T0399029

VOLUME SIX HUNDRED AND FIFTY FIVE

METHODS IN ENZYMOLOGY

mRNA 3' End Processing and Metabolism

METHODS IN ENZYMOLOGY

VOLUME SIX HUNDRED AND FIFTY FIVE

METHODS IN ENZYMOLOGY

mRNA 3' End Processing and Metabolism

Edited by

BIN TIAN

Gene Expression and Regulation Program,
Center for Systems and Computational Biology,
The Wistar Institute, Philadelphia, PA, United States

Academic Press is an imprint of Elsevier
50 Hampshire Street, 5th Floor, Cambridge, MA 02139, United States
525 B Street, Suite 1650, San Diego, CA 92101, United States
The Boulevard, Langford Lane, Kidlington, Oxford OX5 1GB, United Kingdom
125 London Wall, London, EC2Y 5AS, United Kingdom

First edition 2021

ISBN: 978-0-12-823573-7
ISSN: 0076-6879

For information on all Academic Press publications
visit our website at https://www.elsevier.com/books-and-journals

Publisher: Zoe Kruze
Developmental Editor: Federico Paulo S. Mendoza
Production Project Manager: James Selvam
Cover Designer: Alan Studholme

Typeset by SPi Global, India

Contents

Contributors

Wei Shen Aik
Department of Biological Sciences, Columbia University, New York, NY, United States

Stefan L. Ameres
Institute of Molecular Biotechnology (IMBA), Vienna BioCenter (VBC); Max Perutz Labs, University of Vienna, Vienna BioCenter (VBC), Vienna, Austria

Pooja Bhat
Institute of Molecular Biotechnology (IMBA), Vienna BioCenter (VBC); Vienna BioCenter PhD Program, Doctoral School of the University at Vienna and Medical University of Vienna, Vienna, Austria

Thomas R. Burkard
Institute of Molecular Biotechnology (IMBA), Vienna BioCenter (VBC), Vienna, Austria

Zhiqiang Cai
Department of Biochemistry and Biophysics, School of Medicine, Howard Hughes Medical Institute, University of Pennsylvania, Philadelphia, PA, United States

Steven L. Coon
Intramural Research Program, Eunice Kennedy Shriver National Institute of Child Health and Human Development, National Institutes of Health, Bethesda, MD, United States

Angela M. Costa
Tumor and Microenvironment Interactions Group—i3S—Instituto de Investigação e Inovação em Saude, Universidade do Porto; INEB-Instituto Nacional de Engenharia Biomédica, Porto, Portugal

Sven Danckwardt
Posttranscriptional Gene Regulation; Institute for Clinical Chemistry and Laboratory Medicine; Centre for Thrombosis and Hemostasis (CTH), University Medical Centre Mainz, Mainz; German Centre for Cardiovascular Research (DZHK), Berlin, Germany

Zbigniew Dominski
Integrative Program for Biological and Genome Sciences; Department of Biochemistry and Biophysics, University of North Carolina at Chapel Hill, Chapel Hill, NC, United States

Ahmet Doymaz
Department of Chemistry, Hunter College, City University of New York, New York, NY, United States

Gideon Dreyfuss
Department of Biochemistry and Biophysics, School of Medicine, Howard Hughes Medical Institute, University of Pennsylvania, Philadelphia, PA, United States

Nathan D. Elrod
Department of Biochemistry and Molecular Biology, The University of Texas Medical Branch at Galveston, Galveston, TX, United States

Mervin M. Fansler
Cancer Biology and Genetics Program, Memorial Sloan Kettering Cancer Center; Tri-Institutional Training Program in Computational Biology and Medicine, Weill-Cornell Graduate College, New York, NY, United States

Yipeng Gao
Graduate Program in Quantitative and Computational Biosciences; Department of Medicine, Baylor College of Medicine, Houston, TX, United States

Austin E. Gillen
Division of Hematology, University of Colorado School of Medicine, Aurora, CO, United States

Raeann Goering
Department of Biochemistry and Molecular Genetics; RNA Bioscience Initiative, University of Colorado Anschutz Medical Campus, Aurora, CO, United States

R. Samuel Herron
Department of Pathology, University of Pittsburgh, School of Medicine, Pittsburgh, PA, United States

Veronika A. Herzog
Institute of Molecular Biotechnology (IMBA), Vienna BioCenter (VBC), Vienna, Austria

Hun-Way Hwang
Department of Pathology, University of Pittsburgh, School of Medicine, Pittsburgh, PA, United States

James R. Iben
Intramural Research Program, Eunice Kennedy Shriver National Institute of Child Health and Human Development, National Institutes of Health, Bethesda, MD, United States

Madeline K. Jensen
Department of Biochemistry and Molecular Biology, The University of Texas Medical Branch at Galveston, Galveston, TX, United States

Torben Heick Jensen
Department of Molecular Biology and Genetics, Aarhus University, Aarhus, Denmark

Ping Ji
Department of Biochemistry and Molecular Biology, The University of Texas Medical Branch at Galveston, Galveston, TX, United States

Frida Esther Kleiman
Department of Chemistry, Hunter College, City University of New York, New York, NY, United States

Bingnan Li
SciLifeLab, Department of Microbiology, Tumor and Cell Biology, Karolinska Institutet, Solna, Sweden

Qingshun Q. Li
Key Laboratory of the Ministry of Education for Coastal and Wetland Ecosystems, College of the Environment and Ecology, Xiamen University, Xiamen, Fujian, China; Graduate College of Biomedical Sciences, Western University of Health Sciences, Pomona, CA, United States

Tianwei Li
Intramural Research Program, Eunice Kennedy Shriver National Institute of Child Health and Human Development, National Institutes of Health, Bethesda, MD, United States

Wei Li
Division of Computational Biomedicine, Department of Biological Chemistry, School of Medicine, University of California, Irvine, CA, United States

Ai Lin
Department of Biochemistry and Molecular Biology, The University of Texas Medical Branch at Galveston, Galveston, TX, United States; Department of Etiology and Carcinogenesis, National Cancer Center/Cancer Hospital, Chinese Academy of Medical Sciences and Peking Union Medical College, Beijing, China

Juncheng Lin
Key Laboratory of the Ministry of Education for Coastal and Wetland Ecosystems, College of the Environment and Ecology, Xiamen University, Xiamen, Fujian, China

Zhandong Liu
Department of Pediatrics, Baylor College of Medicine; Jan and Dan Duncan Neurological Research Institute, Texas Children's Hospital, Houston, TX, United States

Richard J. Maraia
Intramural Research Program, Eunice Kennedy Shriver National Institute of Child Health and Human Development, National Institutes of Health, Bethesda, MD, United States

Sueli Marques
SciLifeLab, Department of Microbiology, Tumor and Cell Biology, Karolinska Institutet, Solna, Sweden

William F. Marzluff
Integrative Program for Biological and Genome Sciences; Department of Biochemistry and Biophysics, University of North Carolina at Chapel Hill, Chapel Hill, NC, United States

Sandy Mattijssen
Intramural Research Program, Eunice Kennedy Shriver National Institute of Child Health and Human Development, National Institutes of Health, Bethesda, MD, United States

Christine Mayr
Cancer Biology and Genetics Program, Memorial Sloan Kettering Cancer Center; Tri-Institutional Training Program in Computational Biology and Medicine, Weill-Cornell Graduate College, New York, NY, United States

Sibylle Mitschka
Cancer Biology and Genetics Program, Memorial Sloan Kettering Cancer Center, New York, NY, United States

Alexandra Moreira
Gene Regulation, i3S—Instituto de Investigação e Inovação em Saúde; IBMC-Instituto de Biologia Molecular e Celular; ICBAS-Instituto de Ciências Biomédicas Abel Salazar, Universidade do Porto, Porto, Portugal

Michael Robert Murphy
Department of Chemistry, Hunter College, City University of New York, New York, NY, United States

Takayuki Nojima
Sir William Dunn School of Pathology, University of Oxford, Oxford, United Kingdom; Medical Institute of Bioregulation, Kyushu University, Fukuoka, Japan

Anton Ogorodnikov
Posttranscriptional Gene Regulation; Institute for Clinical Chemistry and Laboratory Medicine; Centre for Thrombosis and Hemostasis (CTH), University Medical Centre Mainz, Mainz, Germany

Maria Jose Oliveira
Tumor and Microenvironment Interactions Group—i3S—Instituto de Investigação e Inovação em Saude; INEB-Instituto Nacional de Engenharia Biomédica; Faculdade de Medicina, Universidade do Porto, Porto, Portugal

Andrea Pauli
Research Institute of Molecular Pathology (IMP), Vienna BioCenter (VBC), Vienna, Austria

Vicent Pelechano
SciLifeLab, Department of Microbiology, Tumor and Cell Biology, Karolinska Institutet, Solna, Sweden

Manfred Schmid
Department of Molecular Biology and Genetics, Aarhus University, Aarhus, Denmark

Yongsheng Shi
Department of Microbiology and Molecular Genetics, School of Medicine, University of California, Irvine, CA, United States

Jihae Shin
Department of Microbiology, Biochemistry, and Molecular Genetics, Center for Cell Signaling, Rutgers New Jersey Medical School, Newark, NJ, United States

Guramrit Singh
Department of Molecular Genetics, Center for RNA Biology, The Ohio State University, Columbus, OH, United States

Byung Ran So
Department of Biochemistry and Biophysics, School of Medicine, Howard Hughes Medical Institute, University of Pennsylvania, Philadelphia, PA, United States

Lindsey V. Soles
Department of Microbiology and Molecular Genetics, School of Medicine, University of California, Irvine, CA, United States

Yadong Sun
Department of Biological Sciences, Columbia University, New York, NY, United States

J. Matthew Taliaferro
Department of Biochemistry and Molecular Genetics; RNA Bioscience Initiative, University of Colorado Anschutz Medical Campus, Aurora, CO, United States

Michael Tellier
Sir William Dunn School of Pathology, University of Oxford, Oxford, United Kingdom

Bin Tian
Department of Microbiology, Biochemistry, and Molecular Genetics, Center for Cell Signaling, Rutgers New Jersey Medical School, Newark, NJ; Program in Gene Expression and Regulation, Center for Systems and Computational Biology, The Wistar Institute, Philadelphia, PA, United States

Liang Tong
Department of Biological Sciences, Columbia University, New York, NY, United States

Eric J. Wagner
Department of Biochemistry and Molecular Biology, The University of Texas Medical Branch at Galveston, Galveston, TX, United States

Jingwen Wang
SciLifeLab, Department of Microbiology, Tumor and Cell Biology, Karolinska Institutet, Solna, Sweden

Ruijia Wang
Department of Microbiology, Biochemistry, and Molecular Genetics, Center for Cell Signaling, Rutgers New Jersey Medical School, Newark, NJ, United States

Joana Wilton
Graduate Program in Areas of Basic and Applied Biology (GABBA) PhD Program, ICBAS-Instituto de Ciências Biomédicas Abel Salazar; Gene Regulation, i3S—Instituto de Investigação e Inovação em Saúde, Universidade do Porto; IBMC-Instituto de Biologia Molecular e Celular, Porto, Portugal

Guifen Wu
Department of Molecular Biology and Genetics, Aarhus University, Aarhus, Denmark

Hari Krishna Yalamanchili
Department of Pediatrics, Baylor College of Medicine; Jan and Dan Duncan Neurological Research Institute, Texas Children's Hospital; USDA/ARS Children's Nutrition Research Center, Department of Pediatrics, Baylor College of Medicine, Houston, TX, United States

Xiao-Cui Yang
Integrative Program for Biological and Genome Sciences, University of North Carolina at Chapel Hill, Chapel Hill, NC, United States

Congting Ye
Key Laboratory of the Ministry of Education for Coastal and Wetland Ecosystems, College of the Environment and Ecology, Xiamen University, Xiamen, Fujian, China

Zhongxia Yi
Department of Molecular Genetics, Center for RNA Biology, The Ohio State University, Columbus, OH, United States

Yoseop Yoon
Department of Microbiology and Molecular Genetics, School of Medicine, University of California, Irvine, CA, United States

Preface

Almost all protein-coding transcripts in eukaryotic cells are decorated with a poly(A) tail, which plays critical roles in aspects of mRNA metabolism. Cleavage and polyadenylation (CPA) of precursor RNA is the 3′ end processing mechanism that defines the 3′ end of mature transcripts through endonucleolytic cleavage and poly(A) tail addition, two coupled reactions that take place co-transcriptionally. CPA is also intimately connected to transcriptional termination. Biochemical methods over the last more than three decades have elucidated the molecular underpinning of CPA machinery as we know it today. About two decades ago, bioinformatic analyses of cDNA sequences revealed surprisingly high levels of heterogeneity at the 3′ ends of transcripts, a phenomenon known as alternative polyadenylation (APA). It is now increasingly clear that, due to regulation of CPA, genes express APA isoforms differently in various cell types and under distinct conditions. In this volume of *Methods in Enzymology*, 19 chapters cover a broad range of techniques used to interrogate 3′ ends of RNAs, to identify protein factors involved in 3′ end processing and its regulation, and to unravel the consequences of APA. The chapters fall broadly into four groups, as summarized below.

Since its inception about a dozen years ago, deep sequencing has fundamentally changed the way research questions are being pursued. This is particularly pertinent for APA studies, as regulation of CPA often has global effects. Several chapters in this volume describe cutting-edge sequencing methods to precisely identify 3′ ends and to quantitatively examine APA isoforms, including the polyA-Click-seq (PAC-seq) method developed by Eric Wagner and colleagues, PAS-seq 2 by Yongsheng Shi and colleagues, TREND-seq by Sven Danckwardt and Anton Ogorodnikov, and QPAT-seq by Qingshun Q. Li and colleagues. In addition, Vicent Pelechano and colleagues describe TIF-Seq2, a method they developed to examine both the 5′ and 3′ ends of mRNA; Steven L. Coon, Richard J. Maraia, and colleagues describe SM-PAT-seq, which measures the poly(A) tail length using the PacBio long-read sequencing technology; Torben Jensen and colleagues describe an approach to sequence both poly(A)+ and poly(A)− RNAs; Hun-Way Hwang and R. Samuel Herron describe the cTag-PAPERCLIP method they have developed to interrogate 3′ ends in select cells.

Several chapters in this volume focus on the bioinformatics side of 3′ end research. Zhandong Liu and colleagues describe the PolyA-miner method they have developed to examine APA using 3′ end sequencing data. Stefan L. Ameres and colleagues describe a method named 3′GAmES to analyze APA using sequencing data that inherently contain 3′ end information. Wei Li and Yipeng Gao describe DaPars for 3′ end analysis using regular RNA-seq data and its updated version, scDaPars, for single-cell RNA-seq (scRNA-seq) data analysis. Matthew Taliaferro and colleagues describe LATRAT, a program that also interrogate 3′ end regulation in both bulk RNA-seq and scRNA-seq data.

The third group of chapters in this volume describe biochemical and molecular biology approaches to study 3′ end processing and mRNA metabolism. Frida Kleiman and colleagues describe biochemical methods to examine polyadenylation and deadenylation, as well as measurement of the poly(A) tail length. The chapter by Liang Tong and colleagues describes a biochemical strategy to study histone 3′ end processing machinery. Gideon Dreyfuss and colleagues describe a method to examine the molecular basis of regulation of CPA by U1 snRNP. The chapter by Alexandra Moreira and colleagues describe an integrative approach where APA and gene expression are compared in differentiation of macrophages and their crosstalk with cancer cells is analyzed. Guramrit Singh and Zhongxia Yi describe RIPiT-seq, a method that interrogates interactions between mRNAs with their binding proteins.

Lastly, two chapters discuss methods to perturb APA isoforms in order to examine their functions. Christine Mayr and colleagues describe engineering of genomic sequence by the CRISPR/Cas9 system to selectively express an APA isoform of interest. Bin Tian and colleagues describe a CRISPR/dCas9-based method to change relative expression of APA isoforms through blocking RNA polymerase II elongation.

The production of this book coincides with the COVID-19 pandemic. Many authors overcome difficulties of various kinds to put together their excellent work that will be of great value to the community. I am exceptionally grateful for their time and effort. I would also like to thank Paulo Mendoza, Shellie Bryant, Zoe Kruze, and the team at Elsevier for their support along the process of publishing this book. Paulo, in particular, assisted me in all aspects of book editing and helped authors with all sorts of issues. I could not image how the book could be put together as smoothly as it has been without Paulo's quick response and skillful assistance.

Bin Tian, PhD
Gene Expression and Regulation Program
Center for Systems and Computational Biology
The Wistar Institute
Philadelphia, PA, United States

Application and design considerations for 3′-end sequencing using click-chemistry

Madeline K. Jensen[a], Nathan D. Elrod[a], Hari Krishna Yalamanchili[b,c,d], Ping Ji[a], Ai Lin[a,e], Zhandong Liu[b,c], and Eric J. Wagner[a,*]

[a]Department of Biochemistry and Molecular Biology, The University of Texas Medical Branch at Galveston, Galveston, TX, United States
[b]Department of Pediatrics, Baylor College of Medicine, Houston, TX, United States
[c]Jan and Dan Duncan Neurological Research Institute, Texas Children's Hospital, Houston, TX, United States
[d]USDA/ARS Children's Nutrition Research Center, Department of Pediatrics, Baylor College of Medicine, Houston, TX, United States
[e]Department of Etiology and Carcinogenesis, National Cancer Center/Cancer Hospital, Chinese Academy of Medical Sciences and Peking Union Medical College, Beijing, China
*Corresponding author: e-mail address: ejwagner@utmb.edu

Contents

Methods in Enzymology, Volume 655
ISSN 0076-6879
https://doi.org/10.1016/bs.mie.2021.03.012

Abstract

Over the past 15 years, investigations into alternative polyadenylation (APA) and its function in cellular physiology and pathology have greatly expanded due to the emergent appreciation of its key role in driving transcriptomic diversity. This growth has necessitated the development of new technologies capable of monitoring cleavage and polyadenylation events genome-wide. Advancements in approaches include both the creation of computational tools to re-analyze RNA-seq to identify APA events as well as targeted sequencing approaches customized to focus on the 3′-end of mRNA. Here we describe a streamlined protocol for polyA-Click-seq (PAC-seq), which utilizes click-chemistry to create mRNA 3′-ends sequencing libraries. Importantly, we offer additional considerations not present in our previous protocols including the use of spike-ins, unique molecular identifier primers, and guidance for appropriate depth of PAC-seq. In conjunction with the companion chapter on PolyA-miner (Yalamanchili et al., 2021) to computationally analyze PAC-seq data, we provide a complete experimental pipeline to analyze mRNA 3′-end usage in eukaryotic cells.

1. Introduction

The ascendancy of next-generation sequencing (NGS) has ushered in the field of functional genomics as a powerful tool to understand gene expression regulatory networks. Prior to these technologies, mechanistic investigation was conducted on a case-by-case basis and, while significant insight was attained, our realization of the broad nature of biological phenomena was limited. A vivid example of the NGS impact is on our appreciation of how pervasive alternative splicing is in the human genome. The initial views of alternative splicing posited that differentially included exons were limited to specific examples such as the case of Sxl function in *Drosophila* sex determination (Inoue, Hoshijima, Sakamoto, & Shimura, 1990). Mechanistic studies of alternative splicing in mammals steadily grew to include model substrates such as α-tropomyosin (Gooding & Smith, 2008), cardiac troponin T (Cooper & Ordahl, 1985), c-src (Black, 1992), and FGF-R2 (Wagner & Garcia-Blanco, 2002). These model systems advanced our understanding of alternative splicing regulation but fell short to illustrate exactly how widespread differential splicing is. Only with the arrival of RNA-seq coupled with effective computational analyses were we able to reveal that alternative splicing is not the exception but rather the rule where >95% of multi-exon human genes present evidence of differential exon inclusion (Pan, Shai, Lee, Frey, & Blencowe, 2008).

Following a similar circuitous path to that of splicing, the appreciation for the broad diversity of cleavage and polyadenylation events within the human genome required the power of NGS techniques. Initially, cleavage and polyadenylation was thought to be a constitutive event required for every protein coding gene, with the exception of histone mRNA. Similar to the example of Sxl alternative splicing regulation, studies of how the immunoglobulin M heavy chain gene switches from a membrane-bound to secreted form as well as the heterogeneity reported at the 3′-end of the dihydrofolate reductase gene were among the first reported examples of differential polyadenylation in humans (Alt et al., 1980; Setzer, McGrogan, Nunberg, & Schimke, 1980; Takagaki, Seipelt, Peterson, & Manley, 1996). These early publications revealed the potential to generate diversity at the 3′-end where distinct mRNAs can be created from a single gene with altered 3′UTR length/content or even changes in C-terminal protein coding potential. The pervasiveness of regulating the site of cleavage polyadenylation, or alternative polyadenylation (APA), only began to be understood with the appearance and analyses of databases containing expressed sequence tags (ESTs) (Beaudoing, Freier, Wyatt, Claverie, & Gautheret, 2000; Gautheret, Poirot, Lopez, Audic, & Claverie, 1998; Tian, Hu, Zhang, & Lutz, 2005). But not until the arrival of NGS was it discovered the >70% of human genes undergo APA (Tian et al., 2005). Moreover, the impact of NGS approaches has gone beyond just cataloging APA events and has been instrumental to reveal the importance of APA to both human physiology and diseased states. Currently, there are NGS-based research programs investigating the role of APA in multiple biological and pathophysiological contexts (reviewed in (Gruber & Zavolan, 2019)) including but not limited to: cancer (Masamha et al., 2014), fibrotic diseases (Weng et al., 2020), intellectual dysfunction (Alcott et al., 2020), cellular reprogramming (Brumbaugh et al., 2018), and diabetes (Garin et al., 2010).

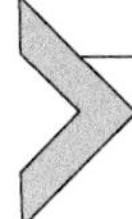

2. Next-generation sequencing tools to detect and measure alternative polyadenylation

The use of RNA-seq has allowed for routine measurement of gene expression as well as alternative exon inclusion. Within any standard RNA-seq dataset, a subset of the reads will be localized near the 3′-ends of genes, thus potentially allowing for the identification and quantification of individual APA events. So, while no specialized sequencing technology would be required to generate RNA-seq that is competent for APA detection, it would be necessary to develop computational tools to extract APA events from the data. Annotation-based tools that rely on previously identified/predicted poly(A) sites (Gruber et al., 2016; Wang, Nambiar, Zheng, & Tian, 2018; You et al., 2015) have been developed including QAPA (Ha, Blencowe, & Morris, 2018) and Roar (Grassi, Mariella, Lembo, Molineris, & Provero, 2016). These tools are powerful but have a potential limitation as annotations are not always completely comprehensive in a given species. To address this issue, there have been a collection of de-novo algorithms developed, notably: TAPAS (Arefeen, Liu, Xiao, & Jiang, 2018), APAtrap (Ye, Long, Ji, Li, & Wu, 2018), and DaPars (Xia et al., 2014), which are all capable of identifying APA events in the absence of annotation. The major advantage to utilizing a computational algorithm to identify and quantify APA events in RNA-seq data is that these tools can analyze datasets retrospectively without the need for any customized sequencing library procedures. Further, these tools allow for accessing a vast repository of sequencing data available in existing databases such as the Cancer Genome Atlas (TCGA) or the Genotype Tissue-Expression (GTEx) consortium for APA re-analysis.

Despite the effectiveness of detecting APA events from RNA-seq data, there are some potential limitations. First, these computational tools allow for the inference of poly(A) sites as opposed to definitive identification. The second limitation is that many of these tools are more effective with higher sequencing depth as only a subset of the sequencing reads near the 3′-ends of genes will be informative. Given the need for higher depth, this creates a concern for cost and could thus limit the number of samples analyzed. In response to this, there have been an astounding 18 distinct 3′-end sequencing approaches developed to enrich for sequence regions near the poly(A) tail in order to study APA on a genome scale including commonly used tools like 3'READS+ (Zheng, Liu, & Tian, 2016), PAS-seq (Shepard et al., 2011), and 3′-seq (Lianoglou, Garg, Yang, Leslie, & Mayr, 2013). With each of these 3′-end sequencing approaches, enriching for the poly(A) tail using an oligo(dT)-based capture or priming is a key component. These techniques differ in other aspects of library preparation such as fragmentation or adapter ligations. While these approaches have not been widely adopted for large consortium sequencing efforts, they offer several advantages over standard RNA-seq. The most important of which is that the vast majority of the total sequencing reads will be localized to the 3′-ends of the genes and, depending on read length, will also contain non-templated adenosines revealing the true location of the poly(A) tail. Similar to identifying APA events in RNA-seq data, 3′-end sequencing approaches have also spurred the creation of computational resources in order to analyze the data. These pipelines were typically developed in parallel with the 3′-end sequencing library protocol but have also been developed independently. In our hands, we have utilized both DPAC (Routh, 2019) and PolyA-miner (Yalamanchili et al., 2020) to analyze 3′-end sequencing datasets.

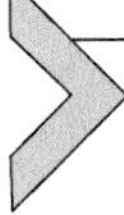

3. PolyA-ClickSeq library preparation and sequencing

3.1 PolyA-Clickseq workflow

The initial creation of the ClickSeq protocol was as a facile alternative to standard RNA-seq library preparations but with the added features of removing fragmentation and ligation steps as well as reducing artifactual recombination events that are commonly observed in library preparations (Jaworski & Routh, 2018; Routh, Head, Ordoukhanian, & Johnson, 2015). This issue is particularly relevant in the investigations of viral infection where distinguishing natural from artifactual recombination can be important. The specific innovation of Clickseq is that fragmentation is not done using the conventional approaches (e.g., chemical or enzymatic) but rather occurs through stochastic incorporation of dideoxy nucleotides containing an azide group at the 3′ position during the reverse transcription reaction. This same component is featured in PAC-seq and a schematic of the basic workflow of PAC-seq is provided in Fig. 1. While ClickSeq reverse transcription is initiated via random priming, PAC-seq reverse transcription (RT) is designed to initiate at the poly(A) tail. From that point onwards, the two library preparations are similar. The RT primer used in PAC-seq contains an oligo(dT) region (21Ts) downstream of a portion of the Illumina p7 sequence (primer sequences in Fig. 2). The oligo(dT) primer can either be anchored or non-anchored, the latter of which would allow annealing anywhere within the poly(A) tail. The reverse-transcription step of PAC-seq is carried out in the presence of 3′-azido-2′,3′-dideoxynucleotides (3′-Azido-ddNTPs) at a specific ratio relative to standard dNTPs (1:5) such that the average RT-generated fragment is ~100 nucleotides (nts) in length with a range of 50 to 400 nts. The RT fragments, which are purified after treatment with RNase H, will possess a 3′-azido group at their terminus and be used in a click-chemistry based ligation reaction with an adaptor that contains a 5′-hexynyl group (Fig. 2). Upon purification of the click-ligated product, the library is amplified through PCR before it

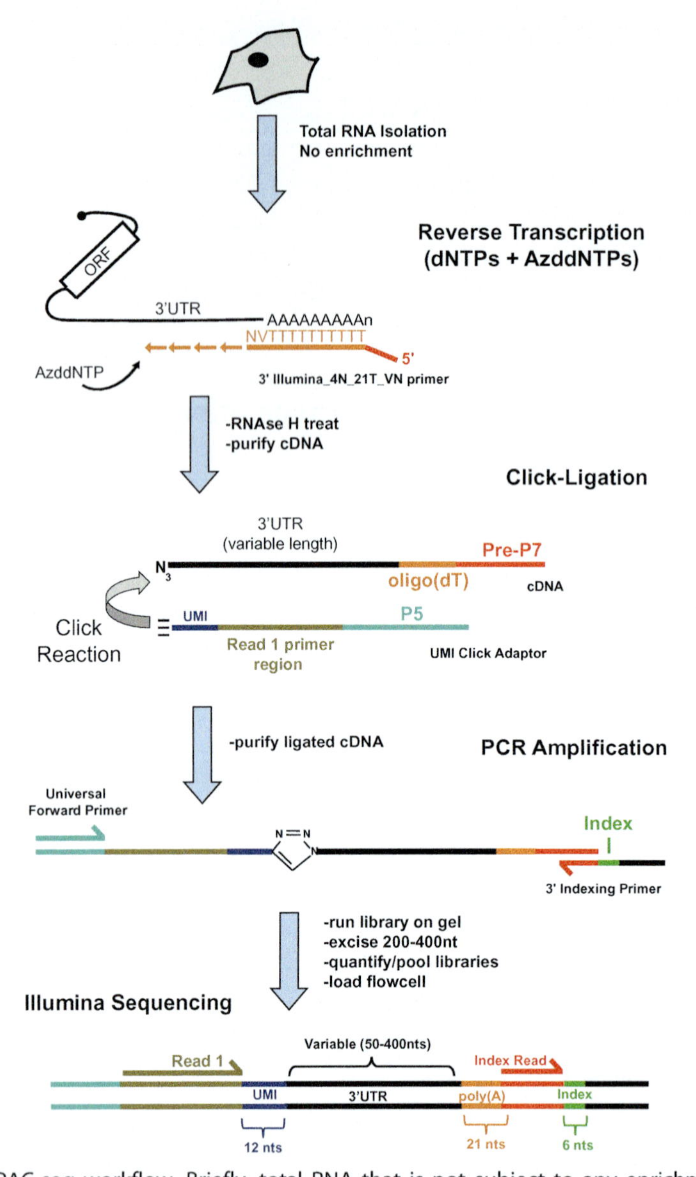

Fig. 1 PAC-seq workflow. Briefly, total RNA that is not subject to any enrichment or fragmentation is used as a template for reverse transcription. The reverse transcription reaction is oligo(dT)-primed and the dNTPs are spiked-in with azido nucleotides at a ratio such that stochastic termination occurs on average at 100nts. The cDNA products are treated with RNase H and purified prior to being subjected to a click chemistry-based ligation of the UMI click adapter, which contains the binding site for the Illumina read one primer, the unique molecular identifier region, and the P5 sequence. Following purification, the ligated cDNA is subject to PCR amplification using a universal forward primer and a 3′ indexing primer that contains a unique index for each PAC-seq library allowing for multiplexing. The library is then run on a gel, purified, quantified, and multiple libraries are pooled before being loaded onto a flowcell. The sequencing reaction involves two reads: Read 1 will first sequence the UMI before going right into the 3′UTR region that will be of variable length (we do not do a 'read 2' as this would be into the poly(A) tail); the index read follows to sequence the barcode/index. Figured derived and reproduced with permission from Oxford Press.

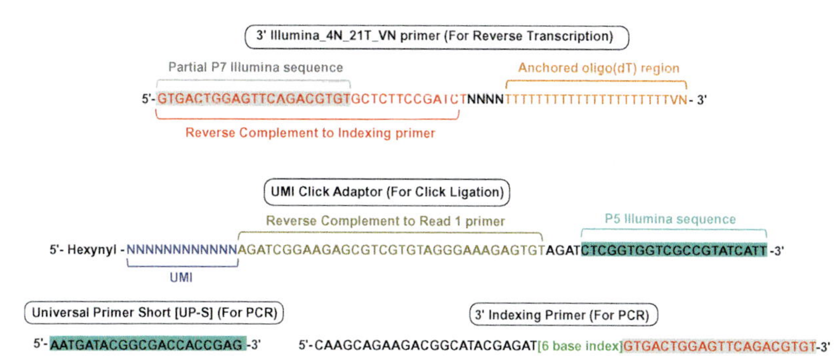

Fig. 2 Table of primers used in PAC-seq. The specific primers used in PAC-seq are shown. In each case, important functional regions are highlighted and shown.

is quantified, pooled, and loaded onto an Illumina sequencing flowcell. The sequencing reaction initiates from the p5 region, thus will sequence directly into the 3′UTR and, in most cases, eventually into the poly(A) region derived from the RT primer. A second primed sequencing reaction will sequence the index region allowing for multiplexing multiple PAC-seq libraries into a single flowcell. As noted above, there are multiple computational tools available to analyze PAC-seq datasets. These tools require common features of read trimming, mapping to a genome, and quantification. In this chapter, we will not describe the analysis pipeline as this will be explained in detail in the following chapter that describes PolyA-miner (Yalamanchili et al., 2021). Our group has currently adopted PolyA-miner as the preferred tool to analyze PAC-seq datasets.

3.2 Additional considerations for PAC-seq provided in this protocol

Below, we present the basic protocol for preparing PAC-seq libraries. We note that previously published manuscripts describing the development of Click-seq as a method of making standard RNA-seq libraries represent the basis of the approach described here with the exception of components unique to PAC-seq (Jaworski & Routh, 2018; Routh et al., 2015). Moreover, while we have published versions of the PAC-seq procedure in three papers from our group (Chu et al., 2019; Elrod, Jaworski, Ji, Wagner, & Routh, 2019; Routh et al., 2017), here we have added several additional features and considerations as well as experimental ‘tips’ to this protocol. We believe that description of these other aspects has become important because, as with any protocol, our experience with PAC-seq has evolved to uncover potential limitations. These four considerations are described broadly below and have populated both the protocol and discussions that follow.

3.2.1 Synergy of PAC-seq and PolyA-miner downstream analysis

We have coordinated the description of this protocol with our recently released computational analysis package, called PolyA-miner (Yalamanchili et al., 2020), which is described in the companion chapter (Yalamanchili et al., 2021). We note that other computational pipelines exist that would be suitable for PAC-seq analysis, but we believe that PolyA-miner is a highly effective resource to analyze PAC-seq datasets possessing several advantages. It can estimate the true effect of 3'UTR shortening or lengthening events by comprehending all APA dynamics including non-proximal/non-distal APA changes irrespective of magnitude. Other key advantages include but are not limited to rigorous removal of mispriming events (e.g., internal priming at adenosine-rich regions), effective 'de-noising' to filter out low confidence poly(A) site events, mapping novel APA sites that are otherwise undetected using reference-based approaches and less susceptibility to inherent data variations. We have coordinated aspects of the PAC-seq protocol in this chapter with analysis features described in the PolyA-miner chapter and make note of these instances. We encourage the user to examine the contents of both of the chapters when they carry out PAC-seq and analysis using PolyA-miner. However, we note that PolyA-miner is not a requisite to analyze PAC-seq and other approaches can be utilized or developed.

3.2.2 Considerations for unique molecular identifiers (UMIs)

The need to amplify NGS libraries using PCR is an obligatory requirement in nearly all sequencing preparations. As a general rule, efforts must be made to keep the number of PCR cycle numbers low to reduce PCR duplication events. By design, the initial cDNA created by PAC-seq will contain an unnatural cycloaddition product at the site of click addition that will require polymerase transversal during PCR amplification. As such, the initial cycles of PCR could be potentially less efficient necessitating additional cycles to amplify the library thereby creating concerns for PCR duplication. Here, we include a specific description of library preparation using primers that contain unique molecular identifiers (UMI) in order to reduce these potential PCR duplication events. The presence of UMIs within PAC-seq primers does not substantively change the protocol to prepare libraries. However, the downstream data/read processing steps need to be modified to handle UMI regions. In brief, UMIs are trimmed from the sequence and attached to the read header prior to alignment and later used to deduplicate the data set. In the accompanying chapter by Yalamanchili et al. (2021), we have provided the necessary computational resources and discussion to handle UMI usage in the PAC-seq analysis.

3.2.3 Use of spike-in normalization controls

The use of spike-in normalization controls has become increasingly apparent as a necessary component to all genome-wide analyses using next generation sequencing (Chen et al., 2015). In the case of PAC-seq, there are several potential sources both commercial and homemade for spike-in normalization. The key requisites are that the spike-ins must contain a poly(A) tail in order to be reverse transcribed, the spike-ins must be present at ~1% of the total experimental materials, and there must be minimal (preferably zero) overlap of the spike-in sequences with the sequences being analyzed by PAC-seq. There are multiple options of spike-ins and two potential entry points where these can be used in the PAC-seq protocol. There are several commercially available products available and we have utilized both Lexogen (cat# 051.01) or Illumina (cat# 20030697). Both of these products contain poly(A)+RNA and are generated from sequences not present in annotated genomes with accompanying manufacturer instructions for computational analysis. Alternatively, and an approach that we favor, the spike-in addition of RNA or cells from a divergent species can be used. We prefer the use of a different species as reagents for normalizations because, in addition to being more cost effective, this approach allows the user to add cells at an early step within the protocol thus allowing for spike-ins to be subjected to the exact same pipeline of purification and library preparation as the experimental cells. This method will ensure that all RNA species are treated equally. In our hands, we routinely spike-in *Drosophila* S2 cells into an experiment focusing on human cells and in cases where S2 cells are the center of the experiment, we spike-in with human cells.

3.2.4 Discussions on barcoding and necessary sequencing depth

One omission from previous descriptions of PAC-seq was clear guidance as to how much sequencing depth is necessary per library when multiplexing samples into a single flowcell. This protocol will now provide recommendations for sequencing depth and how many samples can be analyzed. These recommendations are based upon experience and an extensive analysis provided in the PolyA-miner chapter (Yalamanchili et al., 2021) that empirically determined optimal reading depth to identify annotated PASs as well as

APA events. Moreover, we feel that discussion of sequencing depth has risen in importance because of two reasons. First, the strength of PAC-seq is that reads are localized to the very 3′-ends of genes thereby presenting the opportunity to barcode/index many more samples into a flowcell since the number of reads required to capture PAS usage is much lower than standard RNA-seq approaches. However, this advantage is counterbalanced by a weakness that is similar to many other oligo(dT)-primed methods where the reverse transcription reaction will be subject to mispriming events (i.e., internal priming). The RT primer used in PAC-seq will anneal to any continuous stretches of internally encoded adenosines or even within interrupted stretches of adenosines. Internal priming is difficult to prevent or assess experimentally and must therefore be computationally addressed. Imposing an internal priming computational 'filter' is relatively trivial from a coding perspective but will have significant impact on the final number of reads used to map PAS usage. Our experience using PAC-seq has uncovered the realization that mispriming could be a significant concern. Below, we revisit this concern to make practical recommendations of the depth necessary to make PAC-seq effective, which will ultimately impact the number of samples barcoded and read together in a single flowcell.

3.3 Equipment

3.3.1. Nano Drop 2000c (Thermo Scientific)
3.3.2. Centrifuge 5424 (Eppendorf)
3.3.3. ThermoMixer C (Eppendorf)
3.3.4. T100 Thermocycler (Bio-Rad)
3.3.5. E-gel iBASE Version 1.4.0 and E-gel EX 2% Agarose Gel (Invitrogen, Cat. No. G402002)
3.3.6. ChemiDoc Touch Imaging System (Bio-Rad)
3.3.7. Qubit Fluorometer (Life Tech)
3.3.8. Nextseq 500 Illumina Sequencer (Illumina)
3.3.9. Rare-earth magnet stand (Ambion AM10055 or NEB S1506S)
3.3.10. 0.2 mL PCR Strip Magnetic Separator 8 or 12 Strip (Permagen)

3.4 Chemicals

3.4.1 Deoxyribonucleotide set (dNTPs) (10 mM in water) (any company)
3.4.2 3′-Azido-2′,3′-dideoxynucleotides (AzddNTPs) (10 mM each in water) (Trilink Biotechnologies, N-4007, N-4008, N-4009, N-4014)

3.4.3 Reverse transcriptase—Superscript III (ThermoFisher 18080085)
3.4.4 RNaseOUT (ThermoFisher 10777019)
3.4.5 RNase H (New England Biolabs M0297L)
3.4.6 Click adapter (IDT)
3.4.7 Click catalyst- Cu-TBTA (Lumiprobe Cat. No. 21050)
3.4.8 DMSO (Sigmaaldrich D8418-50ML)
3.4.9 L-Ascorbic Acid (Vitamin C) (Sigmaaldrich A92902-25G)
3.4.10 1 M HEPES pH 7.2 (Sigmaaldrich H3375-25G)
3.4.11 OneTaq DNA Polymerase 2× Master Mix (New England Biolabs, M0482)
3.4.12 E-gel 1 kb plus DNA ladder (Invitrogen, Cat. No. 10488090)
3.4.13 Zymo DNA Clean and Concentrator-5 (Zymo Research, D4013)
3.4.14 Zymo Gel DNA Recovery Kit (Zymo Research, D4007)
3.4.15 Qubit dsDNA HS Assay Kit (Thermo Fisher Scientific Q32851)
3.4.16 Sera-mag SpeedBeads (Fisher #09-981-123)
3.4.17 PEG-8000 (Amresco 0159)
3.4.18 0.5 M EDTA, pH 8.0 (Amresco E177)
3.4.19 1.0 M Tris, pH 8.0 (Amresco E199)
3.4.20 Tween 20 (Amresco 0777)

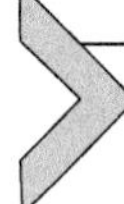

4. PAC-seq protocol

4.1 RNA isolation and quality control assessments (3 h to overnight)

1.1. Typically, we isolate total RNA from a cell culture model system where a single well of a six-well plate contains cells grown to ~80% confluency. In order to improve replicates, we recommend first trypsinizing cells in order to get an accurate cell count prior to RNA isolation. Routinely, we use $\sim 1.5 \times 10^6$ mammalian cells (e.g., HEK 293 T cells) per isolation. Once this number is noted, centrifuge cells and remove the supernatant

Tip: If spike-in normalization is desired, and this is recommended, then this is the first of two possible steps where this can be achieved. Once an accurate cell count has been attained, then add 1% of that number of a different species cells to the cell suspension just prior to centrifugation. In our hands, we add 1% of Drosophila S2 cells that were harvested alongside of the mammalian cells.

1.2. Our typical method of RNA isolation is to use commercially available guanidine-based methods (e.g., TRIzol) but column-based methods are perfectly suitable. Standard phenol/chloroform-based extraction of the RNA is then used followed by precipitation using ethanol and sodium acetate

1.3. Total RNA should be resuspended in RNase-free water and analyzed to determine A260/A280 (>1.8) and RNA Integrity Number (>8.0). If possible, we recommend resuspending total RNA at a concentration of 200–1000 ng/μL

Tip: High quality total RNA is the preferred starting material for PAC-seq library construction. Having said this, we believe that some degree of degradation may be tolerated based upon the fact that ultimately the last <400 nts of the 3′UTR will be sequenced. Importantly, we have not systematically tested the impact of mild sample degradation thus cannot provide strict guidance of acceptable levels meaning this would need to be tested empirically by the user.

4.2 Reverse transcription (2 h)

2.1. Create a stock solution of deoxyribonucleotides containing both dNTPs and 3′-Azido-ddVTPs. The stock solution will have a 1:5 ratio of 3′-Azido-ddVTPs (1 mM final): dNTP (5 mM final). For less than 10 samples, we recommend mixing 10 μL 10 mM dNTPs, 2 μL 10 mM 3′-Azido-ddATP, 2 μL 10 mM 3′-Azido-ddCTP, 2 μL 10 mM 3′-Azido-ddGTP, and 4 μL H_2O. The stock solution can be a larger volume depending on the number of samples

Tip: The initial use of 3′-Azido-ddVTPs was intended if the RT reaction was to use a non-anchored oligo(dT) primer in order to prevent stochastic termination within reverse transcription of the poly(A) tail (Routh et al., 2017). Since that initial design, we have also used anchored oligo(dT) primers, which would allow for immediate reverse transcription into the 3'UTR region therefore abrogating the need to omit 3′-Azido-ddTTP. In our hands, either 3′-Azido-ddVTP or 3′-Azido-ddNTP for anchored primed RT reactions works equally as well.

2.2. *Optional step*: If spike-ins are to be used, this represents a second possible entry point of addition. In this case, spike-ins would be RNA from either commercially available source or a distinct species. In either case, we recommend using 1% by mass of spiked-in RNA with respect to the total RNA added in step 2.1. For example, if 2 μg of total RNA is used, then 20 ng of spike-in RNA should be added

2.3. The total volume of the reverse transcription reaction is 20 μL, but this is assembled in two steps with an incubation of the first 13 μL preceding the addition of the final 7 μL. To set up the first part of the reverse transcription reaction: add 2 μL 1:5 3′-Azido-ddVTP:dNTP stock solution, 0.5 μL 3′ Illumina_4N_21T primer (stock is 100uM), and the remainder of the remaining 10.5 μL is composed of total RNA and RNase-free H_2O in the high-profile 0.2 mL PCR 8-tube strips. Note that the sequences of primers used in PAC-seq are provided in Fig. 2

Tip: PAC-seq is highly sensitive and we have published successful library construction down to 125 ng of total RNA. The user should keep in mind though that low levels of RNA could necessitate the use of additional PCR cycles to amplify a library thus there is a delicate balance between input RNA levels and the potential for higher PCR duplication rate. We note that UMI-containing primers should mitigate this concern and descriptions of these primers are included in this protocol. If RNA is not limiting, we routinely use 2 μg of total RNA as the input for reverse transcription, which results in a reasonable number of PCR cycles to amplify libraries.

2.4. Incubate the first part of the reverse transcription reaction for 5 min at 65 °C to denature the RNA and immediately cool on ice for at least 1 min. This will allow annealing of the 3′ Illumina_4N_21T primer to poly(A) RNA

2.5. It may be necessary to briefly spin the first part of the reverse transcription reaction prior to adding: 4 μL 5 × Superscript First Strand Buffer, 1 μL 0.1 M DTT, 1 μL RNase OUT Recombinant Ribonuclease, and 1 μL Superscript III Reverse Transcriptase on ice for a final reaction volume of 20 μL. Upon addition of all components, mix the reaction to ensure homogeneity

2.6. Incubate the 20 μL reaction in the thermal cycler for 10 min at 25 °C, followed by 40 min at 50 °C, and then followed by 15 min at 75 °C. The reaction can then be held at 4 °C

2.7. To remove residual RNA, add 0.5 μL RNase H and incubate for an additional 20 min at 37 °C and then 10 min at 80 °C to inactivate the RNase H

Tip: Remove a 1 mL aliquot of previously prepared Sera-mag Speedbeads (see next step for preparation) from storage to allow to come to RT. Vortex for 30 s between uses.

4.3 Azido-terminated cDNA purification (1 h)

3.1 There are multiple methods to purify cDNA away from reagents associated with the reverse transcription. We have used both column-based and Sera-mag Speedbeads. Our preference is to use Sera-mag Speedbeads as we have found them effective in terms of cost and yield. To prepare 50 mL Sera-mag Speedbeads slurry: Suspend stock Sera-mag SpeedBeads and transfer 1 mL to a 1.5 mL microtube and drawn the beads to magnet on a magnet stand, wash the beads three times with TE buffer (10 mM Tris-HCl, 1 mM EDTA pH 8.0), the beads are suspended in 1 mL TE buffer. Add the following reagents to a 50 mL conical: 9 g PEG-8000, 10 mL 5 M NaCl (or 2.92 g), 500 μL 1 M Tris-HCl, 100 μL 0.5 M EDTA and fill conical to ~49 mL using sterile ddH2O, mix conical for about 3–5 min until PEG 8000 goes into solution, then add 27.5 μL Tween 20 and transfer 1 mL the Speedbeads in TE to the 50 mL conical. Mix well and aliquot 1 mL the beads slurry into 1.5 mL microtubes and store in a light proof box for up to 6 months at 4 °C

3.2 Add 36 μL of the above room temperature Sera-mag bead slurry (1.8 x volume, to increase the efficiency of binding smaller fragments ~100 bp) to the 20 μL reaction from step 2.7 and mix. Incubate this mixture for 5 min at room temperature in the high-profile 0.2 mL PCR 8-tube strips with occasional and additional mixing

3.3 After the 5 min incubation, place the 8-tube strip into a magnetic separator for 1 min to pulldown Sera-mag beads. Using a pipette, discard the supernatant without disturbing the pellet

Tip: It is not critical to remove all supernatant volume at this step because the beads will be washed in subsequent steps. Be sure to avoid pipetting out Sera-mag beads by accident because these will contain the cDNA.

3.4 Using 200 μL of 80% ethanol, wash the Sera-mag beads by gently pipetting six times to create a suspension. Pulldown the resuspended beads using the magnetic separator. Repeat the wash two additional times. After the final wash, be careful to remove all 80% ethanol as any residual ethanol could negatively impact subsequent steps

3.5 Elute the cDNA by adding 10 μL 50 mM HEPES, pH 7.2 to the Sera-mag bead pellet. Pulldown the beads using the magnetic separator and transfer the supernatant to a new 0.2 mL high-profile PCR 8-tube strip

4.4 Click-ligation (1 h)

4.1 Dilute the azido-terminated cDNA in DMSO and a large molar excess of the click-adapter using the following volumes: 10 μL of the azido-terminated cDNA created in step 3.5, 20 μL 100% DMSO, and 3 μL of the 5 μM UMI-click-adapter in water. (Note: This should be done at room temperature and EDTA will chelate copper required in click-reaction and so must be minimized). The sequence of the UMI-click-adapter is provided in Fig. 2

4.2 To create the chemical catalyst and accelerant mixture for the click reaction, combine 0.4 μL of 50 mM Vitamin C with 2 μL click catalyst- Cu-TBTA. These volumes are for an individual sample and a stock solution should be prepared for multiple samples. There should be a visible color change from light blue to colorless as the Vitamin C reduces copper from the cupric to cuprous state. Wait 30–60 s to ensure the copper ions are fully reduced

Tip: The 50 mM Vitamin C reagent should be prepared as a larger working stock solution and divided into 250 μL aliquots to be stored at −20 °C for no more than 6 months. After thawing and use, dispose of any remaining reagent.

4.3 Add 2.4 μL of the chemical catalyst mixture in step 4.2 to the DNA solution created in step 4.1. Incubate at 37 °C for 30 min

4.4 To increase the efficiency of the click-ligation, repeat steps 4.2 and 4.3. Therefore an additional 4.8 μL of chemical catalyst mixture (2.4 μL × 2) will be added to the ligation.

4.5 Click-ligated cDNA purification (30 min)

5.1 The UMI-adaptor that is click-ligated to the cDNA needs to be purified away from the click-ligation reaction components. To do so, add 64 μL of the Sera-mag bead slurry (1,1.6) to the mixture in step 4.3 and pipette to mix. Incubate this mixture for 5 min at room temperature

5.2 Pellet the beads using the magnet separator and discard the supernatant without disturbing the pellet. Wash the pellet with 200 μL 80% ethanol twice as done previously. When the washes are complete, resuspend the pellet in 20 μL 10 mM Tris, pH 7.4. Pipette to mix the solution and incubate for 2 min at room temperature. Pellet using the magnet separator when the incubation is complete and transfer the 20 μL supernatant to a new tube

4.6 PCR amplification and library purification (2.5 h)

6.1 The PCR reaction should be placed in a 0.2 mL high-profile PCR 8-tube strip and is 50 μL total volume that contains: 10 μL click-ligated DNA from step 5.2, 2.5 μL 5 μM 3' Indexing Primer (1 barcode per sample), 2.5 μL 5 μM Universal Primer Short [UP-S], 10 μL H_2O, and 25 μL 2 × One Taq Standard Buffer Master Mix. Sequences of primers are provided in Fig. 2

Tip: Only half of the click-ligated DNA is used in this step and we recommend storing the remaining 10 μL in case more/less PCR cycles are needed to amplify the library.

6.2 Place the PCR reaction into the thermal cycler and run the following program: 94 °C for 4 min; 53 °C for 30s; 68 °C for 10 min; [94 °C for 30 s, 53 °C for 30s; 68 °C for 2 min] (10–18 cycles); 68 °C for 5 min; hold at 4 °C

Tip: The number of PCR cycles can vary from library to library. We recommend starting with 18 cycles and if the library is very prominent during gel purification (step 6.6 below), then repeat the PCR with less cycles using the remaining 10uL of unused click-ligated DNA. The goal is to generate a library that is just barely visible in the gel to ensure minimal amounts of PCR duplication events.

6.3 To purify the library away from residual PCR reaction components, add 60 μL the Sera-mag bead slurry and pipette to mix. Incubate at room temperature for 5 min

6.4 Pulldown the beads using a magnetic separator. Carefully pipette out and discard the supernatant without disturbing the pellet

6.5 Wash the beads with 80% ethanol twice and then resuspend in 20 μL 10 mM Tris, pH 7.4. Pipette to mix and incubate for 2 min at room temperature. Pulldown the beads using the magnetic separator and transfer the supernatant to a new PCR tube strip. This is the library that will be run on the gel for purification

6.6 Load 10 μL *E*-gel 1 kb plus DNA ladder marker and the 20 μL of the eluted library from step 6.5 into the 2% E-gel EX agarose gel in E-gel iBASE. Run the E-gel EX 1–2% program 7 for 10 min and take an image for records on a ChemiDoc Touch Imaging System for SYBR Gold stain. Crack open precast gel cassette and excise the gel using a fresh razor blade with the desired size of libraries ranging from 200 to 350 bp

Tip: This is an important step to qualitatively determine if your libraries are suitable for sequencing. A proper library should be visible but not highly prominent/strong. If the library is not visible, more PCR cycles may be necessary, and we recommend repeating the PCR using the residual click-ligated DNA from step 5.2. If the library is too prominent, we recommend repeating the PCR with less cycles. In addition, the library should appear as a homogeneous 'smear' on the gel ranging from 200 to 400 in size. If there is any clear banding within the smear, this may reflect specific species being over-amplified. We would generally not recommend using this type of library for sequencing and instead, the user should consider repeating the PCR with less cycles or re-purification of RNA. We have supplied an example of an optimal smeared library versus suboptimal banded library in Fig. 3.

6.7 Transfer the excised gel to a new 1.5 mL tube and mix with 3 × volume of Agarose dissolving buffer (ADB) from the Zymo gel DNA recovery kit. Incubate the gel for approximately 10 min at 50 °C with an occasional shake to melt the gel completely

6.8 Purify the individual library with the Zymo DNA clean column using the manufacturer's protocol. In brief, apply melted agarose in ADB to Zymo DNA clean column, and centrifuge for 30–60 s at 14,000 rpm, wash with 200 μL ethanol-containing wash buffer centrifuge for 30–60 s at 14,000 rpm, and repeat for two washes. To elute, centrifuge for 60 s at 14,000 rpm into fresh, non-stick Eppendorf tubes using 6–12 μL (depending on how prominent the library was on the gel) 10 mM Tris-HCl pH 7.4 or ddH2O

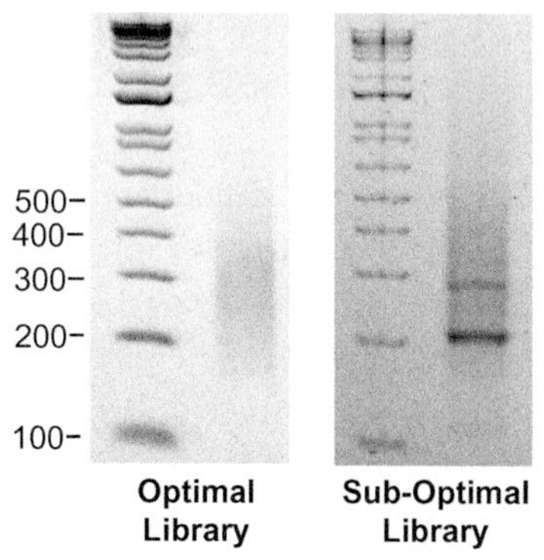

Fig. 3 Example of optimal versus suboptimal PAC-seq library preparation. On the left side is an image of a PAC-seq library that is considered optimal because of its uniform smearing between 200 and 400 nucleotides and the fact that the library is not overly prominent on the gel. On the right is a sub-optimal library because it possesses very prominent banding and is not uniform in terms of smearing.

4.7 Quantification and pooling of PAC-seq libraries (1.5 h)

7.1 Set up the required number of 0.5 mL Qubit assay tubes for standards and samples. The Qubit dsDNA HS Assay requires 2 standards. (Note: Use only thin-wall, clear, 0.5 mL PCR tubes. Acceptable tubes include Qubit assay tubes (Cat. No. Q32856) or Axygen PCR-05-C tubes (part no. 10011–830).

Tip: Do not label the side of the tube as this could interfere with the sample read. Label the lid of each standard tube correctly. Calibration of the Qubit Fluorometer requires the standards to be inserted into the instrument in the right order.

7.2 Prepare the Qubit working solution by diluting the Qubit dsDNA HS Reagent 1:200 in Qubit dsDNA HS buffer. Use a clean plastic tube each time you prepare Qubit working solution. (Note: The final volume in each tube must be 200 μL. Each standard tube requires 190 μL of Qubit working solution, and each sample tube requires 198–199 μL. Prepare sufficient Qubit working solution to accommodate all standards and samples.)

7.3 Add 190 μL of Qubit working solution to each of the tubes used for standards. Add 10 μL of each Qubit standard to the appropriate tube, then mix by vortexing 2–3 s. Be careful not to create bubbles

7.4 Add Qubit working solution to individual assay tubes so that the final volume in each tube after adding sample is 200 μL. (Note: We always use 2 μL sample to mix with 198 μL of Qubit working solution in each assay tube.)

7.5 Allow all tubes to incubate at room temperature for 2 min

7.6 On the Home screen of the Qubit 3.0 Fluorometer, press DNA, then select dsDNA High Sensitivity as the assay type. The "Read standards" screen is displayed. Press Read Standards to proceed

7.7 Insert the tube containing Standard #1 into the sample chamber, close the lid, then press Read Standard. When the reading is complete (~3 s), remove Standard #1. Insert the tube containing Standard #2 into the sample chamber, close the lid, then press Read Standard. When the reading is complete, remove Standard #2. The instrument displays the results on the Read Standard screen. Then press Run Samples

7.8 On the assay screen, select the sample volume and units: a) Press the + or − buttons on the wheel to select the sample volume added to the assay tube. b) From the dropdown menu, select the units (ng/μL) for the output sample concentration

7.9 Insert a sample tube into the sample chamber, close the lid, then press Read tube. When the reading is complete (~3 s), remove the sample tube. The instrument displays the results on the assay screen

7.10 Repeat step 7.9. until all samples have been read. Write down the concentration of all the sample libraries in ng/μL and convert this value from ng/μL to nM using following formula:

$$concentration\ in\ nM = \frac{concentration\ in\ ng/\mu L}{660\frac{g}{mol} \times average\ library\ size\ in\ bp} \times 10^6$$

4.8 Pooling of libraries for sequencing (<1 h)

8.1. Determine the common concentration to dilute the libraries for sequencing, 2–4 nM for each library is the preferred starting concentration for the denaturation and dilution guidelines

8.2. Calculate the dilution of the libraries using the following equation: (C1) (V1) = (C2)(V2) (C1 = initial library concentration; V1 = volume of initial library; C2 = concentration of final solution; V2 = desired volume of final diluted library).

8.3. Dilute the libraries according to the calculations above. The libraries are now normalized

Tip: At this point, the user should determine the number of samples that are to be barcoded in a single flowcell and be sure that barcodes are distinct for each library to allow for de-multiplexing. In the case of human cells, we routinely barcode 20–25 samples (maximum of 30) into a single High Density Next-seq 500 v2.5 flowcell with 150 cycles (cat# 200249070). This will provide more than 15 M reads per sample.

8.4. Pool the normalized libraries. Combine equal volumes of each normalized library into a microcentrifuge tube and gently pipette contents up and down 10 times to mix thoroughly. The normalized pool is now ready to be re-quantified, denatured and sequenced

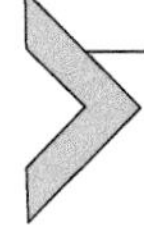

5. Potential limitations of PAC-seq and considerations for depth

PAC-seq was designed to identify and quantify poly(A) site usage genome-wide and is highly effective in achieving this goal. Moreover, we previously published additional computational protocols allowing for

PAC-seq datasets to also be used in conventional differential expression analysis using DEseq (Elrod et al., 2019). Thus, PAC-seq can be used to simultaneously measure changes in gene expression as well as detecting alternative polyadenylation events. PAC-seq will provide only limited capability to detect changes in alternative splicing unless that splicing is accompanied by changes in poly(A) site usage as in the cases of differential terminal exon definition

The primary concern of PAC-seq revolves around its propensity for internal priming and how this impacts the 'effective' depth of sequencing. Stretches of internally encoded adenosine nucleotides will serve as priming locations for the oligo(dT) primer used in the reverse transcription reaction. These events are difficult to avoid experimentally but, as mentioned earlier, can be dealt with computationally by instituting a filtering step in the analysis pipeline. The challenge of this computational step though is to determine how restrictive to make this filter. In other words, how many consecutive adenosines constitutes an 'A-stretch' and how many interruptions of the A-stretch can be expected to disrupt internal priming. This dilemma is important to consider because the stringency of this filter will significantly impact the number of reads that contribute to the final analyzed dataset. As a vivid example, we recently utilized PolyA-miner to re-analyze previously published PAC-seq data (Chu et al., 2019) and determined that as much as 83% of the reads were potentially subject to internal priming using a more restrictive computational filtering step (Yalamanchili et al., 2020). In addition, the downstream goals of a PAC-seq experiment should also be considered when determining required depth. For example, if the desire is to compare APA events between two PAC-seq datasets based upon a previously annotated PAS database, then less reads may be needed. But if the goal is to identify novel polyadenylation sites not present in any annotation, then more reads will be required. As our analyses described in the accompany chapter (Yalamanchili et al., 2021) conclude: the lower end of the recommended PAC-seq depth per sample is ~10 M reads while the ability to resolve novel PASs begins to level off around 25–30 M reads. Thus, the optimal sequencing depth is ~20 M per sample, which translates to ~20–25 barcoded samples into a single flowcell assuming a minimal of 400 M reads passing quality filters.

Acknowledgments

We thank Elizabeth Jaworski and Andrew Routh for helpful suggestions and for their efforts to further optimize PAC-seq. We acknowledge funding support from the Cancer Prevention Research Institute of Texas [RP170387 to Z.L.]; Houston endowment, Chao endowment,

Huffington foundation (to Z.L.); Funded in part with federal funds from the United States Department of Agriculture (USDA/ARS) under Cooperative Agreement No. 58-3092-0-001 and NRI Zoghbi Scholar Award (to H.K.Y.); NIH, National Institute of General Medical Sciences [R01-GM134539 to E.J.W.]; NIH, National Cancer Institute [R03-CA223893 to P.J.].

References

Alcott, C. E., Yalamanchili, H. K., Ji, P., van der Heijden, M. E., Saltzman, A., Elrod, N., et al. (2020). Partial loss of CFIm25 causes learning deficits and aberrant neuronal alternative polyadenylation. *eLife*, *9*, e50895.

Alt, F. W., Bothwell, A. L., Knapp, M., Siden, E., Mather, E., Koshland, M., et al. (1980). Synthesis of secreted and membrane-bound immunoglobulin mu heavy chains is directed by mRNAs that differ at their 3′ ends. *Cell*, *20*(2), 293–301.

Arefeen, A., Liu, J., Xiao, X., & Jiang, T. (2018). TAPAS: Tool for alternative polyadenylation site analysis. *Bioinformatics*, *34*(15), 2521–2529.

Beaudoing, E., Freier, S., Wyatt, J. R., Claverie, J.-M., & Gautheret, D. (2000). Patterns of variant polyadenylation signal usage in human genes. *Genome Research*, *10*(7), 1001–1010.

Black, D. L. (1992). Activation of c-src neuron-specific splicing by an unusual RNA element in vivo and in vitro. *Cell*, *69*(5), 795–807.

Brumbaugh, J., Di Stefano, B., Wang, X., Borkent, M., Forouzmand, E., Clowers, K. J., et al. (2018). Nudt21 controls cell fate by connecting alternative polyadenylation to chromatin signaling. *Cell*, *172*(3), 629–631.

Chen, K., Hu, Z., Xia, Z., Zhao, D., Li, W., & Tyler, J. K. (2015). The overlooked fact: Fundamental need for spike-in control for virtually all genome-wide analyses. *Molecular and Cellular Biology*, *36*(5), 662–667.

Chu, Y., Elrod, N., Wang, C., Li, L., Chen, T., Routh, A., et al. (2019). Nudt21 regulates the alternative polyadenylation of Pak1 and is predictive in the prognosis of glioblastoma patients. *Oncogene*, *38*(21), 4154–4168.

Cooper, T. A., & Ordahl, C. P. (1985). A single cardiac troponin T gene generates embryonic and adult isoforms via developmentally regulated alternate splicing. *The Journal of Biological Chemistry*, *260*(20), 11140–11148.

Elrod, N. D., Jaworski, E. A., Ji, P., Wagner, E. J., & Routh, A. (2019). Development of poly(A)-ClickSeq as a tool enabling simultaneous genome-wide poly(A)-site identification and differential expression analysis. *Methods*, *155*, 20–29.

Garin, I., Edghill, E. L., Akerman, I., Rubio-Cabezas, O., Rica, I., Locke, J. M., et al. (2010). Recessive mutations in the INS gene result in neonatal diabetes through reduced insulin biosynthesis. *Proceedings of the National Academy of Sciences of the United States of America*, *107*(7), 3105–3110.

Gautheret, D., Poirot, O., Lopez, F., Audic, S., & Claverie, J. M. (1998). Alternate polyadenylation in human mRNAs: A large-scale analysis by EST clustering. *Genome Research*, *8*(5), 524–530.

Gooding, C., & Smith, C. W. (2008). Tropomyosin exons as models for alternative splicing. *Advances in Experimental Medicine and Biology*, *644*, 27–42.

Grassi, E., Mariella, E., Lembo, A., Molineris, I., & Provero, P. (2016). Roar: Detecting alternative polyadenylation with standard mRNA sequencing libraries. *BMC Bioinformatics*, *17*(1), 423.

Gruber, A. J., Schmidt, R., Gruber, A. R., Martin, G., Ghosh, S., Belmadani, M., et al. (2016). A comprehensive analysis of 3′ end sequencing data sets reveals novel polyadenylation signals and the repressive role of heterogeneous ribonucleoprotein C on cleavage and polyadenylation. *Genome Research*, *26*(8), 1145–1159.

Gruber, A. J., & Zavolan, M. (2019). Alternative cleavage and polyadenylation in health and disease. *Nature Reviews. Genetics*, *20*(10), 599–614.
Ha, K. C. H., Blencowe, B. J., & Morris, Q. (2018). QAPA: A new method for the systematic analysis of alternative polyadenylation from RNA-seq data. *Genome Biology*, *19*(1), 45.
Inoue, K., Hoshijima, K., Sakamoto, H., & Shimura, Y. (1990). Binding of the Drosophila sex-lethal gene product to the alternative splice site of transformer primary transcript. *Nature*, *344*(6265), 461–463.
Jaworski, E., & Routh, A. (2018). ClickSeq: Replacing fragmentation and enzymatic ligation with click-chemistry to prevent sequence chimeras. *Methods in Molecular Biology*, *1712*, 71–85.
Lianoglou, S., Garg, V., Yang, J. L., Leslie, C. S., & Mayr, C. (2013). Ubiquitously transcribed genes use alternative polyadenylation to achieve tissue-specific expression. *Genes & Development*, *27*(21), 2380–2396.
Masamha, C. P., Xia, Z., Yang, J., Albrecht, T. R., Li, M., Shyu, A. B., et al. (2014). CFIm25 links alternative polyadenylation to glioblastoma tumour suppression. *Nature*, *510*(7505), 412–416.
Pan, Q., Shai, O., Lee, L. J., Frey, B. J., & Blencowe, B. J. (2008). Deep surveying of alternative splicing complexity in the human transcriptome by high-throughput sequencing. *Nature Genetics*, *40*(12), 1413–1415.
Routh, A. (2019). DPAC: A tool for differential poly(a)-cluster usage from poly(a)-targeted RNAseq data. *G3 (Bethesda)*, *9*(6), 1825–1830.
Routh, A., Head, S. R., Ordoukhanian, P., & Johnson, J. E. (2015). ClickSeq: Fragmentation-free next-generation sequencing via click ligation of adaptors to stochastically terminated 3′-azido cDNAs. *Journal of Molecular Biology*, *427*(16), 2610–2616.
Routh, A., Ji, P., Jaworski, E., Xia, Z., Li, W., & Wagner, E. J. (2017). Poly(a)-ClickSeq: Click-chemistry for next-generation 3-end sequencing without RNA enrichment or fragmentation. *Nucleic Acids Research*, *45*(12), e112.
Setzer, D. R., McGrogan, M., Nunberg, J. H., & Schimke, R. T. (1980). Size heterogeneity in the 3′ end of dihydrofolate reductase messenger RNAs in mouse cells. *Cell*, *22*(2 Pt 2), 361–370.
Shepard, P. J., Choi, E. A., Lu, J., Flanagan, L. A., Hertel, K. J., & Shi, Y. (2011). Complex and dynamic landscape of RNA polyadenylation revealed by PAS-Seq. *RNA*, *17*(4), 761–772.
Takagaki, Y., Seipelt, R. L., Peterson, M. L., & Manley, J. L. (1996). The polyadenylation factor CstF-64 regulates alternative processing of IgM heavy chain pre-mRNA during B cell differentiation. *Cell*, *87*(5), 941–952.
Tian, B., Hu, J., Zhang, H., & Lutz, C. S. (2005). A large-scale analysis of mRNA polyadenylation of human and mouse genes. *Nucleic Acids Research*, *33*(1), 201–212.
Wagner, E. J., & Garcia-Blanco, M. A. (2002). RNAi-mediated PTB depletion leads to enhanced exon definition. *Molecular Cell*, *10*(4), 943–949.
Wang, R., Nambiar, R., Zheng, D., & Tian, B. (2018). PolyA_DB 3 catalogs cleavage and polyadenylation sites identified by deep sequencing in multiple genomes. *Nucleic Acids Research*, *46*(D1), D315–D319.
Weng, T., Huang, J., Wagner, E. J., Ko, J., Wu, M., Wareing, N. E., et al. (2020). Downregulation of CFIm25 amplifies dermal fibrosis through alternative polyadenylation. *The Journal of Experimental Medicine*, *217*(2).
Xia, Z., Donehower, L. A., Cooper, T. A., Neilson, J. R., Wheeler, D. A., Wagner, E. J., et al. (2014). Dynamic analyses of alternative polyadenylation from RNA-seq reveal a 3′-UTR landscape across seven tumour types. *Nature Communications*, *5*.

Yalamanchili, H. K., Alcott, C. E., Ji, P., Wagner, E. J., Zoghbi, H. Y., & Liu, Z. (2020). PolyA-miner: Accurate assessment of differential alternative poly-adenylation from 3'Seq data using vector projections and non-negative matrix factorization. *Nucleic Acids Research*, *48*(12), e69.

Yalamanchili, H. K., Elrod, N. D., Jensen, M. K., Ji, P., Lin, A., Wagner, E. J., et al. (2021). A Computational pipeline to infer alternative poly-adenylation from 3′ Sequencing data. *Methods in Enzymology, 655*, 185–204.

Ye, C., Long, Y., Ji, G., Li, Q. Q., & Wu, X. (2018). APAtrap: Identification and quantification of alternative polyadenylation sites from RNA-seq data. *Bioinformatics*, *34*(11), 1841–1849.

You, L., Wu, J., Feng, Y., Fu, Y., Guo, Y., Long, L., et al. (2015). APASdb: A database describing alternative poly(a) sites and selection of heterogeneous cleavage sites downstream of poly(a) signals. *Nucleic Acids Research*, *43*(Database issue), D59–D67.

Zheng, D., Liu, X., & Tian, B. (2016). 3'READS+, a sensitive and accurate method for 3′ end sequencing of polyadenylated RNA. *RNA*, *22*(10), 1631–1639.

PAS-seq 2: A fast and sensitive method for global profiling of polyadenylated RNAs

Yoseop Yoon, Lindsey V. Soles, and Yongsheng Shi*
Department of Microbiology and Molecular Genetics, School of Medicine, University of California, Irvine, CA, United States
*Corresponding author: e-mail address: yongshes@uci.edu

Contents

Abstract

Alternative polyadenylation (APA) is a widespread phenomenon in eukaryotes that contributes to regulating gene expression and generating proteomic diversity. APA plays critical roles in development and its mis-regulation has been implicated in a wide variety of human diseases, including cancer. To study APA on the transcriptome-wide level, numerous deep sequencing methods that capture 3′ end of mRNAs have been developed in the past decade, but they generally require a large amount of hands-on time and/or high RNA input. Here, we introduce PAS-seq 2, a fast and sensitive method for global and quantitative profiling of polyadenylated RNAs. Compared to our original PAS-seq, this method takes less time and requires much lower total RNA input due to improvement in the reverse transcription process. PAS-seq 2 can be applied to both APA and differential gene expression analyses.

Methods in Enzymology, Volume 655
ISSN 0076-6879
https://doi.org/10.1016/bs.mie.2021.03.013

1. Introduction

Alternative polyadenylation (APA) allows a single gene to generate multiple transcript isoforms through the use of alternative poly(A) sites (PAS). Approximately 70% of mammalian genes are subjected to APA (Shi, 2012; Tian & Manley, 2017). APA plays important roles in gene regulation (Mittleman et al., 2020; Sandberg, Neilson, Sarma, Sharp, & Burge, 2008), protein localization (Berkovits & Mayr, 2015), cell differentiation (Zhu et al., 2018), immune response (Rogers et al., 1980), development and evolution (Yoon et al., 2019). Aberrant APA regulation has been linked to many diseases, including cancer and neurological disorders (Gruber & Zavolan, 2019).

APA studies have been facilitated by deep sequencing-based methods for transcriptome-wide profiling of polyadenylated RNAs. These methods can not only map PAS globally, but also quantify their relative usage frequency, thereby allowing comparison of differential PAS usage among different cell types, tissues, and biological conditions. For sequencing library construction, most methods use an oligo(dT) primer to selectively reverse transcribe (RT) polyadenylated RNAs (Derti et al., 2012; Fox-Walsh, Davis-Turak, Zhou, Li, & Fu, 2011; Fu et al., 2011; Lianoglou, Garg, Yang, Leslie, & Mayr, 2013; Martin, Gruber, Keller, & Zavolan, 2012; Sanfilippo, Miura, & Lai, 2017; Shepard et al., 2011; Yoon & Brem, 2010; Zhou et al., 2016). While oligo(dT) primer could also bind internal A-rich sequences during RT, a phenomenon called internal priming, it can be mitigated by computational filtering and use of annotated PAS database for data analyses (Derti et al., 2012; Shepard et al., 2011). Several other methods circumvent the problem of internal priming by linking a 3′ adapter for RT to polyadenylated RNAs (Hoque et al., 2013; Jan, Friedman, Ruby, & Bartel, 2011; Zheng, Liu, & Tian, 2016), or enriching polyadenylated RNAs that bind poly(A) binding proteins (Hwang et al., 2016). However, these methods tend to be laborious compared to oligo(dT) primer-based methods. The large number of steps could also introduce biases that may prevent accurate quantification (Spies, Burge, & Bartel, 2013). Finally, most methods developed so far require high RNA input, up to 10–20 μg of total RNA or 1 μg of total RNA.

Here, we introduce PAS-Seq 2, an improved version of PAS-seq (Shepard et al., 2011), that allows library construction from 100 ng of total RNA. The original PAS-seq combines RT and adapter linking into a single step and has been successfully used in many APA studies, but requires 5–10 μg of total RNA input (Brumbaugh et al., 2018; Huang et al., 2017; Wang et al., 2020; Zhu et al., 2018). In PAS-seq 2, we adopted the reverse transcription protocol from SMART-seq 2 (Picelli et al., 2014), which is intended for samples with low RNA input such as single cells. Specifically, PAS-seq 2 uses a template switch oligo (TSO) that contains a locked nucleic acid (LNA) at the 3′ end, thereby enhancing TSO annealing to untemplated 3′ extension of the cDNA. Also, the protocol includes betaine and magnesium chloride in the RT reaction, which together can increase cDNA yield (Picelli et al., 2014). In our hands, PAS-seq 2 libraries can be successfully generated from 100 ng total RNA. This protocol may work with even lower RNA input if the fragmentation step is optimized.

2. Materials

2.1 Solutions

1. 1 M Magnesium Chloride ($MgCl_2$)
2. 100% Ethanol (EtOH)
3. 3 M Sodium acetate (NaOAc)
4. Betaine, 5 M solution (Thermo Scientific AAJ77507UCR)
5. 10 mM dNTP mix
6. 100% isopropyl alcohol
7. RNase-free water
8. Distilled water

2.2 Reagents and equipment

1. TRIzol™ Reagent (Invitrogen 15596026)
2. RNA fragmentation reagents (Invitrogen AM8740)
3. NEBNext® Poly(A) mRNA magnetic isolation module (NEB E7940)
4. NanoDrop®
5. Heat block or water bath
6. −80 °C freezer
7. Temperature-controlled tabletop centrifuge
8. SuperScript™ III first-strand synthesis system (Invitrogen 18080051)
9. AMPure XP (Beckman-Coulter A63880)
10. Magnetic stand
11. Thermal cycler

12. Phusion® high-fidelity PCR master mix with HF buffer (NEB M0531S)
13. RNaseOUT™ recombinant ribonuclease inhibitor (Invitrogen 1077 7019)
14. GeneJET gel extraction kit (Thermo Scientific K0691)
15. GlycoBlue™ Coprecipitant (Invitrogen AM9516)

2.3 Primer sequences

1. Template switch oligo (TSO) containing locked nucleic acid (LNA): CTACACGACGCTCTTCCGATCTCATrGrG+G
2. PAS-seq 2 oligo(dT): GTGACTGGAGTTCAGACGTGTGCTCT TCCGATCTTTTTTTTTTTTTTTTTTTTTV (V: A/C/G)
3. TruSeq universal adapter:
   AATGATACGGCGACCACCGAGATCTACACTCTTTCCC TACACGACGCTCTTCCGATCT
4. TruSeq indexed adapter:
   CAAGCAGAAGACGGCATACGAGAT**[index]**GTGACTGGA GTTCAGACGTGTGCTCTTCCGATC.

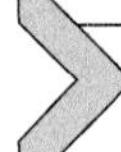

3. Methods

3.1 Isolation of total RNA from biological samples (1–1.5 h (hrs))

1. Isolate total RNA from the biological sample using TRIzol™ Reagent, as per manufacturer's instructions. During the final steps, wash pellet twice instead of once with 75% EtOH. Air-dry and resuspend the pellet with 30 μL RNase-free water. Incubate in a water bath or heat block set at 55–60 °C for 10–15 min. Determine RNA concentration using NanoDrop® and store the samples at −80 °C until use (Fig. 1)

3.2 Isolation of poly(A) RNA (1–1.5 h)

1. Dilute 100 ng—5 μg of total RNA in 50 μL of RNase-free H_2O
2. Isolate poly(A) RNA from total RNA using NEBNext® Poly(A) mRNA magnetic isolation module, as per manufacturer's instructions. Elute the poly(A) RNA using 17 μL of the Tris buffer included in the kit
3. Add 28 μL H_2O to the eluted mRNA, making the final volume of 45 μL
4. Place tube on ice

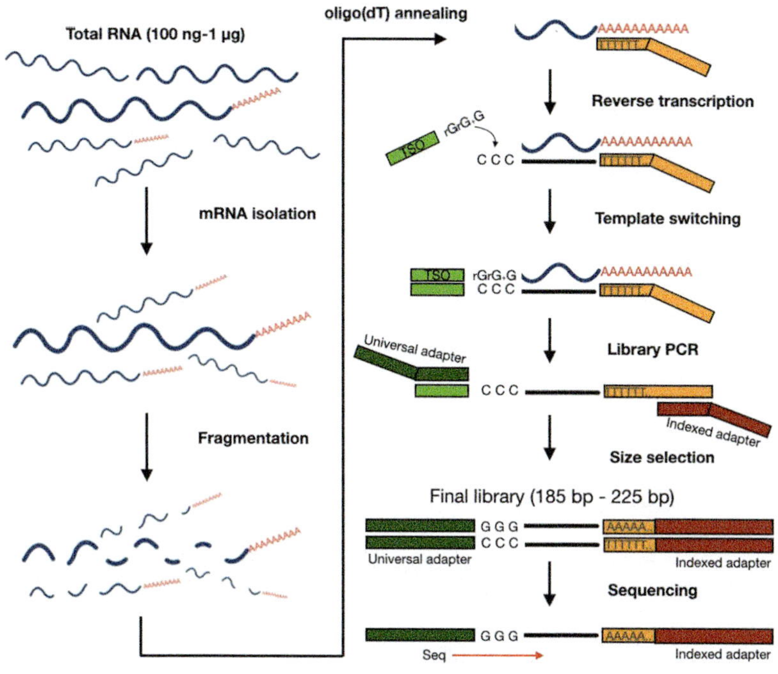

Fig. 1 Overview of PAS-seq 2. TSO: template switch oligo.

3.3 RNA fragmentation (1.5 h)

1. For 1 μg of starting total RNA, add 5 μL of 10 × RNA fragmentation reagent (final volume 50 μL), mix well, and incubate at 70 °C in a thermal cycler for 5 min (min).
2. Add 5.5 μL of Stop solution (10 ×), mix well, and place tube on ice for 2 min
3. Add following mixture to the tube:
   - 144.5 μL H_2O
   - 20 μL 3 M NaOAc
   - 1 μL GlycoBlue™
   - 660 μL ice cold 100% EtOH
4. Precipitate for 1 h at −80 °C
5. Centrifuge at full speed (13,000 rpm) in 4 °C for 30 min
6. Wash pellet with 1 mL of ice cold 75% EtOH. Centrifuge at full speed (13,000 rpm) in 4 °C for 5 min
7. Dissolve the pellet in 5 μL H_2O

3.4 Reverse transcription (3–3.5 h)

1. Set up oligo(dT) primer binding reaction in a PCR tube on ice:

5 µL	Fragmented poly(A) RNA (from step 3.3.7)
2 µL	10 mM dNTP mix
2 µL	10 µM PAS-seq2 oligo(dT) primer
9 µL	**Total**

2. Incubate at 72 °C for 3 min, and put it back to ice
3. Prepare the following mixture in a master mix:

1 µL	SuperScript III reverse transcriptase
4 µL	5 × First-strand buffer
1 µL	100 mM DTT
0.5 µL	RNaseOUT RNase inhibitor
4 µL	5 M Betaine
0.12 µL	1 M $MgCl_2$
0.2 µL	100 µM template switch oligo containing LNA
0.18 µL	RNase-free H_2O
11 µL	**Total**

4. Add 11 µL of mixture from step 3.4.3 to 9 µL of reaction from step 3.4.1 to obtain a final reaction volume of 20 µL. Mix well
5. Incubate the reaction in a thermal cycler with a heated lid, using the following program:
   - 1: 42 °C, 90 min
   - 2: 50 °C, 2 min
   - 3: 42 °C, 2 min
   - 4: Repeat 2–3 for 9 times
   - 5: 72 °C, 15 min
   - 6: 4 °C, hold

*Addition of betaine permits cycle 2 4 by stabilizing reverse transcriptase in higher temperature and promotes unfolding of RNA secondary structures.

6. Add 30 μL of H_2O to obtain a final volume of 50 μL

3.5 Purification of cDNA (0.5 h)

1. Purify 50 μL of cDNA from step 3.4.6. using AMPure XP, as per manufacturer's instructions. Do not over-dry the beads
2. In the final step, elute cDNA with 25 μL of H_2O
3. Transfer 23 μL of the supernatant (cDNA) to a new tube

3.6 PCR amplification and size selection (3.5 h)

1. Prepare the following mixture:

23 μL	cDNA
25 μL	2 × Phusion Hifi PCR master mix
1 μL	10 μM TruSeq universal adapter
1 μL	10 μM TruSeq indexed adapter
50 μL	**Total**

2. Incubate the reaction in a thermal cycler with a heated lid, using the following program:
   - 1: 98 °C, 30 s
   - 2: 98 °C, 10 s
   - 3: 66 °C, 30 s
   - 4: 72 °C, 20 s
   - 5: Repeat step 2–4 19 times
   - 6: 72 °C, 5 min
   - 7: 4 °C, hold
3. Run all PCR products on a 2.5% agarose gel (80 V, 2.5 h).
4. Gel extract 185–225 base-pair (bp) band and submit samples to the sequencing core etc. for quality check prior to sequencing (Fig. 2)
5. Perform sequencing (100 bp single read) of the libraries as per manufacturer's instructions. Add Phix control or multiplex with other libraries to add diversity

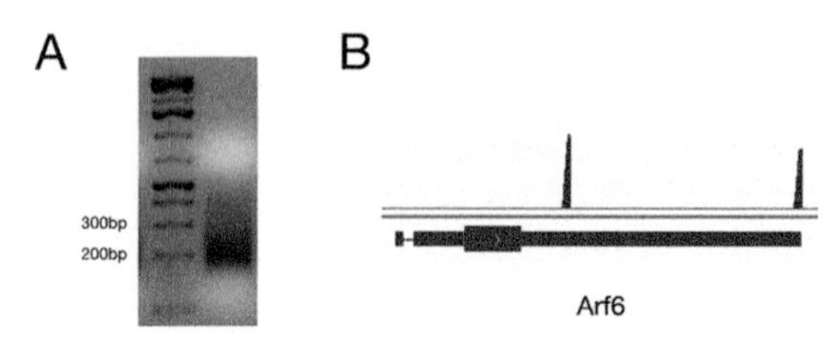

Fig. 2 (A) PAS-seq 2 library. After PCR amplification (Section 3.6), the reaction was resolved on a 2.5% agarose gel. 185–225 bp fragments are selected for sequencing. (B) An example of PAS-seq 2 data. Bigwig file was generated from PAS-seq 2 reads. 3′UTR region of Arf6 gene is shown in the figure. Each peak corresponds alternative cleavage/polyadenylation site for Arf6 gene.

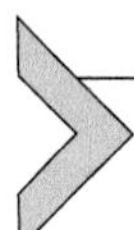

4. Data analysis

1. Trim the sequences by removing the first six nucleotides (CATGGG—from the template switch oligo) and consecutive As (from the oligo(dT) primer sequence) with Cutadapt (Martin, 2011).
2. Align trimmed reads to reference genome using aligner programs such as STAR (Dobin et al., 2013).
3. To filter out internal priming events, remove reads that are mapped to genomic sequences that are immediately followed by 6-consecutive As, or seven As out of 10 nucleotides
4. Compare the 3′ ends of each read to a list of annotated PAS based on published databases (e.g., APADB, PolyA_DB) (Muller et al., 2014; Wang, Nambiar, Zheng, & Tian, 2018). Generate a count table by assigning reads to known PAS if they are mapped within ±40 nt from the site
5. APA analysis is performed with edgeR using the "exon" mode to obtain statistical significance values (false discovery rate (FDR)) for each PAS
6. Genes are considered as APA genes if at least one PAS reaches statistical difference (FDR <0.05) and its percent usage within the gene is different by more than 15 between samples
7. For further bioinformatic analyses, please refer to (Brumbaugh et al., 2018; Shepard et al., 2011).

5. Troubleshooting tips

1. Isolating poly(A) RNA (Section 3.2) from total RNA helps to remove reads from rRNAs, but this protocol works with total RNA as well
2. RNA fragmentation step should be optimized based on the concentration of total RNA input. The current fragmentation protocol is based on

1 μg of total RNA input. Add Stop solution immediately after incubating the samples with RNA fragmentation buffer according to the protocol. Both incubation time and amount of fragmentation buffer can affect average library size. Libraries smaller than 200 bp may contain primer dimers and very short reads

3. Addition of GlycoBlue™ is useful for detecting small pellets during the RNA fragmentation step
4. For gel extraction of PCR amplified libraires, make sure to run the gel sufficiently to remove primer dimers (around 150 bp) from 185 to 225 bp fragments. Check with bioanalyzer to make sure that sample sizes are appropriate and primer dimers are absent from the library
5. PAS-seq 2 reads contain six identical sequences (CATGGG) at the beginning. Therefore, adding spike-in controls such as Phix is recommended. Alternatively, libraries can be multiplexed with other samples to increase diversity at the start of reads
6. While we found that 100 ng of total RNA input was sufficient for generating PAS-seq 2 library, even lower RNA input might be sufficient
7. It is important to perform PAS-seq analysis with biological replicates

Acknowledgments

This study was supported by the following grants: NIH GM090056 and GM128441. We thank the UCI GHTF for sequencing.

References

Berkovits, B. D., & Mayr, C. (2015). Alternative 3' UTRs act as scaffolds to regulate membrane protein localization. *Nature*, *522*(7556), 363–367. https://doi.org/10.1038/nature14321.

Brumbaugh, J., Di Stefano, B., Wang, X., Borkent, M., Forouzmand, E., Clowers, K. J., et al. (2018). Nudt21 controls cell fate by connecting alternative polyadenylation to chromatin signaling. *Cell*, *172*(3), 629–631. https://doi.org/10.1016/j.cell.2017.12.035.

Derti, A., Garrett-Engele, P., Macisaac, K. D., Stevens, R. C., Sriram, S., Chen, R., et al. (2012). A quantitative atlas of polyadenylation in five mammals. *Genome Research*, *22*(6), 1173–1183. https://doi.org/10.1101/gr.132563.111.

Dobin, A., Davis, C. A., Schlesinger, F., Drenkow, J., Zaleski, C., Jha, S., et al. (2013). STAR: ultrafast universal RNA-seq aligner. *Bioinformatics*, *29*(1), 15–21. https://doi.org/10.1093/bioinformatics/bts635.

Fox-Walsh, K., Davis-Turak, J., Zhou, Y., Li, H., & Fu, X. D. (2011). A multiplex RNA-seq strategy to profile poly(A+) RNA: Application to analysis of transcription response and 3′ end formation. *Genomics*, *98*(4), 266–271. https://doi.org/10.1016/j.ygeno.2011.04.003.

Fu, Y., Sun, Y., Li, Y., Li, J., Rao, X., Chen, C., et al. (2011). Differential genome-wide profiling of tandem 3' UTRs among human breast cancer and normal cells by high-throughput sequencing. *Genome Research*, *21*(5), 741–747. https://doi.org/10.1101/gr.115295.110.

Gruber, A. J., & Zavolan, M. (2019). Alternative cleavage and polyadenylation in health and disease. *Nature Reviews. Genetics*, *20*, 599–614. https://doi.org/10.1038/s41576-019-0145-z.

Hoque, M., Ji, Z., Zheng, D., Luo, W., Li, W., You, B., et al. (2013). Analysis of alternative cleavage and polyadenylation by 3′ region extraction and deep sequencing. *Nature Methods*, *10*(2), 133–139. https://doi.org/10.1038/nmeth.2288.

Huang, C., Shi, J., Guo, Y., Huang, W., Huang, S., Ming, S., et al. (2017). A snoRNA modulates mRNA 3′ end processing and regulates the expression of a subset of mRNAs. *Nucleic Acids Research*, *45*(15), 8647–8660. https://doi.org/10.1093/nar/gkx651.

Hwang, H. W., Park, C. Y., Goodarzi, H., Fak, J. J., Mele, A., Moore, M. J., et al. (2016). PAPERCLIP identifies microRNA targets and a role of CstF64/64tau in promoting non-canonical poly(A) site usage. *Cell Reports*, *15*(2), 423–435. https://doi.org/10.1016/j.celrep.2016.03.023.

Jan, C. H., Friedman, R. C., Ruby, J. G., & Bartel, D. P. (2011). Formation, regulation and evolution of *Caenorhabditis elegans* 3'UTRs. *Nature*, *469*(7328), 97–101. https://doi.org/10.1038/nature09616.

Lianoglou, S., Garg, V., Yang, J. L., Leslie, C. S., & Mayr, C. (2013). Ubiquitously transcribed genes use alternative polyadenylation to achieve tissue-specific expression. *Genes & Development*, *27*(21), 2380–2396. https://doi.org/10.1101/gad.229328.113.

Martin, M. (2011). Cutadapt removes adapter sequences from high-throughput sequencing reads. EMBnet.journal. *[S.l.]*, *17*(1), 102226–126089. https://doi.org/10.14806/ej.17.1.200.

Martin, G., Gruber, A. R., Keller, W., & Zavolan, M. (2012). Genome-wide analysis of pre-mRNA 3′ end processing reveals a decisive role of human cleavage factor I in the regulation of 3' UTR length. *Cell Reports*, *1*(6), 753–763. https://doi.org/10.1016/j.celrep.2012.05.003.

Mittleman, B. E., Pott, S., Warland, S., Zeng, T., Mu, Z., Kaur, M., et al. (2020). Alternative polyadenylation mediates genetic regulation of gene expression. *eLife*, *9*. https://doi.org/10.7554/eLife.57492.

Muller, S., Rycak, L., Afonso-Grunz, F., Winter, P., Zawada, A. M., Damrath, E., et al. (2014). APADB: A database for alternative polyadenylation and microRNA regulation events. *Database (Oxford)*, *2014*, bau076. https://doi.org/10.1093/database/bau076.

Picelli, S., Faridani, O. R., Bjorklund, A. K., Winberg, G., Sagasser, S., & Sandberg, R. (2014). Full-length RNA-seq from single cells using smart-seq2. *Nature Protocols*, *9*(1), 171–181. https://doi.org/10.1038/nprot.2014.006.

Rogers, J., Early, P., Carter, C., Calame, K., Bond, M., Hood, L., et al. (1980). Two mRNAs with different 3′ ends encode membrane-bound and secreted forms of immunoglobulin mu chain. *Cell*, *20*(2), 303–312. https://doi.org/10.1016/0092-8674(80)90616-9.

Sandberg, R., Neilson, J. R., Sarma, A., Sharp, P. A., & Burge, C. B. (2008). Proliferating cells express mRNAs with shortened 3′ untranslated regions and fewer microRNA target sites. *Science*, *320*(5883), 1643–1647. https://doi.org/10.1126/science.1155390.

Sanfilippo, P., Miura, P., & Lai, E. C. (2017). Genome-wide profiling of the 3′ ends of polyadenylated RNAs. *Methods*, *126*, 86–94. https://doi.org/10.1016/j.ymeth.2017.06.003.

Shepard, P. J., Choi, E.-A., Lu, J., Flanagan, L. A., Hertel, K. J., & Shi, Y. (2011). Complex and dynamic landscape of RNA polyadenylation revealed by PAS-Seq. *RNA (New York)*, *17*(4), 761–772. https://doi.org/10.1261/rna.2581711.

Shi, Y. (2012). Alternative polyadenylation: New insights from global analyses. *RNA*, *18*(12), 2105–2117. https://doi.org/10.1261/rna.035899.112.

Spies, N., Burge, C. B., & Bartel, D. P. (2013). 3' UTR-isoform choice has limited influence on the stability and translational efficiency of most mRNAs in mouse fibroblasts. *Genome Research*, *23*(12), 2078–2090. https://doi.org/10.1101/gr.156919.113.

Tian, B., & Manley, J. L. (2017). Alternative polyadenylation of mRNA precursors [review]. *Nature Reviews Molecular Cell Biology*, *18*(1), 18–30. https://doi.org/10.1038/nrm.2016.116.

Wang, R., Nambiar, R., Zheng, D., & Tian, B. (2018). PolyA_DB 3 catalogs cleavage and polyadenylation sites identified by deep sequencing in multiple genomes. *Nucleic Acids Research*, *46*(D1), D315–D319. https://doi.org/10.1093/nar/gkx1000.

Wang, X., Hennig, T., Whisnant, A. W., Erhard, F., Prusty, B. K., Friedel, C. C., et al. (2020). Herpes simplex virus blocks host transcription termination via the bimodal activities of ICP27. *Nature Communications*, *11*(1), 293. https://doi.org/10.1038/s41467-019-14109-x.

Yoon, O. K., & Brem, R. B. (2010). Noncanonical transcript forms in yeast and their regulation during environmental stress. *RNA*, *16*(6), 1256–1267. https://doi.org/10.1261/rna.2038810.

Yoon, Y., Klomp, J., Martin-Martin, I., Criscione, F., Calvo, E., Ribeiro, J., et al. (2019). Embryo polarity in moth flies and mosquitoes relies on distinct old genes with localized transcript isoforms. *eLife*, *8*. https://doi.org/10.7554/eLife.46711.

Zheng, D., Liu, X., & Tian, B. (2016). 3'READS+, a sensitive and accurate method for 3′ end sequencing of polyadenylated RNA. *RNA (New York)*, *22*(10), 1631–1639. https://doi.org/10.1261/rna.057075.116.

Zhou, X., Li, R., Michal, J. J., Wu, X. L., Liu, Z., Zhao, H., et al. (2016). Accurate profiling of gene expression and alternative polyadenylation with whole transcriptome termini site sequencing (WTTS-Seq). *Genetics*, *203*(2), 683–697. https://doi.org/10.1534/genetics.116.188508.

Zhu, Y., Wang, X., Forouzmand, E., Jeong, J., Qiao, F., Sowd, G. A., et al. (2018). Molecular mechanisms for CFIm-mediated regulation of mRNA alternative polyadenylation. *Molecular Cell*, *69*(1), 62–74. e64 https://doi.org/10.1016/j.molcel.2017.11.031.

CHAPTER THREE

TRENDseq—A highly multiplexed high throughput RNA 3′ end sequencing for mapping alternative polyadenylation

Anton Ogorodnikov[a,b,c,†] **and Sven Danckwardt**[a,b,c,d,*]

[a]Posttranscriptional Gene Regulation, University Medical Centre Mainz, Mainz, Germany
[b]Institute for Clinical Chemistry and Laboratory Medicine, University Medical Centre Mainz, Mainz, Germany
[c]Centre for Thrombosis and Hemostasis (CTH), University Medical Centre Mainz, Mainz, Germany
[d]German Centre for Cardiovascular Research (DZHK), Berlin, Germany
*Corresponding author: e-mail address: sven.danckwardt@unimedizin-mainz.de

Contents

Abstract

Alternative polyadenylation (APA) is a widespread and highly dynamic mechanism of gene regulation. It affects more than 70% of all genes, resulting in transcript isoforms with distinct 3′ end termini. APA thereby considerably expands the diversity of the transcriptome 3′ end (TREND). This leads to mRNA isoforms with profoundly different physiological effects, by affecting protein output, production of distinct protein isoforms, or modulating protein localization. APA is globally regulated in various conditions, including developmental and adaptive programs. Since perturbations of APA can disrupt biological processes, ultimately resulting in most devastating disorders, querying the APA landscape is crucial to decipher underlying mechanisms, resulting consequences

[†] Present address: Ye Laboratory, Rheumatology Department, University of California San Francisco (UCSF), School of Medicine, San Francisco, CA, United States.

Methods in Enzymology, Volume 655
ISSN 0076-6879
https://doi.org/10.1016/bs.mie.2021.03.022

and potential diagnostic and therapeutic implications. Here we provide a detailed step-by-step protocol for TRENDseq, a method for transcriptome-wide high-throughput sequencing of polyadenylated RNA 3′ ends in a highly multiplexed fashion. TRENDseq exploits linear amplification of the starting material to improve sensitivity while significantly reducing the amount of input material. It thereby represents a powerful tool to study APA in numerous experimental set-ups and/or limited human samples in a highly multiplexed and reproducible manner.

1. Introduction

Next-generation RNA sequencing (RNAseq) has led to the discovery of a perplexingly complex metazoan transcriptome architecture that results from the alternative use of transcription start sites, exon and introns, and polyadenylation sites (Carninci et al., 2005; Reyes & Huber, 2017; Wang et al., 2008). The combinatorial use and incorporation of such elements into mature transcript isoforms significantly expands genomic information and is subject to dynamic spatial and temporal modulation during development and adaptation. Recently, diversification of the transcriptome at the 3′ end by alternative polyadenylation (APA) evolved as an important and evolutionarily conserved layer of gene regulation (Elkon, Ugalde, & Agami, 2013; Tian & Manley, 2017). It results in transcript isoforms that vary at the RNA 3′ end, which can have profoundly different physiological effects. APA in the coding sequence can directly alter protein functions, while variations in the 3′ untranslated region (UTR) can impact the stability, subcellular localization and translation of the mRNA, or direct protein localization in trans through inclusion or exclusion of regulatory RNA elements (Mayr, 2017).

Constitutive RNA 3′ end processing relies on a complex macromolecular machinery that catalyzes endonucleolytic cleavage and polyadenylation (CPA) of pre-mRNA molecules (Danckwardt, Hentze, & Kulozik, 2008). This involves the assembly of four multi-component protein complexes (CPSF, CSTF, CFI and CFII) (Shi et al., 2009) on the pre-mRNA at dedicated, but largely poorly conserved, processing sites (Gruber et al., 2016). Differential expression of individual CPA components (Jenal et al., 2012; Marini, Scherzinger, & Danckwardt, 2021; Masamha et al., 2014; Ogorodnikov et al., 2018) or selective regulation of their binding properties (Danckwardt et al., 2011) can direct the dynamic modulation of APA resulting in transcript isoforms with alternative 3′ ends. In addition, other mechanisms including RNA polymerase II kinetics (Neve et al., 2016; Pinto et al., 2011; Turner et al., 2020) or epigenetic events (Brumbaugh et al., 2018; Ke et al., 2015; Li et al., 2016; Parker et al., 2020; Spies, Nielsen, Padgett, & Burge, 2009; Wood et al., 2008; Yue et al., 2018) regulate APA, illustrating a previously unanticipated complex crosstalk between various cellular processes in the control of transcriptome 3′ end diversity (TREND). Although difficult to detect by standard high throughput profiling techniques (Ha, Blencowe, & Morris, 2018; Hoque et al., 2013; Xia et al., 2014), dynamic changes at the transcriptome 3′ end are widespread (Marini et al., 2021; Mayr & Bartel, 2009; Ozsolak et al., 2010; Sandberg, Neilson, Sarma, Sharp, & Burge, 2008). They affect more than 70% of all genes.

Global APA changes are commonly associated with differentiation and dedifferentiation processes (Turner, Pattison, & Beilharz, 2017). This requires a well-coordinated temporal and spatial interaction between RNA motifs and the multi-component processing complex to ensure that CPA occurs in a timely fashion and at the right position (Gruber & Zavolan, 2019; Lutz & Moreira, 2011). Hence, 3′ end processing is tightly coupled to other co- and posttranscriptional processes (Danckwardt et al., 2008), and controlled by delicate (auto-) regulatory mechanisms (Kamieniarz-Gdula et al., 2019; Wang, Zheng, Wei, Ding, & Tian, 2019). Transcripts that show dynamic regulation at the 3′ end are typically encoded by phylogenetically old genes, which corresponds to the phylogenetic age of most executing APA regulators (Ogorodnikov et al., 2018). Finally, individual components of the CPA machinery are functionally dominant over 'neighboring' factors within and across the multi-component CPA complexes in the control of APA (Ogorodnikov et al., 2018).

While the underlying mechanisms of APA regulation are still being elucidated (Marini et al., 2021), disruption of this process clearly proved to be clinically relevant (Nourse, Spada, & Danckwardt, 2020). For example, APA perturbations are associated with various disorders (Lin et al., 2012; Masamha et al., 2014; Patel, Brophy, Hickling, Neve, & Furger, 2019; Soetanto et al., 2016; Weng et al., 2019; Ye, Zhou, Hong, & Li, 2019) (and refs therein), but they can also possess direct disease eliciting activities, act as oncogenic driver, and thereby mimic genetic alterations (Lee et al., 2018; Ogorodnikov et al., 2018). Importantly, such changes are generally missed in genome profiling endeavors (Stacey et al., 2011). But often they also remain undetected by standard RNAseq technologies

(Ogorodnikov, Kargapolova, & Danckwardt, 2016). However, even when resulting in primarily subtle changes of non-coding RNA sequence elements in the 3′UTR (Geisberg, Moqtaderi, Fan, Ozsolak, & Struhl, 2014), APA perturbations can be functionally significant (Ogorodnikov et al., 2018) and represent unexpectedly potent novel biomarkers (Ogorodnikov et al., 2018; Xia et al., 2014). Aberrant posttranscriptional expansion of the genome complexity can thus result in most devastating consequences; at the same time, it also opens novel diagnostic and therapeutic avenues (Nourse et al., 2020).

The identification, profiling and mapping of polyadenylation events in a transcriptome-wide manner is demanding. It requires the reliable detection of non-templated poly(A) tails in the transcriptome (which must not be confused with genetically encoded adenosine-rich stretches (Ha et al., 2018; Hoque et al., 2013; Jan, Friedman, Ruby, & Bartel, 2011)). In addition, most genes in eukaryotes possess multiple poly(A) sites that can mediate APA (Derti et al., 2012). While a large proportion of APA dynamics (approximately 80%) is found in 3′UTRs (so called "*tandem*" APA (Ogorodnikov et al., 2018)), a significant number of APA affects the body of genes, i.e., the intronic or the coding regions (referred to as "*internal*" APA). The tight coupling of various RNA processing steps (including alternative splicing) can result in APA ultimately leading to a complex co-existence of various mRNA isoforms (Fig. 1). This significantly

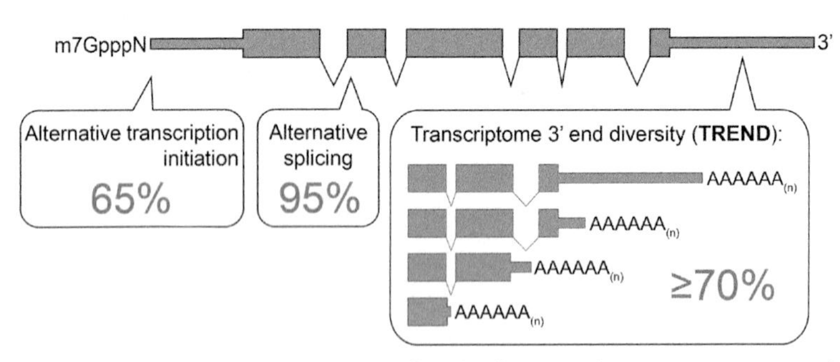

Fig. 1 Alternative polyadenylation diversifies the functional transcriptome complexity. The genome complexity is considerably expanded by co- and posttranscriptional processes. Various mechanisms including alternative polyadenylation (APA) can result in transcriptome 3′ end diversification (TREND) affecting the coding sequence and/or 3′ untranslated regions (UTR). More than 70% of all genes expressed have alternative 3′ ends, thus shaping the transcriptome complexity to a similar extent as alternative splicing (95%) or alternative transcription initiation (65% (Ogorodnikov et al., 2018)).

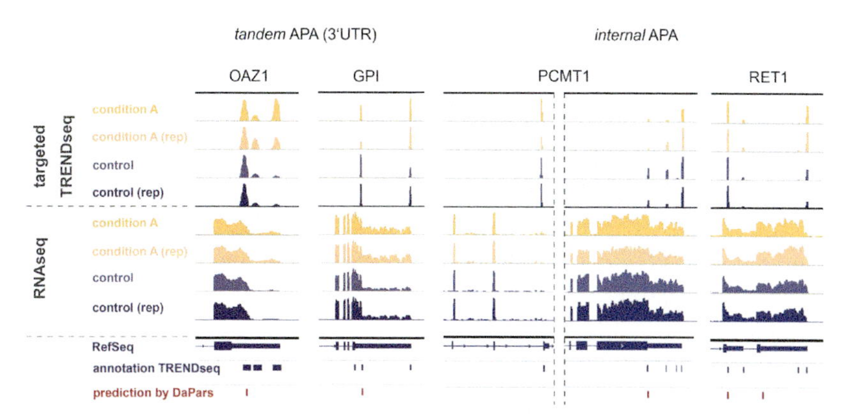

Fig. 2 Comparison of targeted and non-targeted high-throughput RNA sequencing for the definition of the transcriptome 3′ end. Visualization of the transcriptome 3′ end (four selected examples of polyadenylated RNAs) by targeted (TRENDseq, upper four lanes) compared to non-targeted RNAseq (lower four lanes) and subsequent definition of the mRNA 3′ end by DaPars (Xia et al., 2014). Shown are differential changes of the mRNA 3′ end signature comparing two conditions ("condition A" and "control," in biological replicates). The *lower panel* illustrates the respective RefSeq 3′UTR and the annotation of the TRENDseq data (blue) in comparison with a differential use of poly(A) sites by DaPars (pink). Of note, internal APA (illustrated for PCMT1 and Ret1, right two panels) can be reliably detected based on a targeted sequencing approach, despite significantly reduced sequencing depth. *Modified after Ogorodnikov, A., Kargapolova, Y., & Danckwardt, S. (2016). Processing and transcriptome expansion at the mRNA 3′ end in health and disease: Finding the right end.* European Journal of Physiology, 468*(6), 993–1012.*

complicates the analysis of APA based on non-targeted high-throughput sequencing technologies. Furthermore, subtle, yet biologically relevant APA dynamics affecting low abundant RNA isoforms are more likely to be missed by such global approaches. While APA affecting the 3′UTR (or last exon) can be detected reliably based on (already existing) RNAseq data by bioinformatic algorithms (e.g., DaPars (Xia et al., 2014), QAPA (Ha et al., 2018)), the detection of internal APA is demanding, and typically requires targeted sequencing approaches to identify such events in higher resolution (Fig. 2).

In an attempt to systematically profile dynamic APA events in tumorigenic differentiation/dedifferentiation processes, we recently developed a newly designed high-throughput sequencing approach suited to capture polyadenylated transcript 3′ ends of numerous experimental conditions in a highly multiplexed fashion (Ogorodnikov et al., 2016). TRENDseq exploits linear amplification of the starting material improving the complexity of the sequencing-library despite significantly reducing the amount of input RNA, a particularly limited resource in the context of clinical specimens. TRENDseq thereby delivers high quality data on limited starting material (as little as 100 ng total RNA) in highly multiplexed experimental setups (Ogorodnikov et al., 2018).

Here, we provide a detailed step-by-step protocol for TRENDseq. In a proof-of-concept, TRENDseq has been used for cataloging the dynamic nature of the APA landscape transcriptome-wide upon targeting >170 components involved in the definition of RNA 3′ ends (Ogorodnikov et al., 2018). Importantly, this highly multiplexed screening functionally recapitulates known protein complex compositions involved in polyadenylation (Shi & Manley, 2015) (and the cross-talk to other RNA-processing events (Danckwardt et al., 2008)), highlighting throughput, biological relevance, and accuracy of this technique. TRENDseq thereby helps to reliably map and illuminate the role of APA for various biological programs and potential disease mechanisms (Marini et al., 2021). It also permits the identification of novel, clinically most potent diagnostic signatures (Ogorodnikov et al., 2018).

2. TRENDseq protocol—Overview

In order to analyze and map APA in an unbiased and high throughput fashion transcriptome-wide, Next Generation Sequencing (NGS) based technologies are considered state of the art. Despite significant progress on available approaches employing targeted RNA 3′ end sequencing (Berg et al., 2012; Derti et al., 2012; Fu et al., 2011; Shepard et al., 2011; Wang, Dowell, & Yi, 2013; and refs. therein), the capacity of multiplexing based on limited input RNA material is an issue, especially for high-throughput analyses, for example, in the context of numerous experimental conditions in basic (e.g., in screening setups) and translational science (e.g., in cohort studies). To this end, we developed TRENDseq, which makes use of existing NGS platforms for high-throughput analysis in combination with a massive sample multiplexing rate on limited input material (Hashimshony, Wagner, Sher, & Yanai, 2012). To this end, RNA samples are reverse transcribed in presence of a primer containing a T7 promoter,

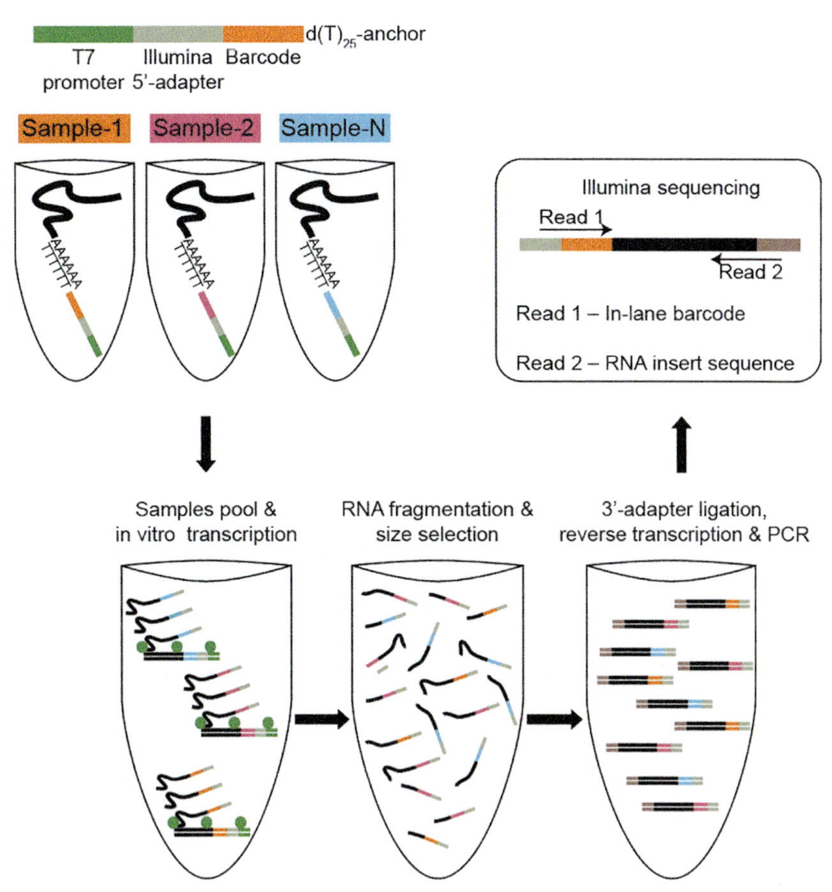

Fig. 3 Workflow of TRENDseq library design for a highly multiplexed transcriptome-wide analysis of alternative polyadenylation (APA). TRENDseq library preparation (see detailed step-by-step protocol) based on barcoded RT-primers with pooling of up to 72 individual experiments in a single library for Illumina Next Generation Sequencing.

an Illumina 5′ adapter, individual in-lane barcode and an anchored oligo-dT stretch (Fig. 3). The in-lane barcode is used to label individual RNA samples for library multiplexing.

Following reverse transcription and second strand cDNA synthesis, samples corresponding to individual RNA inputs are pooled, purified and in vitro transcribed. After in vitro transcription, the obtained amplified aRNA (antisense-RNA) is fragmented and size-selected. This permits a targeted sequencing restricted to the penultimate 25–35 nucleotides upstream of the poly(A) tail of the mRNA (compare Fig. 2) and thereby significantly

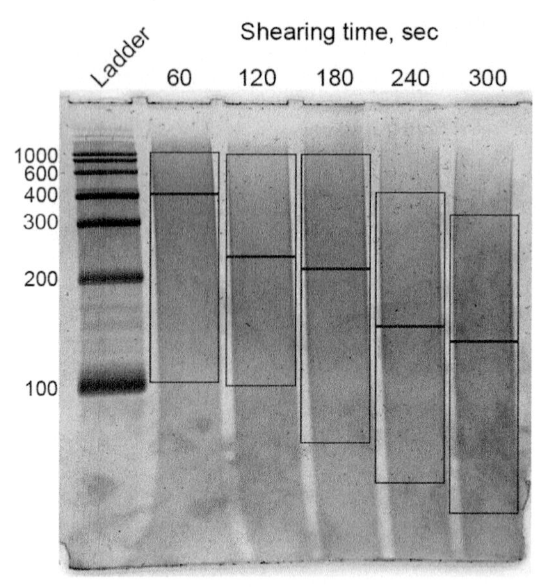

Fig. 4 TRENDseq library synthesis—illustration of aRNA shearing step optimization. aRNA is sheared using a Covaris M220 Focused-Ultrasonicator™ for various durations. The sheared RNA is visualized by the Urea-PAGE showing that longer shearing durations are required to obtain RNA fragments of desired size (100 nt; the regions with maximal signal intensity are marked with horizontal lane, boxes highlight the regions with ~50% of maximal signal intensity).

increases effective read coverage and resolves closely spaced CPA events. To achieve fragmentation in a sequence-unbiased manner, a mechanical fragmentation by ultrasonication is recommended. However the shearing has to be adjusted to produce aRNA fragments of the required size (Fig. 4). In our case, the selected range of 100 nucleotides was estimated as the sum of RNA insert and the size of RT primer (without T7 promoter).

Next, the purified RNA is 5′- and 3′-end dephosphorylated, followed by a specific RNA 5′end phosphorylation using T4 polynucleotide kinase. The RNA is then purified, followed by ligation of a 3′-adapter to the 3′-phosphoryl terminated end of the RNA. The resulting RNA with ligated 3′-adapter is then reverse transcribed using a primer specific to the ligated 3′-adapter, followed by PCR amplification with Illumina-compatible RNA library preparation kits. Thereafter, the amplicon libraries are purified, quality-checked and pooled together in equimolar quantities. Sequencing is typically performed on Illumina HiSeq 2500 or NextSeq 500 platforms with addition of 30% PhiX. Read 1 and Read 2 cover 9 nucleotides and 50 nucleotides respectively, Illumina TruSeq index are sequenced as a dedicated read.

3. Bioinformatic pipeline

For bioinformatic analysis of TRENDseq, we use standard tools for genome analysis in combination with custom constructed pipelines (Fig. 5). Raw sequencing data (FASTQ format) are first demultiplexed using the in-lane RT primer barcode with an average per base quality score above 20, followed by A- and T-stretches trimming as described previously (Hashimshony et al., 2012). Resulting sequences with average length of 25–35 nucleotides are mapped to the human genome using bowtie2 aligner (Langmead & Salzberg, 2012). Mapped reads are filtered from internal priming events using the assembly of a high quality, bona fide poly(A) site/TREND annotation (e.g., obtained by using 3′READS (Hoque et al., 2013)). The number of reads associated with each poly(A) site are calculated using HTSeq (Anders, Pyl, & Huber, 2015) ("htseq-count" command, "intersection-strict" option). They reflect the expression of individual 3′ end RNA isoform(s) and thereby provide deep insights into the landscape of transcriptome 3′ end diversity (TREND). Exemplifying the accuracy of TRENDseq we refer to the high reproducibility of mapped reads (shown in Fig. 2 or Fig. 6), an important prerequisite for applications that go beyond the mere scientific interest (e.g., when applying TRENDseq as a diagnostic tool).

For downstream statistical analysis (e.g., comparing TREND signatures of different experimental conditions) in principle various approaches can be used, with each having advantages and disadvantages. Ideally the type of analysis and statistics applied should be tailored to the type and context of experiment. For the purposes of a proof of principle study, the expression level of the most prevalent transcript isoforms can be examined by Fisher's exact test in comparison to the respective other alternative 3′ end isoforms expressed by the same gene (Ogorodnikov et al., 2018). In this context, we constructed contingency tables including the number of reads of tested isoform and total amount of reads of all the other isoforms of the gene (in a conditional experimental setup, i.e., for knockdown and control samples, respectively).

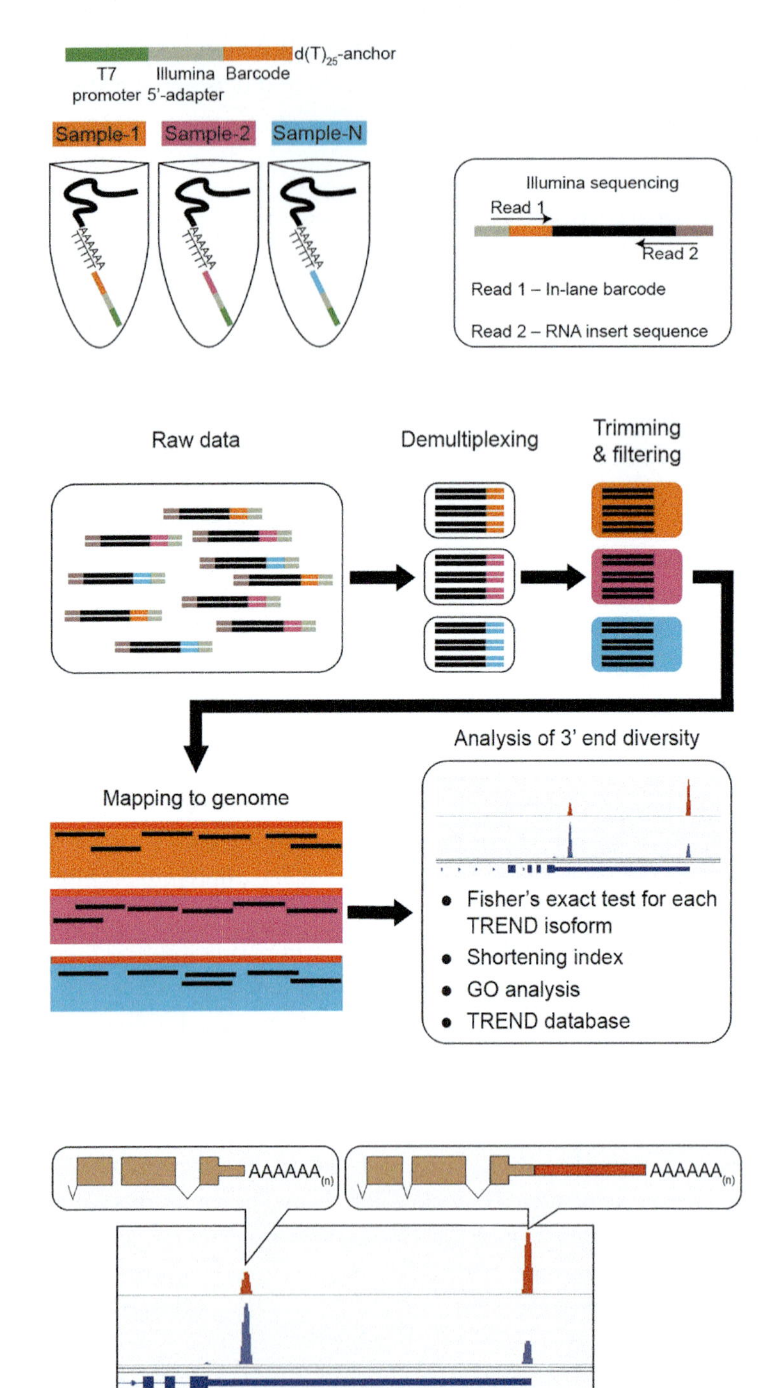

Fig. 5 See figure legend on opposite page.

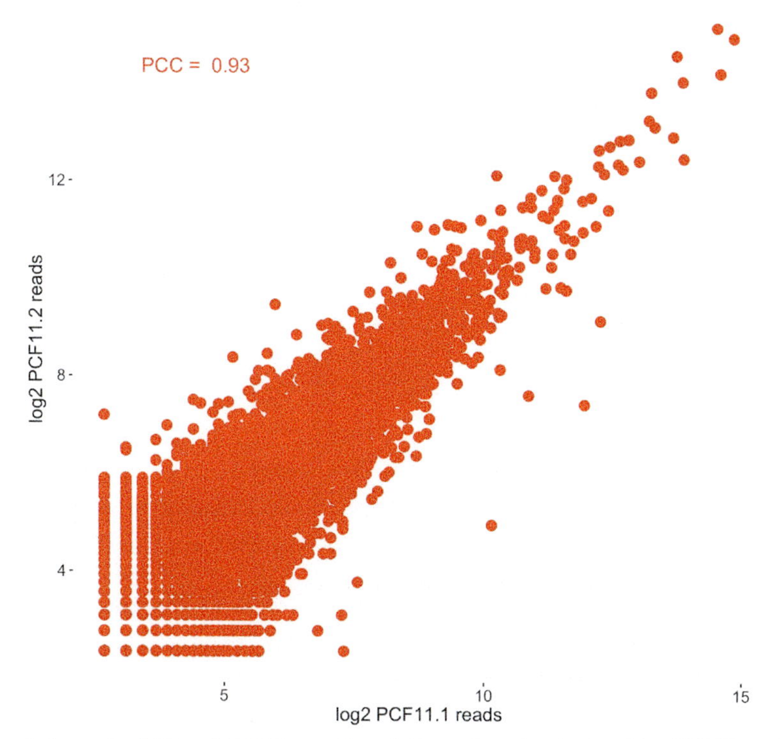

Fig. 6 Reproducibility of TRENDseq. Correlation of reads obtained by TRENDseq in a replicated experiment (depletion of one central component, PCF11, that controls the site of polyadenylation, compare Fig. 7 and (Ogorodnikov et al., 2018)).

Fig. 5—Cont'd Example for bioinformatic processing of TRENDseq data for a highly multiplexed transcriptome-wide analysis of alternative polyadenylation (APA). Bioinformatic analysis of TRENDseq combines the use of standard genome analysis tools and a custom pipeline for demultiplexing, trimming and filtering of the raw reads before mapping to the genome. *Bottom panel*: Schematic representation of short (left) and long (right) TREND-isoforms generated from the same gene by APA (corresponding to the respective peaks in the genome viewer (bottom) for two representative samples (red and blue)). Similar examples can be found in TREND-DB (http://shiny.imbei.uni-mainz.de:3838/trend-db/), a transcriptome-wide atlas of the dynamic landscape of alternative polyadenylation (Marini et al., 2021) based on massive parallel APA studies (Ogorodnikov et al., 2018) (and see Fig. 7).

$$\text{All reads of all CPA's of a gene } j \text{ condition } k:$$

$$A_{j,k} = \sum_{i=1}^{n} r_{i,j,k}$$

$$\text{For shortness let's introduce a substitution}:$$

$$W_{i,j,k} = A_{j,k} - r_{i,j,k}$$

$$\text{Fisher's exact test } p - \text{value}:$$

$$P_{i,j,k} = \frac{\left(r_{i,j,k} + W_{i,j,k}\right)!*\left(r_{i,j,k_0} + W_{i,j,k_0}\right)!*\left(r_{i,j,k} + r_{i,j,k_0}\right)!*\left(W_{i,j,k} + W_{i,j,k_0}\right)!}{r_{i,j,k}!*W_{i,j,k}!*r_{i,j,k_0}!*W_{i,j,k_0}!}$$

Where $r_{i,j,k}$ is number of raw reads for CPA i gene j in condition k (or k_0 for control)

Obtained p-values were adjusted using Benjamini-Hochberg method, and an adjusted p-value ≤ 0.05 filter was applied. To calculate fold-regulation per transcript isoform, the total amount of reads for each gene was normalized to 100%, and the relative percentage of individual isoforms in the experimental sample was divided by the percentage of the same isoform in the control. Further, if one wants to describe the overall tendency of a given gene to express shortened (or lengthened, respectively) transcript isoforms, a proxy of two most significantly (lowest BH-corrected p-value) APA-affected isoforms can be applied. This allows to calculate a "*shortening index*" as the fold-regulation of the shorter isoform normalized to the fold-regulation of the longer transcript isoform of the same gene (see below).

To illustrate throughput, biological relevance, and accuracy of TRENDseq, we refer to a recent proof-of-concept, in which this technique has been used for cataloging the dynamic landscape of APA transcriptome-wide upon RNAi-targeting >170 components involved in the definition of RNA 3′ ends (Fig. 7). Not only did we retrieve the molecular architecture of known protein complex compositions involved in polyadenylation (Shi & Manley, 2015) on a functional level, we were also able to identify novel diagnostic signatures based on APA, outperforming clinically established risk markers for survival in cancer patients (Ogorodnikov et al., 2018). This highlights the accuracy and wide applicability of targeted RNA 3′ end sequencing in biology and beyond.

A
Example for large scale APA screening (>170 experimental conditions)

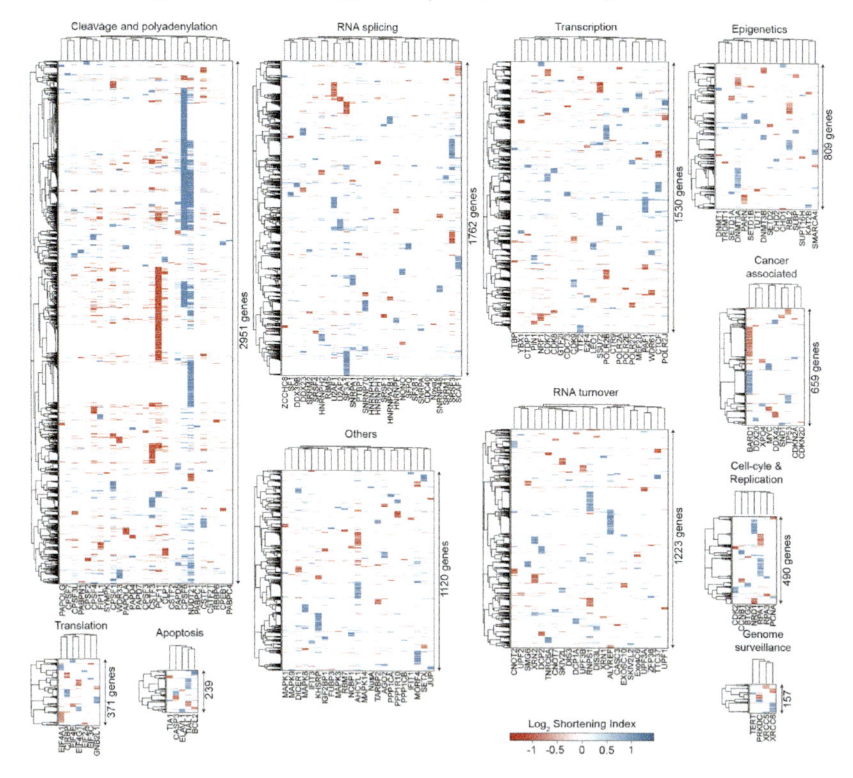

Fig. 7 Example for highly multiplexed analysis of APA in numerous experimental conditions. (A) Heat maps of APA-affected genes obtained by a screening covering more than 170 experimental conditions (knockdown "KD" of various factors with a potential role in APA regulation). For illustration purposes, representation grouped per functional category of depleted APA-regulator (X-axes; hierarchical clustering according to shortening index is based on Pearson's correlation coefficient and complete linkage method). (B) Upon further bioinformatic processing, detailed effects on APA can be visualized per experimental condition (identity of genes and APA-signatures are displayed in detail in the TREND-DB web explorer [http://shiny.imbei.uni-mainz.de:3838/trend-db]). Of note, TRENDseq thereby uncovers a large proportion of differential APA regulation at internal poly(A) sites ("internal" shortening and "internal" lengthening, respectively), modulating the generation of truncated protein isoforms. *Modified after Ogorodnikov, A., Levin, M., Tattikota, S., Tokalov, S., Hoque, M., & Scherzinger, D., et al. (2018). Transcriptome 3′end organization by PCF11 links alternative polyadenylation to formation and neuronal differentiation of neuroblastoma.* Nature Communications, *9(1), 5331. https://doi.org/10.1038/s41467-018-07580-5.*

(Continued)

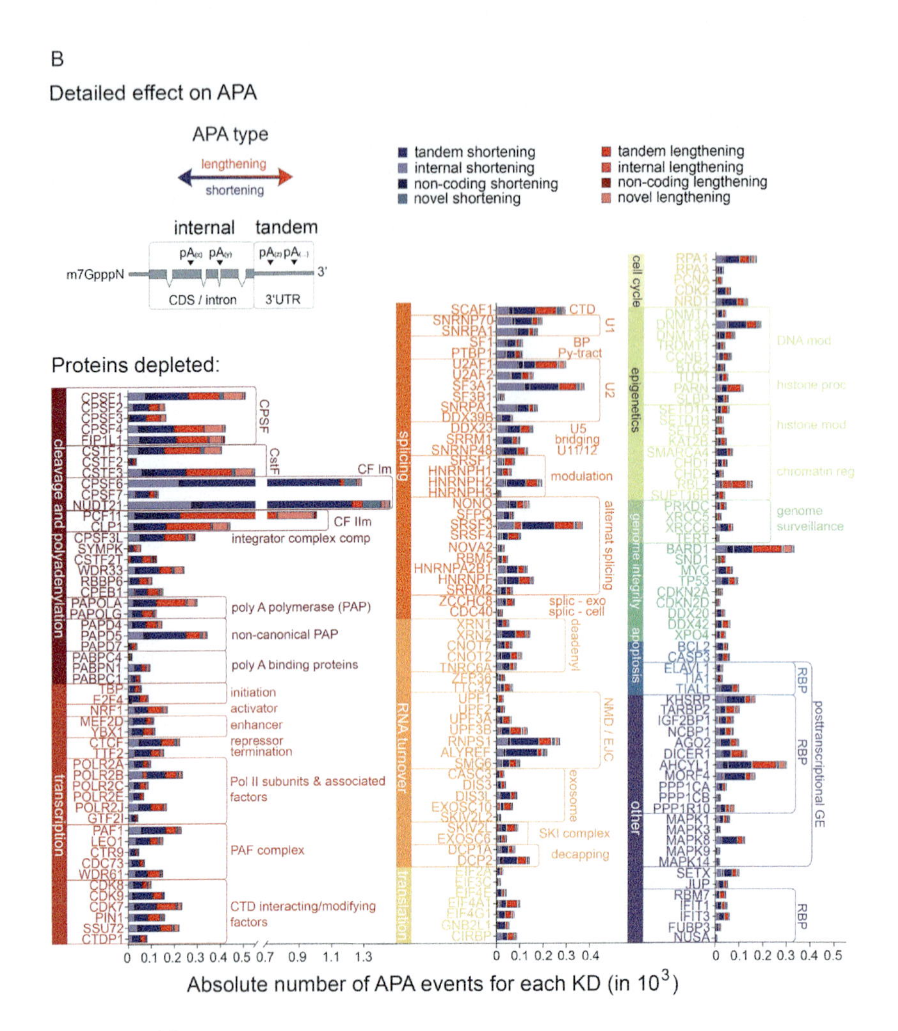

Fig. 7—Cont'd

4. Protocol at glance

Step-by-Step TRENDseq protocol at glance (a detailed protocol including the pipetting scheme is provided in the Appendix):

1. *Preparation of starting material.* RNA extraction from cell culture, tissues or organs (fresh or frozen) is performed according to standard protocols. RNA quality is assayed with an Agilent RNA 6000 Nano Kit (#5067-1512, Agilent Technologies) according to manufacturer's instructions (we recommend using RNA with RIN ≥9.2–9.5 (Fig. 8)).

All following reactions are performed in a PCR cycler (hotlid = 100 °C) in 0.2 μL tubes, if not specified otherwise. Components of MessageAmp II

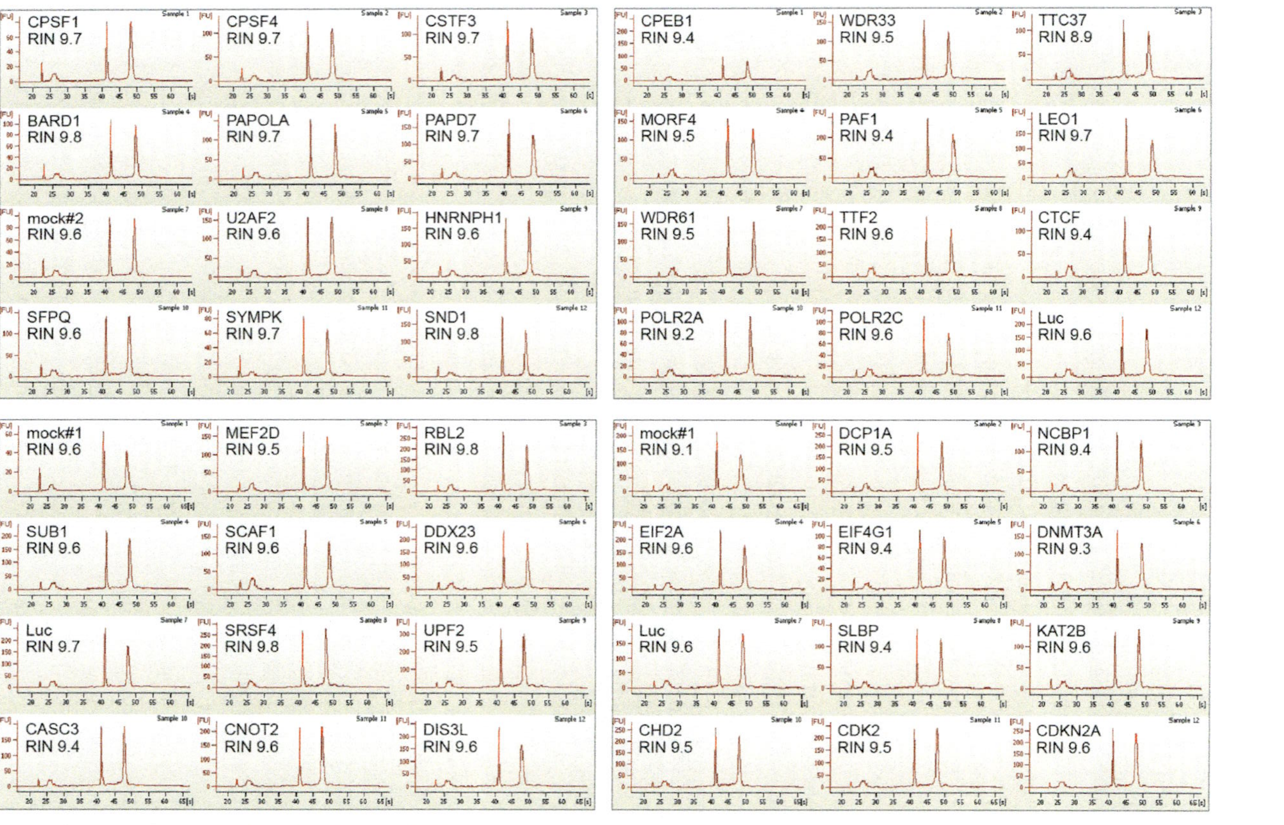

Fig. 8 Exemplified quality control of the starting RNA input using an Agilent Bioanalyzer.

aRNA Amplification Kit (#AM1751, ThermoFisher Scientific) are used, if not specified otherwise. Reverse transcription (RT) primer includes a T7 promoter, Illumina 5′ adapter, individual in-lane barcode and an anchored oligo-dT stretch (for details see Appendix), as described previously (Hashimshony et al., 2012).

2. *cDNA synthesis.* For the cDNA first strand synthesis 100 ng of total RNA input is mixed with 12.5 ng of individual RT primer and 1:2000 ERCC RNA Spike-In (#4456740, ThermoFisher Scientific) in 1.2 μL volume, incubated 10 min at 70 °C and chilled on ice. The following mix is added to comprise the final reaction volume of 2 μL: 10 × First Strand buffer (0.2 μL), dNTP (0.4 μL), RNAse inhibitor and ArrayScrip (0.1 μL each). The mix is incubated at 42 °C for 2 h, and followed by the second strand synthesis reaction immediately (see below).

The second strand cDNA synthesis mix includes 10 × Second Strand buffer (1 μL), dNTP (0.4 μL), First Strand mix (2 μL, full volume of the previous reaction), RNAse H (0.1 μL), DNA polymerase (0.2 μL) and RNase-free water (up to 10 μL). Reaction is carried out at 16 °C for 2 h (open PCR cycler lid). After this step, samples corresponding to individual RNA inputs are pooled together (up to 26 samples) and the cDNA is purified using MessageAmp II Kit components according to the manufacturer's protocol with modifications in the cDNA elution step. Specifically, elution of the cDNA is performed twice with 9 μL of pre-warmed (55 °C) nuclease-free water.

RNA amplification by in vitro transcription (antisense RNA, aRNA, synthesis) is carried out in 40 μL reaction format according to the manufacturer's protocol with 14 h of incubation at 37 °C. Obtained aRNA is purified using MessageAmp II Kit components in accordance with the manufacturer's instructions. Samples are eluted from the column with 100 μL of nuclease-free water.

3. *aRNA shearing* is performed using a Covaris M220 Focused-Ultrasonicator™ with Peak incident power 50 W, Duty Factor 20% and 200 Cycles per Burst (cbp) for 420 s at 7 °C. for size selection, sheared RNA is separated on 6% PAGE in denaturing conditions (7 M urea), and the gel region corresponding to 100 nucleotides is excised, frozen in liquid nitrogen and crushed. The RNA is eluted from the gel by 2 min incubation in 50–100 μL of a buffer containing 100 mM Tris-HCl (pH 8.0), 500 mM NaCl and 1% SDS at room temperature. Size-selected RNA is purified using the miRNeasy Kit (#217004, Qiagen) and eluted from the column with 25 μL of nuclease-free water. A typical outcome of a successful shearing procedure is shown in Fig. 9.
4. *Preparation for linker ligation.* The purified RNA is dephosphorylated with 5 units of Antarctic phosphatase (#M0289S, New England BioLabs) for

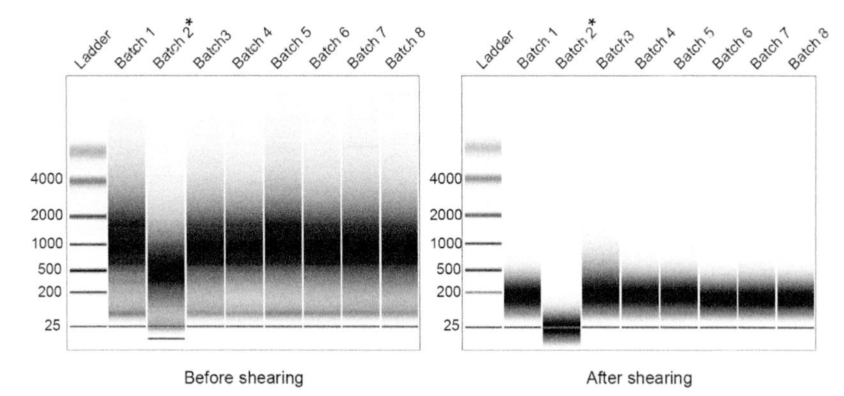

Fig. 9 Exemplified size analysis of aRNA before and after fragmentation. Typical representation (Agilent Bioanalyzer Total RNA Nano chip) of aRNAs before and after shearing using a Covaris M220 Focused-Ultrasonicator™. Shearing (along with the size selection described in the overview of the protocol) results in aRNA fragmentation, which allows TRENDseq analysis to focus on the penultimate 25–35 nucleotides upstream of the poly(A)-site (*technical artifact of Bioanalyzer chip).

30 min incubation according to manufacturer's recommendations. This is followed by RNA 5′ end phosphorylation with T4 Polynucleotide Kinase (#M0201S, New England BioLabs) in the presence of 1 mM ATP (for 60 min) using the manufacturer's protocol. The RNA is then purified using the miRNeasy Kit (#217004, Qiagen), and 30 μL of eluate is concentrated to 10 μL using a vacuum concentrator.

5. *Linker ligation, library generation and quality control.* The following procedures are performed with Illumina TruSeq Small RNA Library Preparation Kits (#15016911—Core solutions, # 15016912—Indexes) if not specified otherwise. RNA obtained from the preceding step and the RNA 3′ adapter (RA3) are incubated for 2 min at 70 °C and immediately placed on ice to prevent annealing. The ligation reaction is performed in HM Ligation buffer (HML) in presence of RNase inhibitor and 200 units of a truncated T4 RNA Ligase 2 (#M0242S, New England BioLabs). After 1 h at 28 °C, 1 μL of stop solution (STP) is added to the reaction and incubation continues for 15 min at the same conditions. The resulting RNA (with ligated 3′ adapter) is reverse transcribed for 1 h at 50 °C using 200 units of SuperScript® II Reverse Transcriptase (#18064014, ThermoFisher Scientific) according to the manufacturer's protocol. The following PCR amplification is performed with the Illumina TruSeq Small RNA Library Preparation Kits with up to 10 cycles (melting: 10 s 98 °C, annealing: 30 s 60 °C, elongation: 30 s 72 °C). Each pooled library is barcoded with the individual TruSeq index. Amplicon libraries are purified applying AMPure XP Beads

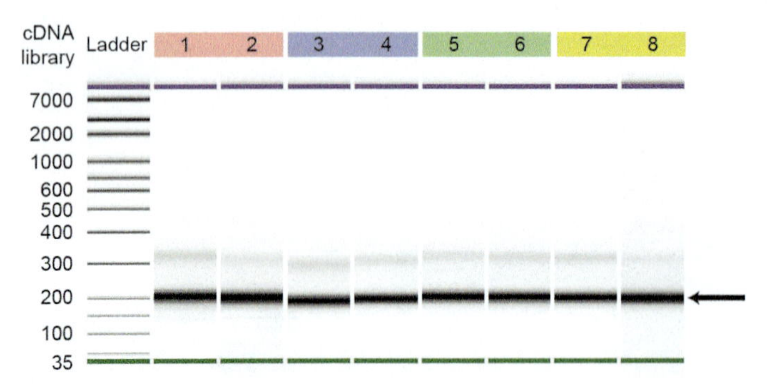

Fig. 10 Size-confirmation of cDNA libraries. The expected size of cDNA (200 base pairs, major band) is highlighted (arrow). Pairing samples (marked with the same color) were labeled with different Illumina indices, pooled in the equimolar quantities and sequenced (Illumina HiSeq 2500).

(#A63880, Beckman Coulter), quality-checked with Agilent 2100 Bioanalyzer (Fig. 10) and pooled together in the equimolar quantities.

6. *Sequencing* is performed on Illumina HiSeq 2500 or NextSeq 500 platforms with addition of 30% PhiX (#FC-110-3001, Illumina). Read 1 and Read 2 cover 9 nucleotides and 50 nucleotides respectively, Illumina TruSeq index is sequenced as a dedicated read.

5. Summary and outlook

The advent of next generation sequencing (NGS) technologies has propelled biomedical sciences and promoted our understanding of gene regulatory mechanisms. NGS has recently revealed APA leading to an enormous diversity at the transcriptome 3′ end on a cellular level, across tissues, developmental stages and species (Derti et al., 2012; Mayr & Bartel, 2009; Ozsolak et al., 2010; Sandberg et al., 2008). Here we present TRENDseq, which combines comprehensiveness and scalability of APA analyses on limited input material with high sensitivity in a cost-effective manner.

Compared to RNAseq, targeted sequencing restricted to the transcriptome 3′ end directly uncovers the variability and perturbations of CPA occurring at the mRNA 3′ end (Fig. 2). This has several advantages; it increases sensitivity and spatial resolution to detect (minor and/or closely spaced) APA events, it reduces the bioinformatic complexity and workload (as the readout is not confounded by other variables that complicate the data processing such as alternative splicing). Further, TRENDseq relies on a technology used for single cell sequencing (Hashimshony et al., 2012). It thereby allows working with minimal input material and opens ample opportunities for multiplexing—while still keeping a high sequencing coverage for a reliable analysis.

TRENDseq has been used for a transcriptome-wide large scale profiling of dynamic APA events in numerous experimental conditions (Ogorodnikov et al., 2018). In this context we depleted more than 170 proteins, including all known factors involved in pre-mRNA 3′ end cleavage and polyadenylation in eukaryotes and selected key factors regulating transcriptional activities, splicing, RNA turnover, and other functions (Batra et al., 2014; Bava et al., 2013; Dermody & Buratowski, 2010; Di Giammartino et al., 2014; Dutertre et al., 2014; Fusby et al., 2016; Hallais et al., 2013; Huang et al., 2012; Kaida et al., 2010; Katahira et al., 2013; Licatalosi et al., 2008; Müller-McNicoll et al., 2016; Neve et al., 2016; Shi et al., 2009) which could directly or indirectly affect APA (Proudfoot, 2016; Tian & Manley, 2017). Combined with a high confidence reference poly(A) annotation obtained by 3′READS (Hoque et al., 2013) this screening resulted in the identification of numerous drivers directing APA (Ogorodnikov et al., 2018; Marini et al., 2021). Most importantly, the functional insights obtained here independently recapitulate the known molecular composition of protein complexes involved in polyadenylation, illustrating biological relevance, accuracy and technical reproducibility of TRENDseq (Fig. 6). The screening also uncovered the decisive role of the transcription termination factor PCF11 for proper neuronal development, which—when perturbed—results in tumor formation (Ogorodnikov et al., 2018). Again, the APA signatures retrieved by TRENDseq closely mirror APA effects determined by other technologies. Furthermore they are predictive, and permit forecasting cancer patient fate (death and high risk). Finally, the dataset obtained from this screening underlies TREND-DB, a user-friendly resource cataloging the dynamic landscape of APA (Marini et al., 2021).

TRENDseq (and complementary technologies (Kargapolova, Levin, Lackner, & Danckwardt, 2017)) will foster disentangling how APA directs dedicated functional programs, how APA perturbations interfere with these and thereby contribute to disease, and whether there are diagnostic signatures and potential therapeutic targets that can be exploited in future (Nourse et al., 2020; Nourse & Danckwardt, 2021).

Acknowledgments

The authors would like to express their gratitude to current and former members of the Danckwardt lab. This work is supported by grants of the DFG (DA 1189/2-1), the GRK 1591, the DFG Priority Program SPP 1935 (Deciphering the mRNP code: RNA bound Determinants of Post-transcriptional Gene Regulation), by the Federal Ministry of Education and Research (BMBF01EO1003), by the EU Framework Programme for Research and Technological Development Horizon 2020 (TICARDIO, Marie Skłodowska-Curie Innovative Training Network), by the Hella Bühler Award for Cancer Research, and by the German Society of Clinical and Laboratory Medicine (DGKL).

Appendix: ERC protocol

TRENDseq—A highly multiplexed high throughput RNA 3′ end sequencing for mapping alternative polyadenylation

key resources table.

Reagent or resource	Source	Identifier
Critical Commercial Assays		
MessageAmp II aRNA Amplification Kit	ThermoFisher Scientific	AM1751
ERCC RNA Spike-In	ThermoFisher Scientific	4,456,740
miRNeasy Kit	Qiagen	217,004
Antarctic phosphatase	New England BioLabs	M0289S
TruSeq Small RNA Library Preparation Kit. Core solutions	Illumina	15,016,911
TruSeq Small RNA Library Preparation Kit. Indexes	Illumina	15,016,912
T4-RNA Ligase 2, truncated	New England BioLabs	M0242S
Agilent RNA 6000 Nano Kit	Agilent Technologies	5067-1512
Agilent High Sensitivity DNA Kit	Agilent Technologies	5067-4627
Qubit® dsDNA HS Assay Kit	ThermoFisher Scientific	Q32851
Oligonucleotides		
Barcoded RT primer	5′-CGATTGAGGCCGGTAATACGACTCACTATA GGGGTTCAGAGTTCTACAGTCCGACGATC[8-bp-barcode] TTTTTTTTTTTTTTTTTTTTTTTTV-3′	

key resources table.—cont'd

Reagent or resource	Source	Identifier
T7 promoter	5′-TAATACGACTCACTATAGGG-3′	
Illumina RA5 adapter (Truseq Small RNA kit)	5′-GUUCAGAGUUCUACAGUCCGACGAUC-3′	
Illumina RA3 adapter (Truseq Small RNA kit)	5′-TGGAATTCTCGGGTGCCAAGG-3′	
Illumina RTP primer (Truseq Small RNA kit)	5′-GCCTTGGCACCCGAGAATTCCA-3′	
Illumina RP1 primer (Truseq Small RNA kit)	5′-AATGATACGGCGACCACCGAGATCTACACG TT-CAGAGTTCTACAGTCCGA-3′	
Illumina RPI[1–48] primers (Truseq Small RNA kit)	5′-CAAGCAGAAGACGGCATACGAGAT[6-bp RPI] GTGACTGGAGTTCCTTGGCACCCGAGAATT CCA-3′	
Read 1 sequencing primer	5′-GTTCAGAGTTCTACAGTCCGACGATC-3′	
Index read primer	5′-TGGAATTCTCGGGTGCCAAGGAACTCCAGT CAC-3′	
Read 2 sequencing primer	5′-GTGACTGGAGTTCCTTGGCACCCGAGAATT CCA-3′	
Illumina P5 adapter	5′-AATGATACGGCGACCACCGAGATCTACAC-3′	
Illumina P7 adapter	5′-CAAGCAGAAGACGGCATACGAGAT-3′	

Materials and equipment

- 2100 Bioanalyzer
- Qubit 2.0 Fluorometer
- Focused-Ultrasonicator (Covaris)
- Thermocycler

Step-by-step method details

1. Preparation of starting material

RNA extraction from cell culture, tissues or organs (fresh or frozen) and RNA quality control.

TIMING **70–90** min.

!CAUTION TriFast (or an analog) reagent is highly toxic. All manipulations with TriFast and plastic waste disposal should be allocated to fume hood. All contact with skin and eyes must be avoided.

i. Add 1 mL of TriFast to 50–100 mg of tissue or up to $10*10^6$ cells. If using a monolayer cell culture lyse cells directly in a cell culture dish (1 mL/10 cm^2). Starting material can be as low as 10^3 cells or 0.32 cm^2 (96 well format) of cells in monolayer. Transfer 1 mL of lysate to 1.5 mL safe seal Eppendorf tube.

NOTE *samples can be stored at −20 °C at this point.*

ii. Add 200 μL of chloroform. Vortex for 60 s

!CAUTION Ensure that the lid of the Eppendorf tube is perfectly closed to avoid leakage of the TriFast-chloroform mix. Leave at room temperature for 10 min. Prepare a fresh 1.5 mL sample tube for the next step.

NOTE *if started with less than 10^5 cells add 20 μg of glycogen RNA grade to the prepared tube.*

iii. Spin the sample at 13.000 g, 4 °C for 5 min. Transfer 500 μL of the upper aqueous phase to a fresh tube.

CRITICAL Do not touch organic phase or interphase.

iv. Add 500 μL (1:1) of isopropanol, vigorously shake the tube and incubate at room temperature for 10 min.

NOTE *incubations with isopropanol at lower temperatures lead to precipitation of salts, which will compromise resulting RNA purity.*

v. Spin the sample at 13.000 g, 4 °C for 10 min. Discard supernatant.

CRITICAL Starting from this point samples should be kept on ice.

vi. Add 1.2 mL of ice cold 75% ethanol, invert the tube 5–10 times and spin at 13.000 g, 4 °C for 10 min. Discard supernatant.

vii. Repeat ethanol wash from step (v). Leave ~200 μL of ethanol in the tube and give a short spin at 4 °C. Remove the remnants of ethanol with 200 μL pipette tip.

viii. To dissolve the pellet, add 40 μL of RNase-free water to the sample and place the tube with the open lid into a thermoblock at 55 °C for 5 min. If started with less than 10^5 cells use 20 μL of water for pellet reconstitution.

ix. Measure RNA concentration using absorption of 260 nm lightwave. Control RNA integrity with Agilent Bioanalyzer (e.g., Agilent RNA 6000 Nano Kit), use RNA with RIN ≥ 9.2–9.5 (Fig. 8).

PAUSE RNA can be stored at −80°C after this point.

All following reactions are performed in a PCR cycler (hotlid = 100 °C) in 0.2 μL tubes, if not specified otherwise. Components of MessageAmp II aRNA Amplification Kit (#AM1751, ThermoFisher Scientific) are used if not specified otherwise. Reverse transcription (RT) primer includes a T7 promoter, Illumina 5′ adapter, individual in-lane barcode and an anchored oligo-dT stretch (Hashimshony et al., 2012).

2. cDNA synthesis and RNA amplification

TIMING **20** h. This step is based on MessageAmp™ II aRNA Amplification Kit Ambion (ThermoFisher Scientific). The following procedure is described for a set of up to 24 samples.

NOTE *all incubations have to be performed in thermocycler in 0.2 mL low-binding tubes, if not specified otherwise.*

i. Prepare dilution of each of the total RNA samples using the formula:

$$C = 1200/N/0.6$$

where C—final concentration of an RNA sample (ng/μL), N—number of samples, 1200—total amount of all RNA samples (ng) that will go into the next steps, 0.6—volume of RNA sample used for cDNA synthesis (μL)

ii. Prepare a mix of barcoded RT-primer (unique for each sample, see table) with ERCC spike ins (per RNA sample):

Reagent	Vx1
$\sum$V	**0.6 μL**
RT-Primer (25 ng/μL)	0.5
Spike in (1:100)	0.1

NOTE *Prepared primers and ERCC spike in mixes can be stored at −80°C up to 6 month.*

iii. Mix the total RNA sample with RT-primer/ERCC mix (from step ii).

Reagent	Vx1
∑V	**1.2 μL**
Total RNA (120 ng)	0.6
RT-primer + Spike in	0.6

Incubate 10 min at 70 °C, move to ice. Quickly spin at max speed for a few seconds to collect as many droplets as possible before the next step, and then return to ice.

iv. Prepare an RT mix (calculations per single RNA sample):

Reagent	Vx1
∑V	**0.8 μL**
10 × First Strand buffer	0.2
dNTP	0.4
RNase Inhibitor	0.1
ArrayScrip	0.1

Gently mix by pipetting and add 0.8 μL to each of 24 samples. Incubate in a thermocycler at 42 °C for 2 h with lid at 45 °C. Move on ice and immediately proceed with the next step.

v. Second strand mix.

Reagent	Vx1
∑V	**10 μL**
H_2O	6.3
cDNA from previous step	2
10 × Second strand buf	1
dNTP	0.4
DNA Polymerase	0.2
RNaseH	0.1

Add 8 μL into each variant.
Incubate 2 h at 16 °C (PCR cycler, open or cool lid, make sure that the lid has cooled down from the previous step).

CRITICAL Immediately proceed to the next step after incubation.

vi. cDNA pool and cleanup.
Pool all **N** samples (10 μL of each from the previous step) that will go to the same IVT. Not recommended to pool more than 24 samples per clean up.
Proceed with Ambion kit. Add (**N***10*2.5) μL cDNA binding buffer. Run through the cDNA clean up column.
Spin 1 min at 10.000 g (12,200 rpm on Minispin), discard flow through.
Add (**N***10*2.5) μL wash buffer, spin as above, discard flow through. Repeat.
Spin additional 1 min at 10,000 g to dry. Move filter to new collection tube.
Elute with 9 μL warm DDW (55 °C), wait 2 min at RT, spin 1.5 min at 10,000 g.
Repeat the step above.

vii. In vitro transcription of pools
Mix:

Reagent	Vx1
∑V	**40 μL**
ds cDNA	16
T7 10 × Reaction Buffer	4
T7 ATP	4
T7 CTP	4
T7 GTP	4
T7 UTP	4
T7 Enzyme Mix	4

Add by 24 μL into each pool of cleaned up cDNA.
Incubate at 37 °C for 14 h, with lid at 100–105 °C. Move to 4 °C.

PAUSE aRNA can be stored at −80°C (until Covaris shearing and further library preparation).

viii. RNA cleanup

Add 60μL DDW to 40μL of IVT reaction to a final volume of 100μL before the cleanup.

Add 350μL sRNA binding buffer, 250μL EtOH, mix by pipetting 3-4 times and load onto spin-column. Immediately spin 1min at 10,000g, discard flow through.

Add 650μL wash buffer, spin as above, discard flow through.

Spin for an additional minute to dry.

Elute with 100μL DDW (55°C), incubate for 2min at RT, spin 1.5min at 10,000g.

Check RNA concentration using Nanodrop. Expected concentrations should be over 100ng/μL.

Bioanalyzer. Take 1.5μL of the sample, keep the rest at −80°C. Heat sample to 70°C for 2min before loading onto Bioanalyzer. Use the Total RNA nano kit for Bioanalyzer.

3. aRNA shearing (using a Covaris M220 Focussed-Ultrasonicator™)

i. Transfer 50μL of the purified aRNA to a covaris tube

ii. Use the following protocol to shear the aRNA to a size range 50–250bp:

Peak Incident Power: 50W.

Duty factor: 20%.

Cycles per Burst: 200cpb.

Time: 420s.

Shearing chamber should be kept at 7°C.

iii. Run RNA sample on 6% polyacrylamide + Urea Gel

Gel Mix:

9mL 6%PAA 7M Urea.

90μL 10% APS (final 0.1%).

9μL TEMED (final 0.1%).

Mix and prepare the gel.

Denature RNA and ladder (1μL) in 50% formamide buffer by incubation at 95°C for 3–5min. Run gel in TAE at 140-180V. Before loading, start prerun for equilibration. After prerun properly flush pockets using a pipette before loading the samples.

A typical result of a successful shearing procedure is shown in Fig. 9.

iv. RNA cleanup (using miRNeasy; Qiagen)

Extraction of RNA from UREA-page (see above):

Cut gel slice at approx. 100 bp.

CRITICAL Avoid touching the gel with hands to minimize contamination.

Place the gel piece into the tube and freeze in liquid nitrogen or at −80 °C.

Crush gel using sterile Eppendorf mortar.

Add 50 μL of elution buffer (100 mM Tris-HCl (pH 8.0); 500 mM NaCl; 1% SDS). Incubate 2 min at RT.

Proceed with Qiagen kit for RNA extraction:

- Add 700 μL of Qiazol from miRNeasy kit. Add 140 μL of chloroform. Incubate 3 min at RT.
- Spin 15 min at 12,000 g at 4 °C.
- Transfer upper phase to a new collection tube.
- Add 1.5 volumes usually 525 μL (check!!) 100% EtOH, mix well by pipetting, and transfer half of the sample to an RNeasy Mini spin column.
- Spin 15 s at ≥8000 g at RT.
- Repeat with the other half of the sample.
- Discard flow-through and add 700 μL Buffer RWT.
- Spin 15 s at ≥8000 g at RT.
- Discard flow-through and add 500 μL Buffer RPE.
- Spin 2 min at 8000 g.
- Transfer column to new collection tube and spin 1 min at max speed.
- Transfer column to new collection tube, and elute separately two times with 25 μL nuclease-free H_2O, spinning 1 min at ≥8000 g. Use the first elution for the downstream protocol. Keep the second elution at −80 °C.
- Size-selected RNA concentrations should be over 5–10 ng/μL.

4. Preparation for linker ligation

i. Phosphatase treatment:

Add 4 μL of the following mix to 16 μL of fragmented mRNA:

10 × phosphatase buffer	2 μL
Antarctic phosphatase	1 μL
RNaseOUT	1 μL

Incubate in a thermal cycler with the following protocol (heated lid 70 °C).

37 °C for 30 min.

65 °C for 5 min.

4 °C indefinite hold.

ii. PNK treatment

Add 30 μL of the following mix to tube from the previous step:

Nuclease-free H_2O	17 μL
10 × Phosphatase buffer	5 μL
ATP (10 mM) Illumina	5 μL
RNaseOUT	1 μL
PNK	2 μL

Incubate in a thermal cycler at 37 °C for 60 min then 4 °C hold (heated lid 42 °C).

iii. Column Cleanup for PNK-treated RNA

Qiagen kit for RNA extraction:

- Add 700 μL of Qiazol from miRNeasy kit. Add 140 μL of chloroform. Incub 3′ t_{room}.
- Spin 15 min at 12,000 g at 4 °C.
- Transfer upper phase to new collection tube.
- Add 1.5 volumes usually 525 μL (check!!) 100% EtOH, mix well by pipetting, and transfer half of the sample to an RNeasy Mini spin column.
- Spin 15 s at ≥8000 g at RT.
- Repeat with the other half of the sample.
- Discard flow-through and add 700 μL Buffer RWT.
- Spin 15 s at ≥8000 g at RT.
- Discard flow-through and add 500 μL Buffer RPE.
- Spin 2 min at 8000 g.
- Transfer column to <u>new collection tube</u> and spin 1 min at max speed.
- Transfer column to <u>new collection tube</u>, and elute with 30 μL nuclease free H_2O, spinning 1 min at ≥8000 g.
- Speed-vac sample to 10 μL (approx. 15 min). Use 5 μL for further steps. Keep the remaining 5 μL at −80 °C.

5. Library generation
   i. 3′-adapter Ligation
      Pre-heat the thermal cycler to 70 °C.
      - Set up the ligation reaction on ice using the following:

RNA 3′ Adapter (RA3)	1 μL
Phosphatase and PNK-treated RNA in Nuclease-free Water	5 μL

      - Gently pipette the entire volume up and down 6–8 times to mix thoroughly, then centrifuge briefly.
      - Incubate the tube on the pre-heated thermal cycler at 70 °C for 2 min and then immediately place the tube on ice. It is very important to keep RNA 3′-adapter on ice after the 70 °C incubation to prevent secondary structure formation!
      - Pre-heat the thermal cycler to 28 °C.
      - Add 4 μL of the following mix to each sample and gently pipette the entire volume up and down 6–8 times to mix thoroughly, then centrifuge briefly:

5 × HM Ligation Buffer (HML) Illumina	2 μL
RNase Inhibitor	1 μL
T4 RNA Ligase 2, truncated	1 μL

      - Incubate the tube on the pre-heated thermal cycler at 28 °C for 1 h (no heated lid).
      - With the reaction tube remaining on the thermal cycler, add 1 μL Stop Solution (STP Illumina) and gently pipette the entire volume up and down 6–8 times to mix thoroughly. Continue to incubate the reaction tube on the thermal cycler at 28 °C for 15 min, and then place the tube on ice.
      - Add 3 μL DDW to each sample
   ii. RT with Illumina RTP primer
      Dilute 25 mM dNTP mix 1:1 in DDW. Mix by pipetting and keep on ice. Required volume 0.5 μL per sample.

Pre-heat the thermal cycler to 70 °C.

- Gently pipette the following mix up and down 6–8 times to mix thoroughly, then centrifuge briefly.

Adapter-ligated RNA	6 µL
RNA RT Primer (RTP)	1 µL

NOTE The remaining 5′ and 3′ adapter-ligated RNA may be stored at −80 °C.

- Incubate the tube on the pre-heated thermal cycler at 70 °C for 2 min and then immediately place the tube on ice.
- Pre-heat the thermal cycler to 50 °C.
- Add 5.5 µL of the following mix to each sample and gently pipette the entire volume up and down 6–8 times to mix thoroughly, then centrifuge briefly.

5 × First Strand Buffer	2 µL
12.5 mM dNTP mix	0.5 µL
100 mM DTT	1 µL
RNase Inhibitor	1 µL
SuperScript II RT	1 µL

Incubate the tube in the pre-heated thermal cycler at 50 °C for 1 h and then place the tube on ice.

iii. PCR amplification

- Add RPI and RPIX primer to the sample before adding PML mix

  2 µL of each to **12.5** µL RT Rx. for **full PCR reaction**

  1 µL of each to **6.25** µL RT Rx. for **half PCR reaction**
- Add **33.5** µL of the PCR mix to full reaction (12.5 µL of RT Rx. + 4 µL of primers)
- Add **16.75** µL of the PCR mix to half reaction (6.25 µL of RT Rx. + 2 µL of primers) prepared in the following order:

Using PML Master Mix	Full Rx./0.5 Rx.
Ultra Pure Water	8.5/4.25 μL
PML Master Mix	25/12.5 μL

- Gently pipette the entire volume up and down 6–8 times to mix thoroughly, then centrifuge briefly and place the tube on ice. The total volume should now be 50/25 μL
- PCR-amplify the reaction in the thermal cycler using the following conditions:
  30 s at 98°.
  12* cycles of:
  10 s at 98 °C.
  30 s at 60 °C.
  30 s at 72 °C.
  10 min at 72 °C.
  Hold at 4 °C.
  up to 15 cycles if necessary, down to 11 if starting with full 10 ng.
  PAUSE Sample can be stored at −20 °C after this point.

iv. Bead Cleanup of PCR—1st round
- Prewarm beads to RT
- If half PCR reaction was performed add 25 μL of H_2O
- Vortex AMPure XP Beads until well dispersed, then add 50 μL to the 50 μL PCR reaction. Mix the entire volume up 10 times to mix thoroughly
- Incubate at RT for 15 min
- Place on magnetic stand at least 5 min, until the liquid appears clear
- Remove and discard 95 μL of the supernatant
- Add 200 μL freshly prepared 80% EtOH
- Incubate at least 30 s, then remove and discard supernatant without disturbing beads
- Add 200 μL freshly prepared 80% EtOH
- Incubate at least 30 s, then remove and discard supernatant without disturbing beads
- Air dry beads for 15 min

- Resuspend with 32.5 μL Resuspension Buffer. Pipette entire volume up and down 10 times to mix thoroughly
- Incubate at RT for 2 min
- Place on a magnet for 5 min or until the liquid appears clear
- Transfer 30 μL of supernatant to a new tube

v. Bead Cleanup of PCR—2nd round

- Repeat as above, but add 39 μL beads (removing 65 μL) and elute in 12.5 μL Resuspension Buffer
  PAUSE Sample can be stored at −20 °C after this point.
- Check concentration of DNA by Qubit
- Run each sample on a high sensitivity DNA chip on Bioanalyzer (Fig. 10)

6. Sequencing

Sequencing is performed on Illumina HiSeq 2500 or NextSeq 500 platforms with addition of 30% PhiX (#FC-110-3001, Illumina). Read 1 and Read 2 cover 9 nucleotides and 50 nucleotides respectively, Illumina TruSeq index is sequenced as a dedicated read.

References

Anders, S., Pyl, P. T., & Huber, W. (2015). HTSeq- -A Python framework to work with high-throughput sequencing data. *Bioinformatics*, *31*(2), 166–169. https://doi.org/10.1093/bioinformatics/btu638.

Batra, R., Charizanis, K., Manchanda, M., Mohan, A., Li, M., Finn, D. J., et al. (2014). Loss of MBNL leads to disruption of developmentally regulated alternative polyadenylation in RNA-mediated disease. *Molecular Cell*, *56*(2), 311–322.

Bava, F.-A., Eliscovich, C., Ferreira, P. G., Minana, B., Ben-Dov, C., Guigo, R., et al. (2013). CPEB1 coordinates alternative 3'-UTR formation with translational regulation. *Nature*, *495*(7439), 121–125.

Berg, M. G., Singh, L. N., Younis, I., Liu, Q., Pinto, A. M., Kaida, D., et al. (2012). U1 snRNP determines mRNA length and regulates isoform expression. *Cell*, *150*(1), 53–64. https://doi.org/10.1016/j.cell.2012.05.029.

Brumbaugh, J., Di Stefano, B., Wang, X., Borkent, M., Forouzmand, E., Clowers, K. J., et al. (2018). Nudt21 controls cell fate by connecting alternative polyadenylation to chromatin signaling. *Cell*, *172*(1), 106–120.e121.

Carninci, P., Kasukawa, T., Katayama, S., Gough, J., Frith, M. C., Maeda, N., et al. (2005). The transcriptional landscape of the mammalian genome. *Science*, *309*(5740), 1559–1563.

Danckwardt, S., Gantzert, A. S., Macher-Goeppinger, S., Probst, H. C., Gentzel, M., Wilm, M., et al. (2011). p38 MAPK controls prothrombin expression by regulated RNA 3′ end processing. *Molecular Cell*, *41*(3), 298–310.

Danckwardt, S., Hentze, M. W., & Kulozik, A. E. (2008). 3′ end mRNA processing: Molecular mechanisms and implications for health and disease. *The EMBO Journal*, *27*(3), 482–498.

Dermody, J. L., & Buratowski, S. (2010). Leo1 subunit of the yeast Paf1 complex binds RNA and contributes to complex recruitment. *Journal of Biological Chemistry*, *285*(44), 33671–33679. https://doi.org/10.1074/jbc.M110.140764.

Derti, A., Garrett-Engele, P., Macisaac, K. D., Stevens, R. C., Sriram, S., Chen, R., et al. (2012). A quantitative atlas of polyadenylation in five mammals. *Genome Research*, *22*(6), 1173–1183. https://doi.org/10.1101/gr.132563.111.

Di Giammartino, D. C., Li, W., Ogami, K., Yashinskie, J. J., Hoque, M., Tian, B., et al. (2014). RBBP6 isoforms regulate the human polyadenylation machinery and modulate expression of mRNAs with AU-rich 3′ UTRs. *Genes & Development*, *28*(20), 2248–2260. https://doi.org/10.1101/gad.245787.114.

Dutertre, M., Chakrama, F. Z., Combe, E., Desmet, F.-O., Mortada, H., Espinoza, M. P., et al. (2014). A recently evolved class of alternative 3′-terminal exons involved in cell cycle regulation by topoisomerase inhibitors. *Nature Communications*, *5*, 3395.

Elkon, R., Ugalde, A. P., & Agami, R. (2013). Alternative cleavage and polyadenylation: Extent, regulation and function. *Nature Reviews. Genetics*, *14*(7), 496–506. https://doi.org/10.1038/nrg3482.

Fu, Y., Sun, Y., Li, Y., Li, J., Rao, X., Chen, C., et al. (2011). Differential genome-wide profiling of tandem 3' UTRs among human breast cancer and normal cells by high-throughput sequencing. *Genome Research*, *21*(5), 741–747. https://doi.org/10.1101/gr.115295.110.

Fusby, B., Kim, S., Erickson, B., Kim, H., Peterson, M. L., & Bentley, D. L. (2016). Coordination of RNA polymerase II pausing and 3′ end processing factor recruitment with alternative polyadenylation. *Molecular and Cellular Biology*, *36*(2), 295–303.

Geisberg, J. V., Moqtaderi, Z., Fan, X., Ozsolak, F., & Struhl, K. (2014). Global analysis of mRNA isoform half-lives reveals stabilizing and destabilizing elements in yeast. *Cell*, *156*(4), 812–824. https://doi.org/10.1016/j.cell.2013.12.026.

Gruber, A. J., Schmidt, R., Gruber, A. R., Martin, G., Ghosh, S., Belmadani, M., et al. (2016). A comprehensive analysis of 3′ end sequencing data sets reveals novel polyadenylation signals and the repressive role of heterogeneous ribonucleoprotein C on cleavage and polyadenylation. *Genome Research*, *26*(8), 1145–1159.

Gruber, A. J., & Zavolan, M. (2019). Alternative cleavage and polyadenylation in health and disease. *Nature Reviews Genetics*, *20*, 599–614. https://doi.org/10.1038/s41576-019-0145-z.

Ha, K. C. H., Blencowe, B. J., & Morris, Q. (2018). QAPA: A new method for the systematic analysis of alternative polyadenylation from RNA-seq data. *Genome Biology*, *19*(1), 45. https://doi.org/10.1186/s13059-018-1414-4.

Hallais, M., Pontvianne, F., Andersen, P. R., Clerici, M., Lener, D., Benbahouche, N. E. H., et al. (2013). CBC–ARS2 stimulates 3′-end maturation of multiple RNA families and favors cap-proximal processing. *Nature Structural & Molecular Biology*, *20*(12), 1358–1366.

Hashimshony, T., Wagner, F., Sher, N., & Yanai, I. (2012). CEL-Seq: Single-cell RNA-Seq by multiplexed linear amplification. *Cell Reports*, *2*(3), 666.

Hoque, M., Ji, Z., Zheng, D., Luo, W., Li, W., You, B., et al. (2013). Analysis of alternative cleavage and polyadenylation by 3′ region extraction and deep sequencing. *Nature Methods*, *10*(2), 133–139. https://doi.org/10.1038/nmeth.2288.

Huang, Y., Li, W., Yao, X., Lin, Q.-j., Yin, J.-w., Liang, Y., et al. (2012). Mediator complex regulates alternative mRNA processing via the Med23 subunit. *Molecular Cell*, *45*(4), 459–469.

Jan, C. H., Friedman, R. C., Ruby, J. G., & Bartel, D. P. (2011). Formation, regulation and evolution of Caenorhabditis elegans 3'UTRs. *Nature*, *469*(7328), 97–101. https://doi.org/10.1038/nature09616.

Jenal, M., Elkon, R., Loayza-Puch, F., van Haaften, G., Kuhn, U., Menzies, F. M., et al. (2012). The poly(A)-binding protein nuclear 1 suppresses alternative cleavage and polyadenylation sites. *Cell*, *149*(3), 538–553.

Kaida, D., Berg, M. G., Younis, I., Kasim, M., Singh, L. N., Wan, L., et al. (2010). U1 snRNP protects pre-mRNAs from premature cleavage and polyadenylation. *Nature*, *468*(7324), 664–668.

Kamieniarz-Gdula, K., Gdula, M. R., Panser, K., Nojima, T., Monks, J., Wiśniewski, J. R., et al. (2019). Selective roles of vertebrate PCF11 in premature and full-length transcript termination. *Molecular Cell*, *74*(1), 158–172.e159. https://doi.org/10.1016/j.molcel.2019.01.027.

Kargapolova, Y., Levin, M., Lackner, K., & Danckwardt, S. (2017). sCLIP—An integrated platform to study RNA–protein interactomes in biomedical research: Identification of CSTF2tau in alternative processing of small nuclear RNAs. *Nucleic Acid Research*, *45*(10), 6074–6086.

Katahira, J., Okuzaki, D., Inoue, H., Yoneda, Y., Maehara, K., & Ohkawa, Y. (2013). Human TREX component Thoc5 affects alternative polyadenylation site choice by recruiting mammalian cleavage factor I. *Nucleic Acids Research*, *41*(14), 7060–7072. https://doi.org/10.1093/nar/gkt414.

Ke, S., Alemu, E. A., Mertens, C., Gantman, E. C., Fak, J. J., Mele, A., et al. (2015). A majority of m6A residues are in the last exons, allowing the potential for 3′ UTR regulation. *Genes & Development*, *29*(19), 2037–2053. https://doi.org/10.1101/gad.269415.115.

Langmead, B., & Salzberg, S. L. (2012). Fast gapped-read alignment with bowtie 2. *Nature Methods*, *9*(4), 357–359. https://doi.org/10.1038/nmeth.1923.

Lee, S.-H., Singh, I., Tisdale, S., Abdel-Wahab, O., Leslie, C. S., & Mayr, C. (2018). Widespread intronic polyadenylation inactivates tumour suppressor genes in leukaemia. *Nature*, *561*(7721), 127–131. https://doi.org/10.1038/s41586-018-0465-8.

Li, W., Park, J. Y., Zheng, D., Hoque, M., Yehia, G., & Tian, B. (2016). Alternative cleavage and polyadenylation in spermatogenesis connects chromatin regulation with post-transcriptional control. *BMC Biology*, *14*(1), 6. https://doi.org/10.1186/s12915-016-0229-6.

Licatalosi, D. D., Mele, A., Fak, J. J., Ule, J., Kayikci, M., Chi, S. W., et al. (2008). HITS-CLIP yields genome-wide insights into brain alternative RNA processing. *Nature*, *456*(7221), 464–469.

Lin, Y., Li, Z., Ozsolak, F., Kim, S. W., Arango-Argoty, G., Liu, T. T., et al. (2012). An in-depth map of polyadenylation sites in cancer. *Nucleic Acids Research*, *40*(17), 8460–8471. https://doi.org/10.1093/nar/gks637.

Lutz, C. S., & Moreira, A. (2011). Alternative mRNA polyadenylation in eukaryotes: An effective regulator of gene expression. *Wiley Interdisciplinary Reviews: RNA*, *2*(1), 22–31. https://doi.org/10.1002/wrna.47.

Marini, F., Scherzinger, D., & Danckwardt, S. (2021). TREND-DB—A transcriptome-wide atlas of the dynamic landscape of alternative polyadenylation. *Nucleic Acids Research*, *49*(D1), D243–D253. https://doi.org/10.1093/nar/gkaa722.

Masamha, C. P., Xia, Z., Yang, J., Albrecht, T. R., Li, M., Shyu, A. B., et al. (2014). CFIm25 links alternative polyadenylation to glioblastoma tumour suppression. *Nature*, *510*(7505), 412–416.

Mayr, C. (2017). Regulation by 3'-untranslated regions. *Annual Review of Genetics*, *27*(51), 171–194.

Mayr, C., & Bartel, D. P. (2009). Widespread shortening of 3'UTRs by alternative cleavage and polyadenylation activates oncogenes in cancer cells. *Cell*, *138*(4), 673–684.

Müller-McNicoll, M., Botti, V., de Jesus Domingues, A. M., Brandl, H., Schwich, O. D., Steiner, M. C., et al. (2016). SR proteins are NXF1 adaptors that link alternative RNA processing to mRNA export. *Genes & Development*, *30*(5), 553–566.

Neve, J., Burger, K., Li, W., Hoque, M., Patel, R., Tian, B., et al. (2016). Subcellular RNA profiling links splicing and nuclear DICER1 to alternative cleavage and polyadenylation. *Genome Research*, *26*(1), 24–35.

Nourse, J., & Danckwardt, S. (2021). A novel rationale for targeting FXI: Insights from the hemostatic microRNA targetome for emerging anticoagulant strategies. *Pharmacology & Therapeutics*, *218*, 107676. https://doi.org/10.1016/j.pharmthera.2020.107676.

Nourse, J., Spada, S., & Danckwardt, S. (2020). Emerging roles of RNA 3′-end cleavage and polyadenylation in pathogenesis, diagnosis and therapy of human disorders. *Biomolecules*, *10*, 915.

Ogorodnikov, A., Kargapolova, Y., & Danckwardt, S. (2016). Processing and transcriptome expansion at the mRNA 3′ end in health and disease: Finding the right end. *European Journal of Physiology*, *468*(6), 993–1012.

Ogorodnikov, A., Levin, M., Tattikota, S., Tokalov, S., Hoque, M., Scherzinger, D., et al. (2018). Transcriptome 3'end organization by PCF11 links alternative polyadenylation to formation and neuronal differentiation of neuroblastoma. *Nature Communications*, *9*(1), 5331. https://doi.org/10.1038/s41467-018-07580-5.

Ozsolak, F., Kapranov, P., Foissac, S., Kim, S. W., Fishilevich, E., Monaghan, A. P., et al. (2010). Comprehensive polyadenylation site maps in yeast and human reveal pervasive alternative polyadenylation. *Cell*, *143*(6), 1018–1029. https://doi.org/10.1016/j.cell.2010.11.020.

Parker, M. T., Knop, K., Sherwood, A. V., Schurch, N. J., Mackinnon, K., Gould, P. D., et al. (2020). Nanopore direct RNA sequencing maps the complexity of Arabidopsis mRNA processing and m6A modification. *eLife*, *9*. https://doi.org/10.7554/eLife.49658, e49658.

Patel, R., Brophy, C., Hickling, M., Neve, J., & Furger, A. (2019). Alternative cleavage and polyadenylation of genes associated with protein turnover and mitochondrial function are deregulated in Parkinson's, Alzheimer's and ALS disease. *BMC Medical Genomics*, *12*(1), 60. https://doi.org/10.1186/s12920-019-0509-4.

Pinto, P. A. B., Henriques, T., Freitas, M. O., Martins, T., Domingues, R. G., Wyrzykowska, P. S., et al. (2011). RNA polymerase II kinetics in polo polyadenylation signal selection. *The EMBO Journal*, *30*(12), 2431–2444. https://doi.org/10.1038/emboj.2011.156.

Proudfoot, N. J. (2016). Transcriptional termination in mammals: Stopping the RNA polymerase II juggernaut. *Science*, *352*(6291).

Reyes, A., & Huber, W. (2017). Alternative start and termination sites of transcription drive most transcript isoform differences across human tissues. *Nucleic Acids Research*, *46*(2), 582–592. https://doi.org/10.1093/nar/gkx1165.

Sandberg, R., Neilson, J. R., Sarma, A., Sharp, P. A., & Burge, C. B. (2008). Proliferating cells express mRNAs with shortened 3′ untranslated regions and fewer microRNA target sites. *Science*, *320*(5883), 1643–1647.

Shepard, P. J., Choi, E. A., Lu, J., Flanagan, L. A., Hertel, K. J., & Shi, Y. (2011). Complex and dynamic landscape of RNA polyadenylation revealed by PAS-Seq. *RNA*, *17*, 761–772.

Shi, Y., Di Giammartino, D. C., Taylor, D., Sarkeshik, A., Rice, W. J., Yates, J. R., 3rd, et al. (2009). Molecular architecture of the human pre-mRNA 3′ processing complex. *Molecular Cell*, *33*(3), 365–376.

Shi, Y., & Manley, J. L. (2015). The end of the message: Multiple protein-RNA interactions define the mRNA polyadenylation site. *Genes & Development*, *29*(9), 889–897. https://doi.org/10.1101/gad.261974.115.

Soetanto, R., Hynes, C. J., Patel, H., Humphreys, D. T., Evers, M., Duan, G. W., et al. (2016). Role of miRNAs and alternative mRNA 3′-end cleavage and polyadenylation of their mRNA targets in cardiomyocyte hypertrophy. *Biochimica et Biophysica Acta*, *1859*, 744–756. https://doi.org/10.1016/j.bbagrm.2016.03.010.

Spies, N., Nielsen, C. B., Padgett, R. A., & Burge, C. B. (2009). Biased chromatin signatures around polyadenylation sites and exons. *Molecular Cell*, *36*(2), 245–254. https://doi.org/10.1016/j.molcel.2009.10.008.

Stacey, S. N., Sulem, P., Jonasdottir, A., Masson, G., Gudmundsson, J., Gudbjartsson, D. F., et al. (2011). A germline variant in the TP53 polyadenylation signal confers cancer susceptibility. *Nature Genetics*, *43*(11), 1098–1103. http://www.nature.com/ng/journal/v43/n11/abs/ng.926.html. - supplementary-information.

Tian, B., & Manley, J. L. (2017). Alternative polyadenylation of mRNA precursors. *Nature Reviews Molecular Cell Biology*, *18*(1), 18–30. https://doi.org/10.1038/nrm.2016.116. http://www.nature.com/nrm/journal/v18/n1/abs/nrm.2016.116.html. - supplementary-information.

Turner, R. E., Harrison, P. F., Swaminathan, A., Kraupner-Taylor, C. A., Curtis, M. J., Goldie, B. J., et al. (2020). Genetic and pharmacological evidence for kinetic competition between alternative poly(A) sites in yeast. *bioRxiv*. https://doi.org/10.1101/2020.12.01.407171. 2020.2012.2001.407171.

Turner, R. E., Pattison, A. D., & Beilharz, T. H. (2017). Alternative polyadenylation in the regulation and dysregulation of gene expression. *Seminars in Cell & Developmental Biology*, *75*, 61–69. https://doi.org/10.1016/j.semcdb.2017.08.056.

Wang, L., Dowell, R. D., & Yi, R. (2013). Genome-wide maps of polyadenylation reveal dynamic mRNA 3′-end formation in mammalian cell lineages. *RNA*, *19*(3), 413–425. https://doi.org/10.1261/rna.035360.112.

Wang, E. T., Sandberg, R., Luo, S., Khrebtukova, I., Zhang, L., Mayr, C., et al. (2008). Alternative isoform regulation in human tissue transcriptomes. *Nature*, *456*(7221), 470–476.

Wang, R., Zheng, D., Wei, L., Ding, Q., & Tian, B. (2019). Regulation of Intronic Polyadenylation by PCF11 Impacts mRNA Expression of Long Genes. *Cell Reports*, *26*(10), 2766–2778.e2766. https://doi.org/10.1016/j.celrep.2019.02.049.

Weng, T., Ko, J., Masamha, C. P., Xia, Z., Xiang, Y., Chen, N.-Y., et al. (2019). Cleavage factor 25 deregulation contributes to pulmonary fibrosis through alternative polyadenylation. *The Journal of Clinical Investigation*, *129*(5), 1984–1999. https://doi.org/10.1172/JCI122106.

Wood, A. J., Schulz, R., Woodfine, K., Koltowska, K., Beechey, C. V., Peters, J., et al. (2008). Regulation of alternative polyadenylation by genomic imprinting. *Genes & Development*, *22*(9), 1141–1146.

Xia, Z., Donehower, L. A., Cooper, T. A., Neilson, J. R., Wheeler, D. A., Wagner, E. J., et al. (2014). Dynamic analyses of alternative polyadenylation from RNA-seq reveal a 3′-UTR landscape across seven tumour types. *Nature Communications*, *5*, 5274. https://doi.org/10.1038/ncomms6274.

Ye, C., Zhou, Q., Hong, Y., & Li, Q. Q. (2019). Role of alternative polyadenylation dynamics in acute myeloid leukaemia at single-cell resolution. *RNA Biology*, *16*(6), 785–797. https://doi.org/10.1080/15476286.2019.1586139.

Yue, Y., Liu, J., Cui, X., Cao, J., Luo, G., Zhang, Z., et al. (2018). VIRMA mediates preferential m6A mRNA methylation in 3′UTR and near stop codon and associates with alternative polyadenylation. *Cell Discovery*, *4*(1), 10. https://doi.org/10.1038/s41421-018-0019-0.

CHAPTER FOUR

QPAT-seq, a rapid and deduplicatable method for quantification of poly(A) site usages

Juncheng Lin[a], Congting Ye[a], and Qingshun Q. Li[a,b,*]

[a]Key Laboratory of the Ministry of Education for Coastal and Wetland Ecosystems, College of the Environment and Ecology, Xiamen University, Xiamen, Fujian, China
[b]Graduate College of Biomedical Sciences, Western University of Health Sciences, Pomona, CA, United States
*Corresponding author: e-mail address: qqli@westernu.edu

Contents

Abstract

Alternative polyadenylation (APA) is an essential regulatory mechanism for gene expression. The next generation sequencing provides ample opportunity to precisely delineate APA sites genome-wide. Various methods for profiling transcriptome-wide poly(A) sites were developed. By comparing available methods, the ways for adding sequencing adaptors to fit with the Illumina sequencing platform are different. These methods have identified more than 50% genes that undergo APA in eukaryotes. However, due to the unbalanced PCR during library preparation, accurate quantification of poly(A) sites is still a challenge. Here, we describe an updated poly(A) tag sequencing

Methods in Enzymology, Volume 655
ISSN 0076-6879
https://doi.org/10.1016/bs.mie.2021.04.002

method that incorporates unique molecular identifier (UMI) into the adaptor for removing quantification bias induced by PCR duplicates. Hence, quantification of poly(A) site usages can be achieved by counting UMIs. This protocol, quantifying poly(A) tag sequencing (QPAT-seq), can be finished in 1 day with reduced cost, and is particularly useful for application with a large number of samples.

1. Introduction

In eukaryotes, nascent RNA undergoes polyadenylation to synthesize mature message RNA (mRNA) for protein translation (Tian & Manley, 2017). This crucial process is maintained by a large protein complex, which binds to *cis*-elements, cleaves and adds a poly(A) tail at a specific site of the precursor mRNA. The location of polyadenylation site is not restricted to the end of genes, but also spread among other regions of genes (e.g., 5′ UTR, non-terminal exons and introns). This phenomenon is termed as alternative polyadenylation (APA). Early study identified that APA governs the production of different mRNA isoforms of mu heavy-chain gene, which regulates the membrane localization of IgM (Danner & Leder, 1985). To date, it has been reported that >50% genes, either in mammals and plants, receive poly(A) tail at two or more locations (Chen et al., 2017; Fu et al., 2016, 2019). In mammals, APA was demonstrated to regulate coding sequence length, nuclear export, stability, localization and translation efficiency of mRNA (de Morree et al., 2019; Jia et al., 2017; Mayr, 2017). These molecular variations could further function in diseases, cell differentiation and tumorigenesis, and so on (Gruber & Zavolan, 2019). APA also shows a conspicuous heritable effect on the biological processes of plants, such as development, stress responses, symbiosis (Deng & Cao, 2017; Hunt, 2014; Pan et al., 2016) and epigenetics (Lin et al., 2020). In summary, APA plays an important role in regulating gene expression and functions in many essential biological processes of eukaryotes.

1.1 Identification of poly(A) sites by next generation sequencing

In line with RNA-seq, genome-wide profiling of polyadenylation landscape is helpful for depicting cellular responses of organisms under different internal or external environmental stimulations. It is also useful for tracing the effect of proteins in mRNA 3′ end processing after genetic manipulation. The feature of a poly(A) tail provides a convenient platform for the enrichment of mRNA or poly(A)$^+$ mRNA fragments by using oligo(dT) primers or its attached beads. After capturing poly(A)$^+$ mRNA, it is easily to be reverse transcribed into cDNA for downstream studies. Hence, various high-throughput sequencing-based approaches were developed for typing polyadenylation (or APA) profiles in eukaryotes. Due to the low coverage of typical RNA-seq on 3′ end of transcripts, mRNA is routinely sheared into small fragments for enriching poly(A)$^+$ tags. Most of 3′ end sequencing methods adopt poly(A)$^+$ tag capturing and directly reverse transcribing by oligo(dT), and simultaneously introduce 3′ adaptor for next-generation sequencing (NGS). The key difference between various methods is the way for adding 5′ adaptors for NGS. For example, 3′READS+ (Zheng, Liu, & Tian, 2016) employs a ligation for introducing a 5′ adaptor; PAT-seq (Ma, Pati, Liu, Li, & Hunt, 2014; Wu et al., 2011) adopts template switching ability of M-MLV reverse transcriptase for adding the 5′ adaptor; and PAC-seq (Routh et al., 2017) utilizes the "click-ligation" between azido-terminated cDNA and alkyne-functionalized 5′ adaptor. High-throughput poly(A) tag sequencing has been broadly applied to elucidate the role of APA in various biological studies (Fu et al., 2016; Sanfilippo, Wen, & Lai, 2017; Singh et al., 2018; Zheng et al., 2018).

1.2 Challenges

Methods of poly(A) tag sequencing differ in time, cost, library production, and initial RNA input, among other considerations. Engaging too many purification steps in a workflow will decrease the concentration of library, and certainly increase time and costs. If the initial RNA input is too low, the procedure requires more cycles for PCR amplification to reach a certain amount of tag library for sequencing. Under this situation, highly expressed genes will be over-amplified, result in large amounts of products over-represented in the final sequencing library. This will reduce the qualified reads output and disturb the quantification of poly(A) tags from other, particularly low expressing, genes. Actually, current widely used poly(A) tag sequencing methods are not able to remove the duplicates generated by PCR. The utility of unique molecular identifier (UMI) in RNA-seq, typically in single-cell RNA-seq, has fulfilled a function of reducing PCR duplicates and quantifying transcripts (Chen et al., 2018; Hong & Gresham, 2017). Here, we describe a method, QPAT-seq, for precisely quantifying poly(A) tags in 3′ end high-throughput sequencing (Fig. 1).

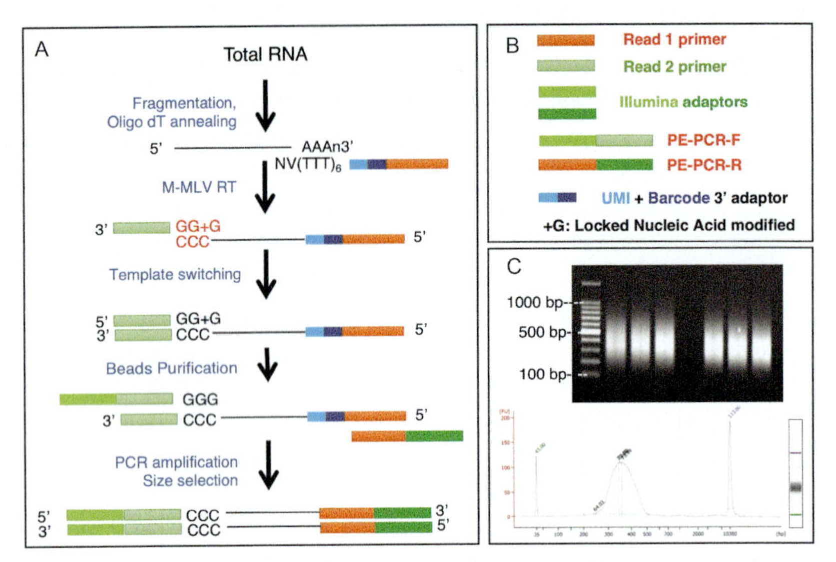

Fig. 1 Scheme of QPAT-seq. (A) The workflow of QPAT-seq. (B) Description of primers. (C) An example electrophoresis of 6 PCR products on a 2% agarose gel (upper) or size range of gel purified product imaged on Bioanalyzer.

QPAT-seq is a template switching-based 3′ end sequencing method, and introduces UMI to the native poly(A)$^+$ tags for quantifying the number of transcripts in the source sample. In addition, the intact workflow of QPAT-seq only needs 1 day to reach the final product before Illumina sequencing.

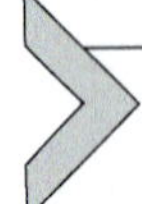

2. QPAT-seq library preparation

2.1 Equipment

(1) Water bath.
(2) Thermal cycler.
(3) Magnetic stand.
(4) Mini-centrifuge.
(5) Agilent bioanalyzer.
(6) Qubit.
(7) Nanodrop.
(8) DNA electrophoresis system.

2.2 Reagents

(1) Oligo $(dT)_{25}$ beads (NEB).
(2) 5 × first strand buffer (contains Mg^{2+}, Takara).
(3) SMARTScribe™ reverse transcriptase (Takara).
(4) RNase inhibitor (NEB).
(5) KAPA HiFi PCR Mix (Roche).
(6) DNA purification beads (Vazyme).
(7) Nuclease free H_2O.
(8) Absolute ethanol.
(9) Tris.
(10) HCl.
(11) LiCl.
(12) EDTA.

2.3 Primers

(1) 3′ adaptor:
5′ACACTCTTTCCCTACACGACGCTCTTCCGATCT(*Bn*)NNNN NNNNTTTTTTTTTTTTTTTTTTTTVN-3′

(*Bn*) indicates 6–8 nt user defined barcodes for distinguish samples.

(2) 5′ adaptor:
5′CGGTCTCGGCATTCCTGCTGAACCGCTCTTCCGATCTGG+ G3′.

+G indicates the last G is a locked nucleic acid (LNA) modified to increase template switching efficiency and stabilize the oligo during storage.

(3) PCR primer 1:
5′AATGATACGGCGACCACCGAGATCTACACTCTTTCCCTAC ACGACGCTCTTCCGATC*T3′

* indicates phosphorothioate modification between C and T.

(4) PCR primer 2:
5′CAAGCAGAAGACGGCATACGAGATCGGTCTCGGCATTCCT GCTGAACCGCTCTTCCGATC*T3′

3. Protocol

3.1 Enrichment of poly(A)$^+$ tags

3.1.1. Aliquot 15 μL of oligo$(dT)_{25}$ beads (NEB) for each sample. Wash the beads twice with 100 μL of binding buffer (20 mM Tris-HCl pH 7.5,

1.0 M LiCl and 2 mM EDTA, RNase free), and remove the supernatant on a magnetic stand. Resuspend the beads in 50 μL of binding buffer. Keep at room temperature until the next step.

3.1.2. Aliquot 1–12 μL (100 ng–4 μg) qualified total RNA (e.g., Bioanalyzer RNA Integrity Number (RIN) >9, *Note 1*) in a thin wall PCR tube. Add 3 μL of 5 × 1st strand buffer come with the reverse transcriptase. Bring up the volume to 15 μL by RNase free H_2O.

3.1.3. Pre-heat thermal cycler to 94 °C with the lid temperature on 105 °C. Heat the total RNA mix at 94 °C for 4 min to fragment RNA. Chill on ice immediately for 2 min.

3.1.4. Add the fragmentated RNA to the washed bead solution. Gently pipette up and down to mix well. Incubate the mix at room temperature for 5 min. Collect beads using the magnetic stand, discard supernatant.

3.1.5. Wash the beads twice with 100 μL of Washing Buffer directly on the magnetic stand (10 mM Tris-HCl pH 7.5, 0.15 M LiCl, 1 mM EDTA, RNase free). Remove the supernatant from the beads. Add 15 μL of RNase-free water and heat the beads at 80 °C for 3 min to elute the poly(A) tagged mRNA.

3.1.6. Place tube on magnetic stand immediately. Remove beads with the magnetic stand, save 14 μL of the supernatant (easier to aliquot than 15 μL, and this will avoid to carry over the beads) in a new thin wall tube.

3.2 Template switching based cDNA synthesis

3.2.1. Add 1 μL of 50 μM 3′ oligo(dT) adaptor (*Note 2*) to each purified poly$(A)^+$ mRNA. Heat the mix at 65 °C for 5 min, then chill on ice for 2 min. Repeat once for the heating and chilling step.

3.2.2. Add 5 μL 5 × 1st strand buffer, 2.5 μL 10 mM each dNTP mix, 1 μL 20 mM DTT, 0.5 μL RNase inhibitor, and 1 μL SMARTScribe™ transcriptase (Takara) (*Note 3*) to make a 25 μL RT mix (*Note 4*). Do RT reaction at 42 °C for 2 h.

3.2.3. Add 1 μL SMARTScribe™ transcriptase and 1 μL 50 μM 5′ adaptor. Gently pipette to mix the solution. Do template switching reaction for 2 h at 42 °C. Inactivate the enzyme at 70 °C for 15 min. This is the cDNA for next step.

3.3 Library preparation

3.3.1. Take DNA purification beads out from fridge. Vortex it to homogenize the beads solution. Aliquot 25 μL beads to a 1.5 mL tube for each sample. Keep it on room temperature for at least 30 min. Add the cDNA to the 25 μL beads (*Note 5*), and gently pipette up and down to make sure the cDNA is mixed well with beads. Incubate for 8 min at room temperature.

3.3.2. Separate beads using the magnet stand (*Note 6*), and discard supernatant. Wash beads twice with 200 μL fresh 80% ethanol. Open the lid and dry for 5 min at room temperature (*Note 7*). Elute cDNA with 36 μL nuclease free water at room temperature for 2–5 min. Then, put on magnetic stand to concentrate beads at least 2 min. Finally, only aliquot 35 μL eluted product to a new thin wall PCR tube to avoid beads carried over. Take half for PCR, other products can be stored in −20 °C for further use (*Note 8*).

3.3.3. The following step will use KAPA HiFi PCR kit (Roche) to amplify the cDNA product and adding Illumina sequencing adaptor. Add 0.75 μL 10 μM PCR primer 1, 0.75 μL 10 μM PCR primer 2, 0.75 μL dNTP mix supplied with the kit, 0.5 μL KAPA HiFi enzyme, and 5 μL KAPA PCR buffer. The final volume is 25 μL. Amplify the mix on thermal cycler with the following program: 95 °C 3 min; 98 °C 20 s, 60 °C 15 s, 72 °C 30 s, 18 cycles; 72 °C 1 min.

3.3.4. Bring up the PCR product with nuclease free water to 100 μL. Add 60 μL warmed DNA beads (room temperature) and mix well by gently pipetting. Incubate at room temperature for 5 min. Use a magnetic stand to concentrate beads. Save supernatant to a new 1.5 ml tube. Add another 20 μL DNA beads and mix well. Incubate at room temperature for 8 min. Use magnetic stand to discard supernatant and collect beads. Wash beads as described in step 3.3.2 and elute the final poly(A) tag library with 20 μL nuclease free water.

3.3.5. The library can be checked by Bioanalyzer and Qubit, and processed for Illumina sequencing (*Note 9*). Sequenced library showed high base quality, and the unique mapping rate is approximal to 95% (Fig. 2).

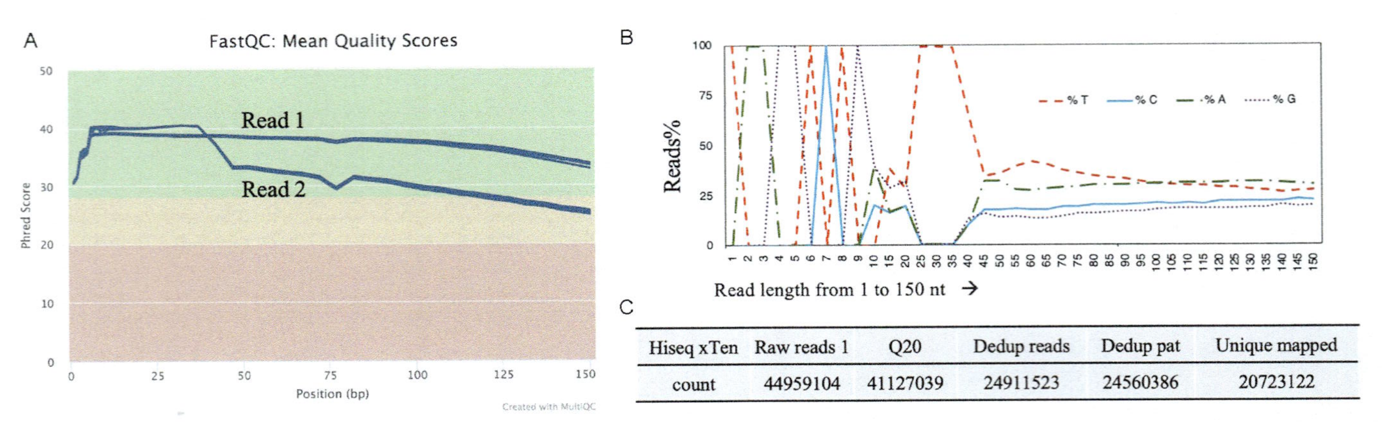

Hiseq xTen	Raw reads 1	Q20	Dedup reads	Dedup pat	Unique mapped
count	44959104	41127039	24911523	24560386	20723122

Fig. 2 An example of QPAT-seq from Illumina HiSeq xTen platform. (A) Read qualities of the paired ends. (B) Per base sequence content of reads. (C) A read count of one sample from the xTen platform. Dedup: removed PCR duplication. Pat: valid poly(A) tags.

4. Discussion

QPAT-seq is able to remove PCR duplicates based on UMIs on each sequenced read. Data mining process (removing PCR duplicates, etc.) can be adopted from the pipeline of single cell mRNA-seq, such as 10× scRNA-seq (Zheng et al., 2017) or Drop-seq (Macosko et al., 2015). To date, large scale poly(A) tag sequencing data accosting tissues or cell types is valuable for understanding the role of APA in eukaryotes. The time and cost saving features of QPAT-seq will provide an opportunity for promoting large scale projects. However, it may not work well with very low amount of RNA input. Perhaps artifacts produced by template switching reverse transcription can be predominantly accumulated and presented as artificial large molecules when the amount of RNA input is too low (Turchinovich et al., 2014). Thus, check library size by Bioanalyzer is recommended before sequencing.

5. Summary

This chapter describes an efficient and low bias quantizable method for profiling polyadenylation landscape for eukaryotic tissues. Careful attentions to fragmentation timing and size selection when using beads will be beneficial for generating high quality libraries. These should be tested before handling large scale samples by individual investigators. Moreover, since the cluster density of sequencing flowcell chips are not the same among different sequencing platform, investigators should carefully consider which platform to sequence the libraries.

6. Notes

(1) For plants, RIN is lower than that of cultured cells or tissues of mammals. See application note at https://www.agilent.com/cs/library/applications/5990-8850EN.pdf. Gel electrophoresis and Nanodrop testing are also recommended for RNA quality assessment.

(2) For plants, 18 nt T is enough for capturing poly(A) tags. For mammals, the poly(A) tail is longer than that of plants, and internal poly(A) may be more common than plants. We recommended to use a longer oligo(dT) to reduce internal priming effect.

(3) Any M-MLV type reverse transcriptase would be fine.

(4) Read user manual of the kit, adjust the volume of mix according to the amount of RNA input.
(5) Volume proportion of cDNA:beads is 1:1.
(6) This may take a minute or two, because of the viscosity of the bead solution, and should quickly go to the next washing step to avoid excessive drying.
(7) *Do not dry excessively.*
(8) The saved half fraction is a backup for further PCR amplification or troubleshooting.
(9) We have tested QPAT-seq on Illumina HiSeq 2500 and Illumina xTen platforms with 13 pM loading concentration.

Acknowledgments

We thank other lab members for testing the protocol. This work was supported in part by a grant from National Nature Science Foundation of China (32000448 to J.L.), and an intramural grant from Western University of Health Sciences (to Q.Q.L.).

References

Chen, W., Jia, Q., Song, Y., Fu, H., Wei, G., & Ni, T. (2017). Alternative polyadenylation: Methods, findings, and impacts. *Genomics, Proteomics & Bioinformatics, 15*(5), 287–300.

Chen, W., Li, Y., Easton, J., Finkelstein, D., Wu, G., & Chen, X. (2018). UMI-count modeling and differential expression analysis for single-cell RNA sequencing. *Genome Biology, 19*(1), 70. https://doi.org/10.1186/s13059-018-1438-9.

Danner, D., & Leder, P. (1985). Role of an RNA cleavage/poly(A) addition site in the production of membrane-bound and secreted IgM mRNA. *Proceedings of the National Academy of Sciences of the United States of America, 82*(24), 8658–8662. https://doi.org/10.1073/pnas.82.24.8658.

de Morree, A., Klein, J. D. D., Gan, Q., Farup, J., Urtasun, A., Kanugovi, A., et al. (2019). Alternative polyadenylation of Pax3 controls muscle stem cell fate and muscle function. *Science, 366*(6466), 734–738. https://doi.org/10.1126/science.aax1694.

Deng, X., & Cao, X. (2017). Roles of pre-mRNA splicing and polyadenylation in plant development. *Current Opinion in Plant Biology, 35*, 45–53.

Fu, H., Wang, P., Wu, X., Zhou, X., Ji, G., Shen, Y., et al. (2019). Distinct genome-wide alternative polyadenylation during the response to silicon availability in the marine diatom Thalassiosira pseudonana. *The Plant Journal, 99*(1), 67–80. https://doi.org/10.1111/tpj.14309.

Fu, H., Yang, D., Su, W., Ma, L., Shen, Y., Ji, G., et al. (2016). Genome-wide dynamics of alternative polyadenylation in rice. *Genome Research, 26*(12), 1753–1760. https://doi.org/10.1101/gr.210757.116.

Gruber, A. J., & Zavolan, M. (2019). Alternative cleavage and polyadenylation in health and disease. *Nature Reviews. Genetics, 20*(10), 599–614. https://doi.org/10.1038/s41576-019-0145-z.

Hong, J., & Gresham, D. (2017). Incorporation of unique molecular identifiers in TruSeq adapters improves the accuracy of quantitative sequencing. *BioTechniques, 63*(5), 221–226. https://doi.org/10.2144/000114608.

Hunt, A. G. (2014). The Arabidopsis polyadenylation factor subunit CPSF30 as conceptual link between mRNA polyadenylation and cellular signaling. *Current Opinion in Plant Biology*, *21*, 128–132.

Jia, X., Yuan, S., Wang, Y., Fu, Y., Ge, Y., Ge, Y., et al. (2017). The role of alternative polyadenylation in the antiviral innate immune response. *Nature Communications*, *8*, 14605.

Lin, J., Hung, F.-Y., Ye, C., Hong, L., Shih, Y.-H., Wu, K., et al. (2020). HDA6-dependent histone deacetylation regulates mRNA polyadenylation in Arabidopsis. *Genome Research*, *30*, 1407–1417.

Ma, L., Pati, P. K., Liu, M., Li, Q. Q., & Hunt, A. G. (2014). High throughput characterizations of poly(A) site choice in plants. *Methods*, *67*(1), 74–83.

Macosko, E. Z., Basu, A., Satija, R., Nemesh, J., Shekhar, K., Goldman, M., et al. (2015). Highly parallel genome-wide expression profiling of individual cells using nanoliter droplets. *Cell*, *161*(5), 1202–1214. https://doi.org/10.1016/j.cell.2015.05.002.

Mayr, C. (2017). Regulation by 3′-untranslated regions. *Annual Review of Genetics*, *51*(1), 171–194. https://doi.org/10.1146/annurev-genet-120116-024704.

Pan, H., Oztas, O., Zhang, X., Wu, X., Stonoha, C., Wang, E., et al. (2016). A symbiotic SNARE protein generated by alternative termination of transcription. *Nature Plants*, *2*, 15197. https://doi.org/10.1038/nplants.2015.197.

Routh, A., Ji, P., Jaworski, E., Xia, Z., Li, W., & Wagner, E. J. (2017). Poly(A)-ClickSeq: Click-chemistry for next-generation 3-end sequencing without RNA enrichment or fragmentation. *Nucleic Acids Research*, *45*(12), e112. https://doi.org/10.1093/nar/gkx286.

Sanfilippo, P., Wen, J., & Lai, E. C. (2017). Landscape and evolution of tissue-specific alternative polyadenylation across Drosophila species. *Genome Biology*, *18*(1), 229. https://doi.org/10.1186/s13059-017-1358-0.

Singh, I., Lee, S.-H., Sperling, A. S., Samur, M. K., Tai, Y.-T., Fulciniti, M., et al. (2018). Widespread intronic polyadenylation diversifies immune cell transcriptomes. *Nature Communications*, *9*(1), 1716. https://doi.org/10.1038/s41467-018-04112-z.

Tian, B., & Manley, J. L. (2017). Alternative polyadenylation of mRNA precursors. *Nature Reviews. Molecular Cell Biology*, *18*(1), 18–30. https://doi.org/10.1038/nrm.2016.116.

Turchinovich, A., Surowy, H., Serva, A., Zapatka, M., Lichter, P., & Burwinkel, B. (2014). Capture and amplification by tailing and switching (CATS). An ultrasensitive ligation-independent method for generation of DNA libraries for deep sequencing from picogram amounts of DNA and RNA. *RNA Biology*, *11*(7), 817–828. https://doi.org/10.4161/rna.29304.

Wu, X., Liu, M., Downie, B., Liang, C., Ji, G., Li, Q. Q., et al. (2011). Genome-wide landscape of polyadenylation in Arabidopsis provides evidence for extensive alternative polyadenylation. *Proceedings of the National Academy of Sciences of the United States of America*, *108*(30), 12533–12538. https://doi.org/10.1073/pnas.1019732108.

Zheng, D., Liu, X., & Tian, B. (2016). 3′READS+, a sensitive and accurate method for 3′ end sequencing of polyadenylated RNA. *RNA*, *22*(10), 1631–1639. https://doi.org/10.1261/rna.057075.116.

Zheng, G. X. Y., Terry, J. M., Belgrader, P., Ryvkin, P., Bent, Z. W., Wilson, R., et al. (2017). Massively parallel digital transcriptional profiling of single cells. *Nature Communications*, *8*(1), 14049. https://doi.org/10.1038/ncomms14049.

Zheng, D., Wang, R., Ding, Q., Wang, T., Xie, B., Wei, L., et al. (2018). Cellular stress alters 3′UTR landscape through alternative polyadenylation and isoform-specific degradation. *Nature Communications*, *9*(1), 2268. https://doi.org/10.1038/s41467-018-04730-7.

CHAPTER FIVE

Using TIF-Seq2 to investigate association between 5′ and 3′mRNA ends

Bingnan Li*, Sueli Marques, Jingwen Wang, and Vicent Pelechano*
SciLifeLab, Department of Microbiology, Tumor and Cell Biology, Karolinska Institutet, Solna, Sweden
*Corresponding author: e-mail address: vicent.pelechano@scilifelab.se

Contents

Abstract

The development of high-throughput technologies has revealed pervasive transcription in all genomes that have been investigated so far. This has uncovered a highly interleaved transcriptome organization involving thousands of overlapping coding and non-coding RNA isoforms that challenge our traditional definitions of genes and functional regions of the genome. In this chapter, we discuss the application of an improved Transcript Isoform Sequencing approach (TIF-Seq2) able to concurrently determine the start and end sites of individual RNA molecules. We exemplify its use for the investigation of the human transcriptome and show how it is especially well suited to discriminate between overlapping molecules and accurately define their boundaries.

Methods in Enzymology, Volume 655
ISSN 0076-6879
https://doi.org/10.1016/bs.mie.2021.03.017

1. Introduction

Transcription is a complex process, producing a myriad of coding and non-coding RNA isoforms, which can have distinct functions, localization and life cycle. These isoforms often differ in their transcription start sites (TSSs), poly(A) sites (PASs) and splicing (de Klerk & Hoen, 2015; The FANTOM Consortium and the RIKEN PMI and CLST (DGT) et al., 2014, Tian & Manley, 2017). Genome-wide methods for the study of transcription have revolutionized biological research, allowing a thorough understanding of gene expression, transcriptional regulation and gene function. However, most of the existing technologies collect cumulative information without an accurate resolution of the individual RNA molecules. This might produce inaccurate annotations of transcript isoforms that could result in misinterpretation of gene expression data. As an example, RNA-Seq reads originating from upstream partially overlapping regulatory transcripts could be wrongly assigned to downstream coding isoforms (Proudfoot, 1986; Van Werven et al., 2012; Wang et al., 2020). Single-end approaches such as CAGE or poly(A) site sequencing have been able to define boundaries of the transcriptomes, but they cannot study the different combinations of TSS and PAS (Gruber et al., 2016; The FANTOM Consortium and the RIKEN PMI and CLST (DGT) et al., 2014). And although long-read sequencing can indeed reveal more accurately transcription complexity, it is still limited by its high cost, low throughput and lower resolution when reading boundary regions. Here, we provide detailed information regarding the development and implementation of an improved Transcript Isoform Sequencing (TIF-Seq2), which enables to concurrently determine the start and end sites of individual RNA molecules and to discriminate between overlapping molecules (Wang et al., 2020).

We have previously developed a Transcript Isoform Sequencing approach (TIF-Seq) and used it to investigate the *Saccharomyces cerevisiae* transcriptome (Pelechano, Wei, Jakob, & Steinmetz, 2014; Pelechano, Wei, & Steinmetz, 2013). Now, we have developed an optimized version, TIF-Seq2, specially well suited for the investigation of more complex transcriptomes (Wang et al., 2020). To demonstrate its applicability, we investigated the transcriptome of a chronic myeloid leukemia (CML) cell line in response to imatinib treatment. We identified thousands of known and unannotated transcript isoforms and accurately defined the boundaries of lowly expressed intergenic transcripts. By focusing on overlapping transcription units, we showed the common existence of short overlapping upstream transcripts that may lead to misinterpretation of RNA-Seq and CAGE gene expression data. We also identified complex transcriptional events involving gene-promoter rewiring and potentially leading to the generation of transcriptionally fused proteins. In addition to K562, we have also applied TIF-Seq2 to investigate the influence of the nuclear exosome shaping the expression of coding-genes in human cells (Wu et al., 2020). Besides human and yeast, we have also used TIF-Seq2 to investigate the complex transcriptome organization of *Arabidopsis thaliana* (Thomas et al., 2020).

TIF-Seq2 procedure is generally divided into four main steps: RNA oligo capping, production of full-length cDNA, intramolecular circularization and sequencing library construction (Fig. 1). For TIF-Seq2, the following points have been optimized:

- We have altered the sequencing library structure and decoupled the grafting and sequencing primer binding parts of the classical Illumina adapter. This allows the direct reading of 5′ and 3′ boundaries. Thus, all reads that can be mapped to the genome enable the identification of boundary sites. Additionally, this precludes the necessity of performing a strict library size selection required in our previous implementation (Pelechano et al., 2014).

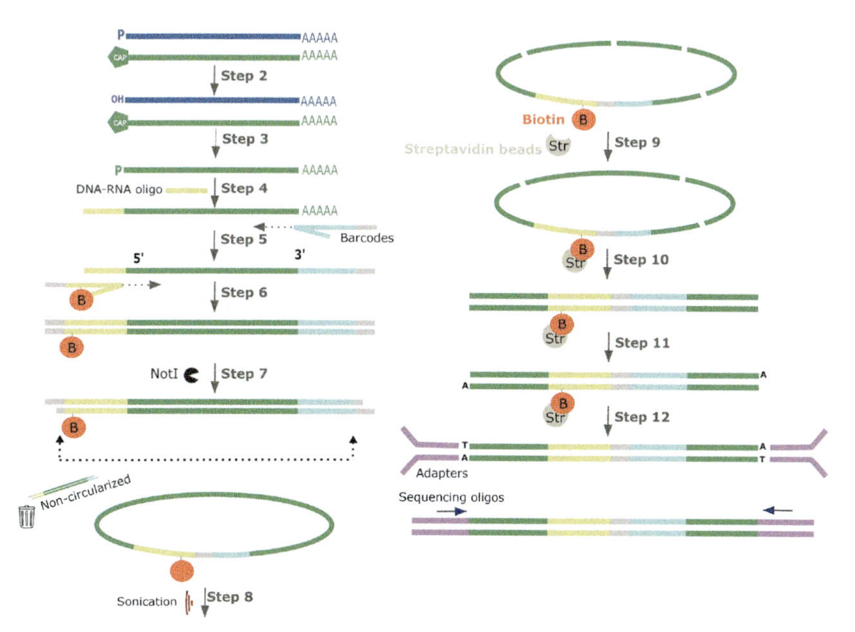

Fig. 1 Schematic representation of Tif-Seq2 protocol.

Table 1 Oligonucleotides list.

Name	Sequence	Production	Barcode	Concentration
TIF2-RNA	TCAGACGTGTGCTCTTCCGATCTrNrNrWrNrNrWrNrN	HPLC	N/A	100 μM
TIF2-Fw1	TATAGCGGCCGCGCCAATGTGAC[BtndT]GGAGTTCA GACGTGTGCTCTTCCGATC	0.05, HPLC	ATTGGC	5 μM
TIF2-Fw2	TATAGCGGCCGCATCTTGTAGTGAC[BtndT]GGAGTTC AGACGTGTGCTCTTCCGATC	0.05, HPLC	TACAAG	5 μM
TIF2-Fw3	TATAGCGGCCGCACAGTGGTGAC[BtndT]GGAGTTCA GACGTGTGCTCTTCCGATC	0.05, HPLC	CACTGT	5 μM
TIF2-Fw4	TATAGCGGCCGCATGTGAAAGTGAC[BtndT]GGAGTTC AGACGTGTGCTCTTCCGATC	0.05, HPLC	TTTCAC	5 μM
TIF2-Fw5	TATAGCGGCCGCCGA[BtndT] GTGTGACTGGAGTTCAGACGTGTGCTCTTCCGATC	0.05, HPLC	ACATCG	5 μM
TIF2-Fw6	TATAGCGGCCGCATTGACCAG[BtndT] GACTGGAGTTCAGACGTGTGCTCTTCCGATC	0.05, HPLC	TGGTCA	5 μM
TIF2-Fw7	TATAGCGGCCGCAGACA[BtndT]CACAGATCGTGACT GGAGTTCAGACGTGTGCTCTTCCGATC	0.05, HPLC	GATCTG	5 μM
TIF2-Fw9	TATAGCGGCCGCAGACA[BtndT]CAAGTCAAGTGACT GGAGTTCAGACGTGTGCTCTTCCGATC	0.05, HPLC	TTGACT	5 μM
TIF2-Fw10	TATAGCGGCCGCCTCA[BtndT]GATATAGTTCCGTGAC TGGAGTTCAGACGTGTGCTCTTCCGATC	0.05, HPLC	GGAACT	5 μM
TIF2-Fw11	TATAGCGGCCGCCTCA[BtndT]GATATGTCAGTG ACTGGAGTTCAGACGTGTGCTCTTCCGATC	0.05, HPLC	TGACAT	5 μM

TIF2-Fw12	TATAGCGGCCGCCTCA[BtndT]GATATGTCCGCG TGACTGGAGTTCAGACGTGTGCTCTTCCGATC	0.05, HPLC	GCGGAC	5 μM
TIF2-RT1	TAGTTCAGTCTTCAGTACCTCGTGCGGCCG CGCCAATACACTCTTTCCCTACACGACGCT CTTCCGATCTTTTTTTTTTTTTTTVN	0.05, HPLC	ATTGGC	1 μM
TIF2-RT2	TAGTTCAGTCTTCAGTACCTCGTGCGGCCG CCTTGTAACACTCTTTCCCTACACGACGCT CTTCCGATCTTTTTTTTTTTTTTTVN	0.05, HPLC	TACAAG	1 μM
TIF2-RT3	TAGTTCAGTCTTCAGTACCTCGTGCGGCCG CACAGTGACACTCTTTCCCTACACGACGCT CTTCCGATCTTTTTTTTTTTTTTTVN	0.05, HPLC	CACTGT	1 μM
TIF2-RT4	TAGTTCAGTCTTCAGTACCTCGTGCGGCCG CGTGAAAACACTCTTTCCCTACACGACGCT CTTCCGATCTTTTTTTTTTTTTTTVN	0.05, HPLC	TTTCAC	1 μM
TIF2-RT5	TAGTTCAGTCTTCAGTACCTCGTGCGGCCG CCGATGTACACTCTTTCCCTACACGACGCT CTTCCGATCTTTTTTTTTTTTTTTVN	0.05, HPLC	ACATCG	1 μM
TIF2-RT6	TAGTTCAGTCTTCAGTACCTCGTGCGGCCG CTGACCAACACTCTTTCCCTACACGACGCT CTTCCGATCTTTTTTTTTTTTTTTVN	0.05, HPLC	TGGTCA	1 μM
TIF2-RT7	TAGTTCAGTCTTCAGTACCTCGTGCGGCCG CCAGATCACACTCTTTCCCTACACGACGCT CTTCCGATCTTTTTTTTTTTTTTTVN	0.05, HPLC	GATCTG	1 μM
TIF2-RT9	TAGTTCAGTCTTCAGTACCTCGTGCGGCCG CAGTCAAACACTCTTTCCCTACACGACGCT CTTCCGATCTTTTTTTTTTTTTTTVN	0.05, HPLC	TTGACT	1 μM

Continued

Table 1 Oligonucleotides list.—cont'd

Name	Sequence	Production	Barcode	Concentration
TIF2-RT10	TAGTTCAGTCTTCAGTACCTCGTGCGGCCG CAGTTCCACACTCTTTCCCTACACGACGCT CTTCCGATCTTTTTTTTTTTTTTTTTVN	0.05, HPLC	GGAACT	1 μM
TIF2-RT11	TAGTTCAGTCTTCAGTACCTCGTGCGGCCG CATGTCAACACTCTTTCCCTACACGACGCT CTTCCGATCTTTTTTTTTTTTTTTTTVN	0.05, HPLC	TGACAT	1 μM
TIF2-RT12	TAGTTCAGTCTTCAGTACCTCGTGCGGCCG CGTCCGCACACTCTTTCCCTACACGACGCT CTTCCGATCTTTTTTTTTTTTTTTTTVN	0.05, HPLC	GCGGAC	1 μM
TIF2-RT13	TAGTTCAGTCTTCAGTACCTCGTGCGGCCG CATCACGACACTCTTTCCCTACACGACGCT CTTCCGATCTTTTTTTTTTTTTTTTTVN	0.05, HPLC	CGTGAT	1 μM
TIF2-RT14	TAGTTCAGTCTTCAGTACCTCGTGCGGCCG CTTAGGCACACTCTTTCCCTACACGACGCT CTTCCGATCTTTTTTTTTTTTTTTTTVN	0.05, HPLC	GCCTAA	1 μM
TIF2-RT15	TAGTTCAGTCTTCAGTACCTCGTGCGGCCG CACTTGAACACTCTTTCCCTACACGACGCT CTTCCGATCTTTTTTTTTTTTTTTTTVN	0.05, HPLC	TCAAGT	1 μM
TIF2-RT16	TAGTTCAGTCTTCAGTACCTCGTGCGGCCG CGATCAGACACTCTTTCCCTACACGACGCT CTTCCGATCTTTTTTTTTTTTTTTTTVN	0.05, HPLC	CTGATC	1 μM
TIF2-RT17	TAGTTCAGTCTTCAGTACCTCGTGCGGCCG CTAGCTTACACTCTTTCCCTACACGACGCT CTTCCGATCTTTTTTTTTTTTTTTTTVN	0.05, HPLC	AAGCTA	1 μM

TIF2-RT18	TAGTTCAGTCTTCAGTACCTCGTGCGGCCG CGGCTACACACTCTTTCCCTACACGACGCT CTTCCGATCTTTTTTTTTTTTTTTTTVN	0.05, HPLC	GTAGCC	1 μM
TIF2-RT19	TAGTTCAGTCTTCAGTACCTCGTGCGGCCG CGTAGAGACACTCTTTCCCTACACGACGCT CTTCCGATCTTTTTTTTTTTTTTTTTVN	0.05, HPLC	CTCTAC	1 μM
TIF2-RT20	TAGTTCAGTCTTCAGTACCTCGTGCGGCCG CGTGGCCACACTCTTTCCCTACACGACGCT CTTCCGATCTTTTTTTTTTTTTTTTTVN	0.05, HPLC	GGCCAC	1 μM
TIF2-Rv	TAGTTCAGTCTTCAGTACCTCGT	0.05, HPLC	N/A	5 μM
PCRgraftP5	AATGATACGGCGACCACCGAGATCTACAC	0.05, HPLC	N/A	5 μM
PCRgraftP7	CAAGCAGAAGACGGCATACGAGAT	0.05, HPLC	N/A	5 μM
TIF2-Fork Fw	AATGATACGGCGACCACCGAGATCTACACA CACCTGCCGGTCACC*T	0.05, HPLC	N/A	1 μM
TIF2-Fork Rv	[Phos]GGTGACCGGCAGGTGTATCTCGTATGCCGT CTTCTGCTTG	0.05, HPLC	N/A	1 μM
SeqR1 + 15 T	ACACTCTTTCCCTACACGACGCTCTTCCGA TCTTTTTTTTTTTTTTTT	0.05, HPLC	N/A	100 μM
SeqR2	GTGACTGGAGTTCAGACGTGTGCTCTTCCGATCT	0.05, HPLC	N/A	100 μM
SeqINDX	GATCGGAAGAGCACACGTCTGAACTCCAGTCAC	0.05, HPLC	N/A	100 μM
Tif3RvRT	GATCGGAAGAGCGTCGTGTAGGGAAAGAGTGT	0.05, DSL	N/A	100 μM

Notes: All oligos are HPLC purified. [BtndT] refers to Biotin dT; [Phos] refers to 5′ phosphorylation; bases preceded by 'r' refer to ribonucleotides; N refers to (A, C, G or U) and W to (A or T). Sequences are represented in 5′–3′ orientation. Illumina oligonucleotide sequences © 2006–2021 Illumina, Inc. All rights reserved.

- By omitting a strict library size selection and starting to sequence from the extracted boundaries sites, we reduce the loss of reads containing only one informative boundary (e.g., only 5′ or 3′).
- To enable the direct sequencing of the poly(A) site and minimize the decrease in quality caused by sequencing through homopolymeric regions (Wilkening et al., 2013), we used a modified read1 sequencing primer with additional 15Ts (seqR1 + 15T) to start reading after the fragment of the poly(A) tail included in the TIF-Seq2 library.
- As a consequence of decoupling the grafting and sequencing primer binding sites, we also modified the primers used during the Illumina sequencing approach. In standard Illumina sequencing, there is always concordance between the strand of the grafting and the sequencing primer binding region. However, in TIF-Seq2 only half of the clusters will have that organization. While the other half of the clusters will present a reverse complementary organization of the sequencing primer binding regions while keeping the same grafting primer region organization. To decode both kinds of clusters simultaneously, we added the sequencing oligos corresponding to read1 and read2 together and used the information in the indexing regions to decode the orientation of each cluster. Specifically, using a NextSeq 500 sequencer we loaded a mix of oligos (i.e., seqR1 + 15T and seqR2) to the positions 7 and 8 on the sequencing cartridge.

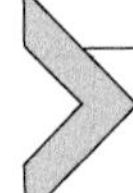

2. Protocol

2.1 Before you begin

2.1.1 Timing: 1 h

1. Prepare the following buffers:
   - **(a)** EB buffer: 10 mM Tris-HCL, pH 8.0
   - **(b)** Bind & Wash buffer (B&W) 2×: 10 mM Tris-Cl pH 7.5, 1 mM EDTA pH 8.0 and 2 mM NaCl
   - **(c)** dA tailing buffer: 2 mM dATP in NEB2 buffer
2. Prepare the annealing reaction of Tif2-Fork Fw and Rv by:
   - **(a)** Mixing equal volumes of the equimolar primers, 15 μM each
   - **(b)** Incubating at 95 °C for 5 min
   - **(c)** Placing on the bench to slowly cool to room temperature

   This will be the 15 × stock, which needs to be freshly diluted before use to 1 ×.

2.1.2 Key resources table

Reagent or resource	Source	Identifier
Chemicals, peptides, and recombinant proteins		
Turbo DNA free kit	Thermo Fisher Scientific	AM1907
Ribolock RNase inhibitor 40,000 U/mL	Thermo Fisher Scientific	EO0382
Alkaline Phosphatase, Calf Intestinal (CIP)	New Englands Biolabs	M0290
Cap-clip acid pyrophosphatase	Cell Script	C-CC15011H
T4 RNA ligase 1 (ssRNA ligase)	New Englands Biolabs	M0204S
DMSO	Thermo Fisher Scientific	85190
dNTP set, 100 mM solution	Thermo Fisher Scientific	R0181
Trehalose 1.57 M	Sigma-Aldrich	T9531
RNase H 5000 U/mL	New Englands Biolabs	M0297S
RNase cocktail enzyme mix	Thermo Fisher Scientific	AM2286
Terra PCR direct polymerase mix	Takara	639271
T4 DNA ligase 2,000,000 U/mL	New Englands Biolabs	M0202T
Plasmid safe ATP-dependent DNase	Lucigen	E310K
Dynabeads M-280 Streptavidin	Thermo Fisher Scientific	11206D
Klenow fragment (3′–5′ *exo*-) 5000 U/mL	New Englands Biolabs	M0212S
NEBNext quick ligation reaction buffer 5 ×	New Englands Biolabs	B6058S
Phusion high-fidelity PCR master mix with HF buffer	New Englands Biolabs	M0531S
Phenol solution. saturated with 0.1 M citrate buffer, pH 4.3 ± 0.2	Sigma-Aldrich	P4682
Chloroform:Isoamyl alcohol 24:1	Sigma-Aldrich	C0549-1PT
Sodium acetate buffer solution, pH 5.3	Sigma-Aldrich	S7899
Glycoblue coprecipitant (15 mg/mL)	Thermo Fisher Scientific	AM9515
70% Ethanol	VWR	83801.360

Continued

—cont'd

Reagent or resource	Source	Identifier
Ampure XP beads	Beckman Coulter	A63881
Ethanol absolute ≥99.5%	VWR	83813.360
Nuclease-free water, not DEPC treated	Thermo Fisher Scientific	AM9937
Qubit RNA HS assay kit	Thermo Fisher Scientific	Q32852
Qubit 1 × DNA HS assay kit	Thermo Fisher Scientific	Q33230
Tris(hydroxymethyl) aminomethane acetate salt, 99%	Fisher Scientific	434500250
Hydrochloric acid 35–38%	Sigma-Aldrich	V800202
ATP solution, Tris buffered, 100 mM	Thermo Fisher Scientific	R1441
Dynabeads™ M-280 Streptavidin	Thermo Fisher Scientific	11206D
RNAClean XP	Beckman Coulter	A63987
UltraPure™ 0.5 M EDTA, pH 8.0	Thermo Fisher Scientific	15575020
Sodium chloride, ACS reagent, ≥99.0%	Sigma-Aldrich	S9888
dATP solution	Thermo Fisher Scientific	R0141
NEBuffer 2	New Englands Biolabs	B7002S
High sensitivity DNA Kit	Agilent	5067-4626
RNA 6000 nano kit	Agilent	5067-1511
Deposited data		
Raw data deposited at GEO		GSE140912
Browsable TIF-Seq2 tracks for K562	SciLifeLabData Repository	https://doi.org/10.17044/scilifelab.13293344.v1
Software and algorithms		
https://github.com/jingwen/TIFseq2	GitHub	
https://github.com/PelechanoLab/TIFseq2	GitHub	
Oligonucleotides		
See Table 1		

2.1.3 Materials and equipment

- Thermoblock
- Microcentrifuge
- PCR machine
- Magnets for PCR tubes and for 1.5 mL Eppendorf tubes
- Covaris sonicator ME220
- QuBit Fluorometer
- Agilent Bioanalyzer 2100
- Chemical hood

2.2 Step-by-step method details

We describe below the protocol for performing TIF-Seq2 in human samples. We recommend processing at least 4 samples, but ideally 6 or 8, in parallel to achieve enough complexity during the multiplexing step. We applied the same protocol to investigate the *Arabidopsis thaliana* transcriptome and expect it would be widely applicable.

2.2.1 Step 1 removal of DNA from RNA samples

Day1, timing: 30 min

1. Starting with 7.5 μg of total RNA (which will be split in 3 for 1st strand synthesis), prepare the following mix and incubate the samples for 20 min at 37 °C

Turbo DNAse buffer 10 ×	5 μL
Turbo DNAse 2 U/μL	3 μL
Ribolock	1 μL
Sample RNA	variable
RNAse-free water	up to 50 μL
Total	**50 μL**

2. Add 6 μL of DNase inactivation reagent (Invitrogen) and flick the tubes. Incubate the tubes for 2–3 min at room temperature
3. Centrifuge at 16,000 *g* × 2 min RT (room temperature) and recover the supernatant. Keep a small aliquot (e.g., 0.5 μL)

2.2.2 Step 2 phosphatase treatment

Timing: 3h

1. Remove the 5′ phosphate of the fragmented and non-capped (5′ tri-phosphate) molecules by setting up the following 100 μL reaction in a 1.5 mL tube

CutSmart buffer 10 ×	10 μL
CIP 10 U/μL	3 μL
Ribolock	1 μL
DNAse-treated sample	50 μL
RNAse-free water	36 μL
Total	**100 μL**

2. Incubate 37 °C 30 min
3. To remove the phosphatase:
   - **(a)** Increase the sample volume with RNase-free water to 300 μL, and transfer it to a 1.5 mL tube containing 300 μL of water-saturated phenol
   - **(b)** Mix the phases by flicking the tube
   - **(c)** Centrifugation 10,000 *g* × 3 min RT
4. Transfer the aqueous phase to a new Eppendorf tube (1.5 mL) containing 300 μL of water-saturated phenol to perform a second extraction
   - **(a)** Mix the phases by flicking the tube
   - **(b)** Centrifugation 10,000 *g* × 3 min RT
5. Add it to an equal volume of phenol:chloroform:isoamyl alcohol 24:25:1 (300 μL).
   - **(a)** Mix the phases by flicking the tube
   - **(b)** Centrifugation 10,000 *g* × 3 min RT
6. Ethanol-precipitate the aqueous phase by adding
   - **(a)** 2.5 volumes (with respect to the sample volume) of 100% (vol/vol) ethanol
   - **(b)** a one-tenth volume of 3 M sodium acetate
   - **(c)** 1 μL of Glycoblue

Mix and incubate the precipitation mixture at −20 °C for a minimum of 30 min.

7. Centrifugation 14,000 *g* × 30 min at 4 °C

8. Wash the pellet with 300 μL of ice-cold 70% (vol/vol) ethanol
9. Centrifugation 14,000 $g \times$ 10 min at 4 °C
10. Remove the remaining ethanol, air-dry for 3 min and resuspend the pellet in 16.5 μL of RNase-free water

2.2.3 Step 3 5′cap removal

Timing: 3 h

1. Remove the 5′ cap of phosphatase-treated RNA molecules by setting up the following 20-μL reaction in a 1.5-mL tube and incubate at 37 °C 60 min

Phosphatase-treated samples	16.5 μL
Cap-Clip Acid Pyrophosphatase buffer	2 μL
Cap-Clip Acid Pyrophosphatase (5 U/μL)	0.5 μL
Ribolock	1 μL
Total	**20 μL**

2. To remove the Cap-clip enzyme:
   - **(a)** Increase the sample volume with RNase-free water to 300 μL, and transfer it to a 1.5 mL tube containing 300 μL of phenol:chloroform: isoamyl alcohol 24:25:1 (300 μL).
   - **(b)** Mix the phases by flicking the tube
   - **(c)** Centrifugation 10,000 $g \times$ 3 min RT
3. Ethanol-precipitate the aqueous phase by adding
   - **(a)** 2.5 volumes (with respect to the sample volume) of 100% (vol/vol) ethanol
   - **(b)** a one-tenth volume of 3 M sodium acetate
   - **(c)** 1 μL of Glycoblue

Mix and incubate the precipitation mixture at −20 °C for a minimum of 30 min.

4. Centrifugation 14,000 $g \times$ 30 min at 4 °C
5. Wash the pellet with 300 μL of ice-cold 70% (vol/vol) ethanol
6. Centrifugation 14,000 $g \times$ 10 min at 4 °C
7. Remove the remaining ethanol, air-dry for 3 min and resuspend the pellet in 4.7 μL of RNase-free water

2.2.4 Step 4 single-stranded RNA ligation

Timing: 18 h

1. Set up the following 10-μL ligation reaction and incubate it overnight (16 h) at 16 °C:

Cap-Clip treated sample	4.7 μL
100 μM DNA-RNA oligo TIF2-RNA	1 μL
T4 RNA ligase buffer 10 ×	1 μL
T4 RNA ligase (10 U/μL)	2.1 μL
DMSO	1 μL
Ribolock	0.1 μL
ATP 100 mM	0.1 μL
Total	**10 μL**

Day 2, timing: 1 h

1. Add 40 μL RNAse-free water to increase the volume. Purify the sample using 1.8 × volumes of Ampure XP beads according to the manufacturer's instructions:
   a. Add 18 μL of RNACleanXP beads and mix the sample by pipetting up and down
   b. Let the sample bind the beads by incubating the mixture at room temperature for 5 min
2. Place the 0.2 mL tubes in the magnet and let the beads settle (~2 min)
   a. Remove the supernatant
   b. Perform two consecutive washes with 200 μL of 70% (vol/vol) ethanol (made fresh).
3. Remove the ethanol and allow the beads to slightly dry for 1 min
   a. Elute the sample in 30 μL of RNase-free water
4. Run an Agilent RNA Bioanalyzer comparison of the input (aliquot from total RNA, Step 1) and the ligated RNA sample (from Step 25). If substantial degradation is observed (e.g., more intense 18S than 26S or an excess of short-length RNAs), discard the sample

2.2.5 Step 5 reverse transcription

Timing: 3 h

1. Allocate primers for different samples. To enable sample differentiation, it is important that barcodes indicating the 3′ side and the 5′ side of each

mRNA molecule (TIF2-RT and TIF2-Fw, respectively) are selected so that all of them can be identified allowing two mismatches (Hamming distance greater than 2). The identity of the 5′ and 3′ ends is decoded during the demultiplexing (see bioinformatic analysis below). As an example:

 a. To multiplex four samples we use the following combination: Sample 1 (Tif2-RT9 + Tif2-FW2), Sample 2 (Tif2-RT15 + Tif2-FW6), Sample 3 (Tif2-RT13 + Tif2-FW5) and Sample 4 (Tif2-RT19 + Tif2-FW7).
 b. To multiplex six samples we use: Sample 1 (Tif2-RT10 + Tif2-FW2), Sample 2 (Tif2-RT13 + Tif2-FW3), Sample 3 (Tif2-RT14 + Tif2-FW5), Sample 4 (Tif2-RT15 + Tif2-FW6), Sample 5 (Tif2-RT19 + Tif2-FW7) and Sample 6 (Tif2-RT20 + Tif2-FW9)

2. Split each sample in three replicates and set up the following 12 μL reaction for each:

Total ligated RNA	10 μL
TIF2-RT (1 μM)	1 μL
dNTPs (10 mM)	1 μL
Total	**12 μL**

3. Denature the samples at 65 °C for 5 min. Afterward, place the samples on ice
4. To each tube, add 11 μL of mixture containing the following:

First strand buffer 5 ×	5 μL
DTT (100 mM)	2 μL
Trehalose (1.57 mM)	3 μL
Ribolock	1 μL
Total	**11 μL**

5. Mix the sample and incubate it for 2 min at 42 °C
6. While the samples are in the PCR machine, add 2 μL of SuperScript III reverse transcriptase (Invitrogen) to each sample
 42 °C for 50 min
 50 °C for 30 min
 55 °C for 30 min

Inactivate at 70 °C for 15 min
Hold the samples at 4 °C

7. Remove the template RNA and excess of rRNA by adding 1 μL of the following mix and by incubating at 37 °C for 30 min

RNaseH (5 U/μL)	0.5 μL
RNase cocktail	0.5 μL
Total	**1 μL**

8. Purify the sample using 2 × volumes of Ampure XP beads according to the manufacturer's instructions
   a. Add 48 μL of Ampure XP beads and mix the sample by pipetting up and down
   b. Let the sample bind the beads by incubating the mixture at room temperature for 5 min
   c. Place the 0.2-mL tubes in the magnet and let the beads settle (~2 min)
9. Remove the supernatant and perform two consecutive washes with 200 μL of 70% (vol/vol) ethanol (made fresh)
10. Remove the ethanol and allow the beads to slightly dry for 1 min

Pause Point: samples can be kept at −20 °C.

2.2.6 Step 6 generation of second strand by PCR

Timing: 3 h

1. Perform a 25 μL PCR, using as template 10 μL of each RT sample, and keep the remaining full-length cDNA sample stored at −20 °C as a backup. Set up the following 25 μL reaction

cDNA template	10 μL
Water	1 μL
2xTERRA direct PCR buffer	12.5 μL
TERRA enzyme	0.5 μL
TIF-Fw oligo(5 μM)	0.5 μL
TIF-Rv oligo(5 μM) (same for all samples)	0.5 μL
Total	**25 μL**

2. Run a PCR with the following cycling conditions
   98 °C for 2 min.
   16 cycles of 98 °C for 20 s, 60 °C 30 s and 68 °C 5 min (+10 s/cycle)
   72 °C for 5 min
   Hold the samples at 4 °C
3. Purify the samples using 1.0 × volumes of Ampure XP beads according to the manufacturer's instructions and as described here
   a. Add 25 μL of Ampure XP beads and mix the sample by pipetting up and down
   b. Let the sample bind the beads by incubating the mixture at room temperature for 5 min
   c. Place the 0.2 mL tubes in the magnet and let the beads settle (~2 min)
4. Remove the supernatant and perform two consecutive washes with 200 μL of 70% (vol/vol) ethanol (made fresh)
5. Remove the ethanol and allow the beads to slightly dry for 1 min
6. Elute the sample in 13 μL of EB buffer (10 mM Tris-HCl, pH 8.0)
7. Quantify the produced full-length cDNA by using the Qubit with the dsDNA HS assay kit. Usually ~300 ng of full-length cDNA is obtained.

2.2.7 Step 7 production of sticky ends

During this step, samples produced in parallel are pooled together at equimolar amounts. Please note that as described in STEP 5 each sample should have a unique combination of barcodes associated with their 5′ and 3′ sides. Performing this process in a pool enables to measure the rate of artefactual chimera formation and simplifies all downstream manipulations.

Day 3, timing: 3h

1. Use ~200 ng of full-length cDNA from each sample
2. Increase the full-length cDNA sample volume to 17 μL with water and set up the following 20 μL reaction

Full-length cDNA sample	17 μL
CutSmart buffer	2 μL
*Not*I HF enzyme (20 U/μL)	1 μL
Total	**20 μL**

CRITICAL: NotI enzyme will also cut any molecules where its recognition sequence (5′-GCCGGCCGC-3′) is present. Bioinformatic analysis of the genome of interest should be performed to identify the potential depletion of a specific gene of interest.

3. Incubate the sample for 1 h at 37 °C in a thermocycler block with a heated lid
4. Inactivate the sample by incubating it at 65 °C for 20 min, and add 20 μL of water
5. Purify the sample using 1.8 × volumes of Ampure XP beads according to the manufacturer's instructions and as described here:
   a. Add 72 μL of Ampure XP beads and mix the sample by pipetting up and down
   b. Let the sample bind the beads by incubating the mixture at room temperature for 5 min
   c. Place the 0.2-mL tubes in the magnet and let the beads settle (~2 min)
6. Remove the supernatant and perform two consecutive washes with 200 μL of 70% (vol/vol) ethanol (made fresh)
7. Remove the ethanol and allow the beads to slightly dry for 1 min
8. Elute the sample in 30 μL of EB buffer (10 mM Tris-HCl, pH 8.0)
9. Quantify the cDNA by using the Qubit with the dsDNA HS assay kit

2.2.8 Step 8 circularization of full-length cDNA

Timing: 18 h

1. Set up the following 600-μL reaction (Pool the samples here in a way that max concentration of cDNA is 1 ng/μL per reaction. If needed, run several reactions):

Sticky end full-length cDNA	variable
Water	variable
T4 DNA ligase buffer 10 ×	60 μL
T4 DNA ligase (2000 U/μL)	20 μL
Total	**600 μL**

2. Incubate the sample overnight at 16 °C (minimum of 16 h)

Day 4, timing: 3h

1. Incubate the samples for 5 min at room temperature to close any remaining DNA nicks in the circularized full-length cDNA
2. Treat the samples with Plasmid Safe to remove any remaining non-circularized full-length cDNA. For every 100 μL system, add 0.5 μL Plasmid Safe and 1 μL ATP 100 mM

3. Incubate for 1 h at 37 °C and 30 min at 70 °C for deactivation
4. Remove the enzyme by phenol-chloroform extraction
   a. To a 2-mL Eppendorf tube, add equal volumes of sample and phenol:chloroform:isoamyl alcohol at a 25:24:1 ratio (600 μL each)
   b. Mix the phases by flicking the tube
   c. Centrifugation 10,000 g × 5 min at 25 °C
   d. Isopropanol-precipitate the sample by adding
      - 2.5 volumes (with respect to the sample volume) of 100% Ethanol
      - a one-tenth volume of 3 M sodium acetate
      - 1 μL of Glycogen Blue
   e. Mix and incubate the sample at −20 °C for a minimum of 30 min
5. Precipitate the sample by centrifugation for 30 min at 14,000 g at 4 °C
6. Wash the pellet with 300 μL of 70% (vol/vol) ice-cold ethanol
7. Centrifuge the sample for 10 min at 14,000 g at 4 °C
8. Remove the remaining ethanol, and resuspend the pellet in 130 μL of water (or 10 mM Tris-HCl (pH 8), as recommended by Covaris)
9. Quantify the cDNA by using the Qubit with the dsDNA HS assay kit

2.2.9 Step 9 sonication of circularized DNA

Timing: 1 h

1. Transfer the sample to a Covaris microTUBE, 6 × 16 mm
2. Sonicate the circularized DNA using the Covaris instrument with the following conditions:

   Duration **240 s**
   Peak Power **30**
   Duty Factor **10**
   200 cycles/burst
   Average Power 3 W
3. Purify the sample using 1.0 × volumes of Ampure XP beads according to the manufacturer's instructions and as described here:
   a. Add 130 μL of Ampure XP beads and mix the sample by pipetting up and down
   b. Let the sample bind the beads by incubating the mixture at room temperature for 5 min
   c. Place the 0.2-mL tubes in the magnet and let the beads settle (~2 min)

CRITICAL: Ampure purification is essential to remove from the sample biotinylated fragments that are too short to be informative (i.e., shorter than 200 bp as we are selecting only longer fragments here).

4. Remove the supernatant and perform two consecutive washes with 200 µL of 70% (vol/vol) ethanol
5. Remove the ethanol and let the beads slightly dry for 1 min. Elute the sample in 20 µL of EB and transfer it to a 0.2-mL tube
6. Check the concentration using the Qubit fluorometer with the dsDNA HS assay kit (optimally, it should be ~10 ng per sample)

CRITICAL: Do not pause sample preparation after sonication, but proceed immediately to the next step.

CRITICAL: Sonication seems to fragment more frequently in the poly(A) tail region, and thus the relative AT content of the genome could affect the specific fragmentation.

2.2.10 Step 10 binding samples to streptavidin beads

Timing: 1 h

1. Prepare M-280 streptavidin beads (Invitrogen) according to the manufacturer's instructions and as described here:
   a. Transfer 20 µL of beads into a 0.2-mL tube
   b. Place the 0.2-mL tubes in the magnet and let the beads settle (~2 min).
   c. Remove the supernatant and perform two consecutive washes with 200 µL of 1 × B&W buffer (5 mM Tris-HCl (pH 7.5), 0.5 mM EDTA (pH 8.0) and 1 M NaCl).
   d. Resuspend the beads in 20 µL of 2 × B&W buffer (10 mM Tris-HCl (pH 7.5), 1 mM EDTA (pH 8.0) and 2 M NaCl).
2. Add to the beads 20 µL of sonicated circularized full-length cDNA and let the samples bind by incubating the beads in a rotator wheel for 30 min at room temperature

Please note that the wheel rotation will not mix the 40-µL bead solution, but rather prevent the beads from settling down.

3. Place the 0.2 mL tubes in the magnet and let the beads settle (~2 min)
4. Remove the supernatant and perform four consecutive washes with 200 µL of 1 × B&W buffer
5. Perform one wash with 200 µL of EB
6. Resuspend the beads in 20.75 µL of water

CRITICAL: At this point, the samples of interest remain bound to the beads owing to the streptavidin-biotin interaction.

2.2.11 Step 11 end repair of DNA fragments

Timing: 1 h

1. Set up the following 25-μL reaction:

Sample beads	20.75 μL
Buffer End Repair	2.5 μL
End Repair Enzymes mix	1.25 μL
Total	**25 μL**

2. Incubate the sample for 30 min at 20 °C in a PCR block
3. Place the 0.2-mL tubes in the magnet and let the beads settle (~2 min). Remove the supernatant
4. Perform washes with 200 μL of 1 × B&W buffer
5. Perform one wash with 200 μL of EB
6. Resuspend the beads in 20.5 μL of water

2.2.12 Step 12 addition of a protruding adenine to the DNA fragments

Timing: 1 h

1. Set up the following 25-μL reaction:

Sample beads	20.5 μL
dA tailing buffer 10 × (see Recipes section)	2.5 μL
Klenow Fragment Exo (5 U/μL)	1.5 μL
Total	**25 μL**

2. Incubate the sample for 30 min at 37 °C in a PCR block
3. Place the 0.2-mL tubes in the magnet and let the beads settle (~2 min)
4. Remove the supernatant. Perform one wash with 200 μL of 1 × B&W buffer
5. Perform one wash with 200 μL of EB
6. Resuspend the beads in 20 μL of water

2.2.13 Step 13 ligation of Illumina adapters

Timing: 1.5 h

1. Set up the following 50 μL reaction:

Sample beads	20 μL
Water	16 μL
TIF2-Fork pre-annealed (1 μM) (freshly diluted from stock solution (15 μM))	1 μL
Quick Ligation buffer 5 ×	10 μL
T4 DNA ligase (2000 U/μL)	3 μL
Total	**50 μL**

2. Incubate the sample for 60 min at RT (or overnight at 16 °C)
3. Place the 0.2 mL tubes in the magnet and let the beads settle (~2 min)
4. Perform three consecutive washes with 200 μL of 1 × B&W buffer, letting the beads settle each time
5. With the last wash, change the beads containing the sample to a new 0.2 mL tube and let the beads settle (~2 min). Remove the supernatant
6. Perform one wash with 200 μL 1 × B&W buffer
7. Perform one wash with 200 μL of EB
8. Resuspend the beads in 20 μL of water

PAUSE: samples should be immediately processed in the next step. Unused samples can be stored at 4C for days or at −20 °C for months (however, storing at −20 °C may damage the integrity of the bead matrix).

2.2.14 Step 14 library PCR amplification

Timing: 1.5 h

1. PCR-amplify the libraries. Prepare the following mix:

Sample beads	20 μL
Water	16 μL
Phusion HF PCR Mastermix 2 ×	25 μL
PCR GRAFT P5	0.5 μL
PCR GRAFT P7	0.5 μL
Total	**50 μL**

2. Run a PCR using the following program

Optionally, if the produced amount of library is too low for Illumina sequencing, increase the number of elongation cycles from 18 to up to 23

98 °C for 30 s

16 cycles of 98 °C for 20 s, 65 °C 30 s and 72 °C 30 s

72 °C for 5 min

Hold the samples at 4 °C

3. Bind Streptavidin beads 2 min in magnet. Recover supernatant, which contains PCR products and continue with it in step 5

4. Wash beads 2× with 1 × BW, 1 × EB and resuspend in 20 μL EB

5. Add 0.35 × Ampure beads to the supernatant. Incubate 5 min and bind 2 min to the magnet. Recover supernatant

6. Add 0.45 × Ampure beads to supernatant. Incubate 5 min and bind 2 min to magnet

7. Wash 2 × 200 μL of 70% ethanol (made fresh).

8. Elute in 10 μL water or EB

9. Quantify dsDNA and run bioanalyzer

10. Measure library molarity (nM) by the following function:

$$\textit{Library concentration}/(\textit{Average library size}*660/100000)*10$$

2.2.15 Custom loading protocol for Illumina NextSeq500

1. Proceed to denaturation and dilution of libraries to loading concentration as indicated in NextSeq 500 system protocol A (standard normalization method)

See documentation in https://support.illumina.com/downloads/nextseq-500-denaturing-diluting-libraries-15048776.html.

2. Prepare five 2 mL tubes labeled as following:

a. Tube 7 (Mixture 1) (For position 7 in cartridge)

i. HT1 buffer	1988 μL
ii. SeqR1+15T (100 μM)	6 μL
iii. SeqR2 (100μM)	6 μL

b. Tube 8 (Mixture 1) (For position 8 in cartridge)

i. HT1 buffer	1988 μL
ii. SeqR1+15T (100μM)	6 μL
iii. SeqR2 (100μM)	6 μL

c. Tube 9 (Mixture 2) (For position 9 in cartridge)

i. HT1 buffer	1988 μL
ii. SeqINDX (100μM)	6 μL
iii. Tif3RvRT (100μM)	6 μL

d. Your library name and details—as prepared in step 1
e. Loading Library

3. Mixture 1 (tube 7 and 8) is loaded into respective positions on reagent cartridge in a NextSeq 500 instrument for custom read1 primer and for custom read2 primer
4. Mixture 2 is loaded into the position for custom index primer (#9 on reagent cartridge)
5. Sequencing was carried out in an Illumina NextSeq 500 instrument with stand-alone configuration and custom sequencing oligos. Paired-end sequencing read lengths were set as read1 76 bp, read2 76 bp, index1 6 bp and index2 6 bp

3. Bioinformatic analysis

Here we detail recommended potential bioinformatic analysis to interpret TIF-Seq2 data. Please also check the script repositories and sample data listed in the key resources table and Fig. 2.

3.1 Preprocessing and alignment

- Process raw sequencing data: Use bcl2fastq (v2.20.0) for converting raw images to sequence information (FASTQ files) and demultiplex. We recommend allowing 1 or 2 mismatches in the indexes depending on the sequencing quality
- Trim adapters: Use cutadapt (v1.16) (Kechin, Boyarskikh, Kel, & Filipenko, 2017) to trim TIF-Seq2 sequencing primer (-a AGGTGACC GGCAGGTGT) and Illumina TruSeq adapter (-a AGATCGGAAG
- Extract UMIs: Use UMI-tools (v0.5.4) (Smith, Heger, & Sudbery, 2017) to extract 8 bp of unique molecular identifiers (UMIs) from the 5′-ends and add the UMI sequence to the head of sequence id
- Trim extra poly(A) stretches: Use UMI-tools (v0.5.4) to remove extra A stretches in the 3′-ends caused by poly(A) slippage during PCR amplification

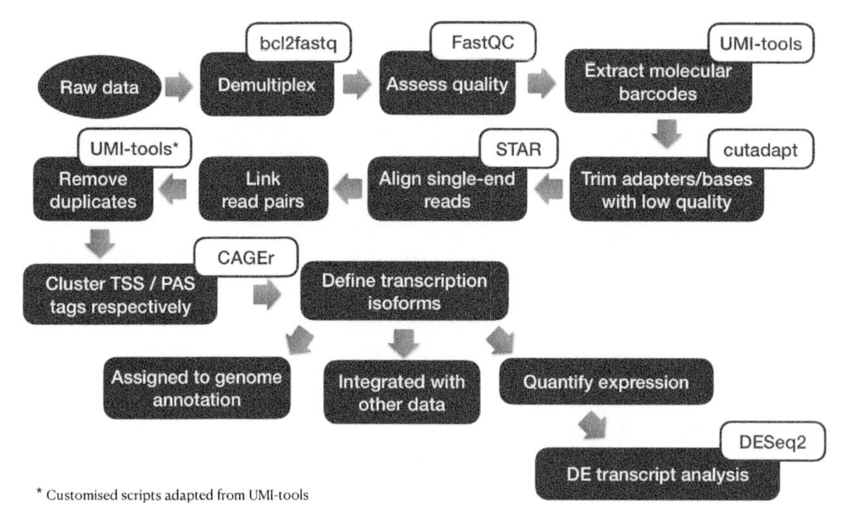

Fig. 2 Bioinformatic analysis workflow.

- Filter short reads: We recommend to keep only reads over 20 bp for alignment
- Align sequencing reads: Use STAR (v2.5.3a) (Dobin et al., 2013) to align 5′-end reads and 3′-end reads separately to the human reference genome hg38, supplying Gencode transcripts as splicing junction annotation. For mapping reads to the human genome, we recommend to adjust the alignment setting as follows: –alignIntronMax 200,000 –alignEndsType Extend5pOfRead1 –alignSJoverhangMin 10. Paired-end reads should then be linked and only uniquely mapped pairs that are on the same chromosome should be kept
- Remove PCR duplicates: We recommend to use a custom script adapted from UMI-tools to remove PCR duplicates from the leftover reads, allowing 1 bp mismatch in the UMIs and 1 bp shifting in the transcription start sites (https://github.com/PelechanoLab/TIFseq2/blob/master/dedup.py)
- An example of mapped TIF-Seq2 tracks is displayed in Fig. 3

3.2 Transcription boundary determination

- Collapse boundaries respectively: We recommend to use a custom python script to extract boundaries of TIF-Seq2 read pairs, collapse the boundary tags and calculate the coverage. (https://github.com/PelechanoLab/TIFseq2/blob/master/boundary.py)

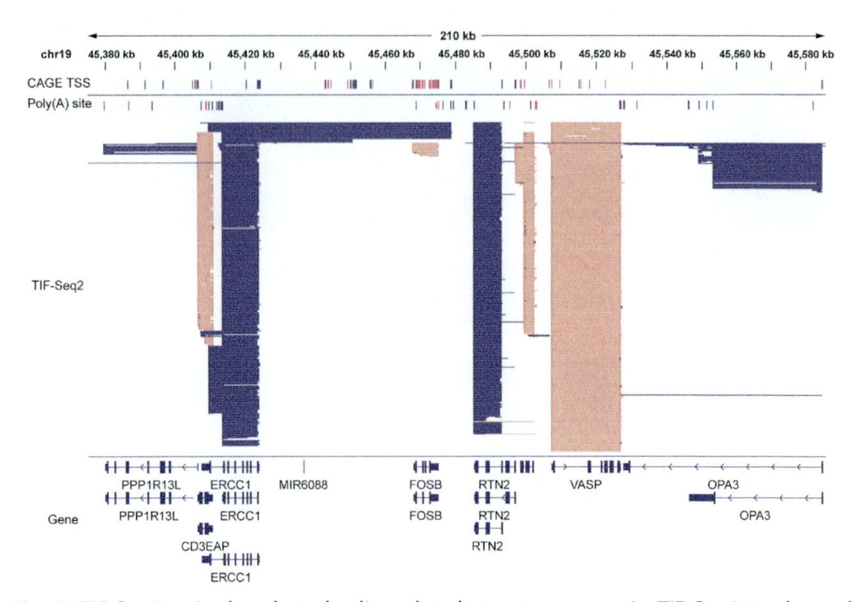

Fig. 3 TIF-Seq2 paired-end reads aligned to human genome. In TIF-Seq2 track, read pairs are labeled as pink lines (the forward strand) and dark blue lines (the reverse strand), representing transcription isoforms (TIFs). CAGE TSS and poly(A) sites (light red in the forward strand and dark blue in the reverse strand) from public repositories validate the TIFs captured by TIF-Seq2.

- Exclude internal priming events: Exclude those 3′-end tags with at least 7 As in the downstream 10 nt sequences from further analysis
- Cluster high-confident boundaries: CAGEr (Haberle, Forrest, Hayashizaki, Carninci, & Lenhard, 2015) can be used to define the cluster of transcripts 5′- or 3′-end tags of TIF-Seq2 respectively. Transcription boundary tags can be normalized to match a power-law distribution. We recommend to exclude low-coverage tags (e.g., those supported by less than 1 normalized counts in multiple samples) before clustering. For humans, we recommend clustering together boundary tags within a 10 bp window. The clusters can be formed into non-overlapping consensus clusters across samples if they are within 10 bp apart

3.3 TIF definition, annotation and quantification

- Form TIFs (transcription isoforms): 5′-end TSS and 3′-end PAS clusters can be linked according to the supporting read pairs from TIF-Seq2. We recommend filtering out pairs with extremely long (>2 Mb) and extremely short (<300 bp) mate-pair distance. To keep a conservative estimate of unannotated transcript isoforms identified, in general, we

recommend to exclude those pairs mapping to different chromosomes. Transcript isoform boundaries (TIFs) can be defined as the connections between TSS clusters and PAS clusters supported by at least four read pairs connecting them across multiple samples (unique molecular events)
- Annotate TIFs: TIFs can be assigned to Gencode v28 annotation features based on their relative distance to the annotated transcripts. For example, TSS distances (d1) and PAS distances (d2) can be calculated between a TIF and its overlapping annotated transcripts. We assign a TIF to the transcript with the least sum of d1 and d2 among all overlapping transcripts, and then we assign it to the gene that harbors the transcript. According to the relative position to their assigned transcripts, the TIFs can be arbitrarily classified as (i) annotated transcripts, if both TSS and PAS are within 200 bp away of annotated transcripts boundaries; (ii) transcripts with new TSS; (iii) transcripts with new PAS; (iv) transcripts with new boundaries, if both TSS and PAS not annotated and (v) intergenic TIFs
- Quantify TIF expression: We measure the expression of TIFs as count of read pairs that link TIF boundaries. We recommend DESeq2 (Love, Huber, & Anders, 2014) for normalization and differential expression analysis

4. Notes

4.1 Expected outcomes

Considering the amount of input RNA indicated in this protocol, the following yields are to be expected:

1. *cDNA yield after second strand synthesis*: in Step 6 after elution with 13μL EB buffer from 1× volume of Ampure beads cleaning, one should obtain 20–50 ng/μL cDNA
2. *cDNA size distribution after second strand synthesis*: in Step 6 for human mRNA a nearly normal distribution of cDNA range from 500 to 7000 bp with a median located within range 2000–3000 bp should be produced. See Fig. 4A as an example of successful cDNA size distribution after second strand synthesis
3. Not1 digestion is not expected to cause any significant loss of cDNA (step 7). An acceptable range when compared to the input amount for Not1 digestion, is maximus loss of 2–10%
4. After PlasmidSafe digestion (second step of day 4), one should anticipate around 3–7% of cDNA left compared to the input for circularization after Not1 digestion. A loss of 93–97% of the cDNA during digestion

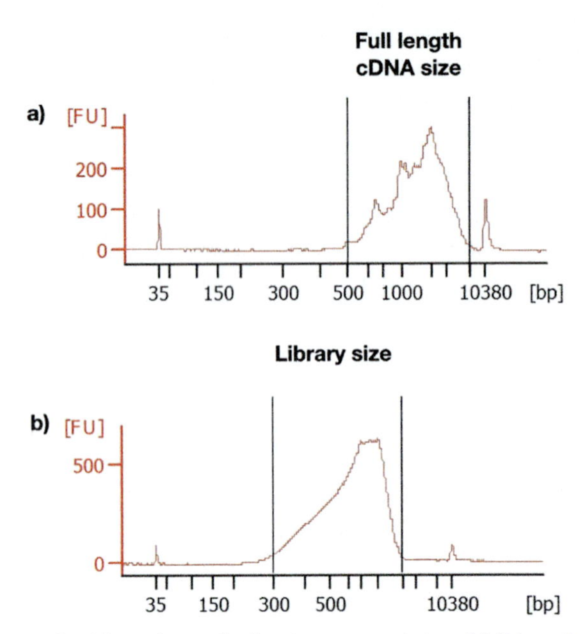

Fig. 4 Representative bioanalyzer plot for the expected size of full-length cDNA and the final library size distribution.

and purification steps can occur. Please note that over digestion could lead to loss of the sample

5. 20–30% cDNA is normally left after sonication and purification with 1 × Ampure XP beads cleaning. The Ampure XP beads should remove the small fragments that are too short to be informative
6. After 11 cycles of final PCR amplification and 0.35–0.8 × Ampure beads purification, one can expect a final library yield of around 60 ng cDNA if 4 samples are multiplexed (15 ng/sample for the final library)
7. Final library size distribution should be between 300 and 1500 bp. See Fig. 4B as an example

4.2 Optimization and troubleshooting

Here we detail some of the problems that might arise during the protocol, their potential origin and suggested solution.

1. After the second strand cDNA synthesis (step 6), too little cDNA amount is produced. This could be caused by:
   a. Failure or inadequate ligation of RNA adapter
   b. Too much sample lost during PCI extraction
   c. Problems with the TERRA PCR, which is rare but possible

In all cases, we recommend performing a control of the mentioned steps and start the protocol from the beginning.

2. Massive loss of cDNA after Not1 digestion (step 7). This could be caused due to inadequate Ampure beads purification (i.e., the beads might have dried up, jeopardizing the capability of eluting cDNA from beads). Alternatively, the Not1 enzyme may be contaminated with other nucleases, in which scenario, replacement of the enzyme would be highly recommended
3. No clear decrease of cDNA level after plasmid safe digestion (second step of day 4). This may be due to either not enough incubation time or problems with the enzyme activity. It is important to observe a clear decrease of DNA amount, as if not, leftover non-circularized cDNA fragments will carry over and lead to the generation of single-end reads surrounding the biotinylation site. If this problem occurs, you could usually observe single-end informative reads combination. That is, only read 1 had an informative reads sequence; meanwhile, read2 is GGGGGG which indicates an empty end. Please see point number 6
4. No clear decrease of cDNA level after sonication and purification with 1 × Ampure XP beads (step 9). This could be caused by inadequate sonication, which would leave long fragments behind. On the other hand, if too little cDNA is left, then an oversonicatiom could have happened. Sonication energy and durance should be adjusted accordingly
5. If too little library is generated by final PCR (step 14), we recommend to:
   a. Confirm that all the steps after sonication are done within one dayframe
   b. Check that the amount of adapter ligation after end repair (step 13) was sufficient
   c. Confirm that the dA tailing buffer contains dATP that often comes in a different tube
   d. Confirm that the adapter has been diluted according to instructions and kept on ice
6. In the case of a massive amount of empty clusters (too many GGGG reads), this would indicate that one end is missing from the final library structure. Possible causes are:
   a. cDNA from short transcripts broken in the middle which would cause one end to be missing. The solution would be to remove these short cDNA fragments by 1 × Ampure beads after sonication

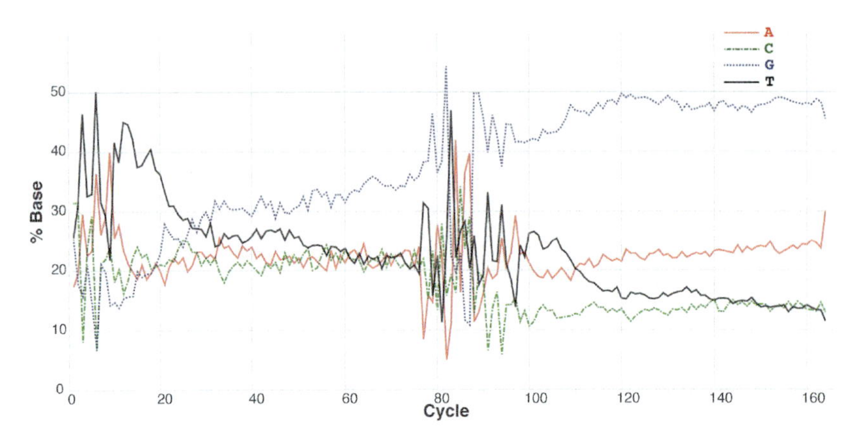

Fig. 5 Percentage of nucleotides in TIF-Seq2 sequencing. Every line represents the distribution of nucleotides A(red), C(green), G(blue) or T(black) in each sequencing cycle. The first 6 cycles are index 1, and cycle 83–88 are index 2. The sequencing cycles of the second reads are cycle 89–164. Please note this refers to the raw nucleotide before any of the recommended quality filters are applied.

b. Inadequate circularization, which is usually caused by insufficient T4 ligase concentration

c. Insufficient digestion by plasmid safe, normally due to insufficient time or low efficacy of the enzyme. Please see problem number 3

However, it is important to note that in TIF-Seq2 we always observe a significant level of empty clusters (GGGG) in the second read (as depicted by the blue line in Fig. 5).

As other poly(A) + sequencing methods which use oligo dT to capture poly(A) tails, TIF-Seq2 also has the potential of internal priming. In order to filter out false positive poly(A) sites caused by internal priming, we performed motif enrichment analysis near putative poly(A) sites according to the composition of adenine (A) in their downstream genomic sequences. Poly(A) associated hexamers (A[*AT*]TAAA) were discovered at 15–30 nt upstream of poly(A) sites with less than 7 As, while no obvious motif was detected near the 3′-end tags with at least 7 As in the downstream. Therefore, we regarded the 3′-end tags with at least 7 As in the downstream 10 nt sequences as internal priming cases, and thus excluded them from downstream analysis. However, the threshold should be evaluated for each experiment

4.3 Advantages of TIF-Seq2

Here we summarize some of the advantages of TIF-Seq2 in respect to alternative approaches.

1. Compared to our first version of Transcript Isoform Sequencing (Pelechano et al., 2013, 2014), the bioinformatic analysis has been rad ically simplified. By altering the reads structure, the output reads sequencer could be directly used for alignment and no sequencing primer or consensus sequence (Not1 site) trimming is now required
2. A higher percentage of useful reads is now obtained as compared to the first version of TIF-Seq. By eliminating the non-informative short fragments using 1 × Ampure XP beads after sonication, the probability of one end only fragments is reduced dramatically. Consequently resulting in a higher read number for the same sequencing investment
3. TIF-Seq2 can define the transcript boundary in a larger throughput and relatively lower cost compared to long-read sequencing
4. TIF-Seq2 links the start and end of the same transcripts, while CAGE and PolyA can only solve one end at a time
5. TIF-Seq2 can better solve the overlapping transcripts when compared to RNAseq

4.4 Limitations of TIF-Seq2

Despite its significant improvement in respect to its previous implementation (Pelechano et al., 2013, 2014), TIF-Seq2 has still several limitations to be accounted for.

1. As any approach involving the amplification of full-length cDNA, long isoforms are in general underrepresented. This length bias facilitates the identification of short isoforms and thus underestimates the abundance of long molecules. It is important to note that this limitation does not affect only TIF-Seq2 but is a general feature of all long-read sequencing approaches or single-cell transcriptome analysis involving the generation of full-length cDNA. However, despite this limitation TIF-Seq2 is still able to quantify relative changes in abundance for the same isoform identified across different conditions. TIF-Seq2 quantification can be further improved by combining it with single-boundary approaches such as poly(A) site quantification (Wang et al., 2020)
2. Despite being based on conventional short-read sequencing instruments with higher throughput, the percentage of useful reads containing perfectly map boundaries is much lower than conventional RNA-Seq approaches. As an example, in RNA-Seq, it is common to have $>80\%$ uniquely mapped reads, while in TIF-Seq2 this percentage is much lower ($\sim 20\%$). On the other hand, a uniquely mapped read pair

is sufficient to define both boundaries of a transcript isoform, something that is not possible with conventional RNA-Seq

3. TIF-Seq2 focuses on the identification of 5′ and 3′ transcript isoform boundaries, and thus it does not offer information of alternative splicing (unless it occurs proximal to the 5′ or 3′ends). Although start and termination sites drive most transcription isoform differences in humans (Reyes & Huber, 2018), to obtain information regarding alternative splicing, we recommend combining TIF-Seq2 with conventional RNA-Seq or long-read sequencing approaches (Wang et al., 2020)
4. Finally, despite our efforts to simplify this method, TIF-Seq2 is an advanced approach. We recommend the user to familiarize with the protocol described above and perform the recommended quality control steps and if possible, a pilot sequencing run

4.5 Safety considerations and standards

- Phenol-Chloroform extractions require proper risk assessment prior to handling of the chemicals, which present several dangers to human health. Gloves, protective clothing and safety glass are necessary as well as handling within a chemical hood. Disposal of corresponding waste should also be performed according to the institution safety regulations
- Ethanol and Isopropanol are flammable and volatile and thus should be handled with caution

5. Summary

Tif-Seq2 is an optimized version of the previous method developed in our lab. This version is better equipped to investigate more complex transcriptomes and better disentangle overlapping transcription, as we are now capable of concurrently determining the start and end sites of individual RNA molecules. It is especially well suited to discriminate between overlapping molecules and accurately define their boundaries. The method has been successfully applied to investigate the transcriptome of human cells as well as yeast and was fundamental to investigate the complex transcriptome organization of *Arabidopsis thaliana*. In this chapter, we present the detailed protocol as well as all the important limitations and troubleshooting tips needed to successfully implement the method to any other kind of sample.

Acknowledgments

We thank Donal Barrett for technical assistance, and all members of the Pelechano, Kutter, Friedländer, Wu and Steinmetz laboratories for discussion. We acknowledge support from SNIC through Uppsala Multidisciplinary Center for Advanced Computational Science (UPPMAX). This work was mainly funded by the Swedish Research Council (VR 2016-01842), a Wallenberg Academy Fellowship (KAW 2016.0123), the Swedish Foundations' Starting Grant (Ragnar Söderberg Foundation), a Joint China-Sweden mobility grant from STINT (CH2018-7750) and Karolinska Institutet (SciLifeLab Fellowship, SFO and KI funds) to V.P.

References

de Klerk, E., & Hoen, P. A. C. (2015). Alternative mRNA transcription, processing, and translation: Insights from RNA sequencing. *Trends in Genetics*, *31*(3), 128–139.

Dobin, A., Davis, C. A., Schlesinger, F., Drenkow, J., Zaleski, C., Jha, S., et al. (2013). STAR: Ultrafast universal RNA-seq aligner. *Bioinformatics*, *29*(1), 15–21.

Gruber, A. J., Schmidt, R., Gruber, A. R., Martin, G., Ghosh, S., Belmadani, M., et al. (2016). A comprehensive analysis of 3′ end sequencing data sets reveals novel polyadenylation signals and the repressive role of heterogeneous ribonucleoprotein C on cleavage and polyadenylation. *Genome Research*, *26*(8), 1145–1159.

Haberle, V., Forrest, A. R. R., Hayashizaki, Y., Carninci, P., & Lenhard, B. (2015). CAGEr: Precise TSS data retrieval and high-resolution promoterome mining for integrative analyses. *Nucleic Acids Research*, *43*(8), e51. https://doi.org/10.1093/nar/gkv054.

Kechin, A., Boyarskikh, U., Kel, A., & Filipenko, M. (2017). cutPrimers: A new tool for accurate cutting of primers from reads of targeted next generation sequencing. *Journal of Computational Biology: A Journal of Computational Molecular Cell Biology*, *24*(11), 1138–1143.

Love, M. I., Huber, W., & Anders, S. (2014). Moderated estimation of fold change and dispersion for RNA-seq data with DESeq2. *Genome Biology*, *15*(12), 550.

Pelechano, V., Wei, W., Jakob, P., & Steinmetz, L. M. (2014). Genome-wide identification of transcript start and end sites by transcript isoform sequencing. *Nature Protocols*, *9*(7), 1740–1759.

Pelechano, V., Wei, W., & Steinmetz, L. M. (2013). Extensive transcriptional heterogeneity revealed by isoform profiling. *Nature*, *497*(7447), 127–131.

Proudfoot, N. J. (1986). Transcriptional interference and termination between duplicated alpha-globin gene constructs suggests a novel mechanism for gene regulation. *Nature*, *322*(6079), 562–565.

Reyes, A., & Huber, W. (2018). Alternative start and termination sites of transcription drive most transcript isoform differences across human tissues. *Nucleic Acids Research*, *46*(2), 582–592.

Smith, T., Heger, A., & Sudbery, I. (2017). UMI-tools: modeling sequencing errors in Unique Molecular Identifiers to improve quantification accuracy. *Genome Research*, *27*(3), 491–499. https://doi.org/10.1101/gr.209601.116.

The FANTOM Consortium and the RIKEN PMI and CLST (DGT), Forrest, A., Kawaji, H., et al. (2014). A promoter-level mammalian expression atlas. *Nature*, *507*, 462–470.

Thomas, Q. A., Ard, R., Liu, J., Li, B., Wang, J., Pelechano, V., et al. (2020). Transcript isoform sequencing reveals widespread promoter-proximal transcriptional termination in Arabidopsis. *Nature Communications*, *11*(1), 2589.

Tian, B., & Manley, J. (2017). Alternative polyadenylation of mRNA precursors. *Nature Reviews. Molecular Cell Biology*, *18*, 18–30.

Van Werven, F. J., Neuert, G., Hendrick, N., Lardenois, A., Buratowski, S., van Oudenaarden, A., et al. (2012). Transcription of two long noncoding RNAs mediates mating-type control of gametogenesis in budding yeast. *Cell*, *150*(6), 1170–1181.
Wang, J., Li, B., Marques, S., Steinmetz, L. M., Wei, W., & Pelechano, V. (2020). TIF-Seq2 disentangles overlapping isoforms in complex human transcriptomes. *Nucleic Acids Research*, *48*(18), e104.
Wilkening, S., Pelechano, V., Järvelin, A. I., Tekkedil, M. M., Anders, S., Vladimir, B., et al. (2013). An efficient method for genome-wide polyadenylation site mapping and RNA quantification. *Nucleic Acids Research*, *41*(5), e65.
Wu, M., Karadoulama, E., Lloret-Llinares, M., Rouviere, J. O., Vaagensø, C. S., Moravec, M., et al. (2020). The RNA exosome shapes the expression of key protein-coding genes. *Nucleic Acids Research*, *48*(15), 8509–8528.

CHAPTER SIX

Single-molecule polyadenylated tail sequencing (SM-PAT-Seq) to measure polyA tail lengths transcriptome-wide

Steven L. Coon, Tianwei Li, James R. Iben, Sandy Mattijssen, and Richard J. Maraia*

Intramural Research Program, Eunice Kennedy Shriver National Institute of Child Health and Human Development, National Institutes of Health, Bethesda, MD, United States

*Corresponding author: e-mail address: maraiar@mail.nih.gov

Contents

Abstract

Polyadenylation of the 3′ end of mRNAs is an important mechanism for regulating their stability and translation. We developed a nucleotide-resolution, transcriptome-wide, single-molecule SM-PAT-Seq method to accurately measure the polyA tail lengths of individual transcripts using long-read sequencing. The method generates cDNA using a double stranded splint adaptor targeting the far 3′ end of the polyA tail for first strand synthesis along with random hexamers for second strand synthesis. This straightforward method yields accurate polyA tail sequence lengths, can identify non-A residues in those tails, and quantitate transcript abundance.

ISSN 0076-6879
https://doi.org/10.1016/bs.mie.2021.04.006

1. Introduction

Polyadenylation of the 3′ end of transcripts is an important mechanism for regulating their stability and processing efficiency (Curinha, Oliveira Braz, Pereira-Castro, Cruz, & Moreira, 2014; Elkon, Ugalde, & Agami, 2013; Tian & Manley, 2013). In eukaryotes, most messenger RNAs are polyadenylated, as are many long non-coding RNAs and primary microRNAs (Nicholson & Pasquinelli, 2019). The length of the polyA tail is dynamically regulated in the nucleus and cytoplasm (Eisen et al., 2020) and affects processes including export, translation, and degradation (Kuhn & Wahle, 2004). In many cases, alternate polyA sites downstream of the terminus of the coding region can also be regulated (Batra, Manchanda, & Swanson, 2015; Tian & Manley, 2013), so it is important to determine the location of the polyA tail in addition to its length.

Methods to accurately measure the length of polyA tails on specific transcripts have been developed only recently and are actively evolving. The first methods utilized high-throughput sequencing on Illumina sequencers. More recently, methods have been developed, including the one detailed here, that take advantage of long-read sequencers such as the Pacific Biosciences Sequel and the Oxford Nanopore machines. In 2014, two methods were published that relied on Illumina sequencers. The first was TAIL-seq (Chang, Lim, Ha, & Kim, 2014), in which reads were generated from the distal end of the polyA tail into the end of the 3′ UTR. Illumina short read sequencers can have difficulty reading through homopolymeric regions such as polyA tails. To deal with this, Chang et al. developed a specialized algorithm to determine the point at which the base calls from the polyA tail transitioned into the 3′ UTR. In 2016, they refined their method with mTAIL-seq which increased the sensitivity (Lim, Lee, Son, Chang, & Kim, 2016). Also in 2014, Subtelny et al. published an alternative approach called PAL-seq (Subtelny, Eichhorn, Chen, Sive, & Bartel, 2014). In this case, rather than directly sequencing the polyA tail, they used a primer extension reaction to incorporate dTTP and biotinylated dUTP, which was then incubated with fluorescent streptavidin. The authors then modified an Illumina sequencer to detect the fluorescent signal which was used to calculate the polyA tail length associated with a specific 3′ UTR sequence. In 2015, Harrison et al. developed PAT-seq which differed from TAIL-seq in a few key steps but still depended on Illumina short reads to directly read the polyA homopolymer (Harrison et al., 2015; Swaminathan, Harrison,

Preiss, & Beilharz, 2019). A new approach was provided in 2018, where the polyA tail was not directly sequenced but inferred by subtracting the length of the non-polyA 3′ UTR sequence from the known insert size of the library; this was called TED-Seq (Woo et al., 2018). In the newest of the Illumina short read-based methods, Poly(A)-Seq is a further refinement of the TAIL-seq, PAL-seq and PAT-seq methods (Yu et al., 2020). Despite the ingenuity of their methodological and algorithmic advancements, the inherent weakness is the underlying technology of Illumina sequencers which is limited for measuring homo-polyA lengths accurately (Chang et al., 2014). Consequently, very significant advances and discoveries were made, new questions arose, the field has grown, and new approaches were needed.

To address this, there are alternative methods which rely on the third generation long-read technologies of Oxford Nanopore and Pacific Biosciences (PacBio). These provide continuous reads that may be as long as tens of thousands of bases but are not as accurate as Illumina sequencers per single read (see below). An advantage of Nanopore sequencers is they can sequence native RNA directly and can provide a direct estimate of polyA tail length as well as information regarding the full-length transcript (Workman et al., 2019). Whereas Nanopore generates reads from just a single pass along the molecule, PacBio sequencers can sequence the same molecule multiple times which greatly improves accuracy. Also in 2019, Legnini et al. devised a way to sequence full-length transcripts along with their polyA tails with high accuracy using the PacBio Sequel in a method they called FLAM-Seq (Legnini, Alles, Karaiskos, Ayoub, & Rajewsky, 2019). That same year, Liu et al. published PAIso-seq, a technique that was similar to FLAM-Seq in that it also sequenced full-length polyadenylated transcripts (Liu, Nie, Liu, & Lu, 2019). Our group developed **S**ingle-**M**olecule **P**oly **A**denylated **T**ail (SM-PAT)-Seq which also takes advantage of the PacBio technology (Mattijssen, Iben, Li, Coon, & Maraia, 2020). The three PacBio-based methods are similar, except that ours does not generate reads for the entire transcript, but focuses on the 3′ UTR.

2. Protocol: Overall strategy

This protocol starts with extraction of high-quality total RNA. Then we use a novel method for targeting and ligating an adapter primer to the end of the polyA tail to prime first strand cDNA synthesis (Fig. 1). Second strand synthesis is primed using random hexamers. Full-length cDNAs are not

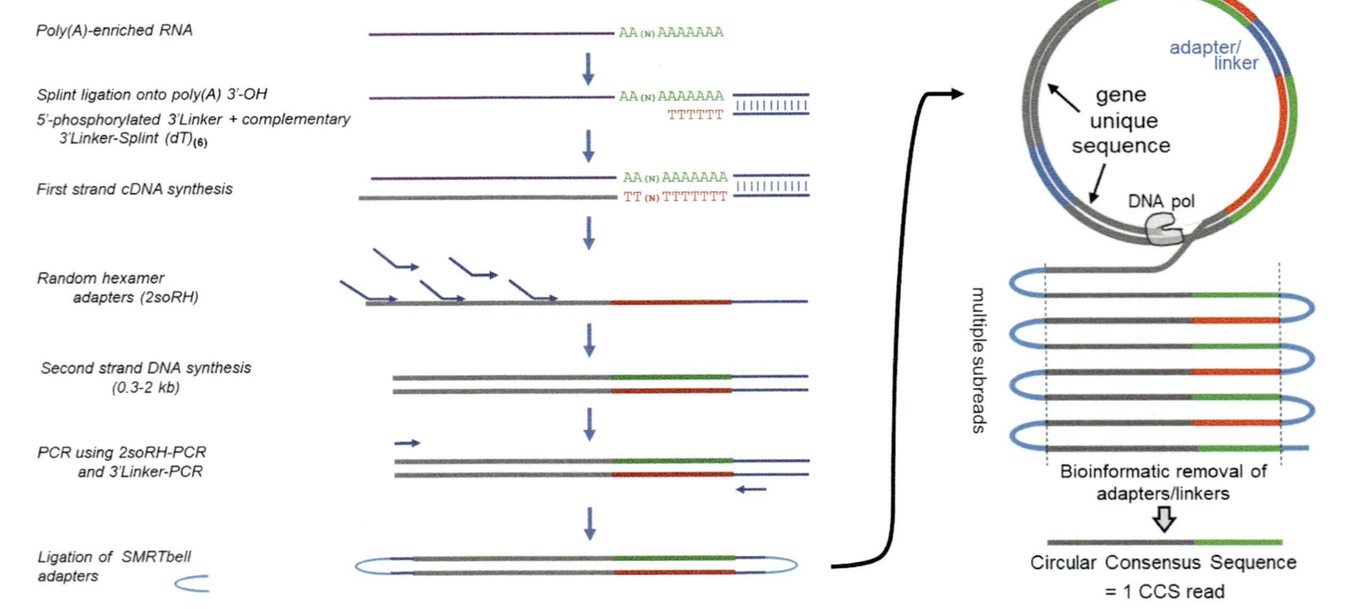

Fig. 1 Schematic of the strategy to construct libraries for determination of polyA tail length and composition. The libraries are sequenced using long-read technology on a Pacific Biosciences Sequel machine which generates highly accurate circular consensus sequences from multiple passes through the sense and antisense strands. *Modified from Mattijssen, S., Iben, J. R., Li, T., Coon, S. L., & Maraia, R. J. (2020). Single molecule poly(A) tail-seq shows LARP4 opposes deadenylation throughout mRNA lifespan with most impact on short tails.* eLife, 9, e59186. *doi:10.7554/eLife.59186 under a Creative Commons CC0 license.*

needed; the primary goal is to get enough information to ascribe the polyA tail to the proper gene transcript. The primer used for first strand synthesis and the random hexamer have sequences appended to them that are used as priming sites for amplification. Thus, the amplified cDNA library contains the full-length polyA tail along with the 3′ end of the related transcript. We intentionally chose to focus on the 3′ end of the transcript, rather than full-length transcripts, to increase the proportion of reads that yielded high quality "circular consensus sequences" (CCSs; i.e., a consensus of base calls from shorter fragments that are read more times for a given unit of time on the PacBio Sequel sequencer as it reads sense then antisense strands around a rolling circle, increasing accuracy). In the end, we were primarily interested in accurately measuring the length of each polyA tail and ascribing it to a transcript from a particular gene.

These cDNAs are used for preparing libraries that can be sequenced on a Pacific Biosciences Sequel machine. As mentioned, the Sequel technology is critical for being able to accurately read through and quantitate the length of polyA tails. The conversion of the cDNA library to a sequencing library uses standard Pacific Biosciences kits, reagents, and protocols.

This protocol also covers general data processing. It explains how to convert the data output files coming off the Sequel to usable polyA tail lengths associated with specific transcripts. The uses of that data and appropriate graphic representations are not covered since they will be specific for individual research studies. We do provide examples of basic data display that can serve a starting point for more-detailed analysis. The protocol is not written for a specific type of sample. The method and data processing are generic for any high-quality total RNA input.

This protocol is the one we used for generating the data in our recent publication (Mattijssen et al., 2020); however, there are potential improvements to the method that are being developed or considered. Since the method is relatively simple and straight-forward, it is easily adaptable. Some potential modifications will be discussed at the end of the protocol.

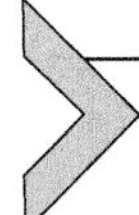

3. Materials

3.1 RNA extraction

1. Equipment (Note, we prefer to use the Maxwell technology for RNA extraction, but this is optional; any method yielding high quality RNA would be acceptable.)

a. Maxwell® 16 Instrument (Promega; Cat.# AS2000) configured with the Maxwell® 16 High Strength LEV Magnetic Rod and Plunger Bar Adaptor (Cat.# SP1070)

2. Reagents

a. Maxwell 16 LEV simplyRNA Cells kit (Promega; Cat.# AS1270)

3.2 Library preparation

1. Equipment
   - a. Thermocycler (e.g., BioRad C1000)
   - b. Qubit Fluorometer (ThermoFisher)
   - c. Bioanalyzer 2100 (Agilent)
   - d. Standard laboratory equipment
     - i. Centrifuge
     - ii. pipettes
2. Reagents
   - a. Custom oligos (5′ > 3′; we had them synthesized by IDT, but it could be done by many vendors). Oligos marked with * were HPLC purified. Dissolve each stock oligo to 100 μM in 10 mM TrisHCl pH ~7.5–8
     - i. *3′Linker**: /5Phos/CTCTGCGTTGATACCACTGCTTGC GGCCGCATTA
       1. This oligo is targeted for ligation to the distal end of the polyA tail to stabilize oligo 3′LinkerSplint at that location
     - ii. *3′LinkerSplint**: /5BiotinTEG/TAATGCGGCCGCAAGCA GTGGTATCAACGCAGAGTTTTTT
       1. This oligo is pre-annealed to the 3′Linker oligo. The terminal T's serve to target the pre-annealed pair to the end of the polyA tail. The number of T's used is 6, but this number was not investigated or optimized. The biotin at the 5′ end is not used in this protocol. This oligo is used for first strand cDNA synthesis
     - iii. *2soRH**: GCCTGAGCTCTGCAGAGATCNNNNNN
       1. This oligo is used for second strand synthesis. It contains random hexamers to target 3′ UTRs universally
     - iv. *2soRH-PCR*: GCCTGAGCTCTGCAGAGATC
       1. This oligo is complementary to oligo 2soRH and used along with oligo 3′Linker-PCR to amplify the double stranded cDNA library
     - v. *3′Linker-PCR*: AAGCAGTGGTATCAACGCAGA
       1. This oligo is complementary to oligo 3′Linker and used along with oligo 2soRH-PCR to amplify the double stranded cDNA library

b. Enzymes (we ordered these from New England Biolabs, so their catalog numbers are provided, but other vendors could be used.)
   i. T4 RNA ligase 2 (M0239S)
   ii. RNase H (M0297S)
   iii. Klenow Fragment 3′–5′ *exo-* (M0212S)
c. RNAClean XP beads (Beckman-Coulter)
d. 10 mM TrisHCl pH 7.5
e. 10 mM Tris pH 8
f. 5 M NaCl
g. Superscript II Reverse transcriptase (ThermoFisher)
h. RiboLock RNase inhibitor (various RNase inhibitors are available from other vendors)
i. Magnetic bead rack for 1.5 mL tubes
j. dNTPs (10 mM; ThermoFisher)
k. Elution Buffer (Qiagen)
l. NEB Buffer 2 (New England Biolabs)
m. KAPA HiFi HotStart Ready Mix (2×) (Roche)
n. TruSeq Stranded mRNA Prep Kit (Illumina)
o. Qubit spectrophotometer using a dsDNA HS Assay kit (ThermoFisher)
p. Pacific Biosciences reagents:
   i. SMRTbell Template Prep Kit 1.0-SPv3 (cat# 100-991-900)
   ii. Sequencing primer v3 (cat# 100-970-100)
   iii. SMRTbell Barcoded Adapter Complete Prep Kit—96 (cat# 100-514-900)
   iv. AMPure PB beads (cat# 100-265-900; do not use other sources when these are called for)
   v. Sequel Binding and Internal Control Kit 3.0 (cat# 101-626-600)
   vi. Sequel Sequencing Kit 3.0, 4 rxn (cat# 101-597-900)
   vii. SMRT Cells 1 M v3 Tray (cat# 101-531-000)
   viii. Sequel Internal Control Complex 3.0 (cat# 101-500-300)

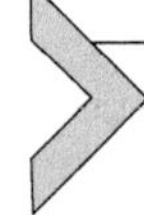

4. Protocol

4.1 RNA extraction

1. Isolate total RNA using a Maxwell 16 LEV simplyRNA purification kit
   a. Depending on the cell type, seed enough cells to yield at least 4 μg of total RNA. For HEK293 cells, this is 2 wells of a 6-well plate, 5×10^5 cells/well. Make sure the cells are not more confluent than 80% at the time of harvest, so they still have room to grow

b. Wash cells with 2 mL PBS per well
c. Lyse cells with 100 μL homogenization buffer per well containing thioglycerol from the Maxwell kit
d. Transfer lysate to a 1.5 mL Eppendorf tube
e. Continue to follow the Maxwell 16 LEV simplyRNA purification kit protocol using 200 μL lysis buffer
f. Load the samples onto the Maxwell® 16 Instrument
g. Use 1 simplyRNA cartridge per 2 wells of a 6-well plate
h. Elute total DNase-treated RNA in 50 μL nuclease-free water into 0.5 mL tubes supplied in the Maxwell kit
i. Proceed immediately, or store at −80 °C

4.2 Library preparation

1. Anneal 3′Linker to 3′LinkerSplint
   a. Mix equimolar amounts of oligo in a solution containing:
      i. 100 pmols of each oligo (= 1 μL of 100 μM stock)
      ii. 50 mM NaCl (= 1 μL of a stock solution of 5 M NaCl)
      iii. 20 mM Tris HCl (pH 7.5) (= 2 μL of 1 M stock)
      iv. 95 μL RNase-free water to bring the total volume to 100 μL
   b. Heat to 80 °C for 2 min
   c. Slowly cool the solution (i.e., in a thermocycler ramp down the temperature to 25 °C at 1 °C/min).
2. Enrich total RNA for polyadenylated transcripts using reagents and protocol from TruSeq stranded mRNA kit (Illumina).
   a. Dilute 4 μg of total RNA to 50 μL with nuclease-free water
   b. Mix RPB by vortexing until well-dispersed (These are essentially oligo-dT magnetic beads.)
   c. Add 50 μL RPB to each sample and mix by pipetting up and down six times
   d. Place in a thermocycler and run a denaturing program
      i. Choose the preheat lid option and set to 100 °C
      ii. 65 °C for 5 min
      iii. Hold at 4 °C
   e. Place on the bench and incubate at room temperature for 5 min
   f. Centrifuge at 280 × *g* for 1 min
   g. Place on a magnetic stand and wait until the liquid is clear (~5 min)
   h. Remove and discard the supernatant
   i. Remove from the magnetic stand

j. Add 200 μL BWB, and then mix by pipetting up and down six times
k. Place on a magnetic stand and wait until the liquid is clear (~5 min)
l. Remove and discard the supernatant
m. Remove from the magnetic stand
n. Elute the polyA-enriched RNA with 41 μL of 10 mM Tris 8 by heating at 75 °C for 2 min

3. Ligate the pre-annealed linker to the polyA-enriched RNA
   a. Mix a solution containing (50.5 μL final volume):
      i. 40 μL of polyA-enriched RNA
      ii. 2.5 μL of annealed 3′Linker solution (= 2.5 pmols of the annealed linker)
      iii. 5 μL T4 RNA ligase 2 10 × buffer
      iv. 2 μL of T4 RNA ligase 2
      v. 1 μL of RiboLock RNase inhibitor
   b. Incubate at 25 °C for 1 h
   c. Perform an RNA cleanup using 100 μL of RNAClean XP beads according to the manufacturer's instructions
   d. Elute in 27.5 μL of Elution Buffer
4. Synthesize first strand cDNA by reverse transcription
   a. Mix a solution by adding the following in order on ice (50 μL final volume):
      i. 26.5 μL of the ligated product from the prior step
      ii. 5 μL of 0.1 M DTT (from Superscript II kit)
      iii. 5 μL of 10 mM dNTPs
      iv. 10 μL of 5 × Superscript II RT buffer
      v. 1 μL of RiboLock RNase Inhibitor
      vi. 2.5 μL of Superscript II RT enzyme
   b. Incubate for 30 min at 42 °C
   c. Incubate for 15 min at 70 °C
   d. Cool to 37 °C
   e. Add 1 μL of RNase H
   f. Incubate for 15 min at 37 °C
   g. Cleanup the cDNA using 100 μL of RNAClean XP beads according to the manufacturer's instructions
      i. Since this is now cDNA, AMPure XP beads (Beckman-Coulter) could also be used. We use RNAClean XP beads because they are slightly less expensive and are already at room temperature
   h. Elute in 25 μL of water

5. Synthesize second strand cDNA
   a. Mix a solution by adding the following in order on ice (30 μL final volume):
      i. 23.5 μL of first strand cDNA from the above purification
      ii. 1 μL of 100 μM oligo 2soRH
      iii. 3 μL of NEB Buffer 2
      iv. 1 μL of 10 mM dNTPs
   b. In a thermocycler heat to 80 °C for 1 min
   c. Cool to 37 °C and hold for 10 min
   d. Add 1.5 μL of Klenow exo -
   e. Incubate for 30 min at 37 °C
   f. Incubate for 20 min at 75 °C
   g. Purify PCR product using 0.6 × (20 μL) AMPure PB beads (as per the Pacific Biosciences protocol).
   h. Elute in 25 μL of Elution Buffer
6. Amplify the cDNA fragments containing the partial 3′UTR + polyA tail
   a. Use custom oligo primers 3′Linker-PCR and 2soRH-PCR (25 μM stocks)
   b. Mix a solution by adding the following (50 μL final volume):
      i. 23 μL of cDNA from the above purification
      ii. 25 μL Kapa HiFi HotStart Ready Mix (2 ×)
      iii. 1 μL of each primer (25 μM stock)
   c. Amplify in a thermocycler using the following program:
      i. Step 1: 95 °C for 45 s
      ii. Step 2: 95 °C for 15 s
      iii. Step 3: 60 °C for 30 s
      iv. Step 4: 72 °C for 1 min
      v. Step 5: 25 total cycles of steps 2–4
      vi. Step 6: 72 °C for 2 min
      vii. Step 7: 4 °C hold
   d. Purify the PCR product using 0.6 × (30 μL) AMPure PB beads (as per the Pacific Biosciences protocol)
   e. Elute in 20 μL of Elution Buffer
7. Quantitate with a Qubit using a dsDNA HS Assay kit
   a. The amount of PCR amplicon should be 1–5 μg in 20 μL
8. Check the size range of fragments on a Bioanalyzer HS DNA chip (you may need to dilute the samples to be within the proper range for this Bioanalyzer chip)

   a. There should be a range of fragment sizes above 500 bp
      i. This can be adjusted by modifying the amount of random hexamer used for second strand synthesis, if desired. For long-read sequencing, fragment length is usually not an issue
      ii. For the concentration of random hexamer used here, we have a range of fragment sizes from 500 to 3000 bp with a peak at about 1700 bp
   b. Use the Bioanalyzer output to calculate the molar concentration (pmols of amplicon per μL)

9. These fragments are converted to a sequencing library for the PacBio Sequel using the protocol detailed in the PacBio document: Preparing Amplicon Libraries using PacBio® Barcoded Adapters for Multiplex SMRT® Sequencing. The protocol is given briefly here; the PacBio protocol should be obtained and followed closely. Each sample will get a barcode in this protocol. This will be used after sequencing to keep track of the sample from which each read was derived. A certain minimum total amount of material from all samples combined will be needed at a later stage in the protocol. This amount of each sample needed will depend on the number of samples and the size range of the fragments

10. The first step is to end-repair the amplicons and ligate a different barcode to the end of each sample. In this protocol, the barcodes are on the SMRTbell adapters. These are the signature "looped" adapters that are ligated to both ends of the library fragments and allow the sequencing to proceed as a rolling circle
   a. Prepare a pre-mix of the required reagents from the PacBio SMRTbell Barcoded Adapter Complete Prep Kit and aliquot an amount for each sample
   b. Add your sample and barcoded adapters
   c. Incubate for the indicated temperatures and times, then hold at 4 °C
   d. The barcoded samples may now be mixed, or multiplexed, for further processing

11. Perform AMPure PB bead purification as prescribed in the protocol. The volume of beads used depends on the average fragment size determined above using the Bioanalyzer. This will purify the desired product away from reaction components and unincorporated adapters

12. Measure the concentration with a Qubit using a dsDNA HS Assay Kit

13. Perform DNA damage repair
   a. Mix DNA with reagents provided in the kit
   b. Incubate at 37 °C for 20 min, then hold at 4 °C

14. Perform exonuclease digestion to remove failed ligation products
    a. Add exonucleases ExoIII and ExoVII from the kit
    b. Incubate at 37 °C for 60 min, then hold at 4 °C
15. Perform a double AMPure PB bead purification as prescribed in the protocol. The volume of beads used depends on the average fragment size determined above using the Bioanalyzer
16. Measure the concentration with a Qubit using a dsDNA HS Assay Kit
17. We also verify the SMRTbell Library (also called Templates) with a Bioanalyzer using a High Sensitivity DNA Analysis Kit

4.3 Sequencing

1. Use the SMRTlink software to set up your PacBio Sequel run
2. You will need to enter the concentration and average size of the library. We use the Qubit for quantitation and the Bioanalyzer to determine the average size. You also tell the SMRTlink software the amount (or final concentration) to be loaded. We used 6 pM. The SMRTlink software then calculates the amount of library to use
3. Anneal the sequencing primer to the SMRTbell Template library. The proper ratio of primer to library fragments is critical. PacBio provides a Binding Calculator (http://calc.pacb.com) to aid in this process
    a. Heat the primer at 80 °C for 2 min, then cool on ice
    b. Mix the multiplexed library with the primer and buffer *from the kit*
    c. Incubate at 20 °C for 30 min, then hold at 4 °C
4. Note that the comments here are relevant for a PacBio Sequel. Essentially the same strategy can be used for the Sequel II, but some final details and the number of flow cells used will need to be adjusted

4.4 Processing sequence data

1. Note that the reads coming from the Sequel are generated by sequencing around a "rolling circle." The reads from the region of interest are sequenced multiple times, alternately from the sense and antisense strands. This generates subreads alternating from forward and reverse strands. These subreads are then aligned to each other and a consensus sequence is generated called a circular consensus sequence (CCS); these are also referred to as HiFi reads. Note that these CCSs are highly accurate, including calling the length of the polyA tail. Briefly before going on, each CCS read represents a single mRNA molecule with a

nucleotide-resolved polyA tail length that will be assigned to a specific gene based on the unique sequence of its 3′ UTR

2. The steps for data processing were carried out on the Biowulf high performance computing cluster at NIH using software obtained from Pacific Biosciences
3. Subread data from each library fragment on a flow cell is converted to a circular consensus sequences (CCS) with the program ccs (v4.0.0) using default settings (–min-passes 3). This step also generates CCS IDs based on the well location and flow cell ID, appended with the "ccs" designation
4. Following CCS generation, the data are separated into the different samples using the program lima (v1.10.0) by providing the barcodes used in library construction and requiring matched barcode pairs (–same).
5. Samples, if run across multiple flow cells, are merged at this point to a single CCS bam per sample using samtools merge (v1.9).
6. CCS read data is extracted in fasta format using samtools for subsequent processing
7. Read data at this step is identified for all subsequent steps by the CCS IDs for tying together polyA tail length measurements and gene assignments
   a. Note that due to the random priming on the Sequel instrument CCSs may be in either forward or reverse directions with an expected approximate 50/50 split
8. For polyA tail length measurements, reads are evaluated in either direction for a polyA stretch abutting the adapter utilized in library construction (CTCTGCGTTGATACCACTGCTT). A one base mismatch is tolerated for the adapter sequence
   a. Length assignment comes from the stretch of A-bases preceding the adapter, while allowing for single non-A bases to be tolerated when they occur bracketed between a minimum of 2 A bases
   b. In regular expression this appears as a capture class ((AA+.)*AA+) followed immediately by adapter sequence
      i. "Regular expression" is a term referring to string-matching logic that is somewhat universal across programming languages. The "capture class" refers to the portion of the expression that is within the parenthesis and is a grouping that gets "captured" or returned by the search. Briefly: a "+" implies the previous term occurs once or more. A "." can match any character. A "*"

means the prior term is matched 0 or more times (the difference from "+" being that 0 is valid).

  ii. Optionally, the composition and rate of these non-A bases may be recorded

9. For gene assignments, the tail and adapter-trimmed CCS reads are mapped against the GENCODE transcriptome for the species with which you are working

  a. We used BLAT for this, but other tools could be used

  b. In our protocol, we accept down to a minimum of 50 bp mapping best scoring alignments. Tail-trimmed CCS reads shorter than this are discarded

  c. We do not utilize isoform information, but only utilize the gene symbol of a successful assignment. Optionally, isoform information or polyadenylation site could be incorporated if that were desired

10. Following determination of both polyA tail length and gene assignment for each CCS read (i.e., each mRNA molecule), the data are connected in tabular format using read IDs for subsequent evaluation in R or other platforms. Optionally, data can also be grouped by gene symbol and sample for gene-by-gene statistics. This table is now the starting point for analysis since for each gene you have a list of the lengths of the associated polyA tails

11. Figs. 2, 3 and 4 show examples of general methods of displaying this type of data [modified from (Mattijssen et al., 2020)]. They show the effect of increasing expression of LARP4, which binds to polyA tails. Fig. 2A shows the most basic method of display. It can be smoothed by averaging across a sliding window (i.e., 10 data points). It is easy to see in this graph that Treatment 2 results in fewer shorter polyA tails with a concomitant increase in the number of longer polyA tails compared to the Control. In Fig. 2B, which displays the same data as the cumulative fraction of reads at or below a certain length, it is easier to see that the average polyA tail increases from Control to Treatment 1 to Treatment 2. Fig. 3 is called a violin plot and graphs the same data from Fig. 2 in a way that facilities showing more detailed information. The y-axis is the median polyA tail length of mRNA transcripts from genes with at least 10 CCS reads, thus the data are plotted on a per gene basis rather than on a per polyA tail basis as in Fig. 2. This shows that the median polyA tail length across genes is increasing with treatment. It is easier to see a shift toward longer lengths across all quantiles of the distribution (y-axis), with a proportionate shift occurring at the longer polyA tails.

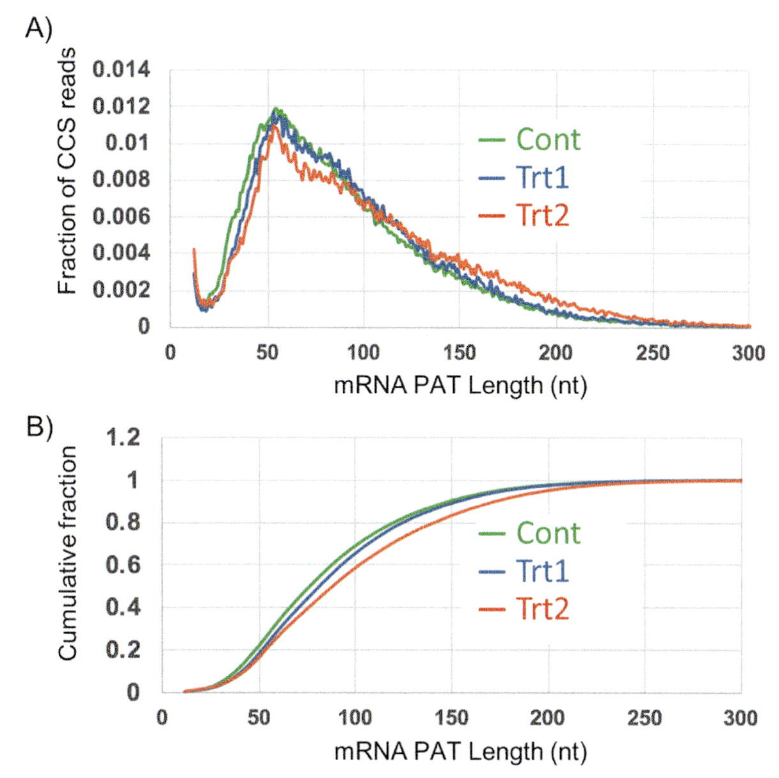

Fig. 2 Graphic presentations of the distribution of polyA tail lengths as a function of experimental treatment. (A) The proportion of the total reads for each treatment observed at each polyA tail length. (B) An alternative presentation of the same data showing the cumulative fraction of reads observed at a given polyA tail length or smaller. Cont, Control; Trt1, Treatment 1; Trt2, Treatment 2; nt, nucleotide; CCS, circular consensus sequence. *Modified from Mattijssen, S., Iben, J. R., Li, T., Coon, S. L., & Maraia, R. J. (2020). Single molecule poly(A) tail-seq shows LARP4 opposes deadenylation throughout mRNA lifespan with most impact on short tails.* eLife, 9, e59186. *doi:10.7554/eLife.59186, under a Creative Commons CC0 license.*

Fig. 4 confirms the data in Figs. 2 and 3, but also shows that the average polyA tail length as a function of categories of gene function based on Gene Ontology terms. This is possible because SM-PAT-Seq attributes polyA tail length to specific gene transcripts and questions can be asked such as whether the treatments effect all classes of transcripts. This graph shows that average polyA tail length increases from Control to Treatment 1 to Treatment 2, except for mitochondrial genes. For abundant genes, the distribution of polyA tails for specific genes can be determined

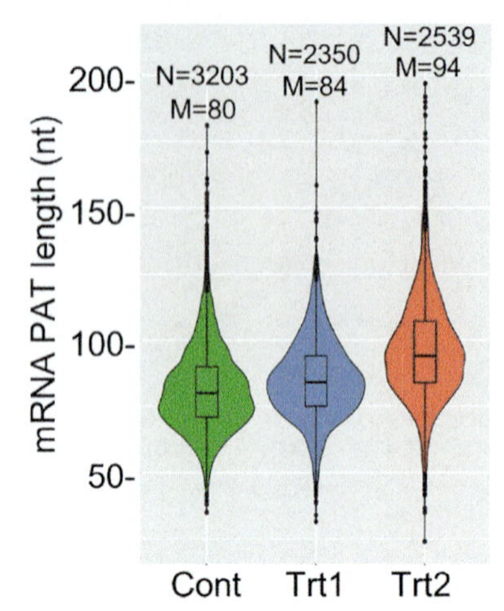

Fig. 3 Violin plots of the median polyA tail lengths determined for genes with ≥10 CCS reads. The width of the violin shows the number of medians at a given polyA length. The box plot inside the violin shows the median plus and minus the first and third quartile. The whiskers show the 95% confidence interval (bottom 5% to top 95%). N, number of genes; M, median length; Cont, Control; Trt1, Treatment 1; Trt2, Treatment 2. *Modified from Mattijssen, S., Iben, J. R., Li, T., Coon, S. L., & Maraia, R. J. (2020). Single molecule poly(A) tail-seq shows LARP4 opposes deadenylation throughout mRNA lifespan with most impact on short tails.* eLife, 9, e59186. *doi:10.7554/eLife.59186 under a Creative Commons CC0 license.*

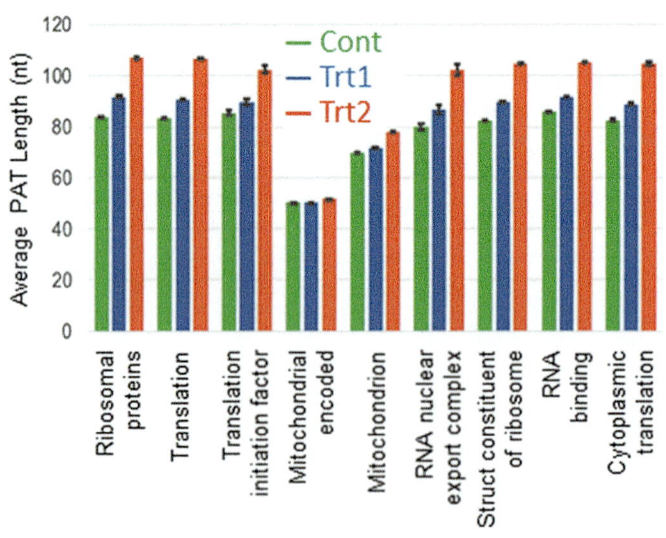

Fig. 4 Histograms showing the average polyA tail length for each treatment as a function of Gene Ontology terms. Cont, Control; Trt1, Treatment 1; Trt2, Treatment 2. This shows that polyA tail length increases from Control to Treatment 1 to Treatment 2 for all classes of molecules, except for mitochondrial encoded genes. *Modified from Mattijssen, S., Iben, J. R., Li, T., Coon, S. L., & Maraia, R. J. (2020). Single molecule poly(A) tail-seq shows LARP4 opposes deadenylation throughout mRNA lifespan with most impact on short tails.* eLife, 9, e59186. *doi:10.7554/eLife.59186 under a Creative Commons CC0 license.*

5. Additional comments

Pacific Biosciences also has a newer, advanced sequencer, the Sequel II. The main advantage with regards to the measurement of polyA tails is that the number of fragments sequenced per flow cell is about eightfold higher, and thus the cost per fragment sequenced is significantly lower. Each investigator needs to determine the number of fragments per sample that need to be sequenced to meet their goals. As mentioned above, there is a trade-off when choosing long-read versus short-read sequencing methods in terms of accuracy versus counts obtained. The Sequel generates about 300,000 reads per flow cell, so it may take multiple flow cells to generate enough data to generate accurate size distributions for any but the most abundant individual genes.

Another advantage of this method is that almost every read is from a unique molecule, with few apparent artifacts from amplification, meaning that quantitation is direct (Mattijssen et al., 2020). This could be due to the low number of reads relative to the complexity of the library. Deeper sequencing, for example, with a Sequel II, will likely be productive in generating more detailed information.

In theory, it should be possible to target specific genes for using this technique. This could be done by using gene specific primers (or pools of primers) for second strand synthesis rather than random hexamers. We have not yet used this modification.

One of the advantages of the accuracy of sequencing through the entire polyA tail is the detection of non-A residues. We see these regularly, but whether these are specifically incorporated, or regulated, is an area of active investigation, as is the question of whether they have a function. While these do not appear to be sequencing artifacts because the Sequel reads them multiple times, it still may be possible that they are derived from library construction, i.e., inaccurate reverse transcriptase, or polymerase. Another potential modification of this method to look at non-A residues could be to investigate terminal uridylation by incorporating one or several As into the splint ligation adaptor used at the beginning of the protocol.

The simplicity and adaptability of the method presented here, along with the advances in sequencing capabilities (i.e., the Sequel II) open new opportunities to investigate polyA tail length and alternative polyadenylation sites in greater detail than before.

Acknowledgments

This work was supported by the Intra-mural Research Program (HD000412-31 PGD) of the Eunice Kennedy Shriver National Institute of Child Health and Human Development, National Institutes of Health (NICHD). This work was conducted at the NICHD Molecular Genomics Core. This work utilized the computational resources of the NIH HPC Biowulf cluster. (http://hpc.nih.gov).

References

Batra, R., Manchanda, M., & Swanson, M. S. (2015). Global insights into alternative polyadenylation regulation. *RNA Biology*, *12*(6), 597–602. https://doi.org/10.1080/15476286.2015.1040974.

Chang, H., Lim, J., Ha, M., & Kim, V. N. (2014). TAIL-seq: Genome-wide determination of poly(A) tail length and 3′ end modifications. *Molecular Cell*, *53*(6), 1044–1052. https://doi.org/10.1016/j.molcel.2014.02.007.

Curinha, A., Oliveira Braz, S., Pereira-Castro, I., Cruz, A., & Moreira, A. (2014). Implications of polyadenylation in health and disease. *Nucleus*, *5*(6), 508–519. https://doi.org/10.4161/nucl.36360.

Eisen, T. J., Eichhorn, S. W., Subtelny, A. O., Lin, K. S., McGeary, S. E., Gupta, S., et al. (2020). The dynamics of cytoplasmic mRNA metabolism. *Molecular Cell*, 77(4), 786–799.e710. https://doi.org/10.1016/j.molcel.2019.12.005.

Elkon, R., Ugalde, A. P., & Agami, R. (2013). Alternative cleavage and polyadenylation: Extent, regulation and function. *Nature Reviews. Genetics*, *14*(7), 496–506. https://doi.org/10.1038/nrg3482.

Harrison, P. F., Powell, D. R., Clancy, J. L., Preiss, T., Boag, P. R., Traven, A., et al. (2015). PAT-seq: A method to study the integration of 3'-UTR dynamics with gene expression in the eukaryotic transcriptome. *RNA*, *21*(8), 1502–1510. https://doi.org/10.1261/rna.048355.114.

Kuhn, U., & Wahle, E. (2004). Structure and function of poly(A) binding proteins. *Biochimica et Biophysica Acta*, *1678*(2–3), 67–84. https://doi.org/10.1016/j.bbaexp.2004.03.008.

Legnini, I., Alles, J., Karaiskos, N., Ayoub, S., & Rajewsky, N. (2019). FLAM-seq: Full-length mRNA sequencing reveals principles of poly(A) tail length control. *Nature Methods*, *16*(9), 879–886. https://doi.org/10.1038/s41592-019-0503-y.

Lim, J., Lee, M., Son, A., Chang, H., & Kim, V. N. (2016). mTAIL-seq reveals dynamic poly(A) tail regulation in oocyte-to-embryo development. *Genes & Development*, *30*(14), 1671–1682. https://doi.org/10.1101/gad.284802.116.

Liu, Y., Nie, H., Liu, H., & Lu, F. (2019). Poly(A) inclusive RNA isoform sequencing (PAIso-seq) reveals wide-spread non-adenosine residues within RNA poly(A) tails. *Nature Communications*, *10*(1), 5292. https://doi.org/10.1038/s41467-019-13228-9.

Mattijssen, S., Iben, J. R., Li, T., Coon, S. L., & Maraia, R. J. (2020). Single molecule poly(A) tail-seq shows LARP4 opposes deadenylation throughout mRNA lifespan with most impact on short tails. *eLife*, *9*, e59186. https://doi.org/10.7554/eLife.59186.

Nicholson, A. L., & Pasquinelli, A. E. (2019). Tales of detailed poly(A) tails. *Trends in Cell Biology*, *29*(3), 191–200. https://doi.org/10.1016/j.tcb.2018.11.002.

Subtelny, A. O., Eichhorn, S. W., Chen, G. R., Sive, H., & Bartel, D. P. (2014). Poly(A)-tail profiling reveals an embryonic switch in translational control. *Nature*, *508*(7494), 66–71. https://doi.org/10.1038/nature13007.

Swaminathan, A., Harrison, P. F., Preiss, T., & Beilharz, T. H. (2019). PAT-Seq: A method for simultaneous quantitation of gene expression, poly(A)-site selection and poly(A)-length distribution in yeast transcriptomes. *Methods in Molecular Biology*, *2049*, 141–164. https://doi.org/10.1007/978-1-4939-9736-7_9.

Tian, B., & Manley, J. L. (2013). Alternative cleavage and polyadenylation: The long and short of it. *Trends in Biochemical Sciences*, *38*(6), 312–320. https://doi.org/10.1016/j.tibs.2013.03.005.

Woo, Y. M., Kwak, Y., Namkoong, S., Kristjansdottir, K., Lee, S. H., Lee, J. H., et al. (2018). TED-Seq identifies the dynamics of poly(a) length during ER stress. *Cell Reports*, *24*(13), 3630–3641.e3637. https://doi.org/10.1016/j.celrep.2018.08.084.

Workman, R. E., Tang, A. D., Tang, P. S., Jain, M., Tyson, J. R., Razaghi, R., et al. (2019). Nanopore native RNA sequencing of a human poly(A) transcriptome. *Nature Methods*, *16*(12), 1297–1305. https://doi.org/10.1038/s41592-019-0617-2.

Yu, F., Zhang, Y., Cheng, C., Wang, W., Zhou, Z., Rang, W., et al. (2020). Poly(A)-seq: A method for direct sequencing and analysis of the transcriptomic poly(A)-tails. *PLoS One*, *15*(6), e0234696. https://doi.org/10.1371/journal.pone.0234696.

CHAPTER SEVEN

3′ End sequencing of pA$^+$ and pA$^-$ RNAs

Guifen Wu†, Manfred Schmid†, and Torben Heick Jensen*
Department of Molecular Biology and Genetics, Aarhus University, Aarhus, Denmark
*Corresponding author: e-mail address: thj@mbg.au.dk

Contents

Abstract

The identity and metabolism of RNAs are often governed by their 5′ and 3′ ends. Single gene loci produce a variety of transcript isoforms, varying primarily in their RNA 3′ end status and consequently facing radically different cellular fates. Knowledge about RNA termini is therefore key to understanding the diverse RNA output from individual

† Equal contribution.

Methods in Enzymology, Volume 655
ISSN 0076-6879
https://doi.org/10.1016/bs.mie.2021.03.020

transcription units. In addition, the 3′ end of a nascent RNA at the catalytic center of RNA polymerase provides a precise and strand-specific measure of the transcription process. Here, we describe a modified RNA 3′ end sequencing method, that utilizes the in vivo metabolic labeling of RNA followed by its purification and optional in vitro polyadenylation to provide a comprehensive view of all RNA 3′ ends. The strategy offers the advantages of (i) nucleotide resolution mapping of RNA 3′ ends, (ii) increased sequencing depth of lowly abundant RNA and (iii) inference of RNA 3′ end polyadenylation status. We have used the method to study RNA decay and transcription termination mechanisms with the potential utility to a wider range of biological questions.

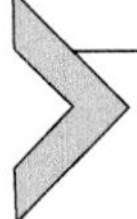

1. Introduction

1.1 RNA 3′ ends

The 3′ ends of cellular RNAs are critically determining their stability and contribute to their subcellular transport and molecular function. This can be exemplified by mRNA poly(A) (pA) tails, which prevent undue transcript degradation by cellular 3′-5′ exonucleases, facilitate RNA nuclear export and affect the cytoplasmic translation process (Nicholson & Pasquinelli, 2019; Tudek, Lloret-Llinares, & Jensen, 2018). However, many stable non-coding RNAs, such as the highly abundant rRNAs, snRNAs, snoRNAs and tRNAs, all lack pA tails (pA^-). Protection of these pA^- 3′ ends against exonucleolysis instead depends on folding of the RNAs into three-dimensional structures or their engagement in shielding RNA:protein complexes (Coy, Volanakis, Shah, & Vasiljeva, 2013; Perumal & Reddy, 2002; Schmid & Jensen, 2018). This is in striking contrast to the plethora of short-lived cellular transcripts, which lack such stabilizing features and instead are removed rapidly, often shortly after their production (Ogami, Chen, & Manley, 2018; Schmid & Jensen, 2018). Interestingly, these unstable RNAs include not only pA^-, but also pA^+ RNAs, where pA tails aid in their targeting for decay (Meola & Jensen, 2017; Ogami et al., 2017; Schmid & Jensen, 2018; Tudek, Lloret-Llinares, et al., 2018).

In eukaryotic cells, both pA^- and pA^+ RNAs are vividly produced from individual transcription units. This can be exemplified by RNA polymerase II (RNAPII)-transcribed protein-coding genes, that are not restricted to generate one single transcript species. The prime product is still a spliced and polyadenylated RNAs, but in addition, early RNAPII transcription is wobbly and prone to termination, which leads to the promiscuous production of unstable pA^- RNAs, primarily within the first ~3 kb of the transcription process (Beckedorff et al., 2020; Elrod et al., 2019; Lykke-Andersen et al., 2020; Tatomer et al., 2019). Moreover, pA sites present in pre-mRNA first introns may also trigger the production of unstable pA^+ RNAs (Chiu et al., 2018; Iasillo et al., 2017; Ogami et al., 2017; Wu et al., 2020). Finally, protein-coding transcription also serves to produce stable non-coding pA^- RNAs, such as snoRNA, scaRNAs and miRNAs, that often reside within pre-mRNA introns (Frankish et al., 2019).

Apart from the mentioned post-transcriptional 3′ ends, cells also contain nascent RNAs, i.e., transcripts undergoing transcription and with their pA^- 3′ ends located within the active center of RNA polymerase. Although these may be of low abundance, they can be enriched by, e.g., RNA polymerase pull-downs or by their brief metabolic labeling and subsequent purification (Churchman & Weissman, 2012; Nojima, Gomes, Carmo-Fonseca, & Proudfoot, 2016; Schwalb et al., 2016). An in-depth analysis of cellular RNA 3′ ends may therefore be used to interrogate the diverse RNA isoform output from single genomic loci alongside its underlying transcription.

1.2 RNA 3′ end sequencing technologies

The advent of next generation sequencing of RNA (RNA-seq) has revolutionized the experimental investigation of RNAs and our resulting understanding of RNA biology. In classical methods short RNA fragments are sequenced, which provide information about the major termini and splicing patterns of abundant cellular transcripts but fall short in distinguishing the patchwork of produced isoforms. Numerous alternatives have therefore been developed, which, together, supply an ever-increasing toolbox for the tailored analysis of specific transcript characteristics relevant to a given biological question. This is also the case for the analysis of RNA 3′ ends, with strategies developed for improving their quantitation and precise mapping to the genome (Ozsolak et al., 2009; Pelechano, Wilkening, Jarvelin, Tekkedil, & Steinmetz, 2012; Routh et al., 2017). Here, we describe one such experimental procedure, that we have employed to examine nuclear RNA decay and transcription termination pathways (Lykke-Andersen et al., 2020; Wu et al., 2020; Zheng, Liu, & Tian, 2016). This method will be described in a step-by-step protocol, including the principal steps of a downstream bioinformatics pipeline to analyze the acquired data.

2. Method overview

The method workflow is outlined in Fig. 1. Briefly, cells of interest are incubated with 4-thiouridine (4sU) for 10 min to allow its incorporation into newly transcribed RNA. After this labeling period, total RNA is

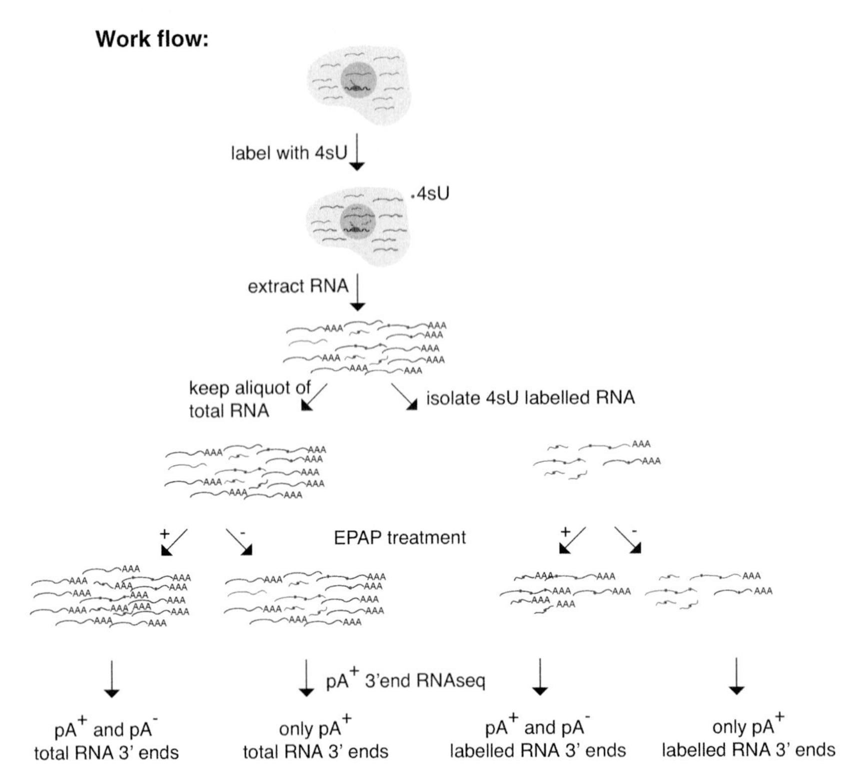

Fig. 1 Workflow of the RNA 3′end sequencing protocol. Cells of choice are metabolically labeled with 4sU. Total RNA is then extracted and 4sU-labeled RNA purified. Aliquots of total and purified 4sU labeled RNA are subjected to EPAP treatment. Resulting RNAs are used for library preparation and analyzed by pA^+ 3′end RNAseq.

harvested and 4sU labeled transcripts are conjugated to biotin and isolated by immunoprecipitation (4sU-IP). The remaining total RNA is subsequently processed in parallel to allow for any comparisons between steady-state and newly made RNA. Aliquots of precipitated total and 4sU-labeled RNA are treated with (or without) *E. coli* pA polymerase (EPAP) to add A-tails to all RNA 3′ OH ends. Finally, rRNAs are depleted from all samples prior to their downstream 3′ end seq library preparation, using the Quant-Seq REV kit from Lexogen (Lexogen Gmbh, Vienna Austria). This technology employs an oligo dT primer for reverse transcription (RT) of RNA and a random primer for second-strand cDNA synthesis, followed by PCR amplification. Libraries can be multiplexed to allow for the sequencing of ~10 samples per lane on a standard Illumina HiSeq machine. Resulting reads are trimmed and filtered to remove adaptor sequences as well as common artefactual T-stretches frequently observed at the start of the raw sequencing reads from Quant-Seq REV libraries. Reads passing these steps are then mapped to the human genome before downstream analysis.

2.1 Metabolic labeling of RNA

RNA-seq experiments, using total cellular RNA preparations, are dominated by abundant RNA species and only allow for steady-state RNA analysis. As a complementary approach, metabolic labeling of RNA, using nucleotide analogs like 4sU, 5-Bromouridine (5BrU) or 5-Ethynyl- uridine (5EU) and followed by its specific purification, allows for a more extensive coverage of low abundant transcripts while also providing information about RNA decay rates (Dolken et al., 2008). Additionally, RNAs derived from brief labeling procedures include a substantial portion of nascent transcripts, the 3′ ends of which mark positions of transcribing RNA polymerase (Schwalb et al., 2016). In the case of HeLa cells, we usually administer 4sU at 500 μM for 10 min by incubation with the growth medium, which allows for downstream analysis of transcription and early RNAs of interest. Note that the 4sU concentration and the labeling time need to be adjusted when using different cell lines and/or addressing different biological questions (see Section 3.8).

2.2 4SU IP

To enrich for newly produced transcripts, the metabolically labeled RNA is then separated from the larger unlabeled pool. In the case of 5BrU-labeling, this is achieved directly via its purification using an anti-BrU antibody. Instead, 4sU- and 5EU-labeled transcripts are first covalently coupled to biotin, using thiol- or click-chemistry, respectively, and then purified on streptavidin resins.

Since 4sU purification is time-consuming and somewhat expensive, we usually validate the experiment prior to this step. This includes assessing relevant protein levels (e.g., after factor depletion) by western blotting and RT-qPCR analyses to confirm expected/predicted RNA phenotypes on total RNA. In addition, total RNA integrity is evaluated using a bioanalyzer. As internal controls for the downstream 4sU IP reaction, we add in vitro transcribed 4sU-labeled, and unlabeled, spike-in RNAs, to the total RNA before its biotin-coupling. An aliquot of each sample is stored for total RNA analysis and the rest is purified. Note, that the amount of labeled RNA is small ($<1\%$ of the total pool) and it is therefore important to carefully adhere to the stringent purification conditions described to obtain clean labeled RNA preparations.

To assess the efficiency of the 4sU purification step, labeled vs. unlabeled spike-in RNA levels are measured by RT-qPCR of both input and IP samples. 4sU labeled spike-in RNAs should preferably be enriched >10-fold over unlabeled spike-in RNAs in the 4sU IP samples as exemplified in Fig. 2A and B. After this step, total and labeled RNAs are treated similarly to allow for their direct comparison in downstream bioinformatics analyses.

2.3 EPAP treatment

To be able to discriminate polyadenylated from non-polyadenylated RNA 3′ ends, samples are split in two, subjecting one aliquot to in vitro polyadenylation by EPAP (EPAP+), while preserving the other as an EPAP- control. This is done because the downstream QuantSeq REV library preparation relies on the production of first strand cDNA using an oligo dT primer, such that only RNAs polyadenylated in vivo are reverse-transcribed from EPAP- samples, while all RNA 3′ OH species, irrespective of their original 3′ end pA status, are reverse-transcribed from EPAP+ samples.

EPAP treatment efficiency can be evaluated by oligo dT-primed RT-qPCR analysis, comparing the abundance of polyadenylated to unadenylated RNA from EPAP+ vs. EPAP-samples. The amplification efficiency of polyadenylated RNA (e.g., *RAB7A* RNA) should remain unchanged after EPAP treatment, while unadenylated RNA (e.g., *U1* RNA) should only be efficiently amplified from EPAP+ samples (Fig. 2C and D).

2.4 Ribosomal RNA depletion

After EPAP treatment, all samples are subjected to rRNA depletion to prevent the highly abundant rRNA 3′ ends to overwhelm the EPAP+ library samples. EPAP- samples are also rRNA-depleted to avoid any potential biases related to this step in the protocol. We usually confirm efficient rRNA depletion on total RNA samples using a Bioanalyzer (Fig. 3A).

2.5 RNA 3′SEQ library preparation

Libraries are then prepared using the QuanSeq REV 3′ mRNA-Seq Library Prep Kit (Lexogen) yielding final cDNA fragments including adapters of ~300 bp (Fig. 3B). Molecular barcodes included in the kit can be used to multiplex 8–12 samples. Sequencing is carried out using 75 nt single-read sequencing on an Illumina HiSeq machine, routinely yielding around 20 mio reads per samples.

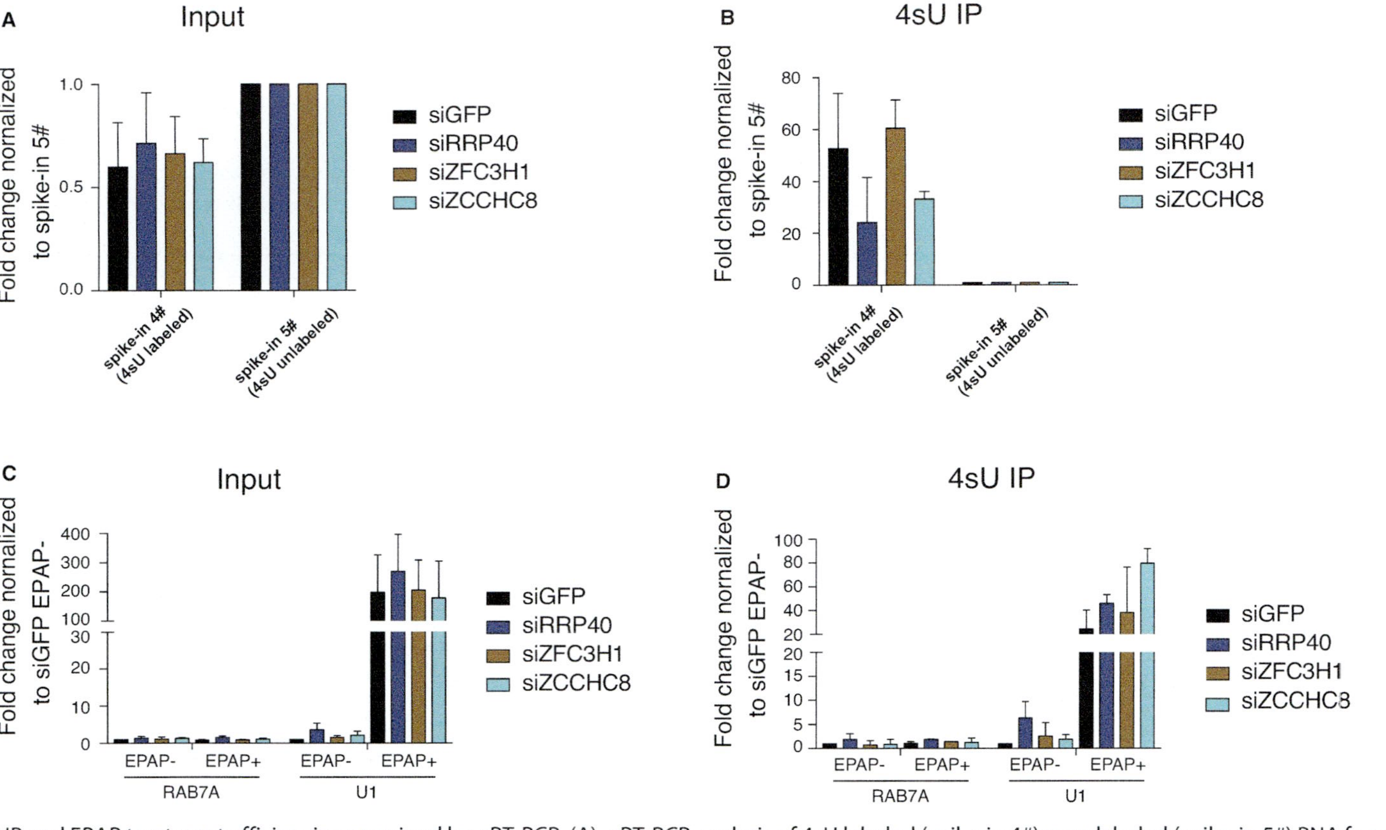

Fig. 2 4sU IP and EPAP treatment efficiencies examined by qRT-PCR. (A). qRT-PCR analysis of 4sU labeled (spike-in 4#) or unlabeled (spike-in 5#) RNA from input samples. cDNA synthesis was primed by an oligo dT primer. Data from biological triplicate experiments of four different samples (siRNA mediated knock-downs as indicated on the right) are shown. (B) As in (A) except that samples from 4sU IPs and no EPAP treatment (EPAP-) are shown. Data from biological duplicate experiments are shown. (C) qRT-PCR analysis of the pA^+ RNA (RAB7A) and the pA^- RNA (U1) from the indicated input samples without (EPAP −) or with (EPAP +) treatment. Data from biological triplicate experiments are shown. (D) As in (C) except that 4sU IP samples are shown. Data from biological duplicate experiments are shown.

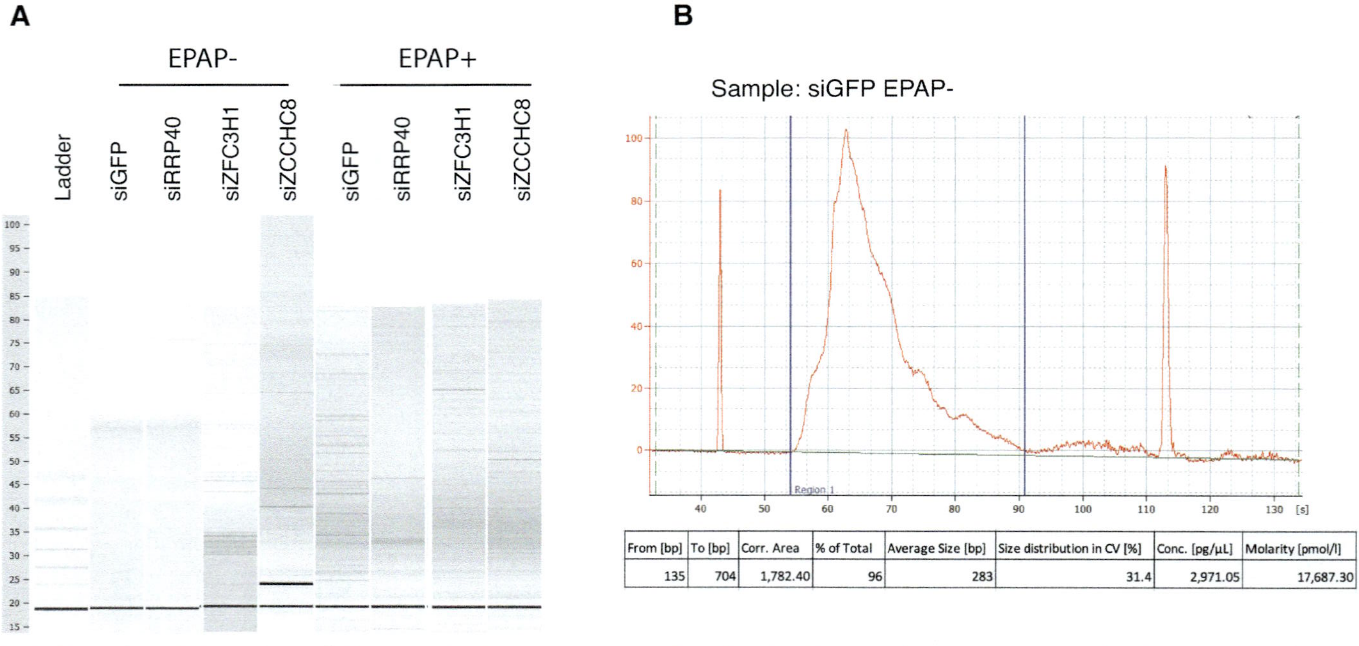

From [bp]	To [bp]	Corr. Area	% of Total	Average Size [bp]	Size distribution in CV [%]	Conc. [pg/μL]	Molarity [pmol/l]
135	704	1,782.40	96	283	31.4	2,971.05	17,687.30

Fig. 3 Bioanalyzer 2100 analysis of rRNA depleted input samples and final libraries. (A) Bioanalyzer gel image of rRNA depleted input samples from Fig. 2. (B) Bioanalyzer electropherogram profile of a final DNA library. The control siGFP EPAP-sample is shown as an example.

2.6 Bioinformatic analysis

Our sequencing is normally performed by a service provider, which also demultiplexes the raw reads. Remaining reads are then trimmed to remove Illumina adaptor sequences as well as any homopolymeric T stretches at read 5′-ends before final filtering for read length and overall quality. Trimmed reads are mapped to the human genome, converted to coverage tracks and normalized to enable comparison of signals between libraries. Since the dT-priming step of the QuantSeq REV protocol will also lead to RNA amplification from internal A-rich sequences, all 3′ ends aligning to adjacent A-rich genomic sequences are removed from further analysis. Library normalization can be performed based on read counts from spike-in RNAs or from high abundance endogenous transcripts like protein-coding RNA, pending the experimental set-up and research question. Normalized data can then be used for any type of downstream computational analysis, such as genome browsing or differential expression and metagene analysis, which will not be described here.

3. Protocol

3.1 Before you begin

3.1.1 Cells

The method described here is in principle suitable to study transcription and RNA metabolism in any cell type. However, exact details of 4sU labeling conditions will have to be established and optimized on a case-by-case basis (see Section 3.8). In the described test case of HeLa cells, we usually start labeling of cells at an 80%–90% confluency in a P15 plate, which gives enough RNA (>400 μg) for a single 4sU IP experiment and its downstream library preparation. Smaller amounts are sufficient for RT-qPCR test experiments, where library preparation is not required. We recommend to always prepare biological duplicates, preferably triplicates, of all samples to account for any experimental variation. Note also, that it is preferable to process all samples in parallel. If this is not feasible, batch-specific effects need to be considered in the downstream analysis and we then recommend to include one sample of all relevant treatments and their controls in each experimental batch.

3.1.2 Spike-in RNAs

The spike-in RNAs, which are added before the biotinylation reaction, can be prepared in advance. Note though, that 4sU-labeled RNA is somewhat unstable so that the quality of spike-in transcripts will decrease during prolonged storage time, even when these are kept at −80 °C.

3.2 Key resources table

Reagent or resource	Source	Identifier
Chemicals, peptides, and recombinant proteins		
4-Thiouridine	Carbosynth	Cat# 13957-31-8
EZ-LinkTM HPDP-Biotin	Thermo Fisher Scientific	Cat# 21341
Chloroform	Sigma-Aldrich	Cat# 366927
Isopropanol	Sigma-Aldrich	Cat# I9516
4sUTP	Jena Bioscience	Cat# NU-1156S
Critical commercial assays		
MEGAscript RNAi Kit	Thermo Fisher Scientific	Cat# AM1626
mMacs streptavidin kit	Miltenyi Biotec	Cat# 130-074-101
miRNeasy Micro kit	QIAGEN	Cat# 217084
E. coli poly(A) polymerase	Invitrogen	Cat# AM2030
RiboLock RNase Inhibitor	Thermo Fisher Scientific	Cat# EO0381
PureLink micro RNA purification kit	Thermo Fisher Scientific	Cat# 12183016
Quant Seq 30 mRNA-Seq REV library prep kit	Lexogen GmbH	Cat# 016
RiboCop rRNA Depletion kit	Lexogen GmbH	Cat# 037.96
EcoRV	Thermo Fisher Scientific	Cat# ER0301
RNase-Free DNase set	QIAGEN	Cat# 79254
TURBO DNA-free kit	Invitrogen	Cat# AM1907
Agilent High Sensitivity DNA Kit	Agilent	Cat# 5067-4626
Qubit dsDNA HS Reagent	Thermo Fisher Scientific	Cat# Q32851

—cont'd

Reagent or resource	Source	Identifier
Recombinant DNA		
pUC19-ERCC-00170 (template for spike in #12)	Schwalb et al. (2016)	N/A
pUC19- ERCC-00002 (template for spike in #9)	Schwalb et al. (2016)	N/A
pUC19-ERCC-00145 (template for spike in #5)	Schwalb et al. (2016)	N/A
pUC19- ERCC-00092 (template for spike in #8)	Schwalb et al. (2016)	N/A
pUC19-ERCC-00136 (template for spike in #4)	Schwalb et al. (2016)	N/A
pUC19-ERCC-00043 (template for spike in #2)	Schwalb et al. (2016)	N/A
Software and algorithms		
bbduk	sourceforge.net/projects/bbmap/	
STAR aligner v2.5.2b	(Dobin et al., 2013) (https://github.com/alexdobin/STAR)	
bedtools v2.25.0	(Quinlan & Hall, 2010) (https://bedtools.readthedocs.io)	
custom python script to mark genomic A-rich positions	https://github.com/manschmi/MS_Metagene_Tools/blob/master/flag_genomic_As_KevinRoy_like.py	

3.3 Materials and equipment

- Nuclease-free H_2O
- 100% Ethanol
- 80% Ethanol
- 75% Ethanol
- Isopropanol
- Chloroform
- TRIzol reagent
- 5M NaCl

- 3 M sodium acetate (pH 5.2)
- Glycogen (20 μg/μL)
- Phenol pH 8.0
- DMF (*N,N*-Dimethylformamide)
- 4-Thiouridine (4sU): Dissolve the 4sU powder in nuclease-free H_2O. The stock concentration is 100 mM. Aliquot and store at −20 °C
- Spike-in plasmids
- *Eco*RV and 10 × FD buffer
- 4sUTP: The stock is 100 mM. Dilute to 75 mM with H_2O before use.
- Biotin-HPDP: Prepare a stock solution at 1 mg/mL in DMF, aliquot and store at −20 °C
- 10 × biotinylation buffer: 100 mM Tris-HCl pH 7.5 (1 mL of 1 M stock solution), 10 mM EDTA (0.2 mL of 500 mM stock solution), nuclease-free H_2O 8.8 mL
- Biotin IP washing buffer: 100 mM Tris-HCl pH 7.5, 10 mM EDTA, 1 M NaCl, 0.1% Tween 20
- DTT (Dithiothreitol): Prepare a 1 M stock in nuclease-free H_2O
- MEGAscript RNAi Kit
- TURBO DNase
- miRNeasy Micro kit
- RNase-Free DNase: Dissolve the lyophilized DNase I in 550 μL nuclease-free H_2O by injecting H_2O into the vail using needle and syringe. Mix gently by inverting the vail (do not vortex), aliquot 20 μL/tube and store at −20 °C
- *E. coli* poly(A) polymerase
- RiboLock RNase Inhibitor
- PureLink micro RNA purification kit
- RiboCop rRNA Depletion Kit
- QuanSeq 3′ mRNA-Seq Library Prep Kit for Illumina REV
- QuantSeq PCR-Add-on kit
- Centrifuge
- Water bath
- Nanodrop™ Spectrophotometer
- Thermomixer
- μMacs columns
- μMacs streptavidin beads
- MACS Multistand (Miltenyi Biotec)
- DynaMag™-2 Magnet
- 2 mL phase lock gel tube
- 1.5 mL low DNA binding Eppendorf tube

3.4 Step-by-step method details

3.4.1 4sU labeling and cell harvesting

3.4.1.1 Timing: 1 h

Note: Protect the 4sU labeling medium and 4sU labeled samples from light.

1. For HeLa cells, we prepare 15 mL per plate of DMEM supplemented with 500 μM 4sU and pre-warm this to 37 °C (4sU labeling medium).
2. Prepare timer, pipettes and falcon tubes for cell harvesting in the hood.
3. Remove medium, add 15 mL 4sU labeling medium and return the plates to the incubator for 10 min. *Note: We suggest to process no more than four plates in parallel, such that labeling times are followed as exactly as possible.*
4. Take out the plates, aspirate the medium.
5. Optional: Scrape some cells from the plates with cell lifter and save them for western blotting analysis.
6. Add 10 mL TRIzol to each plate, pipette up and down to lyse cells, and collect cells in a falcon tube.

Pause point: Cell lysates can be stored at −80 °C.

3.4.2 RNA extraction

3.4.2.1 Timing: 2 h

7. Thaw TRIzol cell lysates on ice.
8. Add 2 mL chloroform and shake vigorously for 20 s.
9. Let stand for 5 min at room temperature.
10. Centrifuge at 12,000 × *g* for 15 min at 4 °C.
11. Transfer aqueous phase into new tube (~6 mL). Add equal volume of isopropanol (~6 mL) and incubate 10 min at room temperature.
12. Centrifuge at 12,000 × *g* for 20 min at 4 °C.
13. Carefully decant the isopropanol and add 5 mL 75% ethanol to the pellet. Turn the tube upside down a few times for efficient washing.
14. Centrifuge at 12,000 × *g* for 5 min at 4 °C.
15. Discard the supernatant, add 1 mL 75% ethanol to the pellet, and transfer the pellet to a 1.5 mL tube using a wide tip.
16. Centrifuge at 12,000 × *g* for 5 min at 4 °C.
17. Completely remove ethanol and air-dry pellet for up to 5 min until no ethanol droplets are visible. Do not over-dry.
18. Resuspend the pellet in 500 μL RNase free H_2O.
19. Measure RNA concentration using, e.g., a Nanodrop™ Spectrophotometer.

Pause point: RNA can be stored at −80 °C.

20. Optional Step: Confirm RNA phenotypes using RT-qPCR analysis or equivalent.

21. Optional: Check RNA quality using a Bioanalyzer, which should reveal two clear bands corresponding to the 18S and 28S rRNA. The RIN (RNA Integrity Number) should be above 9.0, which indicates largely intact RNA.

3.4.3 Spike-in RNA in vitro transcription

3.4.3.1 Timing: 1 day

22. Linearize plasmids using plasmid digestion reactions containing 5 μg plasmid, 25 μL 10× FD buffer, 8 μL *Eco*RV and H_2O to 250 μL. We usually use three templates for reactions without 4sU (templates #5, #9 and #12, see reagent and resource table) and three for reactions with 4sUTP (templates #2, #4 and #8, see reagent and resource table).
23. Incubate 2 h at 37 °C.
24. Add 250 μL phenol pH 8.0 to the reaction, shake thoroughly for 20s. Add 250 μL chloroform, mix thoroughly.
25. Centrifuge at 12,000 × *g* for 5 min at room temperature. Carefully remove the upper aqueous phase to a new tube. Take care not to carry over any phenol during pipetting.
26. Add 1/10 volume of 3 M sodium acetate (pH 5.2) and 1 μL glycogen (20 μg/μL), and add 2.5 volumes of 100% ethanol to the upper aqueous phase. Mix thoroughly.
27. Place the tube at −20 °C overnight for DNA precipitation or place the sample at 80 °C for at least 1 h.
28. Centrifuge the sample at 12,000 × *g* for 20 min at 4 °C.
29. Carefully remove the supernatant without disturbing the DNA pellet and wash pellet with 1 mL of 75% ethanol.
30. Centrifuge at 12,000 × *g* for 5 min at 4 °C.
31. Carefully remove the supernatant and dry pellet up to 5 min until no ethanol is visible.
32. Resuspend the pellet with 20 μL of nuclease-free H_2O.
33. In vitro transcription using MEGAscript RNAi Kit. Prepare for each reaction of unlabeled spike-in RNAs (templates #5, #9 and #12) a mix of 2 μL 75 mM of each ATP, GTP, CTP and UTP, 2 μL 10× reaction buffer, 1 μg linear DNA and 2 μL enzyme mix and H_2O to 20 μL. The reactions for labeled spike-in RNAs (templates #2, #4 and #8) are similar except that they contain only 1.8 μL UTP but also 0.2 μL 75 mM 4sUTP.
34. Mix thoroughly by pipetting up and down and incubate at 37 °C for 4 h.

35. Add 2 μL TURBO DNase to the reaction and mix thoroughly.
36. Incubate at 37 °C for 30 min.
37. Add 10 μL DNase Inactivation Reagent, mix well and incubate 5 min at room temperature.
38. Centrifuge at 10,000 × *g* for 2 min at room temperature.
39. Carefully remove the supernatant to a new tube, add 500 μL TRIzol to the supernatant and 100 μL chloroform. Mix thoroughly for 20s.
40. Centrifuge at 12,000 × *g* for 15 min at 4 °C. Transfer the supernatant to a new tube, add an equal volume of isopropanol and mix well.
41. Centrifuge at 12,000 × *g* for 20 min at 4 °C. Carefully remove the supernatant and add 1 mL 75% ethanol to wash the pellet.
42. Centrifuge at 12,000 × g for 5 min at 4 °C. Carefully remove the supernatant and dry the RNA pellet up to 5 min until no ethanol is visible.
43. Resuspend the pellet with 20 μL nuclease-free H_2O.
44. Measure the RNA concentration with, e.g., a Nanodrop™ Spectrophotometer.
45. Prepare stocks of each spike-in RNA at 600 ng/μL. Mix equal volumes of each spike-in RNA to make a final stock mix.
46. Aliquot the spike-in RNA stock mix into 10 μL per tube and store at −80 °C.

3.4.4 Biotinylation of 4sU labeled RNA and clean-up

3.4.4.1 Timing: 3 h

47. Dilute spike-in RNA to 6 ng/μL with nuclease-free H_2O. Add 4 μL of diluted spike-in RNA to 400 μg total RNA. Supplement with H_2O to a total volume of 500 μL. Mix thoroughly by pipetting up and down. (*Note: use 1.5 mL low DNA binding Eppendorf tube*).
48. Keep a 5 μL aliquot of this RNA mix on ice as total RNA (input) for later use. Heat the remaining RNA for 10 min at 60 °C, then incubate on ice for 2 min.
49. Prepare the biotinylation reaction by adding in this order: 500 μL RNA, 100 μL 10 × biotinylation buffer, 200 μL DMF and 200 μL 1 mg/mL biotin-HDPD and mix immediately by pipetting.
50. Incubate in a thermomixer at 750 rpm at 24 °C for 1.5 h (cover to protect from light).
51. Spin empty 2 mL phase lock tubes at 16,000 rpm for 1 min at room temperature before use.
52. Transfer biotinylation reaction mix to the phase lock gel tube. Add 700 μL chloroform and mix vigorously.

53. Let stand for 3 min at room temperature, and centrifuge at 20,000 rpm for 5 min at room temperature.
54. Transfer the upper phase to a new 2 mL tube.
55. Add 1/10 volume of 5 M NaCl and 1 volume of isopropanol.
56. Mix thoroughly and centrifuge at 12000 g for 30 min at 4 °C.
57. Remove the supernatant carefully and add 1 mL 75% ethanol to wash the pellet.
58. Centrifuge at 12,000 × *g* for 10 min at 4 °C.
59. Remove the supernatant, then do a quick spin and remove the rest of the ethanol carefully with a 200 μL pipette. Air-dry the pellet up to 5 min at room temperature until no ethanol is visible. Do not over-dry.
60. Resuspend RNA in 100 μL of H_2O by pipetting.

Pause point: Biotinylated RNA is stable and can be stored at − 80 °C.

3.4.5 Purification of 4sU RNA

3.4.5.1 Timing: 3 h

61. Equilibrate μMacs streptavidin beads to room temperature for at least 30 min before use.
62. Prepare fresh washing buffer. One needs roughly 7 mL per purification. A 50 mL mix contains 5 mL 1 M Tris-HCl pH 7.5, 1 mL 0.5 M EDTA, 10 mL 5 M NaCl, 500 μL 10% Tween 20 and H_2O to 50 mL.
63. Shake vigorously to mix the buffer. Split in two and heat one aliquot of washing buffer to 65 °C in a water bath and keep the other at room temperature.
64. Heat the biotinylated RNA at 65 °C for 5 min in a thermomixer and then cool on ice for 2 min.
65. Add 100 μL of streptavidin beads to the RNA and incubate in a thermomixer at 400 rpm, 24 °C for 15 min.
66. During this incubation time, place μMacs columns in a magnetic stand and add 900 μL of washing buffer (room temperature) to the columns for equilibration. (*Note: press gently with a finger on the top of the column to make the buffer go through*).
67. Apply the streptavidin beads-RNA mix to the columns, collect the flow-through and reapply once again to columns.
68. Wash columns 3 times with 900 μL 65 °C warm washing buffer.
69. Wash 3 times with 900 μL room temperature washing buffer.
70. Prepare fresh elution buffer of 100 mM DTT by diluting 1 M DTT with nuclease-free H_2O. 200 μL diluted DTT is needed for each reaction.

71. Elute RNA with 100 µL elution buffer.
72. Perform a second round of elution with another 100 µL elution buffer.
73. Proceed directly to RNA cleanup.

3.4.6 RNA cleanup and DNase I digest

3.4.6.1 Timing: 1 h

74. Take out input RNA aliquot from (step 48), add 10 µL 3 M sodium acetate and dilute to 200 µL with nuclease-free H_2O.
75. Take the 4sU labeled RNA (step 71 + 72), add 10 µL 3 M sodium acetate. Treat input RNA and 4sU labeled RNA in the same way from here on.
76. Add 300 µL of 100% ethanol to each sample and mix thoroughly by pipetting up and down.
77. Transfer the sample to a RNeasy MinElute spin column on a 2 mL collection tube (supplied).
78. Centrifuge at 12,000 × *g* for 15 s at room temperature. Discard the flow-through.
79. Add 350 µL buffer RWT (*Note: remember to add isopropanol as described in the RNeasy manual*) into the column and centrifuge at 12,000 × g for 15 s at room temperature. Discard the flow-through.
80. Add 10 µL of DNase I stock solution to 70 µL of buffer RDD (DNase I incubation mix). Mix by gently pipetting up and down.
81. Pipette 80 µL of DNase I incubation mix directly onto the column membrane and incubate at room temperature for 15 min.
82. Add 350 µL of buffer RWT (with isopropanol) into the column and centrifuge at 12,000 × *g* for 15 s at room temperature. Discard the flow-through.
83. Add 500 µL of buffer RPE onto the column. Centrifuge at 12,000 × *g* for 15 s at room temperature. Discard the flow-through.
84. Add 500 µL of 80% ethanol onto the column. Centrifuge at 12,000 × g for 2 min at room temperature. Discard the flow-through.
85. Centrifuge the column at 12,000 × g for 5 min at room temperature to dry the membrane.
86. Place the column in a new 1.5 mL tube. Add 23 µL of nuclease-free H_2O directly on the column membrane. Centrifuge at 12,000 × g for 1 min at room temperature. The expected elution volume is 20–21 µL.
87. Measure RNA concentration by, e.g., a Nanodrop™ Spectrophotometer.

Pause point: RNA samples can be stored at −80°C at this point.

3.4.7 E. coli *poly(A) polymerase (EPAP) treatment*

3.4.7.1 Timing: 1.5 h

88. Take out input and 4sU labeled RNA (step 86). Prepare the EPAP+ and EPAP− reactions: EPAP+ reactions contain 10 μL RNA (corresponding to 1 μg input RNA or roughly 500 ng 4sU RNA), 8 μL 5× EPAP buffer, 4 μL 25 mM $MnCl_2$, 0.4 μL EPAP, 0.4 μL 100 mM ATP, 0.4 μL RiboLock RNase inhibitor and 16.8 μL H_2O. EPAP- reactions are similar except that EPAP enzyme is replaced by H_2O.

89. Incubate at 30 °C for 30 min.

90. For RNA purification with the PureLink micro RNA purification kit add listed reagents in the following order to the 40 μL EPAP reaction: 160 μL H_2O, 200 μL Lysis buffer and 200 μL 100% ethanol.

91. Mix thoroughly by pipetting, transfer it to a PureLink mini column.

92. Centrifuge at 12,000 × *g* for 1 min at room temperature. Discard the flow-through.

93. Add 700 μL wash buffer I. Centrifuge at 12,000 × g for 1 min at room temperature. Discard the flow-through.

94. Wash 2 times by adding 500 μL of wash buffer II onto the column. Centrifuge at 12,000 × *g* for 1 min at room temperature. Discard the flow-through.

95. Centrifuge at 12,000 × *g* for 2 min at room temperature.

96. Put the column on a new tube. Add 30 μL nuclease-free H_2O on the membrane of column. Centrifuge at 12,000 × *g* for 1 min at room temperature and recover the flow through.

97. Measure RNA concentration by Nanodrop™ Spectrophotometer.

Pause point: Store RNA at −80 °C or proceed to rRNA depletion.

3.4.8 rRNA depletion

3.4.8.1 Timing: 2 h

Note: rRNAs are depleted using RiboCop rRNA Depletion Kit.

98. Perform ribo-depletion as described in the manual except that the RNAs are eluted with 8 μL Elution Buffer.

99. Quality control the ribo-depleted RNA by Bioanalyzer before library preparation.

Pause point: Final RNA samples can be stored at −80 °C.

3.4.9 Library generation

3.4.9.1 Timing: 1–2 days

Note: Libraries are made using QuanSeq 3′ mRNA-Seq Library Prep Kit for Illumina (REV).

100. Prepare sequencing libraries as described in the manual.

101. Based on the library size measured by Bioanalyzer and the concentration measured by Qubit, the molarity of the library is calculated. Equimolar amounts of 8–12 libraries are multiplexed. *Note that in our experience, EPAP+ and EPAP− samples should be mixed to avoid sequencing lanes containing only EPAP+ samples* (see Section 3.8).

102. Libraries sequenced on a Illimina HiSeq with single-read 75 nt output. *Note that QuantSeq REV libraries require the use of the kit-specific CSP primer for sequencing on the machine.*

3.4.10 Bioinformatic analysis

3.4.10.1 Timing: 1 day

Note: We usually carried out sequencing at the Next Generation Sequencing Facility at Vienna BioCenter Core Facilities (VBCF, member of the Vienna BioCenter, Austria), which applies demultiplexing as part of their service.

103. Demultiplexed reads are first trimmed and quality-filtered using bbduk script with the command: *bbduk.sh in =<fastq_in> out=<fastq_trimmed> ref=<adapters.fa> k=13 ktrim=r useshortkmers=t mink=5 qtrim=t trimq=10 minlength=20.*

<fastq_in> and *<fastq_trimmed >* are the fastq input and output files and *adapters.fa* contains the sequences to be trimmed from reads. These are the Illumina adapters and a A18 homopolymer.

104. Resulting reads are mapped to a joint index of human genome release GRCh38 and ERCC spike sequences using STAR with the command: *STAR –genomeDir <genome_index> –readFilesIn <fastq_trimmed> –outFilterType BySJout –outFilterMultimapNmax 1 –outFileNamePrefix <output_dir/prefix> –outSAMtype BAM SortedByCoordinate*

<genome_index> is a mapping index for STAR. In our case this is joint index based on the human genome sequence release GRCh38 and ERCC spike-in sequences.

<output_dir/prefix> is the path where output bam files should be placed plus an optional file prefix for the output files.

The option *–outFilterMultimapNmax 1* ensures that only uniquely mapping reads are included in the output bam file.

105. Genomic coverage of 3′ ends is obtained from the reads with the following command for strand-specific coverage:

(a) *bedtools genomecov -ibam <bam> -bg -5 -strand - | awk '!($1 ~ /chr..?$/) | bedtools sort -i - > <plus.bg>*

(b) *bedtools genomecov -ibam <bam> -bg -5 -strand + | awk '!($1 ~ /chr..?$/) | bedtools sort -i - > <minus.bg>*

Note that coverage of alignment 5′ end positions are reported since these mark the RNA 3′ ends (option *-5*) and alignments are reverse-complement relative to the RNA species (*-strand* – for plus strand coverage and *-strand* + for minus strand coverage). The awk command filters positions on major chromosomes and filters away spike-derived reads and reads mapping to various contigs not assigned to a major chromosome.

106. 3′ end positions upstream of genomic A-rich positions are then removed from those bedgraph files:

(a) *bedtools subtract -a <plus.bg> -b <mask_plus.bed> > <plus_Amasked.bg>*

(b) *bedtools subtract -a <minus.bg> -b <mask_minus.bed> > <minus_Amasked.bg>*

mask_plus.bed and *mask_minus.bed* are bed files which contain positions upstream genomic A-rich sequences in GRCh38. A-rich is defined as positions upstream 6 nucleotides comprising only As and Gs and at least 4 As and positions upstream 18 nucleotides with at least 12 As. These criteria are from Roy, Gabunilas, Gillespie, Ngo, and Chanfreau (2016). The mask files are created with custom python scripts available at GitHub (https://github.com/manschmi/MS_Metagene_Tools/blob/master/flag_genomic_As_KevinRoy_like.py).

Note: Further data manipulations from these raw signal bedgraphs are beyond this chapter. Typical downstream steps include normalization of bedgraphs from different samples, genome browsing, metagene analysis and quantification.

3.5 Expected outcomes

Starting from a 80%–90% confluent P15 plate of HeLa cells, labeled for 10 min with 4sU we usually purify 500–750 μg of total RNA. 400 μg of this is then used for 4sU IP experiments, where we usually purify 1–2 μg 4sU labeled RNA. This allows for splitting of samples for EPAP+/EPAP− treatments and rRNA-depletions, yielding sufficient RNA for library preparation.

The library preparation kit includes a qPCR step to determine the optimal final library PCR amplification cycle number. In our experience, sufficient DNA is obtained using 11–17 PCR cycles, which guarantee good sequencing data for low abundant RNA species. If the qPCR reaction indicates that higher cycle numbers are needed, the produced libraries will be less reproducible. In this case, we therefore suggest to optimize the upstream steps to obtain a higher yield of rRNA-depleted RNA, requiring fewer PCR amplification rounds.

When libraries are ready, their concentrations can be measured on a Qubit fluorometer with the Qubit dsDNA HS kit, and their fragment sizes checked by a Bioanalyzer. Generally, final library concentrations range from 2 to 10 μg/mL and average library fragment sizes center around 300 bp. An example Bioanalyzer trace from a final library is shown in Fig. 3B. Note, that we do not use the library molarity output from the Bioanalyzer as we found this measure to be unreliable. Instead, we measure final library concentrations by Qubit and calculate library molarities from these measurements using the molecular weights provided by the Bioanalyzer. Finally, 8–12 different libraries are multiplexed at equal molarity for high throughput sequencing run.

We routinely prepare 25 μL of 20 fmol/μL final QuantSeq REV library mix, which typically allows for multiplexing of 8–12 samples to a lane mix. That is, around 50 fmol of each individual library is needed. The mix is then sequenced on an Illumina HiSeq lane, yielding ~20–30 mio reads per sample. Typically >90% of raw reads from EPAP- library data pass the trimming and filtering steps and can be uniquely aligned to the genome. Fewer, but often >70%, of EPAP+ raw reads pass through. If the number of removed reads is significantly below this expectation, obtained data should be carefully controlled for outlier effects. Finally, during the mapping step some reads will be lost as they map to multiple positions. These are typically ignored. In our datasets ~75% of all reads map to unique genomic positions.

3.6 Advantages

In addition to the general benefit of metabolic labeling to reveal early RNA synthesis patterns as outlined above, a main advantage of the described method is that both steady-state and nascent pA^+/pA^- RNA 3′ ends are measured simultaneously within the same sample. This provides immediate inferences about polyadenylation states, RNA half-lives and transcription activities. The focus on RNA 3′ ends facilitates quantitation and simplifies downstream analysis.

3.7 Limitations

A limitation to the presented method is that genomic A-rich positions are absent from the analysis (~4% of the genome). Moreover, pA^- ends are not measured directly, but are inferred by comparing the EPAP+ and EPAP− samples. Hence, a precise estimation of polyadenylation status is usually not possible for single positions with low-moderate read coverage. On a different note, the QuantSeq REV library preparation kit does—at present—not include an option to use unique molecular identifiers (UMI), preventing the detection, and correction, of PCR overamplification. However, this problem is manageable as long as PCR cycle numbers are kept low during the final library preparation. The related QuantSeq FW+UMI library preparation kit includes UMIs, but since only a fraction of obtained sequencing reads from such libraries reflects RNA 3′ end positions, we do not recommend its use in the present context (see also Section 3.9).

3.8 Optimization and troubleshooting

3.8.1 Starting material

The presented protocol was established for measuring transcription termination and RNA abundance parameters in HeLa cells so 4sU labeling conditions will have to be adjusted to the exact experimental question and biological system. In the case of HeLa cells, we find that certain factor depletions may interfere with 4sU labeling efficiencies and the starting amount of cells and the labeling time will have to be adjusted accordingly in such cases.

3.8.2 4sU labeling

4sU has been reported to interfere with RNA metabolism (Burger et al., 2013) and it is therefore important to avoid unwanted side effects by using too high 4sU concentrations and/or extended labeling times. Conversely, it is necessary to achieve sufficient 4sU incorporation to allow for significant enrichment of 4sU RNA in the downstream purification step. This is especially relevant for long-lived RNAs where 4sU labeled vs. total RNA ratios will be low. Attempting to strike this balance, short labeling times will suffice when information about short-lived and nascent RNAs is needed, whereas longer labeling times are often needed for sufficient enrichment of many protein-coding RNAs. In this relation, a recent sophistication to the 4sU methodology, which is not part of this protocol, is to iodoacetamide-treat 4sU samples before library preparation, a strategy first used in the thiol(SH)-linked alkylation for the metabolic sequencing of RNA (SLAM-seq) protocol (Herzog et al., 2017). This treatment leads to a specific misincorporation at 4sU sites during reverse-transcription yielding T > C conversions in the final reads. Reads derived from 4sU RNA can thus be distinguished from unlabeled RNAs, allowing the detection of both "new" and "old" RNA. Moreover, this approach can be used for a very stringent clean-up of 4sU IP data, i.e., by using only reads containing a T > C misincorporation (Reichholf et al., 2019).

Finally, 4sU and 4sU-labeled RNA are light sensitive and unstable. Samples should therefore be processed within a few days after labeling and protected from light during storage and all incubations.

3.8.3 EPAP sequencing libraries

Some of our early EPAP+ libraries yielded significantly fewer reads than the corresponding EPAP− libraries, which became especially problematic if sequencing lanes contained only EPAP+ libraries. This issue likely relates to the longer and more prevalent A-tails on the RNAs used for library preparation, which can lead to a higher amount of reads with Ts at 5′ end positions in EPAP+ libraries and which in turn disturbs the base calling on Illumina machines. This can be largely alleviated by simply combining EPAP+ and EPAP− samples onto the same sequencing lane.

3.8.4 Low input/overamplification

Low input RNA amounts will require excessive rounds of PCR amplification and accumulation of PCR duplicates, rendering libraries unreliable. It is therefore critical to start library preparation with a sufficient amount of RNA. We aim to use at least 1 μg of total RNA as the starting material.

3.9 Alternative methods/procedures

Alternatives exist for most main steps in the experimental setup: (i) 4sU can be replaced by, e.g., BrU or 5EU where purification strategies are also well-established, (ii) To dually detect pA^+ and pA^- RNA 3′ ends, direct adaptor ligation or G/I-tailing have also been employed to permit cDNA synthesis, and (iii) Numerous alternative RNA 3′ end sequencing strategies for poly(A) + RNAs are available. Concerning sequencing strategies, an obvious alternative is to use the QuantSeq FW + UMI kit instead as described above. Several other alternatives for pA^+ RNA 3′ end sequencing exist.

One noticeable option is to use 3′READS+, which is not based on dT-priming and thus circumvents priming off genomic A rich positions (Zheng et al., 2016). Long-read full-length RNA sequencing approaches, e.g., using PacBio and Oxford Nanopore technologies, may also be used to detect both pA^+ and pA^- RNA 3′ ends and further to assess pA tail lengths. However, the relative low sequencing depths presently provided by these technologies make them less useful for analysis of low abundant RNA.

4. Summary

The 3′ end sequencing method described here has been used in our laboratory to characterize the distinct 3′ end features of mammalian nuclear RNA decay pathways (Wu et al., 2020) and to interrogate transcription termination pathways genome-wide (Lykke-Andersen et al., 2020). We have applied a very similar strategy to analyze immediate phenotypes of depleting pA-binding proteins and mRNA nuclear export factors in the yeast *Saccharomyces cerevisiae* (Tudek et al., 2018). This highlights the broad utility of the experimental strategy, which we find applicable for studying a wide range of biological phenomena.

References

Beckedorff, F., Blumenthal, E., da Silva, L. F., Aoi, Y., Cingaram, P. R., Yue, J., et al. (2020). The human integrator complex facilitates transcriptional elongation by endonucleolytic cleavage of nascent transcripts. *Cell Reports*, *32*(3), 107917. https://doi.org/10.1016/j.celrep.2020.107917.

Burger, K., Muhl, B., Kellner, M., Rohrmoser, M., Gruber-Eber, A., Windhager, L., et al. (2013). 4-thiouridine inhibits rRNA synthesis and causes a nucleolar stress response. *RNA Biology*, *10*(10), 1623–1630. https://doi.org/10.4161/rna.26214.

Chiu, A. C., Suzuki, H. I., Wu, X., Mahat, D. B., Kriz, A. J., & Sharp, P. A. (2018). Transcriptional pause sites delineate stable nucleosome-associated premature polyadenylation suppressed by U1 snRNP. *Molecular Cell*, *69*(4), 648–663.e647. https://doi.org/10.1016/j.molcel.2018.01.006.

Churchman, L. S., & Weissman, J. S. (2012). Native elongating transcript sequencing (NET-seq). *Current Protocols in Molecular Biology*. https://doi.org/10.1002/0471142727.mb0414s98. Chapter 4, Unit 4 14 11-17.

Coy, S., Volanakis, A., Shah, S., & Vasiljeva, L. (2013). The Sm complex is required for the processing of non-coding RNAs by the exosome. *PLoS One*, *8*(6), e65606. https://doi.org/10.1371/journal.pone.0065606.

Dobin, A., Davis, C. A., Schlesinger, F., Drenkow, J., Zaleski, C., Jha, S., et al. (2013). STAR: Ultrafast universal RNA-seq aligner. *Bioinformatics*, *29*(1), 15–21. https://doi.org/10.1093/bioinformatics/bts635.

Dolken, L., Ruzsics, Z., Radle, B., Friedel, C. C., Zimmer, R., Mages, J., et al. (2008). High-resolution gene expression profiling for simultaneous kinetic parameter analysis of RNA synthesis and decay. *RNA*, *14*(9), 1959–1972. https://doi.org/10.1261/rna.1136108.

Elrod, N. D., Henriques, T., Huang, K.-L., Tatomer, D. C., Wilusz, J. E., Wagner, E. J., et al. (2019). The Integrator complex terminates promoter-proximal transcription at protein-coding genes. *Molecular Cell, 76*, 738–752.e7. bio Rxiv, 10.1101/725507 https://doi.org/10.1101/725507.

Frankish, A., Diekhans, M., Ferreira, A. M., Johnson, R., Jungreis, I., Loveland, J., et al. (2019). GENCODE reference annotation for the human and mouse genomes. *Nucleic Acids Research, 47*(D1), D766–D773. https://doi.org/10.1093/nar/gky955.

Herzog, V. A., Reichholf, B., Neumann, T., Rescheneder, P., Bhat, P., Burkard, T. R., et al. (2017). Thiol-linked alkylation of RNA to assess expression dynamics. *Nature Methods, 14*(12), 1198–1204. https://doi.org/10.1038/nmeth.4435.

Iasillo, C., Schmid, M., Yahia, Y., Maqbool, M. A., Descostes, N., Karadoulama, E., et al. (2017). ARS2 is a general suppressor of pervasive transcription. *Nucleic Acids Research, 45*(17), 10229–10241. https://doi.org/10.1093/nar/gkx647.

Lykke-Andersen, S., Zumer, K., Molska, E. S., Rouviere, J. O., Wu, G., Demel, C., et al. (2020). Integrator is a genome-wide attenuator of non-productive transcription. *Molecular Cell, 81*, 514–529.e6. https://doi.org/10.1016/j.molcel.2020.12.014.

Meola, N., & Jensen, T. H. (2017). Targeting the nuclear RNA exosome: Poly(A) binding proteins enter the stage. *RNA Biology, 14*(7), 820–826. https://doi.org/10.1080/15476286.2017.1312227.

Nicholson, A. L., & Pasquinelli, A. E. (2019). Tales of detailed poly(A) tails. *Trends in Cell Biology, 29*(3), 191–200. https://doi.org/10.1016/j.tcb.2018.11.002.

Nojima, T., Gomes, T., Carmo-Fonseca, M., & Proudfoot, N. J. (2016). Mammalian NET-seq analysis defines nascent RNA profiles and associated RNA processing genome-wide. *Nature Protocols, 11*(3), 413–428. https://doi.org/10.1038/nprot.2016.012.

Ogami, K., Chen, Y., & Manley, J. L. (2018). RNA surveillance by the nuclear RNA exosome: Mechanisms and significance. *Noncoding RNA, 4*(1), 8. https://doi.org/10.3390/ncrna4010008.

Ogami, K., Richard, P., Chen, Y., Hoque, M., Li, W., Moresco, J. J., et al. (2017). An Mtr4/ZFC3H1 complex facilitates turnover of unstable nuclear RNAs to prevent their cytoplasmic transport and global translational repression. *Genes & Development, 31*(12), 1257–1271. https://doi.org/10.1101/gad.302604.117.

Ozsolak, F., Platt, A. R., Jones, D. R., Reifenberger, J. G., Sass, L. E., McInerney, P., et al. (2009). Direct RNA sequencing. *Nature, 461*(7265), 814–818. https://doi.org/10.1038/nature08390.

Pelechano, V., Wilkening, S., Jarvelin, A. I., Tekkedil, M. M., & Steinmetz, L. M. (2012). Genome-wide polyadenylation site mapping. *Methods in Enzymology, 513*, 271–296. https://doi.org/10.1016/B978-0-12-391938-0.00012-4.

Perumal, K., & Reddy, R. (2002). The 3′ end formation in small RNAs. *Gene Expression, 10*(1–2), 59–78.

Quinlan, A. R., & Hall, I. M. (2010). BEDTools: A flexible suite of utilities for comparing genomic features. *Bioinformatics, 26*(6), 841–842. https://doi.org/10.1093/bioinformatics/btq033.

Reichholf, B., Herzog, V. A., Fasching, N., Manzenreither, R. A., Sowemimo, I., & Ameres, S. L. (2019). Time-resolved small RNA sequencing unravels the molecular principles of microRNA homeostasis. *Molecular Cell, 75*(4), 756–768.e757. https://doi.org/10.1016/j.molcel.2019.06.018.

Routh, A., Ji, P., Jaworski, E., Xia, Z., Li, W., & Wagner, E. J. (2017). Poly(A)-ClickSeq: Click-chemistry for next-generation 3-end sequencing without RNA enrichment or fragmentation. *Nucleic Acids Research, 45*(12), e112. https://doi.org/10.1093/nar/gkx286.

Roy, K., Gabunilas, J., Gillespie, A., Ngo, D., & Chanfreau, G. F. (2016). Common genomic elements promote transcriptional and DNA replication roadblocks. *Genome Research, 26*(10), 1363–1375. https://doi.org/10.1101/gr.204776.116.

Schmid, M., & Jensen, T. H. (2018). Controlling nuclear RNA levels. *Nature Reviews. Genetics*, *19*(8), 518–529. https://doi.org/10.1038/s41576-018-0013-2.

Schwalb, B., Michel, M., Zacher, B., Fruhauf, K., Demel, C., Tresch, A., et al. (2016). TT-seq maps the human transient transcriptome. *Science*, *352*(6290), 1225–1228. https://doi.org/10.1126/science.aad9841.

Tatomer, D. C., Elrod, N. D., Liang, D., Xiao, M. S., Jiang, J. Z., Jonathan, M., et al. (2019). The integrator complex cleaves nascent mRNAs to attenuate transcription. *Genes & Development*, *33*(21 – 22), 1525–1538. https://doi.org/10.1101/gad.330167.119.

Tudek, A., Lloret-Llinares, M., & Jensen, T. H. (2018). The multitasking polyA tail: Nuclear RNA maturation, degradation and export. *Philosophical Transactions of the Royal Society of London. Series B, Biological Sciences*, *373*(1762). https://doi.org/10.1098/rstb.2018.0169.

Tudek, A., Schmid, M., Makaras, M., Barrass, J. D., Beggs, J. D., & Jensen, T. H. (2018). A nuclear export block triggers the decay of newly synthesized polyadenylated RNA. *Cell Reports*, *24*(9), 2457–2467.e2457. https://doi.org/10.1016/j.celrep.2018.07.103.

Wu, G., Schmid, M., Rib, L., Polak, P., Meola, N., Sandelin, A., et al. (2020). A two-layered targeting mechanism underlies nuclear RNA sorting by the human exosome. *Cell Reports*, *30*(7), 2387–2401. e2385 https://doi.org/10.1016/j.celrep.2020.01.068.

Zheng, D., Liu, X., & Tian, B. (2016). 3'READS+, a sensitive and accurate method for 3′ end sequencing of polyadenylated RNA. *RNA*, *22*(10), 1631–1639. https://doi.org/10.1261/rna.057075.116.

CHAPTER EIGHT

Comprehensive profiling of mRNA polyadenylation in specific cell types *in vivo* by cTag-PAPERCLIP

R. Samuel Herron and Hun-Way Hwang*
Department of Pathology, University of Pittsburgh, School of Medicine, Pittsburgh, PA, United States
*Corresponding author: e-mail addresses: hunway.hwang@pitt.edu; Hunway.Hwang@pitt.edu

Contents

Methods in Enzymology, Volume 655
ISSN 0076-6879
https://doi.org/10.1016/bs.mie.2021.04.003

Abstract

The ability to generate cell-type specific mRNA polyadenylation (pA) maps from complex tissues is crucial for understanding how alternative polyadenylation (APA) is regulated in individual cell types in their physiological environment under different conditions. In this chapter, we discuss cTag-PAPERCLIP, a recently developed method combining the well-established CLIP (crosslinking immunoprecipitation) technique and the Cre-lox system to achieve customized cell-type specific APA profiling from mouse tissue without cell purification or enrichment. In cTag-PAPERCLIP, selective expression of GFP-tagged poly(A) binding protein (PABP-GFP) in the desired cell type is achieved through Cre-mediated activation of a latent knock-in allele of PABP-GFP. Immunoprecipitation of PABP-GFP then allows mRNA 3′ end fragments in the desired cell type to be specifically retrieved from ultraviolet (UV)-irradiated whole tissue lysate. The mRNA fragments are subsequently turned into a cDNA library to provide a comprehensive APA map and an mRNA expression profile of the chosen cell type through deep sequencing.

1. Introduction

Traditionally, mRNA 3′ end profiling is performed through reverse transcription of purified cellular mRNAs with oligo-d(T) primers (hereafter referred to as "3′ end sequencing") (Shi, 2012). Recent technical advances in minimizing reverse transcription from internal adenine-rich mRNA regions substantially improve the fidelity of 3′ end sequencing directly from purified cellular mRNAs (Hoque et al., 2013; Jan, Friedman, Ruby, & Bartel, 2011; Lianoglou, Garg, Yang, Leslie, & Mayr, 2013; Martin, Gruber, Keller, & Zavolan, 2012) (reviewed in Gruber & Zavolan, 2019). Studies utilizing these high-fidelity 3′ end sequencing techniques have provided important insights into APA regulation in different cell types grown in culture and FACS-sorted primary blood cells (Cheng et al., 2020; Singh et al., 2018).

In contrast, application of 3′ end sequencing to primary cells in solid tissues is more challenging because the cell type of interest has to be completely dissociated and isolated from its residing tissue in order to generate highly pure cellular mRNAs for 3′ end profiling. Although this "purification first" approach is technically feasible, it might not be able to fully capture the physiological mRNA 3′ end profiles in intact tissue as the isolation process could induce gene expression changes in cell types sensitive to tissue dissociation (Cardona, Huang, Sasse, & Ransohoff, 2006; Okaty, Sugino, & Nelson, 2011).

cTag-PAPERCLIP is a recently developed method that combines mouse genetics and robust biochemistry of the CLIP technique to generate cell-type specific mRNA 3′ end profiles from solid tissue without prior purification for the cell type of interest (Hwang et al., 2017). In cTag-PAPERCLIP, the cell-type specificity is established by restricting expression of PABP-GFP to the cell type of interest through specific Cre-dependent activation of a latent knock-in allele of PABP-GFP in the transgenic "cTag-PABP" mouse (Fig. 1). PABP-GFP, which naturally binds to mRNA poly(A) tails with high affinity, is then crosslinked to the poly(A) tails in the cell type of interest by exposing the tissue to ultraviolet (UV) light in a UV crosslinker. Immunoprecipitation (IP) of the resulting PABP-GFP: RNA complexes (containing mRNA 3′ ends) formed in the cell type of interest using GFP antibodies then provides the starting materials for high-throughput mRNA 3′ end profiling. Importantly, because the IP can be performed directly from the whole tissue lysate without losing the cell-type specificity, cTag-PAPERCLIP bypasses cell purification altogether and has the technical advantage of preserving sensitive cell types such as brain microglia, which tend to partially activate during tissue dissociation, in their native state (Hwang et al., 2017).

In the original proof-of-principle study, cTag-PAPERCLIP was applied to four major cell types in the mouse brain and revealed distinct APA

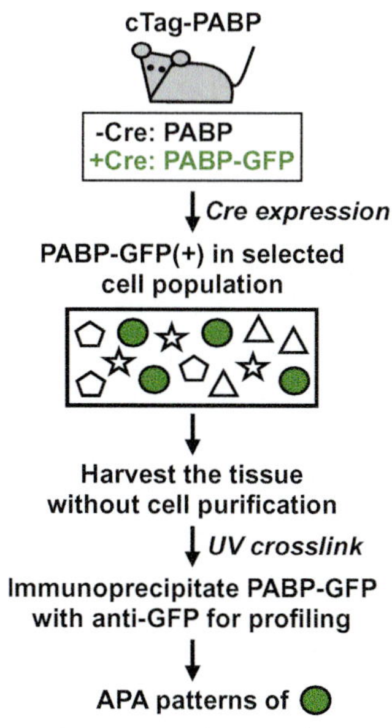

Fig. 1 An illustration showing key steps of cTag-PAPERCLIP in a hypothetical mouse tissue with four different cell types. Each geometric shape represents a cell type.

landscapes in neurons and glia that are directly regulated by RNA-binding proteins specifically expressed in each cell type (Hwang et al., 2017). The compatibility with the Cre/lox system makes cTag-PAPERCLIP highly customizable and the continued expansion of the Cre/lox toolbox provides exciting opportunities to utilize cTag-PAPERCLIP to investigate mRNA 3′ end processing and APA in a large variety of biological contexts beyond the original four cell types in the brain (Challis et al., 2019; Daigle et al., 2018; Korecki et al., 2019). Here, we discuss the modified cTag-PAPERCLIP protocol with improved yield (~35% enhancement compared with the original protocol in Hwang et al., 2017).

2. Materials

2.1 Ultraviolet crosslinking of cTag-PABP mouse tissue

1. Tissue culture grade, Mg^{2+}/Ca^{2+}-free 1 × phosphate-buffered saline (PBS).
2. Sterile 18-gauge needle and syringe.
3. Petri dishes.
4. Plastic tray.
5. UV crosslinker.
6. Refrigerated microcentrifuge.
7. Tissue pulverizer.

2.2 Immunoprecipitation of PABP-GFP

1. Magnetic rack.
2. Dynabeads Protein G.
3. Antibody Binding Buffer: 1 × phosphate-buffered saline (prepare from tissue culture grade Mg^{2+}/Ca^{2+}-free 10 × PBS), 0.02% Tween-20.
4. Mouse monoclonal antibody against GFP, clones 19F7 and 19C8.
5. 1 × PXL Buffer: 1 × phosphate-buffered saline (prepare from tissue culture grade Mg^{2+}/Ca^{2+}-free 10 × PBS), 0.1% SDS, 0.5% NP-40, 0.5% Sodium deoxycholate.
6. (Optional) RNase inhibitor.
7. RQ1 DNase.
8. Incubator/Shaker.
9. RNase A (Molecular Biology Grade).
10. High RNase A solution: 1:100 dilution of RNase A in 1 × PXL Buffer.
11. Low RNase A solution: 1:100,000 or user-optimized dilution of RNase A in 1 × PXL Buffer.

12. Refrigerated microcentrifuge.
13. 5 × PXL Buffer: 5 × phosphate-buffered saline (prepare from tissue culture grade Mg^{2+}/Ca^{2+}-free 10 × PBS), 0.1% SDS, 0.5% NP-40, 0.5% Sodium deoxycholate.
14. 1 × PNK Wash Buffer: 50 mM Tris-HCl, pH 7.4, 10 mM $MgCl_2$, 0.5% NP-40.

2.3 5′ labeling of immunoprecipitated RNA

1. Fast AP.
2. 1 × PNK+EGTA Wash Buffer: 50 mM Tris-HCl, pH 7.4, 20 mM EGTA, 0.5% NP-40.
3. T4 PNK.
4. ^{32}P-γ-ATP.

2.4 SDS-PAGE and nitrocellulose transfer

1. NuPAGE 4 × LDS Sample Buffer.
2. 1 M dithiothreitol (DTT).
3. NuPAGE 8% Bis-Tris Midi Gel.
4. NuPAGE 20 × MOPS SDS Running Buffer.
5. Full-Range Rainbow Molecular Weight Markers.
6. Nitrocellulose membrane.
7. NuPAGE 20 × Transfer Buffer.
8. Autorad Markers.

2.5 RNA isolation and purification

1. Disposable scalpels.
2. Filter paper.
3. Proteinase K solution (20 mg/mL).
4. PK-SDS Buffer: 100 mM Tris-HCl, pH 7.5, 50 mM NaCl, 1 mM EDTA, 0.2% SDS.
5. Saturated Phenol:Chloroform, pH 4.5.
6. Heavy Phase-Lock Gel tubes.
7. Linear acrylamide (5 mg/mL).
8. 3 M NaOAc, pH 5.2.
9. 1:1 mix of ethanol and isopropanol.

2.6 cDNA synthesis and purification I

1. 50 × Denhardt's Solution.
2. Nuclease-free water (RT-PCR grade).

3. 75% ethanol.
4. 8.2 mM dATP/dCTP/dGTP (prepare from 100 mM stock).
5. 8.2 mM BrdUTP (prepare from powder).
6. 5 × RT Buffer/0.1 M DTT/SuperScript III.
7. RNasin Plus (or other RNase inhibitors).
8. Indexed RT primers (Order with PAGE purification):

RT-1 (CGAT): /5Phos/DDDATCGNNNNNNNAGATCGGAAGAG
CGTCGT/iSp18/CACTCA/iSp18/
CAAGCAGAAGACGGCATACGAGATTTTTTTTTTTTTTTTT
TVN

RT-2 (TAGC): /5Phos/DDDGCTANNNNNNNAGATCGGAA
GAGCGTCGT/iSp18/CACTCA/iSp18/
CAA GCAGAAGACGGCATACGAGATTTTTTTTTTTTTTTTT
TVN

RT-3 (AGTC):
/5Phos/DDDGACTNNNNNNNAGATCGGAAGAGCGTCGT/
iSp18/CACTCA/
iSp18/CAA GCAGAAGACGGCATACGAGATTTTTTTTTTTTT
TTTTTVN

RT-4 (GACT):
/5Phos/DDDAGTCNNNNNNNAGATCGGAAGAGCGTCGT/
iSp18/CACTCA/
iSp18/CAA GCAGAAGACGGCATACGAGATTTTTTTTTTTTT
TTTTTVN

9. Anti-BrdU, clone IIB5.
10. 4 × IP Buffer: 1.2 × SSPE, 4 mM EDTA, 0.2% Tween-20.
11. 2 × IP Buffer: 1:2 dilution of 4 × IP Buffer.
12. 1 × IP Buffer: 1:4 dilution of 4 × IP Buffer.
13. RNase H (make 2 U/μL working solution in nuclease-free water).
14. Microspin G-25 column.
15. Nelson Low Salt Buffer: 15 mM Tris-HCl, pH 7.5, 5 mM EDTA.
16. Nelson Stringent Buffer: 15 mM Tris-HCl, pH 7.5, 5 mM EDTA, 2.5 mM EGTA, 120 mM NaCl, 25 mM KCl, 1% Triton X-100, 1% Sodium deoxycholate, 0.1% SDS.

2.7 cDNA purification II and library construction

1. CircLigase Wash Buffer: 33 mM Tris-Acetate, 66 mM KCl, pH 7.8.
2. CircLigase II (includes 10 × Reaction Buffer, 5 M Betaine and 50 mM $MnCl_2$).

3. Phusion Wash Buffer: 50 mM Tris-HCl, pH 8.0.
4. Phusion DNA Polymerase (includes 5 × Phusion HF Buffer).
5. 10 mM dNTPs.
6. 50 × SYBR Green I.
7. DP5-PE primer (Order with PAGE purification): 5′-AATGATACG GCGACCACCGAGATCTACACTCTTTCCCTACACGACGCT CTTCCGATCT
8. DP3-PAT primer: 5′-CAAGCAGAAGACGGCATA
9. 0.2 mL PCR tube compatible with real-time PCR.
10. Thermocycler for real-time PCR.
11. Agencourt AMPure XP Beads.
12. TapeStation.

3. Methods

The entire cTag-PAPERCLIP protocol can be completed in 5 days with a typical 8-h workday schedule:

Day 1: Ultraviolet Crosslinking of cTag-PABP Mouse Tissue (Section 3.1), Immunoprecipitation of PABP-GFP (Section 3.2) and 5′ labeling of immunoprecipitated RNA (Section 3.3)

Day 2: SDS-PAGE and Nitrocellulose Transfer (Section 3.4)

Day 3: RNA Isolation and Purification (Section 3.5)

Day 4: cDNA Synthesis and Purification I (Section 3.6)

Day 5: cDNA Purification II and Library Construction (Section 3.7)

If desired, the protocol can be completed in 3 days with an alternative schedule:

Day 1: Ultraviolet Crosslinking of cTag-PABP Mouse Tissue (Section 3.1), Immunoprecipitation of PABP-GFP (Section 3.2), 5′ labeling of immunoprecipitated RNA (Section 3.3), SDS-PAGE and Nitrocellulose Transfer (Section 3.4)

Day 2: RNA Isolation and Purification (Section 3.5)

Day 3: cDNA Synthesis and Purification I (Section 3.6), cDNA Purification II and Library Construction (Section 3.7)

3.1 Ultraviolet crosslinking of cTag-PABP mouse tissue (1–2 h)

1a. *Tissue processing and UV crosslinking, for small tissue (e.g., the brain cortex)*: Dissect and collect the tissue of interest from the cTag-PABP mouse and keep it in 1 × PBS on ice. Break it into small pieces in 1 × PBS with an 18-gauge needle and syringe. Spread the tissue suspension in a 10-cm petri dish placed on ice in a tray. Put the tray in a UV crosslinker, remove the lid of the petri dish, and irradiate the dish three times at 400 mJ/cm^2 (see Note 1). Rotate the tray between each irradiation to ensure even distribution of UV light energy across the petri dish. When crosslinking is complete, collect the tissue suspension from the petri dish and transfer it to microcentrifuge tubes such that each tube contains sufficient materials to make one cTag-PAPERCLIP library (see Note 2). Spin the microcentrifuge tubes at 4 °C to pellet the crosslinked tissue. Remove the supernatant and proceed to immunoprecipitation. Alternatively, the crosslinked tissue pellets can be frozen at −80 °C until use.

1b. *Tissue processing and UV crosslinking, for large tissue such as the liver or for sensitive cell types such as the brain microglia*: Dissect and collect the tissue of interest from the cTag-PABP mouse and immediately snap-freeze the tissue in liquid nitrogen (see Note 3). Pulverize the snap-frozen tissue and transfer the tissue powder to a pre-chilled 10-cm petri dish on dry ice in a tray (see Note 4). Put the tray in a UV crosslinker, remove the lid of the petri dish, and irradiate the dish three times at 400 mJ/cm^2 (see Note 5). Rotate the tray between each irradiation to ensure an even distribution of crosslinking. When UV crosslinking is complete, transfer the tissue powder to pre-chilled centrifuge tubes such that each centrifuge tube contains sufficient materials to make one cTag-PAPERCLIP library (see Note 6). Proceed to immunoprecipitation. Alternatively, the crosslinked tissue powder can be frozen at −80 °C until use.

3.2 Immunoprecipitation of PABP-GFP (4 h)

1. *Magnetic bead preparation*: Prepare sufficient amount of Dynabeads for all groups (see Note 7). Wash the beads three times with Antibody Binding Buffer assisted by a magnetic rack. When the washing is finished, resuspend the beads in the appropriate amount of Antibody Binding Buffer containing the mixture of two GFP antibodies (see Note 8) and rotate the beads for at least 30 min at room temperature to bind the GFP antibodies.
2. *Prepare crosslinked tissue lysate*: Resuspend each tube of crosslinked tissue in appropriate amount of 1 × PXL Buffer and incubate the tubes on ice for 5–10 min with intermittent vortexes to lyse the tissue (see Note 9). After tissue lysis, add 50 μL of RQ1 DNase to each microcentrifuge tube containing the tissue lysate. Incubate the tubes at 37 °C for 5 min with

continuous shaking at 1200 rpm on an incubator/shaker (see Note 10). After completion of the DNase digestion, put the tubes back on ice briefly before proceeding to RNase digestion. For RNase digestion, add 10–20 μL of High RNase A solution per 1 mL lysate for the High RNase group and add 10 μL of Low RNase A solution per 1 mL lysate for the Low RNase group (see Note 11). Incubate the tubes at 37 °C for 5 min with continuous shaking at 1200 rpm on an incubator/shaker. Clear the lysate by spinning the tubes in a pre-chilled microcentrifuge at ~20,000 *g* for 10 min at 4 °C. Transfer the cleared lysate to new microcentrifuge tubes and leave them on ice.

3. *IP*: Assisted by a magnetic rack, wash the antibody-bound beads three times with 1 × PXL Buffer and divide them evenly in the last wash buffer to new microcentrifuge tubes that corresponds to the number of tissue lysate samples. Remove the last wash buffer from the beads and add the cleared lysate. Rotate the tubes containing the beads and the cleared lysate for 2 h at 4 °C.
4. *Post-IP washes*: Remove the lysate and wash the beads assisted by a magnetic rack with buffers in the following order: twice with 1 × PXL Buffer, twice with 5 × PXL Buffer, and twice with 1 × PNK Wash Buffer (see Note 12).

3.3 5′ labeling of immunoprecipitated RNA (1 h)

1. *Alkaline phosphatase treatment*: Make a master mix in sufficient amount for all tubes—for each reaction, add 51.1 μL water, 6 μL 10 × Fast AP Buffer, 0.6 μL 10% Tween-20, and 2.3 μL Fast AP (see Note 13). Add 60 μL master mix per tube immediately after removal of the last 1 × PNK Wash Buffer on a magnetic rack. Remove the tubes from the magnetic rack and thoroughly mix the contents by brief and gentle vortexes (or finger tapping). Incubate the tubes at 37 °C for 20 min with intermittent shakes (at 1200 rpm for 15 s every 90 s) on an incubator/shaker. When the incubation is complete, remove the reaction mix and wash the beads assisted by a magnetic rack with buffers in the following order: once with 1 × PNK Wash Buffer, once with 1 × PNK +EGTA Wash Buffer, and twice with 1 × PNK Wash Buffer.
2. *PNK 5′ labeling*: Make a master mix in sufficient amount for all tubes—for each reaction, add 49.5 μL water, 6 μL 10 × PNK Buffer, 3 μL T4 PNK, and 1.5 μL ^{32}P-γ-ATP (see Note 14). Add 60 μL master mix per tube immediately after removal of the last 1 × PNK Wash Buffer

on a magnetic rack. Remove the tubes from the magnetic rack and thoroughly mix the contents by brief and gentle vortexes (or finger tapping). Incubate the tubes at 37 °C for 20 min with intermittent shakes (at 1200 rpm for 15 s every 90 s) on an incubator/shaker. When the incubation is complete, remove the reaction mix and wash the beads three times with 1 × PNK Wash Buffer. Leave the beads in the last wash and store at 4 °C overnight.

3.4 SDS-PAGE and nitrocellulose transfer (5–6 h)

1. *Elution*: Make a master mix in sufficient amount for all tubes—for each reaction, add 30 μL 4 × LDS Sample Buffer, 27 μL 1 × PNK Wash Buffer, 3 μL 1 M DTT. Add 60 μL master mix per tube immediately after removal of the last 1 × PNK Wash Buffer on a magnetic rack. Remove the tubes from the magnetic rack and thoroughly mix the contents by brief and gentle vortexes (or finger tapping). Incubate the tubes at 70 °C for 10 min with continuous shaking at 1200 rpm on an incubator/shaker. When the incubation is complete, put the tubes back on a magnetic rack and transfer the solution to a new microcentrifuge tube. Discard the original tubes containing the beads.
2. *SDS-PAGE*: Load each tube (60 μL) in two wells (30 μL/well) of a 20-well 10% gel. Leave an empty lane between different samples. Run the gel at 200 V in the cold room with 1 × MOPS buffer for about 1.5 h (see Note 15).
3. *Transfer*: Assemble a Novex wet transfer apparatus to transfer the PABP-GFP-RNA complexes from the gel to a pure nitrocellulose membrane in 1 × NuPAGE Transfer Buffer supplemented with 10% methanol. Perform the wet transfer in a cold room at 30 V for 80–90 min.
4. *Film exposure*: When the transfer is complete, disassemble the transfer apparatus and rinse the nitrocellulose membrane briefly in 1 × PBS. Drip off excess buffer from the nitrocellulose membrane, enclose it in plastic wrap and expose it to a film in a cassette at −80 °C (see Note 16 and Fig. 2).

3.5 RNA isolation and purification (2–3 h)

1. *Cutting the nitrocellulose membrane*: Mark the corners of the desired area on the nitrocellulose membrane (see Fig. 2) using the film as a guide. The lower margin should be slightly above the band from the High RNase group (see Fig. 2). Cut the area out from the rest of the membrane with a

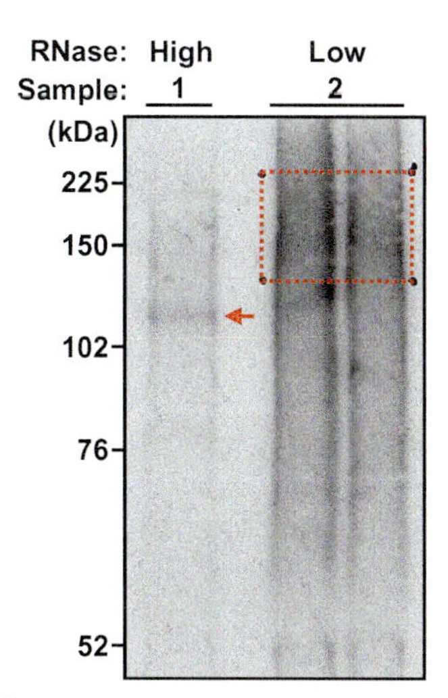

Fig. 2 An autoradiogram from a two-sample cTag-PAPERCLIP experiment. The High RNase group on the left shows a discrete band (denoted by the arrow) consisting of the completely digested PABP-GFP:RNA complex slightly above the 102 kDa marker. (The molecular weight of PABP-GFP is estimated to be ~100 kDa.) In contrast, the Low RNase group on the right shows a desired smear pattern from limited digestion. The dashed lines connecting the dot labels on the original film show the margins of the nitrocellulose membranes used for RNA extraction and sequencing library construction. The lane between samples 1 and 2 was left empty to avoid cross-contamination.

disposable scalpel against a filter paper. Cut the desired piece of membrane further into multiple small pieces against a filter paper and carefully transfer all the pieces into a microcentrifuge tube. Repeat the process for each sample with a new scalpel and a new filter paper each time.

2. *Proteinase K digestion*: Make a master mix in sufficient amount for all tubes—for each reaction; add 10 μL Proteinase K and 200 μL PK-SDS Buffer (see Note 17). Add 210 μL master mix to each tube containing the nitrocellulose membrane pieces and make sure all pieces are submerged in the master mix. Incubate the tubes at 50 °C for 60 min with intermittent shakes (at 1400 rpm for 15 s every 30 s) on an incubator/shaker.
3. *RNA extraction*: Add 240 μL PK-SDS buffer and 450 μL pH 4.5 saturated Phenol:Chloroform to each tube. Vortex each tube vigorously for 20 s. Incubate the tubes at 37 °C for 10 min with continuous shaking at 1400 rpm on an incubator/shaker. During the incubation, spin the Heavy Phase-Lock Gel tubes at 1500 *g* for 30 s. When the incubation is complete, transfer the entire solution from each microcentrifuge tube

to a Heavy Phase-Lock Gel tube and leave the membrane pieces behind. Vortex each Heavy Phase-Lock Gel tube vigorously for 20 s. Spin all Heavy Phase-Lock Gel tubes in a pre-chilled microcentrifuge at full speed (>20,000 *g*) for 5 min at 4 °C. Transfer the aqueous phase (~500 μL) from each Heavy Phase-Lock Gel tube to a new microcentrifuge tube. For each tube, add 5 μL linear acrylamide, 50 μL 3 M NaOAc, and 1 mL ethanol-isopropanol mixture. Thoroughly mix the tubes then store the tubes at −20 °C overnight to precipitate the RNAs.

3.6 cDNA synthesis and purification I (5–6 h)

1. *Magnetic bead preparation, blocking*: Prepare sufficient amount of Dynabeads for all groups (see Note 18). Wash the beads three times with Antibody Binding Buffer assisted by a magnetic rack. When the washing is finished, resuspend the beads in the appropriate amount of Antibody Binding Buffer supplemented with 50 × Denhardt's Solution (see Note 19). Rotate the beads for at least 45 min at room temperature to block non-specific binding.
2. *RNA precipitation*: Spin all microcentrifuge tubes stored at −20 °C overnight from Section 3.5 in a pre-chilled microcentrifuge at full speed (>20,000 *g*) for 20 min at 4 °C to precipitate the RNAs (see Note 20). Carefully remove the supernatant and add 200 μL 75% ethanol to each tube. Spin the tubes at full speed (>20,000 *g*) for 10 min at 4 °C. Repeat the ethanol wash once before drying the RNA pellet (see Note 21). Add 8 μL nuclease-free water to each pellet and incubate the tube at 65 °C for 5 min. After the incubation, do a quick spin to collect all fluids at the bottom and carefully resuspend the RNAs by gently pipetting up and down a few times. Transfer the RNA solution to a new 0.2 mL PCR tube. To prepare the minus reverse transcriptase (RT) control tube, pool 0.5–1 μL RNA solution from all the other groups and bring the volume to 8 μL with nuclease-free water if needed. Leave all samples on ice (see Note 22).
3. *Prepare master mixes for reverse transcription*: Prepare Master Mix I in a microcentrifuge tube for all samples and the minus RT control: for each sample, add 4 μL 5 × RT Buffer, 1 μL 8.2 mM dATP, 1 μL 8.2 mM dCTP, 1 μL 8.2 mM dGTP, and 1 μL 8.2 mM Br-dUTP. Prepare Master Mix IIA in a 0.2 mL PCR tube for all samples (but not the minus RT control): for each sample, add 1 μL 0.1 M DTT, 1 μL RNasin Plus, and 1 μL SuperScript III. Prepare Master Mix IIB in a 0.2 mL PCR tube

specifically for the minus RT control: add 1 μL 0.1 M DTT, 1 μL RNasin Plus, and 1 μL nuclease-free water. Leave all master mixes on ice. Start and hold the thermocycler at 75 °C.

4. *Reverse transcription*: Add 8 μL of Master Mix I and 1 μL of 25 μM indexed RT primer to each sample and the minus RT control (see Note 23). Transfer all the PCR tubes containing the samples and the minus RT control to the thermocycler. Denature at 75 °C for 3 min then ramp down to 48 °C and hold. Transfer Master Mix IIA and IIB to the thermocycler heat block and incubate briefly at 48 °C. Add 3 μL of Mix IIA to each sample and add 3 μL of Mix IIB to the minus RT control. Continue with reverse transcription—45 min at 48 °C, 15 min at 55 °C, 5 min at 85 °C and hold at 4 °C.
5. *Magnetic bead preparation, antibody binding*: Wash the blocked beads from step 1 in Section 3.6 three times with Antibody Binding Buffer assisted by a magnetic rack. Prepare a master mix for all samples and the minus RT control: for each tube, add 40 μL Antibody Binding Buffer, 5 μL 50 × Denhardt's Solution and 5 μL (5 μg) anti-BrdU antibody. Resuspend all the beads in the master mix and rotate the beads for at least 45 min at room temperature to bind the antibodies.
6. *Post-RT clean up*: When the reverse transcription is complete, add 1 μL of RNase H to each PCR tube and incubate the PCR tubes at 37 °C on a thermocycler for 20 min. When the RNase H digestion is complete, add 10 μL nuclease-free water to each PCR tube and do a quick spin down to collect all liquids at the bottom (see Note 24). Prepare the G-25 columns following the manufacturer's instructions. Spin each RT reaction through a G-25 column into a new microcentrifuge tube at 750 *g* for 2 min to remove excess Br-dUTP. Discard the used G-25 columns.
7. *cDNA purification: immunoprecipitation I*: Use a pipet to measure the volume of each cleared RT reaction in the microcentrifuge tube and add nuclease-free water to bring the volume up to 40 μL for each tube. Add 50 μL 2 × IP Buffer and 10 μL 50 × Denhardt's Solution to each tube (see Note 25). Denature the samples at 70 °C for 5 min followed by incubation on ice until use. Wash the antibody-bound beads from step 5 in Section 3.6 three times with 1 × IP Buffer assisted by a magnetic rack and leave the washed beads in 1 × IP Buffer from the last wash. Split half of the washed beads into new microcentrifuge tubes by the number of samples including the minus RT control and store the remaining half of the washed beads at 4 °C. On a magnetic rack, remove the 1 × IP Buffer from the beads and immediately add the denatured sample.

Mix the sample and the beads by brief and gentle vortexes (or finger tapping). Rotate the tubes at room temperature for 45 min.

8. *cDNA purification: post-IP wash I*: Wash the beads assisted by a magnetic rack with buffers in the following order: once with 1 × IP Buffer plus 5 × Denhardt's Solution, twice with Nelson Low Salt Buffer plus 1 × Denhardt's Solution, twice with Nelson Stringent Buffer plus 1 × Denhardt's Solution, and twice with 1 × IP Buffer (see Note 26). Each wash includes a 3-min incubation—the tubes containing the beads and the wash buffer will be rotated for 3 min at room temperature before buffer removal on a magnetic rack (see Note 27).
9. *cDNA purification: heat elution*: After removal of the 1 × IP Buffer from the last wash, immediately add 90 μL Elution Buffer to each tube. Incubate the tubes at 98 °C for 1 min with continuous shaking at 1200 rpm on an incubator/shaker. When the incubation is complete, immediately put the tubes on a magnetic rack. Transfer the eluate to a new microcentrifuge tube and add 10 μL 50 × Denhardt's Solution for a total volume of 100 μL for each tube. Store the samples overnight at 4 °C.

3.7 cDNA purification II and library construction (5–6 h)

1. *cDNA purification: immunoprecipitation II*: Denature the samples at 70 °C for 5 min followed by incubation on ice until use. Split the remaining half of the washed beads into new microcentrifuge tubes by the number of samples including the minus RT control. On a magnetic rack, remove the 1 × IP Buffer from the beads and immediately add the denatured sample. Mix the sample and the beads by brief and gentle vortexes (or finger tapping). Rotate the tubes at room temperature for 45 min (see Note 28).
2. *cDNA purification: post-IP wash II*: Wash the beads assisted by a magnetic rack with buffers in the following order: once with 1 × IP Buffer plus 5 × Denhardt's Solution, twice with Nelson Low Salt Buffer plus 1 × Denhardt's Solution, twice with Nelson Stringent Buffer plus 1 × Denhardt's Solution, and twice with CircLigase Wash Buffer (see Note 29). Each wash includes a 3-min incubation—the tubes containing the beads and the wash buffer will be rotated for 3 min at room temperature before buffer removal on a magnetic rack.
3. *cDNA circularization*: Make a master mix in sufficient amount for all tubes—for each tube, add 12 μL nuclease-free water, 2 μL 10 × CircLigase

Reaction Buffer, 4 μL 5M Betaine, 1 μL 50mM $MnCl_2$, and 1 μL CircLigase II. Add 20 μL master mix per tube immediately after removal of the last CircLigase Wash Buffer on a magnetic rack. Remove the tubes from the magnetic rack and thoroughly mix the contents by brief and gentle vortexes (or finger tapping). Incubate the tubes at 60 °C for 1 h with intermittent shakes (at 1600 rpm for 15 s every 15 s) on an incubator/shaker. When the incubation is complete, wash the beads assisted by a magnetic rack with buffers in the following order: twice with Nelson Low Salt Buffer, twice with Nelson Stringent Buffer, and twice with Phusion Wash Buffer. Each wash includes a 3-min incubation—the tubes containing the beads and the wash buffer will be rotated for 3 min at room temperature before buffer removal on a magnetic rack.

4. *Prepare master mixes for PCR amplification*: Prepare Master Mix I in a microcentrifuge tube for all samples: for each sample, add 37 μL nuclease-free water, 10 μL 5× Phusion HF Buffer, and 1 μL 10 mM dNTPs (see Note 30). Prepare Master Mix II in a 0.2 mL PCR tube for all samples: for each sample, add 0.5 μL DP5-PE (20 μM), 0.5 μL DP3-PAT (20 μM), and 0.5 μL Phusion DNA Polymerase. Make 50× SYBR Green I Solution in a microcentrifuge tube by adding 1 μL SYBR Green I stock solution (10,000×) to 199 μL Phusion Wash Buffer.
5. *cDNA elution and PCR amplification*: After removal of the Phusion Wash Buffer from the last wash, immediately add 48 μL Master Mix I to each tube. Incubate the tubes at 98 °C for 1 min with continuous shaking at 1200 rpm on an incubator/shaker. When the incubation is complete, immediately place the tubes on a magnetic rack. Transfer the eluate into new PCR tubes on ice (see Note 31) and add 1.5 μL Master Mix II and 0.5 μL 50× SYBR Green I Solution to each tube. Start and hold the thermocycler at 98 °C. Transfer the PCR tubes to the thermocycler and start real-time PCR. The initial denaturation is 98 °C for 30 s. The cycling parameters are: 10 s at 98 °C, 15 s at 60 °C, and 20 s at 72 °C. Remove the reaction tube individually from the thermocycler when the fluorescence signal reaches ~1000 (see Note 32).
6. *Post-PCR processing: library purification and quantitation*: Purify the PCR product using Agencourt AMPure XP Beads following manufacturer's instructions (use 90 μL beads per 50 μL PCR reaction). Use TapeStation to measure the concentration of each library.

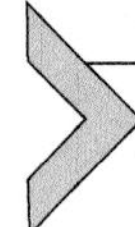

4. Notes

4.1 Ultraviolet crosslinking of cTag-PABP mouse tissue

1. The setting for UV crosslinking is adequate for the brain and the liver in a Stratalinker. Optimization might be necessary if a different crosslinker or a different type of tissue is used. Sufficient cooling during crosslinking is essential.
2. A 1.7 or 2 mL microcentrifuge tube can hold up to 300 mg tissue pellets. We usually put 150–300 mg brain cortex in one microcentrifuge tube, which is regarded as one sample for the subsequent steps. Depending on the cell type of interest, it might be necessary to use more than one tube to have sufficient materials for library construction.
3. For large tissue such as the liver, it can be quickly cut into a suitable size before snap-frozen in liquid nitrogen to facilitate processing.
4. To prevent the tissue powder from thawing during the process, it is critical to pre-chill the pulverizer and petri dishes. Use sufficient dry ice to keep the petri dishes cold during crosslinking.
5. The setting for UV crosslinking is adequate for the brain and the liver in a Stratalinker. Optimization might be necessary if a different crosslinker or a different type of tissue is used.
6. We usually use a 15 mL conical centrifuge to store tissue powder from large tissues and put 300–400 mg tissue powder in one tube, which will be used as one sample for the subsequent steps. Depending on the cell type of interest, it might be necessary to use more tubes to have sufficient materials for library construction.

4.2 Immunoprecipitation of PABP-GFP

7. For each microcentrifuge tube of crosslinked tissue, use 200–300 μL of Dynabeads Protein G. The minimum number of experimental groups is two (one High RNase group and one Low RNase group). If the total volume of beads exceeds the maximum capacity of one microcentrifuge tube, evenly split the beads into two or more tubes.
8. For each group, use 40 μg anti-GFP mixture (20 μg of 19F7 and 20 μg of 19C8) for immunoprecipitation. Resuspend the beads in the original slurry volume (200–300 μL) of Antibody Binding Buffer containing anti-GFP.

9. The ideal volume of 1 × PXL Buffer for tissue lysis is 3 × ~5 × tissue volume. For tissue with RNase activity such as the liver, RNase inhibitors can be added to the tissue lysis buffer to minimize RNA degradation. Lyse the crosslinked tissue in their original storage tubes. For large tissue stored in 15 mL conical centrifuge tubes, it might be necessary to split the tissue lysate into several 1.7 or 2 mL microcentrifuge tubes that fit into the incubator/shaker for subsequent DNase and RNase digestion steps.
10. The incubator/shaker we use is an Eppendorf Thermomixer. The shaking speed might need to be optimized if a different incubator/shaker is used.
11. The optimal dilution for Low RNase A solution needs to be optimized depending on the tissue and cell type of interest.
12. This and all subsequent washes are performed on a magnetic rack in the same volume as the IP reaction (usually 1 mL per tube). To speed up processing, vacuum can be used to remove the wash solutions with precautions to avoid losing beads.

4.3 5′ labeling of immunoprecipitated RNA

13. The volume is based on 300 μL of beads per reaction. It can be scaled down if fewer amounts of beads are used. For example, reduce the total volume to 40 μL if 200 μL beads are used.
14. Radioactive materials are used. Use protection gears and collect all solid/liquid waste according to institutional guidelines for this and all subsequent steps in Sections 3.4–3.7. The volume is based on 300 μL of beads per reaction. It can be scaled down if fewer amounts of beads are used.

4.4 SDS-PAGE and nitrocellulose transfer

15. We usually end the run when the 52 kDa marker (purple) reaches the bottom of the gel.
16. Be sure to use autorad markers because a perfect alignment between the film and the membrane is essential for the subsequent RNA extraction step. The adequate exposure time depends on the amount of starting material, the tissue type and the cell type of interest but usually ranges from several hours to overnight when fresh gamma ATP is used.

4.5 RNA isolation and purification

17. It is a good practice to warm up the PK-SDS Buffer before use as SDS may precipitate from refrigeration.

4.6 cDNA synthesis and purification I

18. Use 50 μL of Dynabeads Protein G per group for two rounds of cDNA purification (25 μL each). Prepare extra 50 μL for the minus RT control.
19. Use of Denhardt's Solution is essential for minimizing non-specific carryover. Please note that the working concentration of Denhardt's solution is 5 × instead of 1 ×. To prepare master mix for 2–4 samples, use 225 μL Antibody Binding Buffer and 25 μL 50 × Denhardt's Solution per sample. For 5–8 samples, use 900 μL Antibody Binding buffer and 100 μL 50 × Denhardt's Solution for the master mix.
20. The RNA pellet might be difficult to visualize, as it is usually small and half-opaque. Use a Radiation Survey Meter to confirm the presence of a pellet and to keep track of it if necessary.
21. It is important not to over-dry the pellet. Air dry for 2–3 min at room temperature or SpeedVac for 45–60 s at room temperature usually is sufficient.
22. It is important to keep all samples and reaction components cold for this and subsequent steps prior to reverse transcription. We use a chilled metal heat block, which sits on ice in a laboratory ice bucket, to hold all the samples and reaction components.
23. The RT primers are compatible with Illumina sequencers. Each primer carries a distinct 4-nt index, so it is possible to retrieve the sequencing reads from individual libraries from a pooled sequencing run.
24. The additional 10 μL water will bring the total volume to ~30 μL, which is above the minimum requirement of 25 μL for the G-25 columns.
25. Alternatively, 50 μL 2 × IP Buffer and 10 μL 50 × Denhardt's Solution can be added to the collection tube prior to spinning the sample through the G-25 column. After spinning, the total volume will be adjusted to 100 μL by adding nuclease-free water.
26. It is essential to include Denhardt's Solution in the indicated wash buffers (5 × for 1 × IP Buffer; 1 × for Nelson Low Salt Buffer and Nelson Stringent Buffer). Make a master mix sufficient for all the tubes. The dilution of 50 × Denhardt's Solution doesn't need to be exact.

For example, for five samples, prepare the master mix by adding 500 μL 50 × Denhardt's Solution to 5 mL 1 × IP Buffer.

27. Only use manual pipetting to remove the wash buffers. Use of a vacuum will result in significant loss of beads and is not recommended.

4.7 cDNA purification II and library construction

28. At least two rounds of cDNA immunoprecipitation are necessary to achieve the desired purity.

29. The two washes with CircLigase Wash Buffer are critical for optimal performance of the CircLigase II.

30. It is important to keep all samples and reaction components cold for this and subsequent steps prior to reverse transcription. We use a chilled metal heat block, which sits on ice in a laboratory ice bucket, to hold all the samples and reaction components.

31. We perform real-time PCR to customize the number of amplification cycles to individual samples in order to minimize over amplification. We use Bio-Rad CFX96™ Touch Real-Time PCR Detection System, which provides real-time tracking of multiple samples simultaneously and allows interruption during PCR amplification for sample retrieval. We use 8-strip PCR tubes and optically clear 8-cap strip from Bio-Rad that are compatible with real-time PCR and we cut the connectors to separate each tube and cap.

32. The optimal level of fluorescence signal needs to be experimentally determined and the number of PCR cycles needed to reach the optimal level depends on the cell type of interest (usually ranging from high teens to high twenties from our experience).

Acknowledgements

We would like to thank Dr. Robert Darnell for his guidance and support in the development of cTag-PAPERCLIP. H.-W.H. was supported by a grant from the National Institutes of Health (NS113861).

References

Cardona, A. E., Huang, D., Sasse, M. E., & Ransohoff, R. M. (2006). Isolation of murine microglial cells for RNA analysis or flow cytometry. *Nature Protocols*, *1*(4), 1947–1951.

Challis, R. C., Kumar, S. R., Chan, K. Y., Challis, C., Beadle, K., Jang, M. J., et al. (2019). Systemic AAV vectors for widespread and targeted gene delivery in rodents. *Nature Protocols*, *14*(2), 379–414.

Cheng, L. C., Zheng, D., Baljinnyam, E., Sun, F., Ogami, K., Yeung, P. L., et al. (2020). Widespread transcript shortening through alternative polyadenylation in secretory cell differentiation. *Nature Communications*, *11*(1), 1–14.

Daigle, T. L., Madisen, L., Hage, T. A., Valley, M. T., Knoblich, U., Larsen, R. S., et al. (2018). A suite of transgenic driver and reporter mouse lines with enhanced brain-cell-type targeting and functionality. *Cell*, *174*(2), 465–480.e22.

Gruber, A. J., & Zavolan, M. (2019). Alternative cleavage and polyadenylation in health and disease. *Nature Reviews Genetics*, *20*(10), 599–614.

Hoque, M., Ji, Z., Zheng, D., Luo, W., Li, W., You, B., et al. (2013). Analysis of alternative cleavage and polyadenylation by 3′ region extraction and deep sequencing. *Nature Methods*, *10*(2), 133–139.

Hwang, H.-W., Saito, Y., Park, C. Y., Blachère, N. E., Tajima, Y., Fak, J. J., et al. (2017). cTag-PAPERCLIP reveals alternative polyadenylation promotes cell-type specific protein diversity and shifts Araf isoforms with microglia activation. *Neuron*, *95*(6), 1334–1349.e5.

Jan, C. H., Friedman, R. C., Ruby, J. G., & Bartel, D. P. (2011). Formation, regulation and evolution of Caenorhabditis elegans 3'UTRs. *Nature*, *469*(7328), 97–101.

Korecki, A. J., Hickmott, J. W., Lam, S. L., Dreolini, L., Mathelier, A., Baker, O., et al. (2019). Twenty-seven tamoxifen-inducible iCre-driver mouse strains for eye and brain, including seventeen carrying a new inducible-first constitutive-ready allele. *Genetics*, *211*(4), 1155–1177.

Lianoglou, S., Garg, V., Yang, J. L., Leslie, C. S., & Mayr, C. (2013). Ubiquitously transcribed genes use alternative polyadenylation to achieve tissue-specific expression. *Genes & Development*, *27*(21), 2380–2396.

Martin, G., Gruber, A. R., Keller, W., & Zavolan, M. (2012). Genome-wide analysis of pre-mRNA 3′ end processing reveals a decisive role of human cleavage factor I in the regulation of 3′ UTR length. *Cell Reports*, *1*(6), 753–763.

Okaty, B. W., Sugino, K., & Nelson, S. B. (2011). A quantitative comparison of cell-type-specific microarray gene expression profiling methods in the mouse brain. *PLoS One*, *6*(1), e16493.

Shi, Y. (2012). Alternative polyadenylation: New insights from global analyses. *RNA*, *18*(12), 2105–2117.

Singh, I., Lee, S.-H., Sperling, A. S., Samur, M. K., Tai, Y.-T., Fulciniti, M., et al. (2018). Widespread intronic polyadenylation diversifies immune cell transcriptomes. *Nature Communications*, *9*(1), 1716.

CHAPTER NINE

A computational pipeline to infer alternative poly-adenylation from 3′ sequencing data

Hari Krishna Yalamanchili[a,b,c], Nathan D. Elrod[d], Madeline K. Jensen[d], Ping Ji[d], Ai Lin[d,e], Eric J. Wagner[d], and Zhandong Liu[a,b,*]

[a]Department of Pediatrics, Baylor College of Medicine, Houston, TX, United States
[b]Jan and Dan Duncan Neurological Research Institute, Texas Children's Hospital, Houston, TX, United States
[c]USDA/ARS Children's Nutrition Research Center, Department of Pediatrics, Baylor College of Medicine, Houston, TX, United States
[d]Department of Biochemistry and Molecular Biology, The University of Texas Medical Branch at Galveston, Galveston, TX, United States
[e]Department of Etiology and Carcinogenesis, National Cancer Center/Cancer Hospital, Chinese Academy of Medical Sciences and Peking Union Medical College, Beijing, China
*Corresponding author: e-mail address: zhandong.liu@bcm.edu

Contents

Abstract

An increasing number of investigations have established alternative polyadenylation (APA) as a key mechanism of gene regulation through altering the length of 3′ untranslated region (UTR) and generating distinct mRNA termini. Further, appreciation for the

Methods in Enzymology, Volume 655
ISSN 0076-6879
https://doi.org/10.1016/bs.mie.2021.04.001

significance of APA in disease contexts propelled the development of several 3′ sequencing techniques. While these RNA sequencing technologies have advanced APA analysis, the intrinsic limitation of 3′ read coverage and lack of appropriate computational tools constrain precise mapping and quantification of polyadenylation sites. Notably, Poly(A)-ClickSeq (PAC-seq) overcomes limiting factors such as poly(A) enrichment and 3′ linker ligation steps using click-chemistry. Here we provide an updated PolyA-miner protocol, a computational approach to analyze PAC-seq or other 3′-Seq datasets. As a key practical constraint, we also provide a detailed account on the impact of sequencing depth on the number of detected polyadenylation sites and APA changes. This protocol is also updated to handle unique molecular identifiers used to address PCR duplication potentially observed in PAC-seq.

1. Introduction

After solving the double helical structure of deoxyribonucleic acid (DNA) and the central dogma, one of the most complex questions that has consumed investigations is to decode the underlying regulation of transcriptional dynamics. Over the past couple decades, contributions from next generation sequencing (NGS) technologies such as ribonucleic acid (RNA)-Seq have significantly advanced our understanding of gene expression and unraveled several fundamental cellular mechanisms. In particular, our understanding of post-transcriptional diversity beyond transcription has greatly progressed with several studies establishing the role of alternative splicing in human diseases (Dargahi et al., 2014; Dredge, Polydorides, & Darnell, 2001; Luco, Allo, Schor, Kornblihtt, & Misteli, 2011). Both neurological and cancer research domains have been expanded through NGS-generated data to uncover novel alternative splicing and cryptic splicing events (Hsieh et al., 2019; Ling, Pletnikova, Troncoso, & Wong, 2015; Tan et al., 2016).

2. Alternative polyadenylation

In eukaryotes, almost all precursor mRNA (pre-mRNA) molecules must undergo cleavage and polyadenylation (Erson-Bensan, 2016) where the 3′ end of pre-mRNA is cleaved, and a stretch of adenosine monophosphate units are added. It is now realized that about 70% of human genes undergo alternative polyadenylation (APA) and generate different 3′ end mRNA transcripts. The length of the 3′ untranslated region (UTR) is altered by cleaving the mRNA at distinct polyadenylation sites and more than 50% of the human genes have three or more polyadenylation sites

(Derti et al., 2012). Alternative polyadenylation plays a crucial role in fundamental biological processes including gene regulation (Erson-Bensan & Can, 2016), mRNA localization (Tushev et al., 2018), cell proliferation (Sandberg, Neilson, Sarma, Sharp, & Burge, 2008), differentiation (Ji, Lee, Pan, Jiang, & Tian, 2009), and senescence (Shen et al., 2019). The significance of alternative polyadenylation is also demonstrated in the development and prognosis of various endocrinal, neurological, oncological and immunological diseases (Chang, Yeh, & Yong, 2017). Alternative polyadenylation, by definition, can direct cis-acting micro-RNA (miRNA) mediated gene regulation by altering the length of 3′ untranslated region (Boutet et al., 2012). Recently, studies have also revealed a novel trans-acting gene regulatory dimension involving competing endogenous RNA (Park et al., 2018). Over the past decade, the aggregate effect of many studies indicate that alternative polyadenylation is potentially pervasive and is associated with a large number of human pathologies (Chang et al., 2017). With the current advances in RNA biology, alternative polyadenylation is emerging as a center stage event in post transcriptional regulation (Ren, Zhang, Zhang, Miller, & Pu, 2020).

2.1 Mapping and quantifying alternative polyadenylation

Initially, expressed sequence tags (EST) were used to map polyadenylation sites (Zhang, Lee, & Tian, 2005). Later, low throughput Affymetrix microarrays based on alternative transcript probes and individual oligonucleotide hybridization were also used to detect APA events (Sandberg et al., 2008). Variations of paired-end ditag (PET) approaches were also used to detect global APA changes (Ng et al., 2006). However, it was not until the use of RNA-Seq technology that the analysis of alternative polyadenylation began to undergo transformation. Several studies successfully demonstrated the potential of RNA-Seq data to infer genome-wide polyadenylation changes (Wang & Tian, 2020; Xia et al., 2014). Despite promising studies, the precise mapping and quantification of polyadenylation sites remains challenging because of the intrinsic coverage noise at the 3′ end. Fig. 1A and B illustrate representative 3′UTR coverage of the genes *TARDBP* and *RBM17* in human brain. The 3′UTR coverage of the gene *TARDBP* suggests that one of the isoforms is expressed significantly higher than the other. The step-like expression pattern is highly suggestive of two polyadenylation sites. In contrast, the coverage of *RBM17* is more noisy/heterogeneous making it more challenging to confidently map and quantify polyadenylation sites with precision.

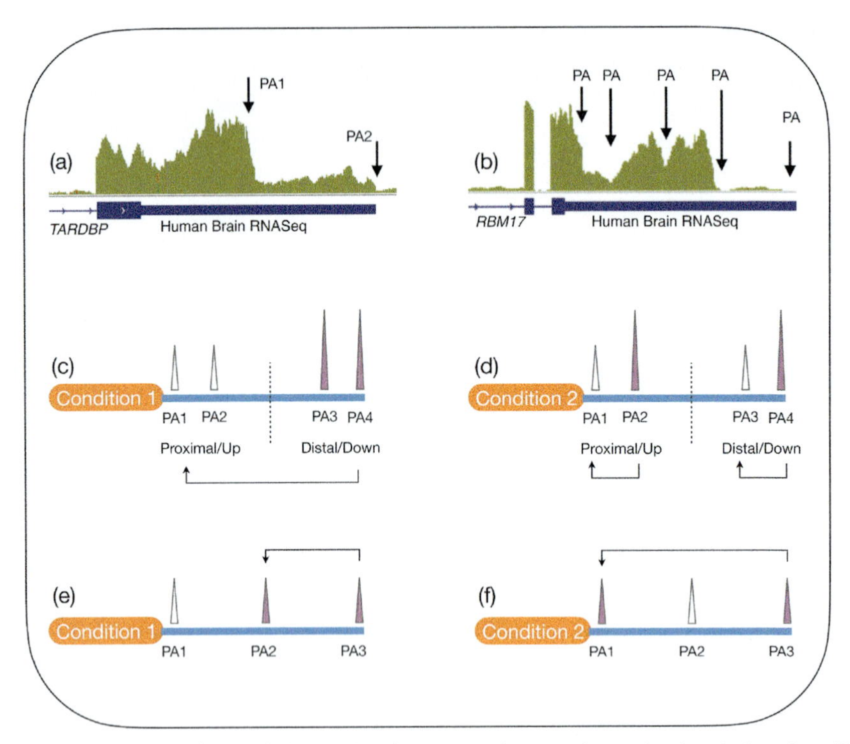

Fig. 1 Illustration of (A and B) RNA-Seq limitations in mapping polyadenylation sites (C and D) limitation of distal to proximal usage approach in APA analysis and (E and F) merit of vector projections in estimating the true effect of 3′UTR shortening.

2.2 Advances in 3′ RNA sequencing techniques

The increasing significance of alternative polyadenylation in human disease context and the limitations of traditional RNA-Seq in accurately identifying polyadenylation sites propelled the development of novel 3′-end RNA sequencing techniques specifically designed to map mRNA cleavage sites. 3′-end sequencing techniques like 3′Seq, polyadenylation sequencing (PA-seq), and poly(A) site sequencing (PAS-seq) use oligo(dT) primer based reverse transcription to capture the 3′-end of mRNA thus allowing the accurate identification of individual polyadenylation sites. However, a majority of 3′-end sequencing techniques suffer from poor base-calling quality and mispriming, where the oligo(dT) primer that is intended to bind the poly(A) tail instead binds a sequence of genomically-encoded adenosines (Chen et al., 2017). Techniques like poly(A)-test RNA-sequencing (PAT-seq) and poly(A)-position profiling (3P-seq) try to minimize mispriming by adding adapters prior to primer annealing, but require complex

RNA manipulation steps and can present challenges for quantification (Derti et al., 2012). Among these various techniques, Poly(A)-ClickSeq (PAC-seq) overcomes common limiting factors like poly(A) enrichment and 3′ linker ligation steps using click-chemistry. PAC-seq was also demonstrated to infer differential gene expression (Elrod, Jaworski, Ji, Wagner, & Routh, 2019). A detailed protocol of PAC-seq is provided in the accompanying chapter "Application and Design Considerations for 3′-End Sequencing using Click-chemistry" (Jensen et al., 2021).

2.3 Computational tools for alternative polyadenylation analysis

With new promising sequencing technologies, complementing novel computational tools are required to realize the true potential of the data. Several computational methods were proposed to analyze conventional RNA-seq data to infer alternative polyadenylation changes. DaPars (Xia et al., 2014) uses a fisher exact test to detect differences in proximal to distal PAS usage. Similarly, QAPA (Ha, Blencowe, & Morris, 2018) uses DEXseq (Smith et al., 2010) and TAPAS (Arefeen, Liu, Xiao, & Jiang, 2018) uses a change point strategy to infer APA changes from regular RNA-seq. However, none of these approaches can specifically identify poly(A) sites and thus will be limited to account for all APA isoforms. Despite the increasing significance of APA and 3′ sequencing techniques, computational tools that are precisely designed for 3′Seq datasets are very limited. Those approaches typically relied on either existing poly-A annotations, ignored novel APA sites, or were limited to proximal and distal polyadenylation sites (Brumbaugh et al., 2018) thus ignoring APA changes involving intermediate poly(A) sites. Enrichment of proximal or distal polyadenylation sites, commonly referred to as proximal to distal usage (PDU) or distal to primal usage (DPU), are computed to infer gene level APA changes (Brumbaugh et al., 2018; Chu et al., 2019). Fig. 1C illustrates a situation where a distal to proximal usage (DPU) would be effective. The usage of the two distal sites tend to decrease (indicated by arrow direction) resulting in overall gene level distal to proximal shift. On the other hand, Fig. 1D illustrates a situation where a DPU approach would fail as the shift in internal polyadenylation sites (indicated by arrow directions) balance out the net distal to proximal shift. With more than 50% of human genes having more than two polyadenylation sites, gene level APA changes are better comprehended by accounting for all polyadenylation sites. However, almost none of the existing approaches abstract all polyadenylation sites in quantifying gene level APA changes. The lack of 3′Seq specific approaches strongly advocate the need for new computational methods to accurately extract the 3′Seq datasets. Here we provide the PolyA-miner protocol, a computational approach that is precisely designed to analyze PAC-seq or other 3′Seq data. With the emerging importance of APA in human diseases, PolyA-miner can significantly accelerate analysis and help decoding underlying APA dynamics.

3. PolyA-miner

PolyA-miner (Yalamanchili et al., 2020) is a novel, de novo differential alternative polyadenylation detection algorithm based on vector projections and iterative non-negative matrix factorization (NMF) (Dhillon & Sra, 2005). In the context of alternative polyadenylation (APA) changes, a gene is abstracted as a matrix where polyadenylation sites are represented as rows and samples as columns. Unlike other approaches, PolyA-miner comprehends all polyadenylation sites including non-proximal and non-distal changes and can distinguish the most distal to most proximal changes from intermediate site changes irrespective of absolute change magnitude. Fig. 1E and F illustrate a situation where the magnitudes of APA change are comparable, but the actual length of shortening is greater in Fig. 1F (PA3–PA1). PolyA-miner uses vector projections to distinguish such changes, and this is essential to estimate the true effect of 3′UTR shortening or lengthening events. The magnitude of alternative polyadenylation change is then computed as the difference of vector projections with respect to a reference point (the most distal or the most proximal polyadenylation site) in an n-dimensional vector space, where n is the number of polyadenylation sites. Use of iterative Consensus NMF clustering makes PolyA-miner less susceptible to intra-sample variation. Statistical significance of the NMF based sample clustering is evaluated as the goodness-of-fit over a null model. Extensive filtering of mispriming and noisy polyadenylation sites makes PolyA-miner an excellent choice to analyze 3′sequencing datasets.

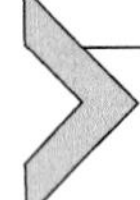

4. Alternative polyadenylation analysis using PolyA-miner

In this chapter we provide an easy-to-follow protocol to analyze PAC-seq data starting from quality control checks, alternative polyadenylation analysis using PolyA-miner and further downstream functional analyses.

For demonstration purposes, here we used a *NUDT21* knock down in a glioblastoma cell line where PAC-seq was used to measure poly(A) site usage. Additional considerations on (1) use of unique molecular identifiers (UMIs) to track PCR duplicates (Marx, 2017) and (2) adequate sequencing depth to reliably map and detect APA changes are also highlighted. Through addressing these considerations, the analysis protocol is also updated from our previous PolyA-miner description (Yalamanchili et al., 2020) to handle unique molecular identifiers. We provide a detailed description of the impact of sequencing depth on the number of detected polyadenylation sites and APA changes. The overall block diagram of the protocol is illustrated in Fig. 2. The protocol consists of five main steps, each step corresponds to a specific phase of the analysis that achieve a certain milestone: (1) Grooming raw reads -UMI marking and trimming, (2) mapping and extracting putative polyadenylation sites, (3) denoising -filtering misprimed and noisy sites, (4) inferring APA changes and (5) downstream data integration and functional analysis.

4.1 Setting up the analysis environment

4.1.1 PolyA-miner is implemented in Python and the source code is freely available at http://www.liuzlab.org/PolyA-miner/. The code will check for the key dependencies and install them. This protocol is conducted with: fastp v0.20.0, bowtie2 v2.3.4.1, samtools v1.7, featureCounts v1.6.0.

Tip: If any of the installations fail, try installing python3.X-dev library.

```
$ apt-get install python3.X-dev.
```

4.1.2 Hardware: A computer or server with access to UNIX/LINUX command environment.

4.1.3 Data processing software: UMI-tools, fastp, bowtie2, samtools, featureCounts.

4.1.4 Python libraries: pandas, cython, pybedtools, scipy, sklearn, statsmodels.

4.1.5 Input files: Raw sequencing reads in fastq format. Bowtie2 reference Index. Annotated polyadenylation sites and gens in bed format. Polyadenylation annotations of model organism organisms are available at PolyA_DB3 (Wang, Nambiar, Zheng, & Tian, 2018).

4.2 Executing PolyA-miner

Sections 4.2.1–4.2.3 describe how to execute PolyA-miner code, list of accepted arguments and an example command for demonstration.

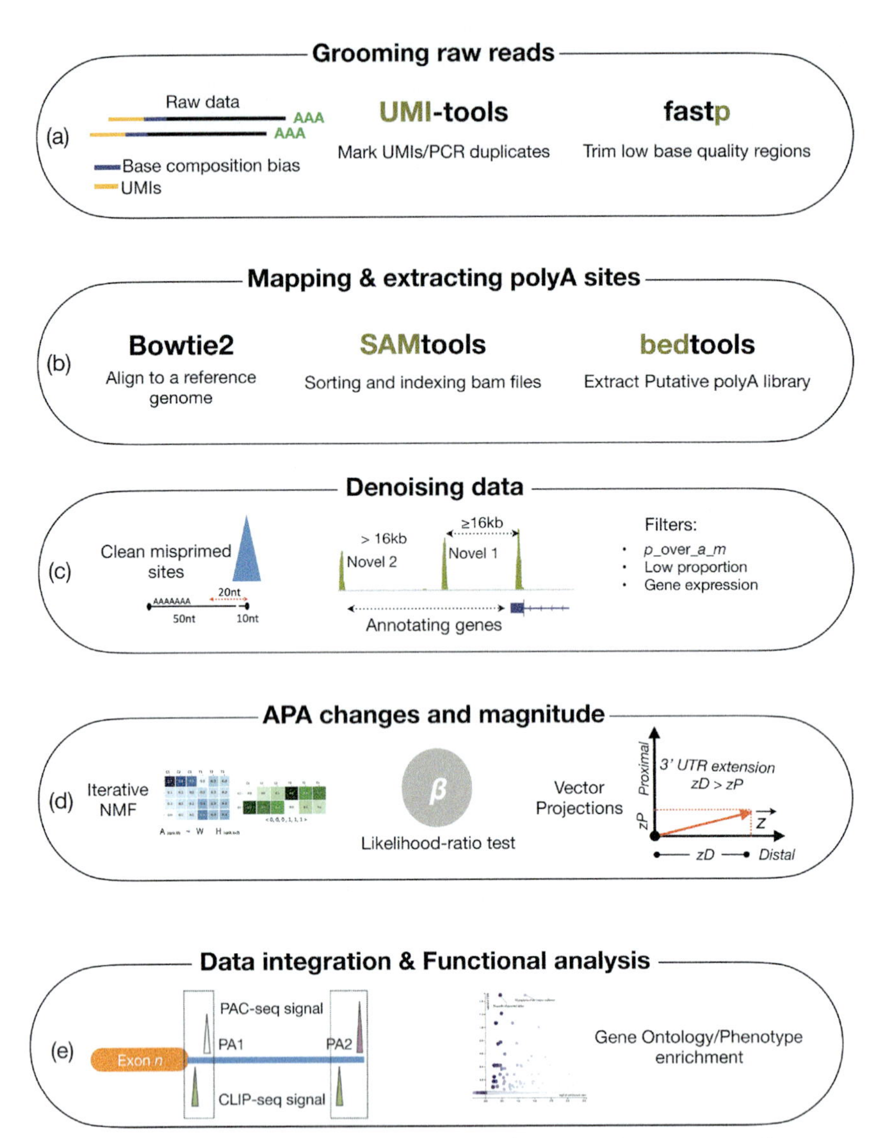

Fig. 2 Block diagram of alternative polyadenylation analysis using PolyA-miner: (A) Grooming reads, (B) mapping and extracting putative polyadenylation sites, (C) denoising data, (D) inferring APA changes and (E) downstream data integration and functional analysis.

4.2.1 Usage

```
$ PolyA-miner.py [-h] -d D -o O -pa PA -index INDEX -fasta FASTA -bed BED
      -c1 C1 [C1 ...] -c2 C2 [C2 ...] [-i I] [-p P]
      [-expNovel] [-ip IP] [-pa_p PA_P] [-pa_a PA_A]
      [-pa_m PA_M] [-apa_min APA_MIN] [-gene_min GENE_MIN]
```

```
[-mdapa MDAPA] [-md MD] [-apaBlock APABLOCK]
[-anchor ANCHOR] [-key KEY]
```

4.2.2 Arguments

```
-h, -help   show this help message and exit
-d D    Base directory of input fastq files
-o O    Output directory
-pa PA    PolyA annotations file
-index INDEX  Reference genome bowtie2 index
-fasta FASTA  Reference fasta sequence
-bed BED   Reference genes bed file
-c1 C1 [C1 ...] Comma-separated list of condition1 fastq files
-c2 C2 [C2 ...] Comma-separated list of condition2 fastq files
-umi UMI   Length of UMI
-i I    No. of NMF iterations
-p P    No. of processors to use
-expNovel   Explore novel APA sites
-ip IP    Internal priming window
-pa_p PA_P   pOverA filter: P
-pa_a PA_A   pOverA filter: A
-pa_m PA_M   pOverA filter: M
-apa_min APA_MIN Min. proportion per APA
-gene_min GENE_MIN Min counts per Gene
-mdapa MDAPA  Cluster distance for annotated polyA sites
-md MD    Cluster distance for de-novo polyA sites
-apaBlock APABLOCK Block size for annotated polyA sites
-key KEY   Output file key
```

4.2.3 Run command

```
$ python3 PolyA-miner/PolyA-miner.py -d Data/ -o PolyA-miner_Analysis/ -pa GRCh38.PolyA_DB.bed -index Bowtie2Index/GRCh38v33 -fasta GRCh38.genome.fa -bed GRCh38v33.Genes.bed -i 50 -p 20 -expNovel -pa_p 0.6 -pa_a 5 -pa_m 3 -ip 30 -key TestRun -c1 Control_S1.fastq.gz,Control_S2.fastq.gz,COntrol_S3.fastq.gz -c2 Knockdown_S1.fastq.gz, Knockdown_S2.fastq.gz, Knockdown_S3.fastq.gz
```

Note: Parameters (*-i, -p -pa_p, -pa_a, -pa_m, -ip*) used above are for demonstration purpose. For faster execution use higher number of processing cores by adjusting the *-p* parameter. Other individual parameters are described in Sections 4.4–4.6. Use appropriate parameters as per specific experimental setup.

4.3 Grooming raw reads

PAC-seq library require additional PCR amplification to offset the effect of click chemistry. However, additional PCR cycle could potentially cause overrepresentation of some fragments (Fu, Wu, Beane, Zamore, & Weng, 2018). To address this bias PAC-seq libraries are updated with unique molecular identifiers (UMIs) to track PCR duplicates (Jensen et al., 2021).

4.3.1 Marking unique molecular identifiers (UMIs)

The PolyA-miner protocol starts with marking PCR duplicates using. UMI-tools (Smith, Heger, & Sudbery, 2017) are used to extract the UMI nucleotides from a read and append it to the read name, which later are used for deduplication. Fig. 3A illustrates a sudden spike in nucleotide content ratio at the 12th nucleotide suggesting possible UMI end. This matches with the UMI barcode length used in generating the data (Jensen et al., 2021). PolyA-miner accepts barcode pattern as the parameter *-umi*. UMI marking is skipped by default.

4.3.2 Trimming low quality nucleotides

Almost all of the next generation sequencing techniques suffer from biased sequence composition at the start of the read due to random hexamer priming while amplifying cDNA (Hansen, Brenner, & Dudoit, 2010). To improve mappability, PolyA-miner trims any adapter contamination and the first four nucleotides using fastp (Chen, Zhou, Chen, & Gu, 2018). Fig. 3B illustrates base composition noise in first 4 nucleotide of the representative data. Reads less than 40 bp are also filtered out to further minimize any ambiguous alignments.

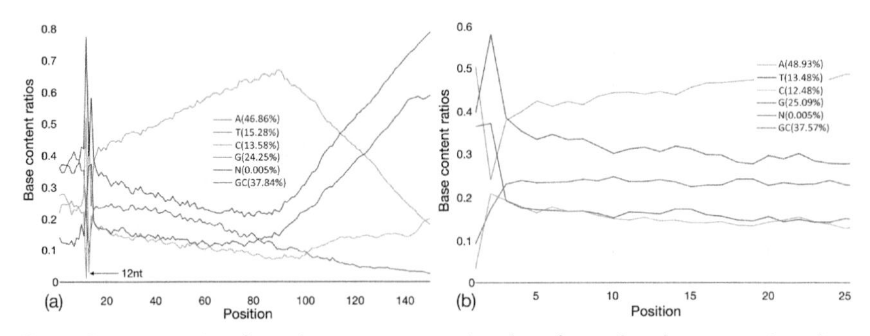

Fig. 3 Representative fastp base content ratio plots from the demonstration data. (A) Raw reads: a sudden spike in nucleotide content pattern at 12th nucleotide hints possible UMI end. (B) UMI marked reads: base composition is relatively noisy for the first four nucleotides.

4.4 Mapping and extracting putative polyadenylation sites

PolyA-miner orchestrate multiple computational tools to map and extract putative polyadenylation sites. PolyA-miner expects reference genome index path as *-index* parameter.

4.4.1 Mapping

Groomed reads are mapped to a reference genome using bowtie2 (Langmead & Salzberg, 2012). Human GRCh38 genome index is used for demonstration here. If the reference index is not available, *bowtie2-build* can be used to builds a Bowtie index from fasta sequence. Bowtie2 outputs Sequence alignment map (SAM) files from *bowtie2* are converted, coordinate sorted and indexed to binary alignment maps (BAMs) using *samtools* (Li et al., 2009). Mapped reads are deduplicated based on the UMI marked read names and the mapping co-ordinates.

4.4.2 Extracting putative polyadenylation sites

PolyA-miner can be executed in both *de-novo* and annotation-based modes. Parameter *-expNovel* will enable *de-novo* mode. By default, PolyA-miner runs in annotation-based mode. In either mode, PolyA-miner requires annotated polyadenylation sites and gens in bed format as *-pa* and *-bed* parameters. Aligned reads are summarized using Bedtools *genomecov* module. Every contiguous non-zero coverage nucleotide positions are extracted as putative polyadenylation sites. All sample-wise feature files are pooled to compute a comprehensive library of polyadenylation site. Any overlapping or sites that fall within a minimal distance *md* (PolyA-miner parameter) as illustrated in Fig. 4A. Parameter *md* should be set based as per the resolution of the sequencing protocol, by default it is 15 nt.

4.5 Denoising -filtering misprimed and noisy polyadenylation sites

Most of the 3′ sequencing protocols, including PAC-seq, use oligo(dT) primers to capture mRNA tails. These oligo(dT) primers potentially induce mispriming, a phenomenon where the primers bind to stretches of adenosines within the body of the mRNA resulting in sequencing reads that do not align with true cleavage sites (Nam et al., 2002). In addition, because our understanding of alternative polyadenylation is not as extensive as gene expression or alternative splicing, annotated APA sites in databases may not be exhaustive. Thus, it may be any computational approach must be capable

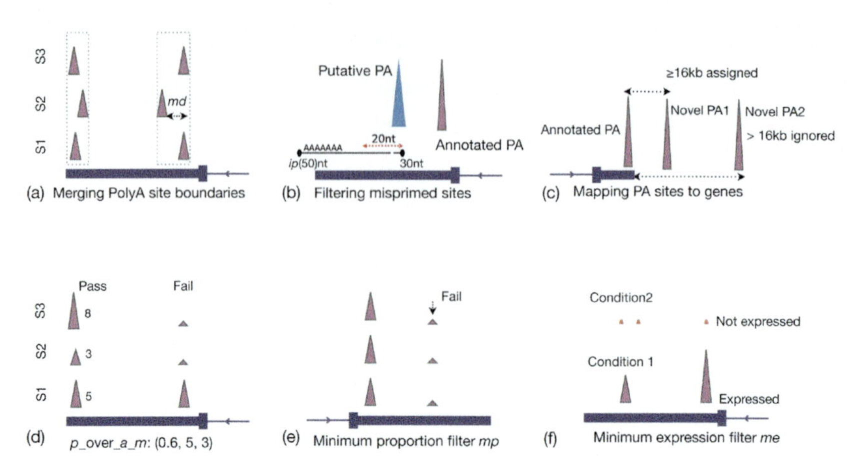

Fig. 4 Denoising putative polyadenylation sites. (A) Merging adjacent polyadenylation features. (B) Filtering misprimed sites. (C) Annotating novel polyadenylation sites. (D) *p_over_a_m* filter. (E) Minimum proportion (*mp*) filter. (F) Minimum expression (*me*) filter.

of identifying novel polyadenylation sites. One drawback of exploring novel sites is that mispriming can overestimate the number of bona fide polyadenylation sites. Underscoring this concern, PolyA-miner previously revealed massive scale of mispriming and noisy polyadenylation sites in 3′sequencing data sets using several heuristics (Yalamanchili et al., 2020).

4.5.1 Filtering misprimed sites

Misprimed sites can be filtered by exploring the downstream base composition (Lee, Yeh, Park, & Tian, 2007). PolyA-miner adopts a conservative if not strict approach in handling mispriming: each putative poly(A) site is extended by a mispriming distance *mpd* toward the 3′ end and scanned for a genomic PolyA feature. Any poly(A) site with more than 65% of adenosines in a sliding window of 20 bp are flagged and filtered out. Fig. 4B illustrates PolyA-miner sliding window approach. Regardless of the percentage of adjacent adenosines, poly(A) sites within *apaBlock* nt from annotated cleavage sites are retained.

4.5.2 Mapping novel polyadenylation sites to genes

Polyadenylation sites retained after filtering misprimed sites are mapped to respective genes. However, often novel polyadenylation sites fall beyond the annotated gene boundaries. PolyA-miner maps novel sites to genes if they are within 16 kb, the longest known 3′UTR length (Miura, Shenker, Andreu-Agullo, Westholm, & Lai, 2013), from respective transcriptional end site (TES) and do not overlap with any other gene (illustrated in

Fig. 4C). After mapping poly(A) sites to genes, sample-wise polyadenylation site counts are computed using featureCounts (Liao, Smyth, & Shi, 2013). Each gene is abstracted as a matrix with APA sites as rows and sample as columns.

4.5.3 Filtering noisy polyadenylation sites

To minimize the untoward effect of sequencing noise in *de-novo* mode PolyA-miner adopts a *p-over-a_m* function to filter noisy polyadenylation sites. This function guarantees that at least "*p*" proportion of samples are with "*a*" number of reads or more with an overall minimum of "*m*" reads. In other words, if *p, a,* and *m* are set to 0.6, 5 and 3 respectively, polyadenylation sites with less than 5 reads in at least 60% of the samples with an overall minimum less than 3 are filtered out (illustrated in Fig. 4D). PolyA-miner accepts parameters *p, a,* and *m* as-*pa_p, -pa_a, -pa_m*. Polyadenylation sites that fall short of a minimum proportion -*mp* of total reads mapped to the respective gene in both the conditions are also pruned (Fig. 4E). Low or non-expressed genes could constrain APA changes. Thus, genes with less than a minimum expression count -*me* in either of the conditions are filtered out (Fig. 4F).

Tip: Higher *p-over-a_m* parameters are recommended for highly consistent hits. On the other hand, if the lower *p-over-a_m* parameters are recommended if intra sample heterogeneity is high.

4.6 Inferring alternative polyadenylation changes

The number of iterations in the consensus NMF module of PolyA-miner could be adjusted by the parameter -*i*. Iterations should increase with the number of samples and heterogeneity, as it could take several iterations to converge. By default, the number of iterations is set to 100. The default value should suffice for most of the datasets. The proximal index of a gene is computed as the $\log_2$ ratio of proximal to distal projections. Detailed description of iterative NMF model and vector projections is given in (Yalamanchili et al., 2020). Higher proximal index suggests a dominant short 3′UTR and vice versa. Fig. 5A illustrates a strong global 3′UTR shortening in *NUDT21* knock down (KD) samples. Proximal indices of *NUDT21* KD samples are higher than controls. This observation is in good agreement with several other *NUDT21* KD studies (Alcott et al., 2020; Chu et al., 2019). PolyA-miner identified a total of 4858 APA changes (Adjusted *P* value ≤ 0.05) from the current *NUDT21* KD demonstration data, out of which 4176 are 3′UTR shortening and 682 are 3′UTR lengthening events. Fig. 5B and C illustrate 3′UTR shortening in *LAMC1* and lengthening in *FOXD1* genes, respectively. Other well established *NUDT21* targets like *VMA21* and *PAK1* are also found to have significant 3′UTR shortening.

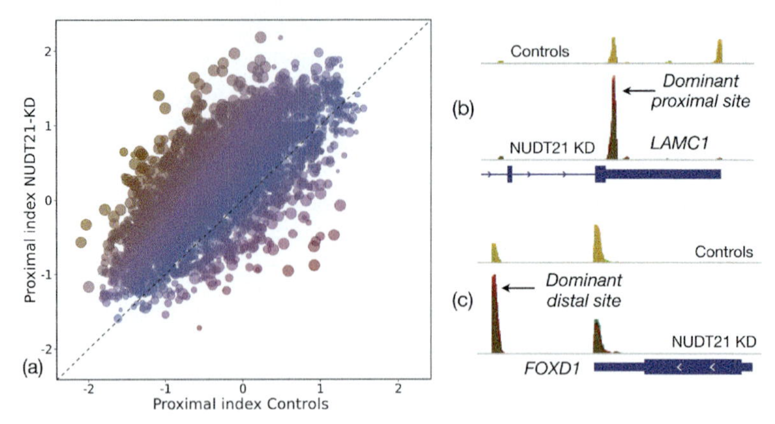

Fig. 5 (A) Overall 3′UTR shrinking effect. Representative examples of (B) 3′UTR shortening and (C) 3′UTR elongation.

4.7 PolyA-miner downstream analysis

Readouts from high throughput sequencing technologies like PAC-seq needs to be validated to make robust biological conclusions. However, it would be extremely challenging to experimentally validate each individual APA change. On the other hand, functional enrichment of high throughput assays plays a significant role in understanding the molecular dynamics of the underlying biological phenomenon (Huang et al., 2007). In this subsection we describe commonly used approaches to substantiate the APA changes inferred using PolyA-miner and infer biological insights.

4.7.1 Integration with CLIP-seq data

RNA processing readouts like alternative splicing and alternative polyadenylation can be substantiated by matching CLIP-seq (Stork & Zheng, 2016) signal near the splice sites or cleavage sites (Zhu et al., 2018), respectively. For demonstration purpose here we used publicly available *NUDT21* CLIP-seq data in GEO database (GSM917661). Processing of CLIP-seq data is described elsewhere (Uhl, Houwaart, Corrado, Wright, & Backofen, 2017). Fig. 6A and B illustrate 3′UTR shrinking in *PAK1* and *VMA21*, respectively. Corresponding *NUDT21* binding (CLIP-seq signal) in the proximity of the polyadenylation sites are marked with arrow heads. CLIP-seq signal substantiate the 3′UTR shrinking of *PAK1* and *VMA21* as the direct consequence of *NUDT21* KD.

4.7.2 Functional enrichment analysis

While there are several functional enrichment tools, in this protocol we use WebGestalt (Yuxing Liao, Wang, Jaehnig, Shi, & Zhang, 2019) a

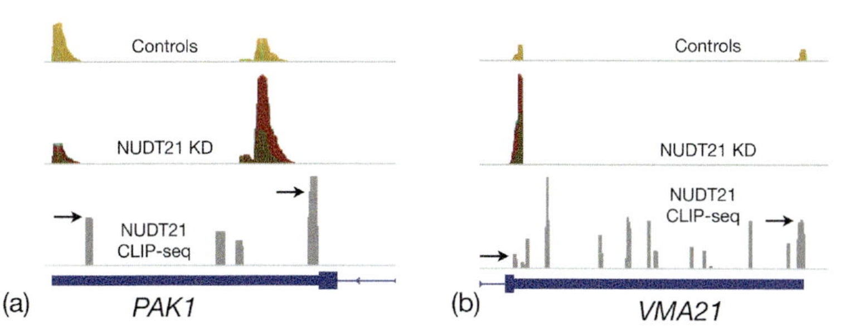

Fig. 6 Integration with CLIP-seq data: CLIP signal overlapping with PAC-seq signal for (A) PAK1 and (B) VMA21.

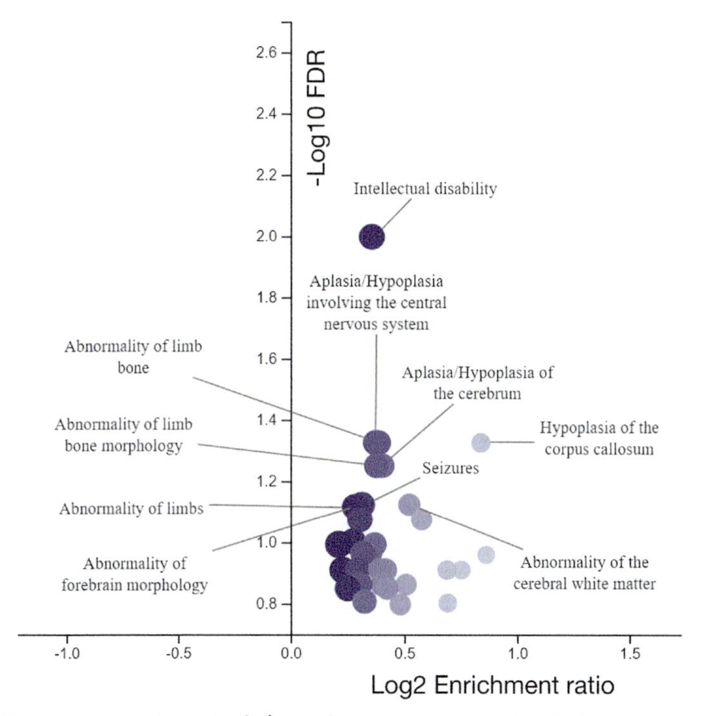

Fig. 7 Phenotype enrichment of 3′UTR shortening gene using WebGestalt.

WEB-based GEne SeT AnaLysis Toolkit to perform the function enrichment of the APA gene. It is completely online and can intuitively help biologists in exploring large sets of genes with almost no coding necessity. Genes with 3′ UTR shortening are enriched for intellectual disability and other developmental phenotypes (Fig. 7). These enrichments agree with the recent advances in *NUDT21* biology (Gennarino et al., 2015).

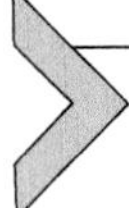

5. Impact of sequencing depth on the number of APA changes detected

Though our understanding of PAC-seq and other 3′ sequencing techniques is advanced the total number of sequencing reads required to reliably capture the bulk of the APA changes is still an open question. Several studies investigated the impact of RNA-seq depth on gene and isoform quantifications (Łabaj et al., 2011). However, we cannot extrapolate any conclusions to 3′ sequencing data. Understanding the relationships between sequencing depth and number of detectable polyadenylation sites is extremely critical for better experimental design and interpreting results. Our chapter on PAC-seq (Jensen et al., 2021) highlights the consideration for sequencing depth. Despite its importance, very little is known on the impact of sequencing depth on the number of detected APA changes. In this section present a saturation experiment to evaluate the impact of sequencing depth on the number of APA changes detected.

Multiple datasets with varying sequencing depth were derived from the *NUDT21* KD data by random sampling without replacement using seqtk (Li, 2012). A total of 8 datasets ranging from 2.5 million to 35 million reads were compiled. APA changes were inferred from each dataset using the PolyA-miner protocol with consistent parameters as shown in Section 4.2.3. Fig. 8A and B illustrate the percentage of detected APA changes and polyadenylation sites across various sequencing depths, respectively.

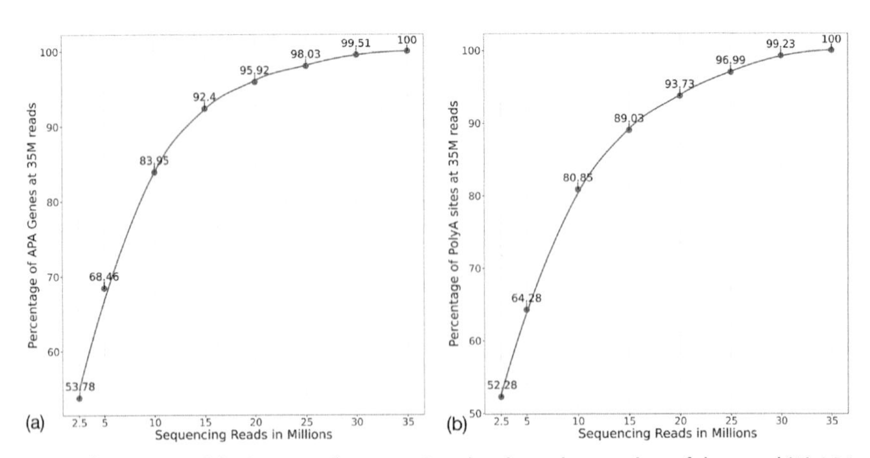

Fig. 8 Illustration of the impact of sequencing depth on the number of detected (A) APA genes and (B) polyadenylation sites.

Only ~54% of APA changes are detected at 2.5M reads when compared to 35M reads. Maximum gain in the number of APA changes is observed at 10M reads and the curve starts to flatten after 15M reads with ~92% (Fig. 8A). A similar trend is also observed for the number of detected poly-adenylation sites (Fig. 8B). With more than 95% of both APA changes and polyadenylation sites identified at 20M reads, we suggest this to be an optimal depth for sequencing even using our conservative mispriming filter.

6. Summary

Researchers investigating alternative polyadenylation using PAC-seq or other 3′ sequencing protocols encounter several challenges in analyzing the data. In this chapter we provide an easy-to-follow PolyA-miner protocol to analyze PAC-seq data. PolyA-miner pipeline offers a single window computational solution to biologists with minimal human intervention. It orchestrates multiple analytical steps from deduplication to read alignment, and denoising to statistical modeling. Ability to comprehend non-proximal to non-distal APA dynamics, extensive filtering of mispriming and noisy polyadenylation sites, and less susceptibility to intra sample variation makes PolyA-miner an excellent choice to analyze 3′sequencing datasets. Though PolyA-miner can identify novel APA sites an instant future extension could be to categorize APA events into tandem, splicing and intronic poly-adenylation events. With the increasing research focus on alternative poly-adenylation, PolyA-miner can significantly accelerate data analysis and help decode alternative polyadenylation dynamics.

Acknowledgments

We thank the members of the Liu and Wagner labs for helpful discussions and suggestions. This work is supported by Cancer Prevention Research Institute of Texas [RP170387 to Z.L.]; Houston endowment, Chao endowment, Huffington foundation (to Z.L.); Funded in part with federal funds from the United States Department of Agriculture (USDA/ARS) under Cooperative Agreement No. 58-3092-0-001 and NRI Zoghbi Scholar Award (to H.K.Y.); NIH, National Institute of General Medical Sciences [R01-GM134539 to E.J.W.]; NIH, National Cancer Institute [R03-CA223893-01 to P.J.].

References

Alcott, C. E., Yalamanchili, H. K., Ji, P., van der Heijden, M. E., Saltzman, A., Elrod, N., et al. (2020). Partial loss of CFIM25 causes learning deficits and aberrant neuronal alternative polyadenylation. *eLife*, *9*, e50895.

Arefeen, A., Liu, J., Xiao, X., & Jiang, T. (2018). TAPAS: Tool for alternative poly-adenylation site analysis. *Bioinformatics*, *34*(15), 2521–2529.

Boutet, S. C., Cheung, T. H., Quach, N. L., Liu, L., Prescott, S. L., Edalati, A., et al. (2012). Alternative polyadenylation mediates microRNA regulation of muscle stem cell function. *Cell Stem Cell, 10*(3), 327–336.

Brumbaugh, J., Di Stefano, B., Wang, X., Borkent, M., Forouzmand, E., Clowers, K. J., et al. (2018). Nudt21 controls cell fate by connecting alternative polyadenylation to chromatin signaling. *Cell, 172*(1–2), 106–120.e21.

Chang, J. W., Yeh, H. S., & Yong, J. (2017). Alternative polyadenylation in human diseases. *Endocrinology and Metabolism, 32*(4), 413–421.

Chen, W., Jia, Q., Song, Y., Fu, H., Wei, G., & Ni, T. (2017). Alternative polyadenylation: Methods, findings, and impacts methods and findings of alternative polyadenylation. *Genomics, Proteomics & Bioinformatics, 15*(5), 287–300.

Chen, S., Zhou, Y., Chen, Y., & Gu, J. (2018). Fastp: An ultra-fast all-in-one FASTQ preprocessor. *Bioinformatics, 34*(17), i884–i890.

Chu, Y., Elrod, N., Wang, C., Li, L., Chen, T., Routh, A., et al. (2019). Nudt21 regulates the alternative polyadenylation of Pak1 and is predictive in the prognosis of glioblastoma patients. *Oncogene, 38*(21), 4154–4168.

Dargahi, D., Swayze, R. D., Yee, L., Bergqvist, P. J., Hedberg, B. J., Heravi-Moussavi, A., et al. (2014). A pan-cancer analysis of alternative splicing events reveals novel tumor-associated splice variants of matriptase. *Cancer Informatics, 13*, 167–177.

Derti, A., Garrett-Engele, P., MacIsaac, K. D., Stevens, R. C., Sriram, S., Chen, R., et al. (2012). A quantitative atlas of polyadenylation in five mammals. *Genome Research, 22*(6), 1173–1183.

Dhillon, I. S., & Sra, S. (2005). Generalized nonnegative matrix approximations with Bregman divergences. In *Advances in neural information processing systems* (pp. 283–290).

Dredge, B. K., Polydorides, A. D., & Darnell, R. B. (2001). The splice of life: Alternative splicing and neurological disease. *Nature Reviews Neuroscience, 2*(1), 43–50.

Elrod, N. D., Jaworski, E. A., Ji, P., Wagner, E. J., & Routh, A. (2019). Development of Poly(A)-ClickSeq as a tool enabling simultaneous genome-wide poly(A)-site identification and differential expression analysis. *Methods, 155*(15 February 2019), 20–29.

Erson-Bensan, A. E. (2016). Alternative polyadenylation and RNA-binding proteins. *Journal of Molecular Endocrinology, 57*(2), F29–F34.

Erson-Bensan, A. E., & Can, T. (2016). Alternative polyadenylation: Another foe in cancer. *Molecular Cancer Research, 14*(6), 507–517.

Fu, Y., Wu, P. H., Beane, T., Zamore, P. D., & Weng, Z. (2018). Elimination of PCR duplicates in RNA-seq and small RNA-seq using unique molecular identifiers. *BMC Genomics, 19*(1), 531.

Gennarino, V. A., Alcott, C. E., Chen, C. A., Chaudhury, A., Gillentine, M. A., Rosenfeld, J. A., et al. (2015). NUDT21-spanning CNVs lead to neuropsychiatric disease and altered MeCP2 abundance via alternative polyadenylation. *eLife, 4*(August 2015), e10782.

Ha, K. C. H., Blencowe, B. J., & Morris, Q. (2018). QAPA: A new method for the systematic analysis of alternative polyadenylation from RNA-seq data. *Genome Biology, 19*(1), 45.

Hansen, K. D., Brenner, S. E., & Dudoit, S. (2010). Biases in Illumina transcriptome sequencing caused by random hexamer priming. *Nucleic Acids Research, 38*(12), e131.

Hsieh, Y. C., Guo, C., Yalamanchili, H. K., Abreha, M., Al-Ouran, R., Li, Y., et al. (2019). Tau-mediated disruption of the spliceosome triggers cryptic RNA splicing and neurodegeneration in Alzheimer's disease. *Cell Reports, 29*(2), 301–316.e10.

Huang, D. W., Sherman, B. T., Tan, Q., Kir, J., Liu, D., Bryant, D., et al. (2007). DAVID Bioinformatics Resources: Expanded annotation database and novel algorithms to better extract biology from large gene lists. *Nucleic Acids Research, 35*(Suppl. 2), W169–W175.

Jensen, M. K., Elrod, N. D., Yalamanchili, H. K., Ji, P., Lin, A., Liu, Z., et al. (2021). Application and design considerations for 3′-end sequencing using click-chemistry. *Methods in Enzymology, 655*, 1–23.

Ji, Z., Lee, J. Y., Pan, Z., Jiang, B., & Tian, B. (2009). Progressive lengthening of 3′ untranslated regions of mRNAs by alternative polyadenylation during mouse embryonic development. *Proceedings of the National Academy of Sciences of the United States of America*, *106*(17), 7028–7033.

Łabaj, P. P., Leparc, G. G., Linggi, B. E., Markillie, L. M., Wiley, H. S., & Kreil, D. P. (2011). Characterization and improvement of RNA-seq precision in quantitative transcript expression profiling. *Bioinformatics*, *27*(13), i383–i391.

Langmead, B., & Salzberg, S. L. (2012). Fast gapped-read alignment with Bowtie 2. *Nature Methods*, *9*(4), 357–359.

Lee, J. Y., Yeh, I., Park, J. Y., & Tian, B. (2007). PolyA_DB 2: mRNA polyadenylation sites in vertebrate genes. *Nucleic Acids Research*, *35*(Suppl. 1), D165–D168.

Li, H. (2012). *seqtk Toolkit for processing sequences in FASTA/Q formats*. GitHub.

Li, H., Handsaker, B., Wysoker, A., Fennell, T., Ruan, J., Homer, N., et al. (2009). The sequence alignment/map format and SAMtools. *Bioinformatics*, *25*(16), 2078–2079.

Liao, Y., Smyth, G. K., & Shi, W. (2013). The Subread aligner: Fast, accurate and scalable read mapping by seed-and-vote. *Nucleic Acids Research*, *41*(10), e108.

Liao, Y., Wang, J., Jaehnig, E. J., Shi, Z., & Zhang, B. (2019). WebGestalt 2019: Gene set analysis toolkit with revamped UIs and APIs. *Nucleic Acids Research*, *47*(W1), W199–W205.

Ling, J. P., Pletnikova, O., Troncoso, J. C., & Wong, P. C. (2015). TDP-43 repression of nonconserved cryptic exons is compromised in ALS-FTD. *Science*, *349*(6248), 650–655.

Luco, R. F., Allo, M., Schor, I. E., Kornblihtt, A. R., & Misteli, T. (2011). Epigenetics in alternative pre-mRNA splicing. *Cell*, *144*(1), 16–26.

Marx, V. (2017). How to deduplicate PCR. *Nature Methods*, *14*(5), 473–476.

Miura, P., Shenker, S., Andreu-Agullo, C., Westholm, J. O., & Lai, E. C. (2013). Widespread and extensive lengthening of 39 UTRs in the mammalian brain. *Genome Research*, *23*(5), 812–825.

Nam, D. K., Lee, S., Zhou, G., Cao, X., Wang, C., Clark, T., et al. (2002). Oligo(dT) primer generates a high frequency of truncated cDNAs through internal poly(A) priming during reverse transcription. *Proceedings of the National Academy of Sciences of the United States of America*, *99*, 6152–6156.

Ng, P., Tan, J. J. S., Ooi, H. S., Lee, Y. L., Chiu, K. P., Fullwood, M. J., et al. (2006). Multiplex sequencing of paired-end ditags (MS-PET): A strategy for the ultra-high-throughput analysis of transcriptomes and genomes. *Nucleic Acids Research*, *34*(12), e84.

Park, H. J., Ji, P., Kim, S., Xia, Z., Rodriguez, B., Li, L., et al. (2018). 3′ UTR shortening represses tumor-suppressor genes in trans by disrupting ceRNA crosstalk. *Nature Genetics*, *50*(6), 783–789.

Ren, F., Zhang, N., Zhang, L., Miller, E., & Pu, J. J. (2020). Alternative polyadenylation: A new frontier in post transcriptional regulation. *Biomarker Research*, *8*(1), 67.

Sandberg, R., Neilson, J. R., Sarma, A., Sharp, P. A., & Burge, C. B. (2008). Proliferating cells express mRNAs with shortened 3′ untranslated regions and fewer microRNA target sites. *Science*, *320*(5883), 1643–1647.

Shen, T., Li, H., Song, Y., Li, L., Lin, J., Wei, G., et al. (2019). Alternative polyadenylation dependent function of splicing factor SRSF3 contributes to cellular senescence. *Aging*, *11*(5), 1356–1388.

Smith, J. J., Deane, N. G., Wu, F., Merchant, N. B., Zhang, B., Jiang, A., et al. (2010). Experimentally derived metastasis gene expression profile predicts recurrence and death in patients with colon cancer. *Gastroenterology*, *138*(3), 958–968.

Smith, T., Heger, A., & Sudbery, I. (2017). UMI-tools: Modeling sequencing errors in Unique Molecular Identifiers to improve quantification accuracy. *Genome Research*, *27*(3), 491–499.

Stork, C., & Zheng, S. (2016). Genome-wide profiling of RNA–protein interactions using CLIP-Seq. *Methods in Molecular Biology*, *1421*, 137–151.

Tan, Q., Yalamanchili, H. K., Park, J., De Maio, A., Lu, H. C., Wan, Y. W., et al. (2016). Extensive cryptic splicing upon loss of RBM17 and TDP43 in neurodegeneration models. *Human Molecular Genetics*, *25*(23), 5083–5093.

Tushev, G., Glock, C., Heumüller, M., Biever, A., Jovanovic, M., & Schuman, E. M. (2018). Alternative 3′ UTRs modify the localization, regulatory potential, stability, and plasticity of mRNAs in neuronal compartments. *Neuron*, *98*(3), 495–511.e6.

Uhl, M., Houwaart, T., Corrado, G., Wright, P. R., & Backofen, R. (2017). Computational analysis of CLIP-seq data. *Methods*, *118–119*, 60–72.

Wang, R., Nambiar, R., Zheng, D., & Tian, B. (2018). PolyA-DB 3 catalogs cleavage and polyadenylation sites identified by deep sequencing in multiple genomes. *Nucleic Acids Research*, *46*(D1), D315–D319.

Wang, R., & Tian, B. (2020). APAlyzer: A bioinformatics package for analysis of alternative polyadenylation isoforms. *Bioinformatics*, *36*(12), 3907–3909.

Xia, Z., Donehower, L. A., Cooper, T. A., Neilson, J. R., Wheeler, D. A., Wagner, E. J., et al. (2014). Dynamic analyses of alternative polyadenylation from RNA-seq reveal a 3′2-UTR landscape across seven tumour types. *Nature Communications*, *5*, 5274.

Yalamanchili, H. K., Alcott, C. E., Ji, P., Wagner, E. J., Zoghbi, H. Y., & Liu, Z. (2020). PolyA-miner: Accurate assessment of differential alternative poly-adenylation from 3′Seq data using vector projections and non-negative matrix factorization. *Nucleic Acids Research*, *48*(12), e69.

Zhang, H., Lee, J. Y., & Tian, B. (2005). Biased alternative polyadenylation in human tissues. *Genome Biology*, *6*(12), R100.

Zhu, Y., Wang, X., Forouzmand, E., Jeong, J., Qiao, F., Sowd, G. A., et al. (2018). Molecular mechanisms for CFIm-mediated regulation of mRNA alternative poly-adenylation. *Molecular Cell*, *69*(1), 62–74.e4.

CHAPTER TEN

Systematic refinement of gene annotations by parsing mRNA 3′ end sequencing datasets

Pooja Bhat[a,b], Thomas R. Burkard[a], Veronika A. Herzog[a], Andrea Pauli[c], and Stefan L. Ameres[a,d,*]

[a]Institute of Molecular Biotechnology (IMBA), Vienna BioCenter (VBC), Vienna, Austria
[b]Vienna BioCenter PhD Program, Doctoral School of the University at Vienna and Medical University of Vienna, Vienna, Austria
[c]Research Institute of Molecular Pathology (IMP), Vienna BioCenter (VBC), Vienna, Austria
[d]Max Perutz Labs, University of Vienna, Vienna BioCenter (VBC), Vienna, Austria
*Corresponding author: e-mail address: stefan.ameres@univie.ac.at

Contents

Abstract

Alternative cleavage and polyadenylation generates mRNA 3′ isoforms in a cell type-specific manner. Due to finite available RNA sequencing data of organisms with vast cell type complexity, currently available gene annotation resources are incomplete, which poses significant challenges to the comprehensive interpretation and quantification of transcriptomes. In this chapter, we introduce 3′GAmES, a stand-alone computational pipeline for the identification and quantification of novel mRNA 3′end isoforms from 3′mRNA sequencing data. 3′GAmES expands available repositories and improves comprehensive gene-tag counting by cost-effective 3′ mRNA sequencing, faithfully mirroring whole-transcriptome RNAseq measurements. By employing R and bash shell scripts (assembled in a Singularity container) 3′GAmES systematically augments cell type-specific 3′ ends of RNA polymerase II transcripts and increases the sensitivity of quantitative gene expression profiling by 3′ mRNA sequencing. Public access: https://github.com/AmeresLab/3-GAmES.git.

Methods in Enzymology, Volume 655
ISSN 0076-6879
https://doi.org/10.1016/bs.mie.2021.03.016

1. Introduction

The biogenesis of eukaryotic messenger RNA (mRNA) comprises multiple processing steps that diversify the fate and function of transcripts in eukaryotes (Moore & Proudfoot, 2009). Among these, 3′ end cleavage and polyadenylation of RNA polymerase II products controls transcription termination and export from the nucleus, as well as cytoplasmic stability, localization and translation of mRNAs (Di Giammartino, Nishida, & Manley, 2011; Tian & Manley, 2017). Alterations in the sites of mRNA 3′ end processing represent a major source for transcript isoform heterogeneity (Tian, Hu, Zhang, & Lutz, 2005; Tian & Manley, 2017): In mammals, at least 70% of genes produce multiple mRNA 3′ isoforms that frequently exhibit tissue- and cell type-specific expression patterns (Derti et al., 2012; Hoque et al., 2013). Generally, poly(A) sites proximal to coding sequences prevail in proliferating cells and distal sites are used in resting cells (Sandberg, Neilson, Sarma, Sharp, & Burge, 2008). Because the 3′ untranslated regions (UTRs) of protein coding mRNAs act as hubs for post-transcriptional gene regulation (Mayr, 2016a), alternative polyadenylation acts as a coordinated mechanism for altering the expression of many genes in a wide variety of biological settings, including T cell activation, embryonic development, or neuronal activation (Flavell et al., 2008; Ji, Lee, Pan, Jiang, & Tian, 2009; Sandberg et al., 2008). Given its overall scope and biological impact, it is not surprising that sequence determinants and processing factors of mRNA 3′ end formation have emerged as common targets for mutations in a variety of human pathologies, such as β-thalassemia, thrombophilia and cancer (Danckwardt, Hentze, & Kulozik, 2008).

With the exception of some histone mRNAs, the formation of eukaryotic mRNA 3′ ends occurs in a tightly coupled two-step process that involves specific sequence elements at the 3′ end of precursor mRNAs (pre-mRNAs) that engage a large multi-protein complex that can constitute more than 80 proteins in humans (Shi et al., 2009). Such sequence elements include a hexameric polyadenylation signal (PAS) "A[A/U]UAAA" that is recognized by the multimeric cleavage and polyadenylation specificity factor (CPSF) and determines the site of cleavage preferentially 3′ of CA dinucleotides at a distance of 10–30 nt downstream of the PAS; and a G/U or U-rich downstream sequence element (DSE) located ~30 nt downstream of the cleavage site that is bound by a heterotrimeric cleavage-stimulating factor

(CstF) to enhance the efficiency of 3′ end processing. Further sequence motifs can contribute in an accessory or compensatory manner to mRNA 3′ end formation—including a U-rich upstream sequence element (USE) or UGUA motif(s) upstream of the PAS—by recruiting or anchoring processing factors (Tian & Graber, 2012). After endonucleolytic cleavage catalyzed by CPSF73, a nuclear poly(A) polymerase (PAP) adds ~250 adenine residues to the 3′ end of transcripts in a template-independent manner (Eckmann, Rammelt, & Wahle, 2011). Notably, the motifs found at individual mRNA 3′ ends exhibit a wide range of sequence variability and spatial organization (Tian & Graber, 2012): More than 14% of expressed sequence tag (EST)-supported human mRNA 3′ ends are associated with degenerate PAS motifs that frequently don't support efficient processing *in vitro*; and a minimal functional PAS can consist of only an A-rich upstream sequence and strong U-rich downstream elements (Beaudoing, Freier, Wyatt, Claverie, & Gautheret, 2000). Furthermore, the frequent presence of multiple PAS per gene combined with regulated expression of factors involved in alternative polyadenylation (APA), splicing, transcription and nucleosomes modulate the choice of PAS usage depending on biological context (Berg et al., 2012; Grosso, de Almeida, Braga, & Carmo-Fonseca, 2012; Tian & Manley, 2017). As a consequence, the *de novo* prediction of mRNA 3′ ends from genomic sequences remains unreliable and requires experimental approaches to account for their cell type and tissue specific emergence.

With the advent of next generation sequencing and the development of dedicated protocols for cDNA library preparation, the qualitative and quantitative assessment of transcriptomes has significantly expanded our view of gene expression in diverse biological settings (Stark, Grzelak, & Hadfield, 2019). However, because of relatively low read coverage at transcript termini, most RNA sequencing approaches fail to precisely inform on poly(A) site usage. As a consequence, several targeted RNA sequencing methods have specifically addressed mRNA 3′ ends by taking advantage of the poly(A) tail, producing an expanding atlas of poly(A) sites in a growing number of organisms (Chen et al., 2017; Herrmann et al., 2020). Among such targeted sequencing approaches, 3′ mRNA sequencing (e.g., QuantSeq) grew beyond a qualitative mRNA 3′ end discovery method and is now widely used as a rapid and cost-effective protocol to quantify gene expression by simple gene-tag counting (Beck et al., 2010; Herzog et al., 2017; Moll, Ante, Seitz, & Reda, 2014; Xiao et al., 2016). But its applicability for comprehensive and quantitative transcriptomics is hampered due to incomplete annotation of tissue- and cell type-specific transcript 3′ ends in available repositories, demanding a computational strategy for the re-evaluation of poly(A) sites in 3′ end sequencing datasets.

In this chapter we discuss the implementation of 3′ GAmES (3′ gene annotation by mRNA end sequencing), a stand-alone pipeline for the systematic refinement of gene annotations by parsing 3′ mRNA sequencing data.

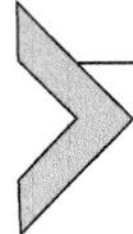

2. Advantages and limitations of 3′ mRNA sequencing approaches

In contrast to whole-transcriptome sequencing approaches, 3′ mRNA sequencing methods generate a single fragment per poly-adenylated transcript (also referred to as tag-read), where tag abundance is proportional to cellular RNA concentration (Fullwood, Wei, Liu, & Ruan, 2009; Moll et al., 2014; Morrissy et al., 2009; Stark et al., 2019). The underlying protocol employs anchored oligo(dT)-based reverse transcription followed by random-priming-based second-strand synthesis and PCR amplification for the generation libraries consisting of short cDNA fragments from the 3′ end of poly-adenylated transcripts (Fig. 1). Because the protocol is devoid of ligation steps, it allows for the rapid generation of amplicons derived from mRNA 3′ ends. When compared to other RNA sequencing protocols, 3′ mRNA sequencing provides several advantages on the experimental and bioinformatic side: (1) By employing anchored oligo(dT) priming for reverse transcription during cDNA generation, it provides a simple and rapid protocol to generate NGS libraries without a need for prior poly(A)-enrichment or ribosomal RNA depletion. (2) The generation of only one fragment per transcript simplifies computational analyses by decreasing the number of reads that span splice junctions and avoiding the need to normalize for gene-length, while maintaining high strand-specificity. (3) Because of the vastly reduced library complexity, lower sequencing depth is sufficient to accurately determine transcript abundance, allowing for a higher level of multiplexing while maintaining a sensitivity similar to that of standard mRNA sequencing approaches. On the other hand, 3′ mRNA sequencing provides only a limited view on transcript context and prevents any insights into expression of isoforms that emerge from differential promoter usage or alternative splicing. Furthermore, the advantages of employing oligo(dT)-based priming in the context of total RNA comes at the cost of producing erroneous tags, that are derived from internal priming events on homopolymeric regions of transcripts. Identification and removal of such

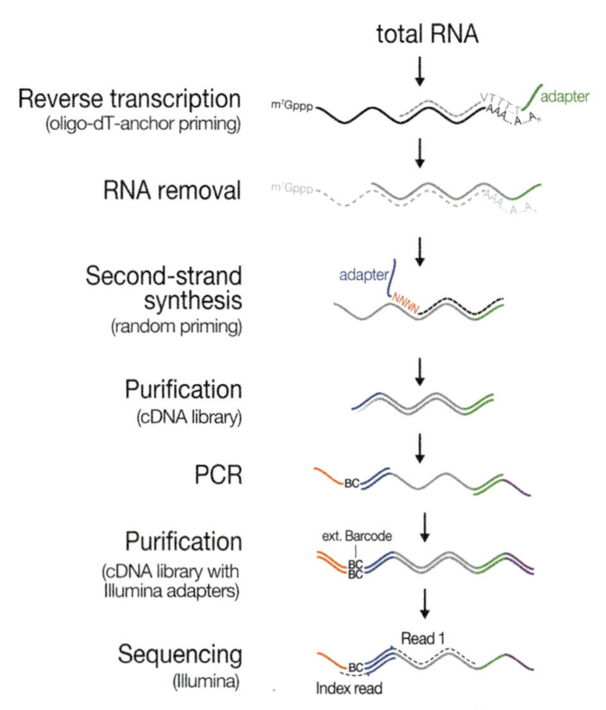

Fig. 1 Schematic overview of cDNA library preparation for 3′ mRNA sequencing using the QuantSeq protocol. Total RNA is subjected to reverse transcription using an anchored oligo(dT) primer that contains an Illumina-specific Read 2 linker sequence. After first strand synthesis the RNA is removed, and second-strand synthesis is initiated by random priming and a DNA polymerase. The random primer contains an Illumina-specific linker sequence. The resulting double-stranded cDNA library is subjected to purification, followed by PCR-based library amplification which introduces the sequences required for cluster generation on Illumina sequencing flow-cells. Multiplexing is possible by introducing a barcode (BC) that is read-out by Index read sequencing. For further details see Moll et al. (2014).

contaminants are prerequisites for the accurate determination of transcript abundance. Finally, the applicability of 3′ mRNA sequencing for transcript quantification strongly depends on accurate genome annotations. While recent efforts led to the generation of several dedicated databases for mRNA 3′ end annotations for a number of organisms (Gruber & Zavolan, 2019), this information has not been systematically integrated into genome annotation resources that are commonly employed for analyzing RNAseq datasets such as RefSeq (O'Leary et al., 2016) or Ensembl (Yates et al., 2019). Moreover, much of the available mRNA 3′ end datasets disregard isoforms that are specifically expressed in unconventional tissues and cell types as well as non-standard model organisms. With the increasing use of 3′ mRNA sequencing strategies in quantitative transcriptomics, the lack of precise annotations with respect to gene-ends pose a significant limitation to the comprehensive and quantitative interpretation of gene expression and warrants a simple and easy-to-employ method to parse datasets for the existence of novel mRNA 3′ isoforms.

3. Implementation of 3′GAmES

3′GAmES is a stand-alone analysis pipeline for the identification and annotation of cell type-specific isoforms from 3′ mRNA sequencing data. In principle, it is compatible with any 3′ mRNA library preparation protocol but it has been optimized and benchmarked on data generated with the commercially available QuantSeq kit (forward kit, Lexogen GmbH) (Moll et al., 2014). As output, 3′GAmES generates a BED-format catalog of coordinates and abundances of 3′ isoforms per gene along with counting windows to support the quantification of transcript abundances on a per-transcript or per-gene basis. The pipeline employs R and Shell scripts compiled in a Singularity container that is available for download on GitHub. Test-datasets are available on GEO (GSE146738) and expected results are discussed below (Section 4). In the following, we provide an overview of the resources and requirements to execute 3′GAmES (Section 3.1), as well as a step-by-step description of sequential processing steps (Section 3.2).

3.1 Resources and requirements to execute 3′GAmES

In the following, we provide an overview of required resources, tools and input files required to execute 3′GAmES (Fig. 2A):

1. The installation of "Singularity"—a software for the execution of contained computational pipelines—is required to execute 3′GAmES (Kurtzer, Sochat, & Bauer, 2017). For installation instructions and support please visit http://singularity.lbl.gov.
2. 3′GAmES can be downloaded from GitHub (https://github.com/AmeresLab/3-GAmES.git). It is available as a Singularity image that contains all the dependencies for running the pipeline.
3. 3′ mRNA sequencing datasets must be provided in a FASTQ (.fq or .fastq) format and stored in a folder named "quantseq." It is also possible to provide a compressed archive file in GNU zip (.gz) format.

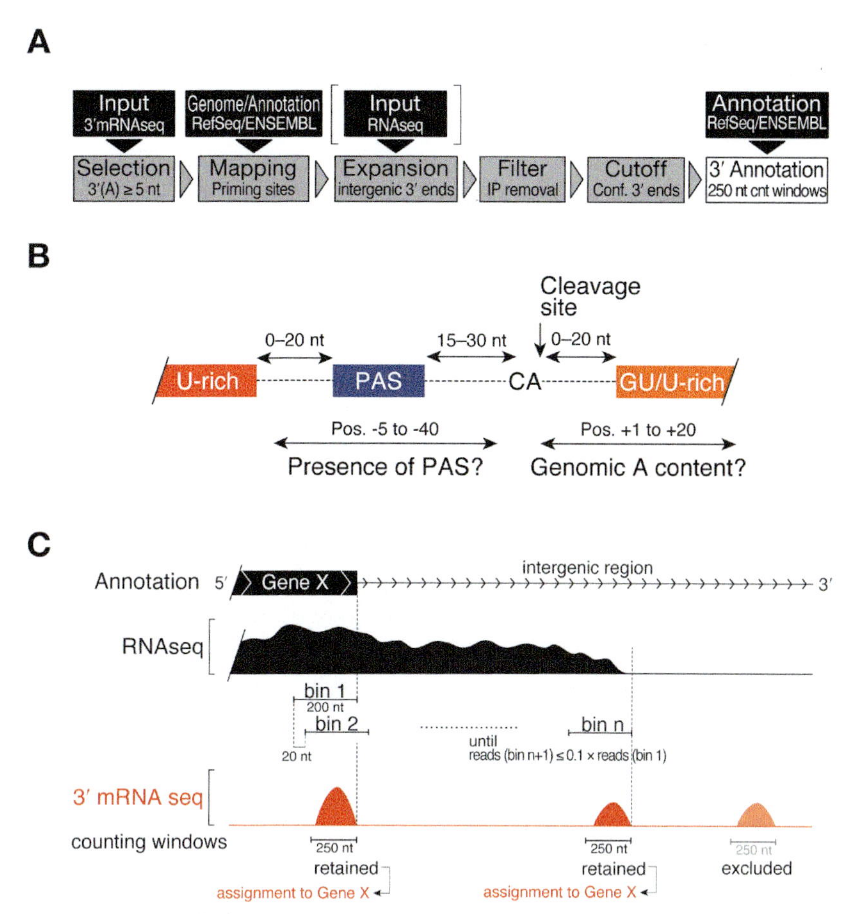

Fig. 2 Overview of 3′GAmES. (A) Steps (gray) and required input files (black) for 3′GAmES. (B) 3′ end formation of mRNAs relies on sequence motifs encoded in the genome, where a poly(A) site (PAS) upstream of a given cleavage site is preceded by a U-rich sequence and succeeded by a GU- or U-rich sequence downstream of the cleavage site. The approximate distance between sequence motifs is indicated. 3′GAmES inspects the presence of a PAS motif upstream (position −5 to −40) and genomic A-content downstream (position +1 to +20) of a putative cleavage site to extract mRNA 3′ ends from 3′ mRNA sequencing datasets. (C) For the annotation of bona-fide mRNA 3′ ends, 3′GAmES assigns reads to overlapping gene annotations. Intergenic 3′ ends are assigned to the closest upstream gene annotation. To this end, 3′GAmES integrates matched RNAseq data by measuring RNAseq read coverage downstream of annotated mRNA 3′ ends using a sliding window approach (200 nt window length with 20 nt step size) over a length of up to 40 kb, unless bins overlap with another sense annotation or if coverage within individual windows drop below 10% of the most upstream window. 3′ ends within concatenated windows will be retained and assigned to upstream annotations.

4. A reference genome sequence must be provided in FASTA (.fa) format and will be used to map 3′ mRNA sequencing data using the previously described alignment software NextGenMap (Neumann et al., 2019; Sedlazeck, Rescheneder, & von Haeseler, 2013).
5. Reference gene annotations matching the provided genome sequence are required to assign detected mRNA 3′ ends to gene annotations and must be provided in Browser Extensible Data (.bed) format. Annotations can be directly downloaded from the UCSC table browser (https://genome.ucsc.edu/cgi-bin/hgTables) or retrieved using the interface biomaRt (Durinck, Spellman, Birney, & Huber, 2009). The following annotations are required for running 3′GAmES; files must be named as indicated in parenthesis:
   a. RefSeq 3′ UTR annotations (refSeq_mrna_utrsPresent.bed)
   b. Ensembl 3′ UTR annotations (proteinCoding_annotatedUTRs.bed)
   c. Ensembl exon annotations (exonInfo_proteinCodingGenes.bed)
   d. Ensembl intron annotations (intronInfo_proteinCodingGenes.bed)
   e. Ensembl transcript annotation (transcriptStartsAndEnds_all.txt)

   Note, that the annotation files must be provided in tab delimited format and must not contain column headers; information must be provided in the following order: chromosome, start coordinate, end coordinate, gene name, score, strand, transcript ID. Additionally, transcript biotype (protein coding, non-coding etc.) should be provided as an eighth column for Ensembl transcript annotations (see [e]).
6. *Optional*: Whole-transcriptome RNA sequencing datasets can be employed to assign intergenic mRNA 3′ ends to upstream annotations. Such datasets must be derived from matched biological samples and mapped to the same reference genome using standard mapping tools (e.g., NextGenMap) (Sedlazeck et al., 2013). Mapped whole-transcriptome datasets must be provided in a Binary Alignment Map (.bam) format and placed in a folder named "rnaseq."

3.2 Stepwise description of 3′GAmES

3′GAmES annotates mRNA 3′ ends in the following six-step procedure (Fig. 2):

1. Reads that overlap oligo(dT)-based reverse transcription priming sites are extracted and processed, followed by alignment to the genome to determine the genomic coordinates of priming sites:

a. Reads that contain 3′ terminal oligo(A) stretches are considered indicative for priming events. We therefore extract reads that end in five or more adenines.

b. Extracted sequences are trimmed at their 3′ end by removing consecutive adenines. Only reads with a length of at least 18 nucleotides are retained to ensure robust mapping to the genome. Retained reads are mapped to the reference genome using NextGenMap (Sedlazeck et al., 2013), which is provided through a tool called SLAMdunk (Neumann et al., 2019). Multimapping to the genome is allowed but limited to a maximum of 100 instances. Only reads mapping with a sequence identity of ≥95% to the reference genome are retained.

2. To extract the genomic coordinates of priming sites, trimmed reads with sufficient experimental support for individual end positions are retained and immediately adjacent sites are merged to account for 3′ end processing inaccuracies.

a. The genomic coordinate corresponding to the 3′-terminal nucleotide of trimmed reads is extracted and considered as the site of priming for reverse transcription. Such recovered priming site-coordinates are filtered based on the number of reads supporting the respective position (threshold must be provided by the user). As a guideline, we recommend a cutoff of >3 reads for a conventional 3′ mRNA sequencing library prepared from a mammalian cell type with an overall sequencing depth of 50 Million reads that was sequenced in SR100 mode.

b. Immediately adjacent priming sites (±1 nt) may reflect the previously described imprecision of mRNA 3′ end processing (Herrmann et al., 2020; Hon et al., 2013). In such cases, the coordinate of the most 3′ site is considered for further analysis.

3. Employing oligo(dT)-based priming in the context of total RNA comes at the cost of producing erroneous tags, that are derived from internal priming events on homopolymeric (A) regions of transcripts. To exclude internal priming events, we assess (i) the presence of a PAS sequence motif upstream of each priming site, as well as (ii) the genomic A-content downstream (Fig. 2B). These parameters are used to extract bona-fide mRNA 3′ ends.

a. The presence of a canonical PAS (AATAAA) or degenerate PAS motifs (ATTAAA, TATAAA, AGTAAA, AATACA, CATAAA, AATATA, GATAAA, AATGAA, AAGAAA, ACTAAA, AAT AGA, AATAAT, AACAAA, ATTACA, ATTATA, AACAAG,

AATAAG) is determined for the genomic sense strand 5–40 nucleotides upstream of the identified priming sites; and the genomic A-content in the genomic sense strand 1–20 nucleotides downstream of priming sites is calculated.

   b. Conditional filtering of priming sites is applied by removing all instances where no PAS motif can be identified and a genomic A-content $\geq$0.24 is detected or where a PAS motif is present and genomic A-content is $\geq$0.36. Note, that those cutoffs recover 83.4% and 78.6% of RefSeq-annotated mRNA 3′ ends in mouse and zebrafish that contain a PAS motif and only 31.7% and 28.1%, respectively, without PAS motif and therefore provide a conservative filter for the *de novo* annotation of mRNA 3′ ends.

   c. All retained priming sites are considered to emerge from bona-fide mRNA 3′ ends and considered for further analysis.

4. To assign bona-fide mRNA 3′ ends to specific genes, we make use of genome annotations available through public resources such as Ensembl and RefSeq (Fig. 2C).

   a. 3′ end coordinates are assigned to genes by sequentially overlapping them with 3′ UTR annotations and exon/intron annotations, which are provided by the user upon extraction from available repositories (i.e., Ensembl and RefSeq).

   b. 3′ end coordinates that remain unassigned are marked as "intergenic" and will be disregarded unless further experimental evidence allows for their unambiguous assignment to upstream transcription units, as specified in Step 5 (see below).

5. *Optional*: If available, whole-transcriptome RNA sequencing datasets can be employed to unambiguously assign intergenic mRNA 3′ ends to upstream annotations (Fig. 2C). To this end, RNAseq data must be derived from matched biological samples.

   a. Continuous read coverage downstream of existing transcript annotations is determined by counting reads in sliding windows (200 nt with step size of 20 nt) starting 200 nt upstream of the annotated 3′ end of the respective gene.

   b. Sliding windows are extended in downstream direction until coverage drops below 10% of the first window or until another sense gene annotation is encountered. The sliding window approach is limited to a maximum of 40 kb.

   c. Intergenic 3′ end coordinates (from Step 4b) that overlap with sliding windows that are supported by whole-transcriptome RNAseq signal are annotated as novel mRNA 3′ ends and assigned to the respective upstream annotation.

6. In a final step, mRNA 3′ ends are filtered to exclude spuriously expressed isoforms, followed by the creation of counting windows for the measurement of transcript abundance in 3′ mRNA sequencing libraries.
   a. For each gene, assigned 3′ ends are ranked according to sequencing counts and excluded unless exceeding 10% of the total counts per gene. All excluded 3′ ends are considered to emerge from spuriously expressed isoforms and discarded.
   b. For each retained 3′ end, a 250 nt upstream counting window is generated. Those are supplemented with equally sized counting windows upstream of all repository (Ensembl)-provided mRNA (and other RNA polymerase II transcript) 3′ ends. Overlapping counting windows are merged. Additional counting windows can be included manually by the user at this stage.

The final output of 3′GAmES is a BED-format catalog of coordinates and abundances of 3′ isoforms per gene along with counting windows for the analysis of relative transcript abundances in 3′ mRNA sequencing datasets.

3.3 Command lines

In order to download 3′GAmES from github, git clone is executed on a local bash shell (Fig. 3A). Subsequent compiling of 3′GAmES is achieved by changing the directory to 3-GAmES/bin and executing a shell script to download dependencies (Fig. 3A). Prior to executing 3′GAmES, the input data should be organized as specified in Section 3.1 (see above). 3′GAmES can subsequently be run by executing the shell script run_3GAmES.sh in the context of the following variables (Fig. 3B):

-a 3′ adapter sequence that was used for 3′mRNA seq library preparation.

-i Input directory containing each on folder named "quantseq" (containing 3′mRNA seq data in .fastq, .fq, or .gz format) and "rnaseq" (optional; containing mapped, sorted and indexed .bam files of matched whole-transcriptome RNAseq data)

-o Output directory, which will contain the resulting output file of 3′GAmES.

-g Genome file in .fasta format.

-t Threshold of the minimal number of reads to define each priming site.

-e Ensembl directory containing the five annotation files specified above (see Section 3.1).

A

```
>[user]$ git clone https://github.com/poojabhat1690/3-GAmES.git
Cloning into '3-GAmES'...
remote: Enumerating objects: 113, done.
remote: Counting objects: 100% (113/113), done.
remote: Compressing objects: 100% (83/83), done.
remote Total 363 (delta 51), reused 70 (delta 24), pack-reused 250
Receiving objects: 100% (363/363), 7.41 miB | 3.65 miB/s, done.
Resolving deltas: 100% (183/183), done.
>[user]$ cd 3-GAmES/bin
>[user]$ ./getDependencies.sh
pulling dependencies for 3´GAmES
INFO: Downloading shub image
583.6miB / 583.6miB [==========================================] 100% 51.5 miB/s 0s
pulling SLAMdunk
INFO: Converting OCI blobs to SIF format
INFO: Starting build...
```

B

```
>[user]$ ./run_3GAmES.sh -a AGATCGGAAGAGCACACGTCTGAACTCCAGTCACNNNNNNATCTCGTATGCCGTCTTC
TGCTTG -i xcondition_input/ -o xcondition_output/ -g danio_rerio/dr11/danRer11.fa
-t 2 -e annotations_all/ -m p -c trialSample -p /scratch/3-GAmES/ -s all
the adapter trimmed has the sequence AGATCGGAAGAGCACACGTCTGAACTCCAGTCACNNNNNNATCTCGTAT
GCCGTCTTCTGCTTG
the input directory containing QuantSeq data is xcondition_input/
the output directory is xcondition_output/
the genome is specified in danio_rerio/dr11/danRer11.fa
the threshold to consider priming sites is 2
the ensembl directory specified is annotations_all/
the rnaseq mode is p
the condition is trialSample
the pipeline is here /scratch/3-GAmES/
Stopping after running all
```

Fig. 3 Example screenshots of bash shell command lines used to download and compile (A) or execute (B) 3′GAmES. Commands that have to be entered by the user are highlighted in bold. For details regarding command lines and user-specified variables (blue) see main text.

-m Mode of counting for RNAseq coverage, derived from bedtools multicov (s: counting on the sense strand, p: paired end reads, S: counting on the antisense strand).

-c User-defined specification of sample (example: timepoint or organism).

-p path to the 3-GAmES folder.

-s Steps after which pipeline will stop (options: "preprocessing," "primingsites," "intergenicends," "all"; default option should be "all").

The approximate computing time of a .fastq input file that contains ~27 million reads is 1 h and 20 min when using 50 GB memory.

4. Application of 3′GAmES and expected results

To benchmark 3′GAmES, we analyzed 3′ mRNA sequencing datasets prepared from total RNA of mouse embryonic stem cells (mESCs, previously published by (Herzog et al., 2017)) and zebrafish embryos (1-day post fertilization; 1 dpf) using QuantSeq 3′ mRNA-Seq Library Prep Kit FWD for Illumina sequencing (sequencing mode: SR100) (Moll et al., 2014). Between 26% and 29% of reads in such libraries contained 3′ terminal A-stretches with a length of at least five nucleotides and were subjected to mapping after A-stretch removal to identify priming sites (Fig. 4A). Among the 204,560 (mESC) and 211,873 (zebrafish) recovered priming sites, 48% (mESC) and 45% (zebrafish) overlapped with annotated genes (Fig. 4B). Filtering based on upstream PAS and downstream GAC retained 26% (mESC) and 25% (zebrafish) of priming sites; and cut-off filtering to eliminate spurious priming events for each gene retained 44% (mESC) and 46% (zebrafish) of filtered priming sites (Fig. 4B). This final list of priming sites featured nucleotide profiles identical to bona-fide mRNA 3′ ends (Fig. 4C), where 53% ($n=12{,}142$ in mESC) and 43% ($n=10{,}742$ in zebrafish) of coordinates overlapped with Ensembl/RefSeq annotations. The 10,788 (mESC) and 14,120 (zebrafish) new annotations were indistinguishable in nucleotide profiles to repository mRNA 3′ ends (Fig. 4D). Overall, 3′GAmES identified 22,930 3′ ends for 10,690 genes in mESC and 24,862 3′ ends for 12,350 genes in zebrafish. Hence, the majority of genes expressed multiple 3′ isoforms (59.8% in mESC and 57.5% in zebrafish; Fig. 4E), where >10% of all genes expressed more than three isoforms, consistent with the previously described heterogeneity of mRNA 3′ ends (Mayr, 2016b).

Finally, we observed an increase in the correlation of 3′ mRNAseq with whole-transcriptome RNAseq by applying 3′GAmES-derived annotations when compared to Ensembl-annotations in mESC (Spearman correlation coefficient, $R=0.76$ vs 0.7) and zebrafish ($R=0.74$ vs 0.67; Fig. 5).

5. Alternative approaches

Apart from 3′GAmES, several tools and databases have used a number of RNA 3′ end cDNA generation protocols to improve on existing annotations. While those tools vary greatly in their utility and applicability, they provide alternative strategies to derive and annotate mRNA 3′ ends (Table 1). In early attempts to systematically annotate transcript ends, poly(A)-position profiling by sequencing (3P-seq) has been employed to map poly(A) sites in *C. elegans* and zebrafish from reads with non-genome-matching 3′ terminal adenine(s) and supporting experimental evidence from conventional RNA sequencing datasets (Jan, Friedman, Ruby, & Bartel, 2010; Ulitsky et al., 2012). But these resources are so-far limited to selected species and developmental stages and do not provide a readily accessible

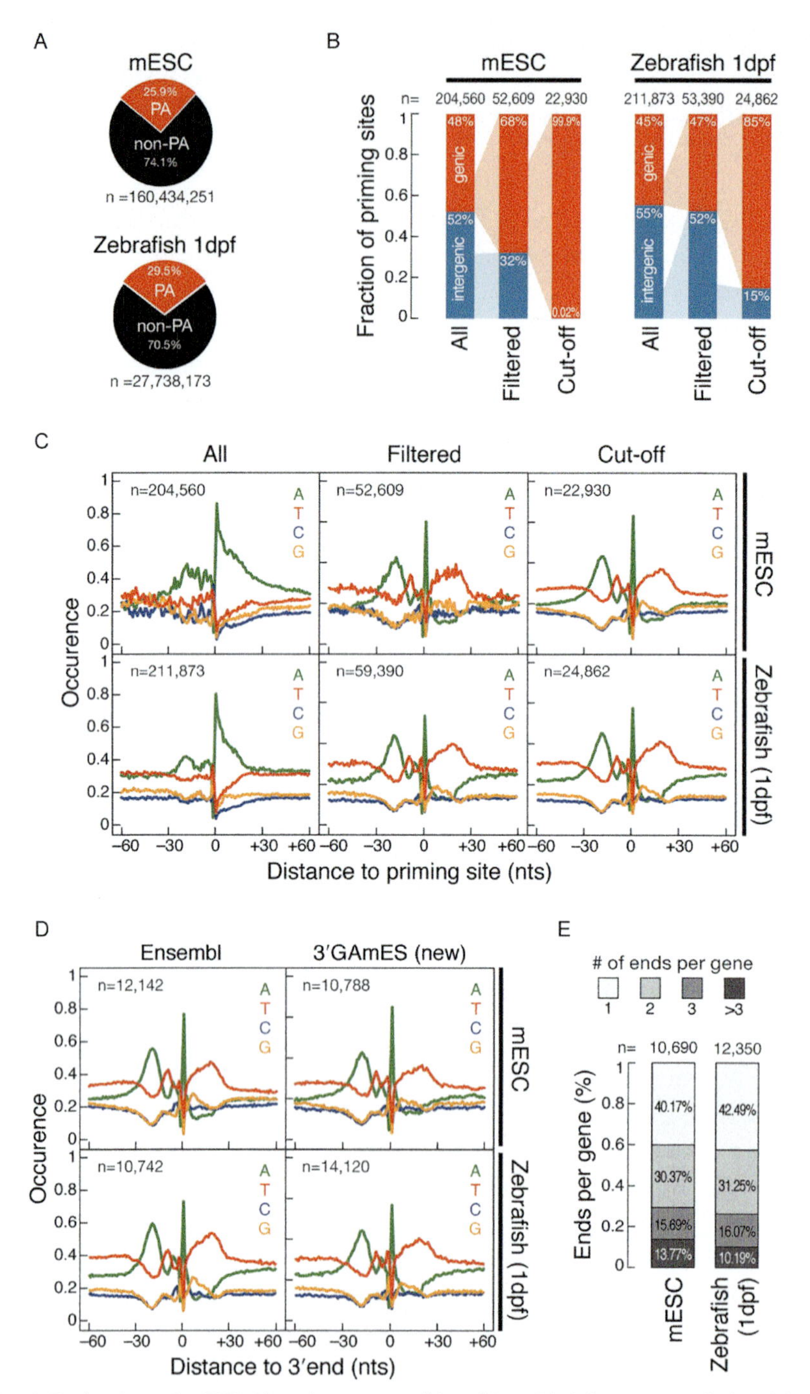

Fig. 4 Evaluation of mRNA 3′ ends recovered by 3′GAmES. (A) Fraction of oligo(A)$_{\geq 5}$-containing reads in 3′ mRNA sequencing libraries prepared from total RNA of mouse embryonic stem cells (mESC) and zebrafish embryos 1 day post fertilization (1 dpf). *(Continued)*

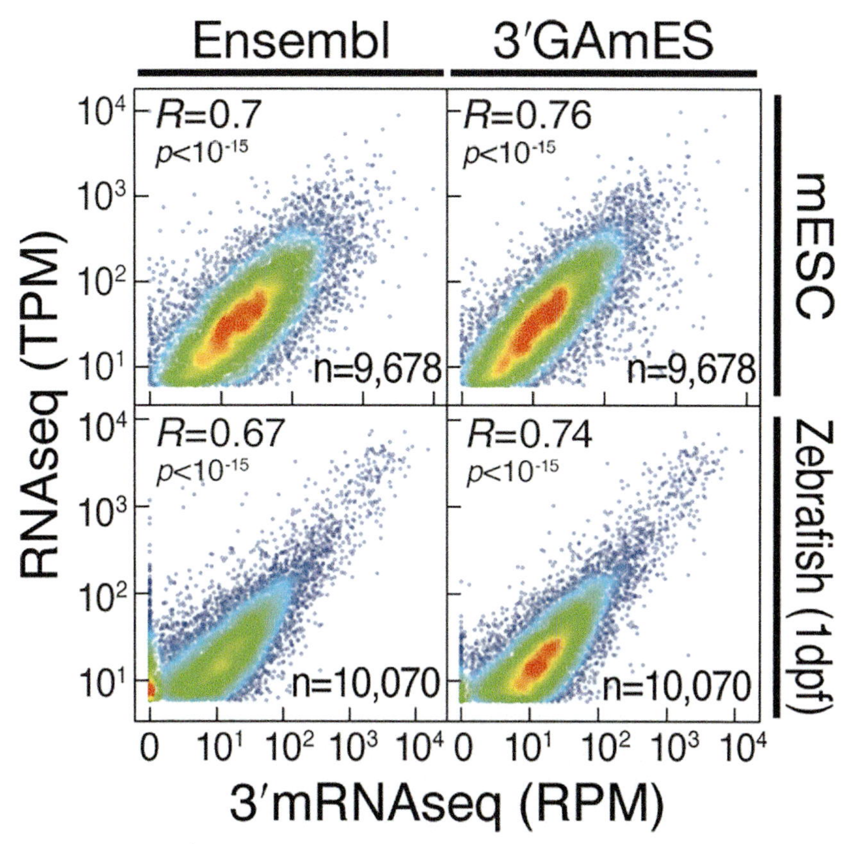

Fig. 5 Impact of 3′GAmES-derived annotations on the quantitative representation of transcripts in 3′mRNA sequencing datasets compared to conventional RNA sequencing approaches. Comparison of gene expression profiles determined by mRNA sequencing (RNAseq in transcripts per million, TPM) and 3′ mRNA sequencing (3′ mRNAseq in reads per million, RPM) in mESC (top) or zebrafish embryos 1 dpf (bottom) relying on Ensembl- (left) or 3′GAmES-derived annotations (right). Number of comparisons (n) were restricted to genes that exhibited >5TPM in RNAseq data. Spearman correlation coefficient (R) and associated *p*-value (p) are indicated.

Fig. 4—Cont'd The total number of inspected reads (n) is indicated. (B) The effect of filtering internal priming events (filtered) and spurious 3′ ends (cut-off) on the absolute number (n) and fraction of intergenic (blue) or intragenic (red) priming sites in processed 3′ mRNA sequencing datasets prepared from mESC (left) or zebrafish embryos 1 dpf. (C) Nucleotide-occurrence around the indicated number of priming sites (n) at individual steps of 3′ mRNA sequencing data-set processing in libraries derived from mESC (top) or zebrafish embryos 1 dpf (bottom). Nucleotide profiles are displayed across the indicated area (±60 nucleotides, nts) relative to the priming site (position 0). (D) Nucleotide-occurrence around the indicated number of 3′GAmES-annotated mRNA 3′ ends that do (left) or do not (right) overlap with Ensembl/RefSeq 3′ UTR annotations in libraries derived from mESC (top) or zebrafish embryos 1 dpf (bottom). The number of 3′ ends is indicated (n). Nucleotide profiles are displayed across the indicated area (±60 nts) relative to the priming site (position 0). (E) Distribution of the number of mRNA 3′ end isoforms per gene, as detected by 3′GAmES, is shown for the indicated number of genes (n) in mESC (left) or zebrafish embryos 1 dpf (right).

Table 1 Comparison of available tools and resources for the annotation of mRNA 3′ ends.

Study	Int. priming filter	RNAseq integration	Quantification of 3′ ends	Tool or resource
Ulitsky et al. (2012)	No	Yes	Yes	Resource
Gruber et al. (2016)	Yes	No	No	Resource
Rot et al. (2017)	Yes	No	Yes	Tool
Herrmann et al. (2020)	Yes	No	Yes	Resource/tool
3′GAmES (this study)	Yes	Yes	Yes	Tool

computational framework. Gruber et al. and Hermann et al. created a flexible computational pipeline to identify mRNA 3′ ends from nine different 3′ end sequencing protocols that includes stringent removal of internal priming sites but does not permit identification of intergenic priming sites based on available RNA sequencing datasets (Gruber et al., 2016; Herrmann et al., 2020). Finally, Rot et al. established a dedicated tool to analyze 3′ mRNA sequencing datasets (using QuantSeq 3′ mRNA-Seq Library Prep Kit REV) which includes a stringent filtering step for internal priming sites but automatically assigns intergenic ends within 10 kb to upstream annotations without probing for continuous transcription (Rot et al., 2017). 3′GAmES provides a readily accessible alternative to complement previously published tools and resources and facilitate the *de novo* generation of counting windows for the quantification of 3′ mRNA sequencing data in organisms with poorly annotated genomes or scarcely-studied cell types. To this end, the tool can in principle be applied to any mRNA 3′ end sequencing strategy that provides reads partially overlapping with poly(A) tails.

6. Conclusion

Among the diversity of RNA sequencing protocols, 3′ mRNA sequencing strategies are becoming increasingly popular as they provide the ability to rapidly sequence large number of samples at a low cost. In this chapter, we describe common challenges associated with the analysis of 3′ mRNA sequencing datasets and presents 3′GAmES, an easy-to-implement pipeline for the refinement and quantification of cell type-specific mRNA 3′ ends. We show that 3′GAmES significantly expands existing genome annotations as illustrated by the systematic identification of 22,930 3′ ends for 10,690 genes in mESC and 24,862 3′ ends for 12,350 genes in zebrafish, where 47% and 57%, respectively, did not overlap with existing 3′ end annotations and were thus newly identified. Such novel annotations enabled more comprehensive gene-tag counting of 3′ mRNA sequencing datasets to faithfully mirror whole-transcriptome sequencing strategies. The application of 3′GAmES therefore exemplifies the power of deriving custom annotations for the systematic assessment of transcript abundance in 3′ mRNA sequencing datasets.

Acknowledgments

Sequencing was performed at the VBCF NGS Unit (https://www.vbcf.ac.at). This work has been supported by Boehringer Ingelheim, the FFG, the FWF START program (Y 1031-B28) and the HFSP Career Development Award (CDA00066/2015) to A.P. and the Austrian Academy of Sciences, the Austrian Science Fund FWF (W1207-B09) and the European Research Council (ERC-PoC-825710 and ERC-CoG-866166) to S.L.A.

References

Beaudoing, E., Freier, S., Wyatt, J. R., Claverie, J.-M., & Gautheret, D. (2000). Patterns of variant polyadenylation signal usage in human genes. *Genome Research*, *10*(7), 1001–1010. https://doi.org/10.1101/gr.10.7.1001.

Beck, A. H., Weng, Z., Witten, D. M., Zhu, S., Foley, J. W., Lacroute, P., et al. (2010). 3′-end sequencing for expression quantification (3SEQ) from archival tumor samples. *PLoS One*, *5*(1), e8768. https://doi.org/10.1371/journal.pone.0008768.

Berg, M. G., Singh, L. N., Younis, I., Liu, Q., Pinto, A. M., Kaida, D., et al. (2012). U1 snRNP determines mRNA length and regulates isoform expression. *Cell*, *150*(1), 53–64. https://doi.org/10.1016/j.cell.2012.05.029.

Chen, W., Jia, Q., Song, Y., Fu, H., Wei, G., & Ni, T. (2017). Alternative polyadenylation: Methods, findings, and impacts. *Genomics, Proteomics & Bioinformatics*, *15*(5), 287–300. https://doi.org/10.1016/j.gpb.2017.06.001.

Danckwardt, S., Hentze, M. W., & Kulozik, A. E. (2008). 3′ end mRNA processing: Molecular mechanisms and implications for health and disease. *The EMBO Journal*, *27*(3), 482–498. https://doi.org/10.1038/sj.emboj.7601932.

Derti, A., Garrett-Engele, P., MacIsaac, K. D., Stevens, R. C., Sriram, S., Chen, R., et al. (2012). A quantitative atlas of polyadenylation in five mammals. *Genome Research*, *22*(6), 1173–1183. https://doi.org/10.1101/gr.132563.111.

Di Giammartino, D. C., Nishida, K., & Manley, J. L. (2011). Mechanisms and consequences of alternative polyadenylation. *Molecular Cell*, *43*(6), 853–866. https://doi.org/10.1016/j.molcel.2011.08.017.

Durinck, S., Spellman, P. T., Birney, E., & Huber, W. (2009). Mapping identifiers for the integration of genomic datasets with the R/Bioconductor package biomaRt. *Nature Protocols*, *4*(8), 1184–1191. https://doi.org/10.1038/nprot.2009.97.

Eckmann, C. R., Rammelt, C., & Wahle, E. (2011). Control of poly(A) tail length. *Wiley Interdisciplinary Reviews. RNA*, *2*(3), 348–361. https://doi.org/10.1002/wrna.56.

Flavell, S. W., Kim, T.-K., Gray, J. M., Harmin, D. A., Hemberg, M., Hong, E. J., et al. (2008). Genome-wide analysis of MEF2 transcriptional program reveals synaptic target genes and neuronal activity-dependent polyadenylation site selection. *Neuron*, *60*(6), 1022–1038. https://doi.org/10.1016/j.neuron.2008.11.029.

Fullwood, M. J., Wei, C.-L., Liu, E. T., & Ruan, Y. (2009). Next-generation DNA sequencing of paired-end tags (PET) for transcriptome and genome analyses. *Genome Research*, *19*(4), 521–532. https://doi.org/10.1101/gr.074906.107.

Grosso, A. R., de Almeida, S. F., Braga, J., & Carmo-Fonseca, M. (2012). Dynamic transitions in RNA polymerase II density profiles during transcription termination. *Genome Research*, *22*(8), 1447–1456. https://doi.org/10.1101/gr.138057.112.

Gruber, A. J., Schmidt, R., Gruber, A. R., Martin, G., Ghosh, S., Belmadani, M., et al. (2016). A comprehensive analysis of 3′ end sequencing data sets reveals novel polyadenylation signals and the repressive role of heterogeneous ribonucleoprotein C on cleavage and polyadenylation. *Genome Research*, *26*(8), 1145–1159. https://doi.org/10.1101/gr.202432.115.

Gruber, A. J., & Zavolan, M. (2019). Alternative cleavage and polyadenylation in health and disease. *Nature Reviews Genetics*, *20*(10), 599–614. https://doi.org/10.1038/s41576-019-0145-z.

Herrmann, C. J., Schmidt, R., Kanitz, A., Artimo, P., Gruber, A. J., & Zavolan, M. (2020). PolyASite 2.0: A consolidated atlas of polyadenylation sites from 3′ end sequencing. *Nucleic Acids Research*, *48*(D1), D174–D179. https://doi.org/10.1093/nar/gkz918.

Herzog, V. A., Reichholf, B., Neumann, T., Rescheneder, P., Bhat, P., Burkard, T. R., et al. (2017). Thiol-linked alkylation of RNA to assess expression dynamics. *Nature Methods*, *14*(12), 1198–1204. https://doi.org/10.1038/nmeth.4435.

Hon, C.-C., Weber, C., Sismeiro, O., Proux, C., Koutero, M., Deloger, M., et al. (2013). Quantification of stochastic noise of splicing and polyadenylation in Entamoeba histolytica. *Nucleic Acids Research*, *41*(3), 1936–1952. https://doi.org/10.1093/nar/gks1271.

Hoque, M., Ji, Z., Zheng, D., Luo, W., Li, W., You, B., et al. (2013). Analysis of alternative cleavage and polyadenylation by 3′ region extraction and deep sequencing. *Nature Methods*, *10*(2), 133–139. https://doi.org/10.1038/nmeth.2288.

Jan, C. H., Friedman, R. C., Ruby, J. G., & Bartel, D. P. (2010). Formation, regulation and evolution of Caenorhabditis elegans 3′UTRs. *Nature*, *469*(7328), 97–101. https://doi.org/10.1038/nature09616.

Ji, Z., Lee, J. Y., Pan, Z., Jiang, B., & Tian, B. (2009). Progressive lengthening of 3′ untranslated regions of mRNAs by alternative polyadenylation during mouse embryonic development. *Proceedings of the National Academy of Sciences*, *106*(17), 7028–7033. https://doi.org/10.1073/pnas.0900028106.

Kurtzer, G. M., Sochat, V., & Bauer, M. W. (2017). Singularity: Scientific containers for mobility of compute. *PLoS One*, *12*(5), e0177459. https://doi.org/10.1371/journal.pone.0177459.

Mayr, C. (2016a). Regulation by 3′-untranslated regions. *Annual Review of Genetics*, *51*(1), 1–24. https://doi.org/10.1146/annurev-genet-120116-024704.

Mayr, C. (2016b). Evolution and biological roles of alternative 3′UTRs. *Trends in Cell Biology*, *26*(3), 227–237. https://doi.org/10.1016/j.tcb.2015.10.012.

Moll, P., Ante, M., Seitz, A., & Reda, T. (2014). QuantSeq 3′ mRNA sequencing for RNA quantification. *Nature Methods*, *11*(12), i–iii. https://doi.org/10.1038/nmeth.f.376.

Moore, M. J., & Proudfoot, N. J. (2009). Pre-mRNA processing reaches back to transcription and ahead to translation. *Cell*, *136*(4), 688–700. https://doi.org/10.1016/j.cell.2009.02.001.

Morrissy, A. S., Morin, R. D., Delaney, A., Zeng, T., McDonald, H., Jones, S., et al. (2009). Next-generation tag sequencing for cancer gene expression profiling. *Genome Research*, *19*(10), 1825–1835. https://doi.org/10.1101/gr.094482.109.

Neumann, T., Herzog, V. A., Muhar, M., von Haeseler, A., Zuber, J., Ameres, S. L., et al. (2019). Quantification of experimentally induced nucleotide conversions in high-throughput sequencing datasets. *BMC Bioinformatics*, *20*(1), 258. https://doi.org/10.1186/s12859-019-2849-7.

O'Leary, N. A., Wright, M. W., Brister, J. R., Ciufo, S., Haddad, D., McVeigh, R., et al. (2016). Reference sequence (RefSeq) database at NCBI: Current status, taxonomic expansion, and functional annotation. *Nucleic Acids Research*, *44*(D1), D733–D745. https://doi.org/10.1093/nar/gkv1189.

Rot, G., Wang, Z., Huppertz, I., Modic, M., Lenče, T., Hallegger, M., et al. (2017). High-resolution RNA maps suggest common principles of splicing and polyadenylation regulation by TDP-43. *Cell Reports*, *19*(5), 1056–1067. https://doi.org/10.1016/j.celrep.2017.04.028.

Sandberg, R., Neilson, J. R., Sarma, A., Sharp, P. A., & Burge, C. B. (2008). Proliferating cells express mRNAs with shortened 39 untranslated regions and fewer MicroRNA target sites. *Science*, *320*(5883), 1643–1647. https://doi.org/10.1126/science.1155390.

Sedlazeck, F. J., Rescheneder, P., & von Haeseler, A. (2013). NextGenMap: Fast and accurate read mapping in highly polymorphic genomes. *Bioinformatics*, *29*(21), 2790–2791. https://doi.org/10.1093/bioinformatics/btt468.

Shi, Y., Giammartino, D. C. D., Taylor, D., Sarkeshik, A., Rice, W. J., Yates, J. R., et al. (2009). Molecular architecture of the human pre-mRNA 3′ processing complex. *Molecular Cell*, *33*(3), 365–376. https://doi.org/10.1016/j.molcel.2008.12.028.

Stark, R., Grzelak, M., & Hadfield, J. (2019). RNA sequencing: The teenage years. *Nature Reviews Genetics*, *20*(11), 631–656. https://doi.org/10.1038/s41576-019-0150-2.

Tian, B., & Graber, J. H. (2012). Signals for pre-mRNA cleavage and polyadenylation. *Wiley Interdisciplinary Reviews. RNA*, *3*(3), 385–396. https://doi.org/10.1002/wrna.116.

Tian, B., Hu, J., Zhang, H., & Lutz, C. S. (2005). A large-scale analysis of mRNA polyadenylation of human and mouse genes. *Nucleic Acids Research*, *33*(1), 201–212. https://doi.org/10.1093/nar/gki158.

Tian, B., & Manley, J. L. (2017). Alternative polyadenylation of mRNA precursors. *Nature Reviews Molecular Cell Biology*, *18*(1), 18–30. https://doi.org/10.1038/nrm.2016.116.

Ulitsky, I., Shkumatava, A., Jan, C. H., Subtelny, A. O., Koppstein, D., Bell, G. W., et al. (2012). Extensive alternative polyadenylation during zebrafish development. *Genome Research*, *22*(10), 2054–2066. https://doi.org/10.1101/gr.139733.112.

Xiao, M., Zhang, B., Li, Y., Gao, Q., Sun, W., & Chen, W. (2016). Global analysis of regulatory divergence in the evolution of mouse alternative polyadenylation. *Molecular Systems Biology*, *12*(12), 890. https://doi.org/10.15252/msb.20167375.

Yates, A. D., Achuthan, P., Akanni, W., Allen, J., Allen, J., Alvarez-Jarreta, J., et al. (2019). Ensembl 2020. *Nucleic Acids Research*, *48*(D1), D682–D688. https://doi.org/10.1093/nar/gkz966.

CHAPTER ELEVEN

Computational analysis of alternative polyadenylation from standard RNA-seq and single-cell RNA-seq data

Yipeng Gao[a,b] and Wei Li[c,*]

[a]Graduate Program in Quantitative and Computational Biosciences, Baylor College of Medicine, Houston, TX, United States
[b]Department of Medicine, Baylor College of Medicine, Houston, TX, United States
[c]Division of Computational Biomedicine, Department of Biological Chemistry, School of Medicine, University of California, Irvine, CA, United States
*Corresponding author: e-mail address: wei.li@uci.edu

Contents

Abstract

Alternative polyadenylation (APA) is a major mechanism of post-transcriptional regulation in various cellular processes including cell proliferation and differentiation. Since conventional APA profiling methods have not been widely adopted, global APA studies are very limited. In this chapter, we summarize current computational methods for analyzing APA in standard RNA-seq and scRNA-seq data and describe two state-of-the-art bioinformatic algorithms DaPars and scDaPars in detail. The bioinformatic pipelines for both DaPars and scDaPars are presented and the application of both algorithms are highlighted.

Methods in Enzymology, Volume 655
ISSN 0076-6879
https://doi.org/10.1016/bs.mie.2021.03.015

1. Introduction

Polyadenylation, which is intimately linked to transcription termination, involves endonucleolytic cleavage of the nascent RNA followed by synthesis of a poly (A) tail on the 3′ terminus of the cleaved product (Sherstnev et al., 2012; Tian & Manley, 2017). It was first reported more than three decades ago that a gene can give rise to transcripts with multiple polyadenylation sites (poly (A) sites) (Edwalds-Gilbert, Veraldi, & Milcarek, 1997), which are defined by surrounding RNA sequence motifs, and that differential usage of these sites can lead to the formation of distinct mRNA isoforms, a phenomenon termed *alternative polyadenylation (APA)* (Edwalds-Gilbert et al., 1997). In fact, approximately 70% of human genes (Derti et al., 2012) are characterized by multiple poly (A) sites and thereby significantly contributing to transcriptome diversity (Tian, Hu, Zhang, & Lutz, 2005). The majority of poly (A) sites are located in the mRNA 3′-untranslated region (3′UTR). As a result, while the protein-coding regions are unaltered in most APA examples, the important *cis*-regulatory elements located in the 3′UTRs, including adenylate-uridylate-rich elements (ARE) and binding sites of microRNAs and RNA binding proteins, are disrupted, resulting in altered mRNA stability, localization and translation efficiency (An et al., 2008; Garneau, Wilusz, & Wilusz, 2007; Hoffman et al., 2016).

Given the functional importance of alternative polyadenylation, identification of poly (A) sites and quantification of differential usage of identified poly (A) sites in a global scale is the critical first step toward understanding the underlying mechanism of 3′UTR-mediated gene regulation. Efforts has been made to address this question through experimental protocols. The earliest global experimental studies of APA used various microarray platforms, in which the APA profile of a gene was calculated by the ratio between average signal intensities of probes targeting the extended region only existed in the longer mRNA isoform and all other probes targeting the same gene (Flavell et al., 2008; Ji, Lee, Pan, Jiang, & Tian, 2009; Sandberg, Neilson, Sarma, Sharp, & Burge, 2008; Shi, 2012). These microarray-based studies revealed widespread and coordinated APA changes during T cell activation (Sandberg et al., 2008), neuronal activity (Flavell et al., 2008) and developmental processes (Ji et al., 2009). While these studies demonstrated the pervasiveness of APA, they are limited by their dependence on annotated poly (A) sites, the inherent technical biases such as cross-hybridization, and the availability of microarrays, especially for less-studied species (Shi, 2012; Xia et al., 2014). More recently, with the advent of high-throughput sequencing, a number of simpler and more quantitative experimental methods have been developed for targeted sequencing of poly (A) junctions (*i.e.*, poly (A) junction sites enrichment followed by high-throughput sequencing) (Hoque et al., 2013; Xia et al., 2014). These poly (A)-specific-sequencing, including DRS (Ozsolak et al., 2010), 3P-Seq (Jan, Friedman, Ruby, & Bartel, 2011; Ulitsky et al., 2012), 3′READs (Hoque et al., 2013), and PAT-seq (Harrison et al., 2015), although powerful in providing the precise locations of poly (A) sites and the global landscape of APA, are hampered by technical issues, such as internal priming artifacts, and high procedure complexity, and thus have not been widely adopted in application (Xia et al., 2014). As a result, the availability of APA-data is still relatively limited with only a few publicly available poly (A) site databases (*i.e.*, APASdb (You et al., 2015), APADB (Müller et al., 2014), and PolyA_DB (Wang, Nambiar, Zheng, & Tian, 2018)) covering a limited number of well-studied organisms such as human, mouse, rice and Arabidopsis (Chen et al., 2020).

RNA-sequencing (RNA-seq) has become the *de facto* method for transcriptome-wide analysis of differential gene expression, primarily because it generates comprehensive, high-quality data that capture quantitative expression levels across the transcriptome (Shi, 2012). Because of the quantitative nature of RNA-seq, it could be powerful to identify and quantify mRNA isoform expression including those resulted from APA. However, only a small portion of the reads can be used to distinguish APA isoforms, and even less reads are mapped to the poly (A) junctions, posing challenges for us to accurately identify poly (A) sites and reliably quantify APA isoforms (Jan et al., 2011). For example, an ultra-deep sequencing study only identified ~40,000 putative poly (A) reads (~0.003%) from 1.2 billion total RNA-seq reads (Xia et al., 2014). Therefore, in order to better understand the scale and the underlying mechanism of APA regulation, *powerful, precise and easy-to-use computational tools are urgently needed to analyze the exponentially growing RNA-seq data in an APA-aware manner.*

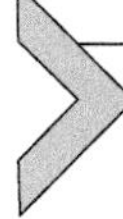

2. Current bioinformatic tools for analyzing APA in RNA-seq data

Current bioinformatic tools for identifying poly (A) sites and/or quantifying differential poly (A) sites usage from RNA-seq data could be separated into four major categories (Chen et al., 2020):

2.1 Methods depending on annotated poly (A) sites

The poly (A)-specific-sequencing data, although limited in number, has provided precise locations of poly (A) sites and the global landscape of APA (Sherstnev et al., 2012). Several bioinformatics tools utilized the publicly available poly (A) site databases constructed by poly (A)-specific-sequencing data for APA profiling (Table 1). For example, a popular bioinformatics tool MISO (mixture-of-isoforms) (Katz, Wang, Airoldi, & Burge, 2010) can readily detect annotated alternative tandem 3′UTRs. However, these methods rely on annotated poly (A) sites and cannot identify novel poly (A) sites beyond poly (A) databases. While the problem has been mitigated to some extent by latest tools Roar (Grassi, Mariella, Lembo, Molineris, & Provero, 2016) and PAQR (Gruber et al., 2018), which generate relatively large poly (A) references from different databases, the poly (A) site databases are still far from complete.

2.2 Methods depending on transcript assembly

Transcript assembly can reveal plausible isoform structures in a given sample which can be used to infer 3′UTRs from RNA-seq data (Chen et al., 2020). The major advantage of these methods (Table 2) is that they are no longer confined by annotated poly (A) sites, which potentially increase the power of these tools (Chen et al., 2020). However, transcript reconstruction from

Table 1 Summary of methods that depend on annotated poly (A) sites to perform APA analysis.

Method	Description of method	Publication links
MISO	Estimates the expression of alternatively spliced exons by a mixture-of-isoforms model	https://www.nature.com/articles/nmeth.1528
Roar	Identifies genes with differential poly (A) site usage using annotated poly (A) sites	https://bmcbioinformatics.biomedcentral.com/articles/10.1186/s12859-016-1254-8
QAPA	Quantifies relative abundance of APA isoform using RNA-seq reads that uniquely map to 3′ UTR sequences refined by annotated poly (A) sites	https://genomebiology.biomedcentral.com/articles/10.1186/s13059-018-1414-4
PAQR	Quantifying poly (A) site usage from activities of oligomeric sequence motifs on poly (A) site choice	https://genomebiology.biomedcentral.com/articles/10.1186/s13059-018-1415-3

Table 2 Summary of methods that depend on transcript assembly to perform APA analysis.

Method	Description of method	Publication links
PASA	Models complete and partial gene structure based on assembled spliced alignment	https://bmcgenomics.biomedcentral.com/articles/10.1186/1471-2164-7-327
Scripture	*De novo* assembly of RNA-seq full-length gene transcriptome using only RNA-Seq reads and the genome sequence	https://www.nature.com/articles/nbt.1633
Cufflinks	Uses aligned reads to map against the genome to assemble the reads into transcripts	https://www.nature.com/articles/nprot.2012.016
3USS	Identifies reconstructed 3′UTRs and detect alternative 3′UTRs by directly comparing with annotated genome	https://academic.oup.com/bioinformatics/article/31/11/1845/2364843#84759371
ExUTR	Predict 3′UTRs based on the intrinsic signals of assembled transcripts, regardless of reference genomes and annotations	https://bmcgenomics.biomedcentral.com/articles/10.1186/s12864-017-4241-1

short-read RNA-seq data itself is still a difficult and unsolved problem (Behr et al., 2013), and it can be even more challenging to use transcript reconstruction to analyze APA. For example, the short 3′UTRs are often embedded within the long ones, and thus the isoforms with short 3′UTRs are commonly overlooked by transcript assembly tools (Xia et al., 2014), such as Cufflinks (Trapnell et al., 2010), which can cause inaccurate assignment of reads to isoforms. Notable, most methods (first three) in this category are transcript assembly not designed specifically for APA analysis.

2.3 Methods using reads that contain a string of untemplated adenosines

RNA-seq reads originated from transcript ends that contain a string of untemplated adenosines (poly (A) reads) can be used to directly pinpoint poly (A) sites (Shi, 2012). Unfortunately, unlike poly (A)-specific-sequencing, RNA-seq data contain very few of those poly (A) reads (0.003% in an ultra-deep sequencing study), which hampers the ability of computational tools (Table 3) to identify poly (A) sites.

Table 3 Summary of methods using poly (A) reads to perform APA analysis.

Method	Description of method	Publication links
KLEAT	Characterizes 3′ UTRs in *de novo* assembled RNA-seq data through direct observation of poly (A) tails	https://www.ncbi.nlm.nih.gov/pmc/articles/PMC4350765/
ContextMAP	Maps poly (A) reads to predict poly (A) sites using a context-based approach	https://journals.plos.org/plosone/article?id=10.1371/journal.pone.0170914

2.4 Methods based on fluctuations in RNA-seq density near the 3′-end

The *main focus* of this chapter will be on methods belongs to this category since they are specifically designed for APA analysis in RNA-seq data, can perform *de novo* identification of poly (A) sites and are not confined by the limited amount of poly (A) reads. The general assumption of these methods (Table 4) is that any major changes in poly (A) site usage among different samples/conditions will results in localized fluctuations in RNA-seq read density near the 3′end of mRNA (Xia et al., 2014). Therefore, poly (A) sites are predicted as the sites of these localized fluctuations which are readily detected through single-nucleotide resolution RNA-seq analysis. The differential poly (A) site usage are then determined by comparing the ratios of expression levels of the isoform with longer 3′UTR over the total transcript abundance among samples (Xia et al., 2014).

3. The DaPars algorithm

In the following section, we will provide a detailed description of the most popular APA analysis tool in RNA-seq data DaPars. Dynamic analysis of Alternative PolyAdenylation from RNA-Seq (DaPars) is a bioinformatics tool that directly performs *de novo* identification of poly (A) sites and quantification of differential poly (A)-site usage among different conditions/samples regardless of prior poly (A) annotations. Although DaPars was initially developed for large-scale genomic project, The Cancer Genome Atlas (TCGA), which only compares between two conditions (*i.e.*, normal and tumor) (Xia et al., 2014), it was later extended to be applicable to any large-scale RNA-seq datasets with multiple conditions. For example, DaPars has been successfully used to systematically identify poly (A) sites

Table 4 Summary of methods based on fluctuations in RNA-seq density to perform APA analysis.

Method	Description of method	Publication links
DaPars	*De novo* identification and quantification of dynamic APA events	https://www.nature.com/articles/ncomms6274
PHMM	Identify potential APA using Poisson hidden Markov model	https://www.ncbi.nlm.nih.gov/pmc/articles/PMC3902974/
GETUTR	Predict poly (A) site by finding local maximum gradients in RNA-seq data with smoothed read coverage	https://pubmed.ncbi.nlm.nih.gov/25899044/
ChangePoint	Detect 3'UTR switching (differential poly (A) site usage) based on a likelihood ratio test	https://academic.oup.com/bioinformatics/article/30/15/2162/2391066
EBChangePoint	Identify 3′ splice sites by an empirical Bayes change-point model	https://academic.oup.com/bioinformatics/article/32/12/1823/1743857
IsoSCM	Identifying alternative 3′UTRs by searching change points in the read density	https://rnajournal.cshlp.org/content/21/1/14
APAtrap	Predict poly (A) sites and differential poly (A) site usage from RNA-seq data	https://academic.oup.com/bioinformatics/article/34/11/1841/4816794
TAPAS	Detect novel poly (A) sites from RNA-seq data	https://academic.oup.com/bioinformatics/article/34/15/2521/4904269

and quantify differential poly (A) site usage in 9475 samples across 53 human tissues using RNA-seq data form the Genotype-Tissue Expression (GTEx) project (Hong et al., 2020).

3.1 Method description

Fig. 1 presents a schematic illustration of the DaPars algorithm. For a given transcript, DaPars first (I) identifies the *de novo* distal poly (A) site (relative to the 5′ end of the mRNA) based on a continuous RNA-seq signal independent of the gene model; then (II) assuming there is an alternative *de novo*

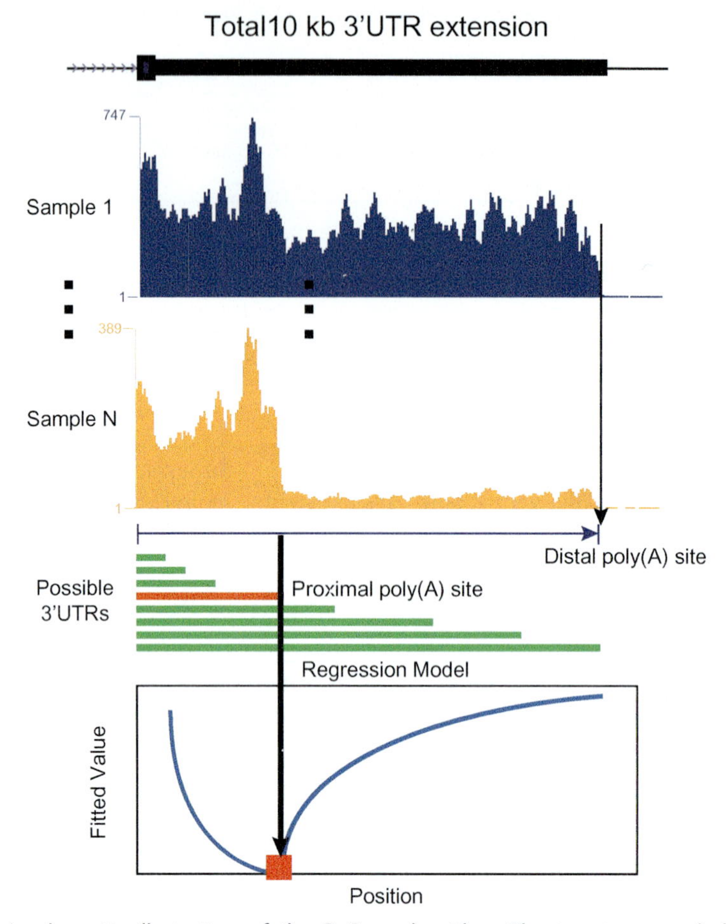

Fig. 1 A schematic illustration of the DaPars algorithm. The top two panel shows RNA-seq coverage 10 kb extended 3′UTR without prior knowledge of poly (A) sites. The y-axis of the bottom panel is the fitted value of the DaPars regression model and the position with the minimum fitted value (red rectangle) corresponds to the predicted proximal APA site (red horizontal bar).

proximal poly (A) site, DaPars forms a linear regression model to jointly infer the exact location of the *de novo* proximal poly (A) site by minimizing the deviation between the observed read density and the expected read density in all RNA-seq samples; finally (III) DaPars quantifies differential poly (A) sites usage as the change in Percentage of Distal poly (A) site Usage Index (PDUI), which is calculated as the proportion of the estimated abundances of transcripts with distal poly (A) sites (longer 3′UTRs) out of all transcripts (with both distal and proximal poly (A) sites), and transcripts favoring distal poly (A) site usage (long 3′UTRs) will have PDUI values near 1, whereas transcripts favoring proximal poly (A) site usage (short 3′UTRs) will have PDUI values near 0.

I. Distal poly (A) site identification from RNA-seq

Given two or more RNA-seq samples, distal poly (A) site of a transcript is defined as the end point of the longest 3′UTR among all samples (Xia et al., 2014). To avoid missing potential distal poly (A) sites that may locate outside of gene annotation, DaPars extends the annotated gene 3′end by up to 10kb before reaching a neighboring gene. RNA-seq data from all samples will be combined along the extended gene model and a 50-bp sliding window will be used to smooth the combined coverage. DaPars then identifies the *de novo* distal poly (A) site as the position in the extended 3′UTR whose coverage is significantly lower than the coverage at the start of the proceeding exon. In addition, since most current RNA-seq data are not strand-specific, DaPars applies additional cutoffs to separate overlapping 3′UTRs from two neighboring "tail-to-tail" genes from different strands to avoid false-positive distal poly (A) site. The distal poly (A) sites identified in this step is independent from gene models (not included in gene model), and therefore will benefit the downstream proximal poly (A) site identification.

II. Proximal poly (A) site identification from RNA-seq

For each transcript with a distal poly (A) site estimated from step (I), DaPars forms a linear regression model to predict the exact location of the *de novo* proximal poly (A) site. The regression model scans each locus along the extended 3′UTRs and minimizes the deviation between the observed read density from RNA-seq and the expected read density based on the two poly (A)-site model (assume there are only two poly(A) sites in the 3′UTR region) in all samples simultaneously. The regression model is as following:

$$\left(W_L^{1,2,3,\ldots,N*}, W_S^{1,2,3,\ldots,N*}, P^*\right) = \frac{argmin}{W_L^{1,2,3,\ldots,N}, W_S^{1,2,3,\ldots,N} \geq 0, 1 < P < L}$$

$$\sum_{i=1}^{N} \left\| C_i - \left(W_L^i I_L + W_S^i I_P\right) \right\|_2^2 \quad (1)$$

where W_L^i and W_S^i are the abundances of transcripts with distal and proximal poly (A) sites for sample i, respectively, and for each P $(1 < P < L)$, W_L^i and W_S^i in all N samples can be estimated using quadratic programming; N is the number of samples; $C_i = [C_{i1}, C_{i2}, \ldots, C_{iL}]^T$ is the read coverage of sample i at single-nucleotide resolution normalized by total sequencing depth; L is the length of the longest 3′UTR from step (I); P is the length of the proximal 3′UTRs to be inferred; I_L and I_P are two indicator functions for long and short 3′UTRs, respectively, such that $I_L = \underset{L}{[1, \cdots, 1]}$ and $I_P = \underset{P, L-P}{[1, \cdots, 1, 0, \cdots, 0]}$; $W_L^{\,i} I_L + W_S^{\,i} I_P$ is the expected read coverage. The locus with the optimal (minimized) fitting value from the above equation is identified as the *de novo* proximal poly (A) site (left vertical arrow in Fig. 1).

III. Quantification of differential poly (A) sites usage

Denote the optimal *de novo* proximal poly (A) site identified in step (II) as P^*, and the corresponding optimal expression levels of transcripts with distal and proximal poly (A) sites as $W_L^{\,i*}$ and $W_S^{\,i*}$. DaPars quantifies the relative poly (A) site usage as the Percentage of Distal poly (A) site Usage Index (PDUI) defined as:

$$PDUI_i = \frac{W_L^{i*}}{W_L^{i*} + W_S^{i*}} \tag{2}$$

The smaller the PDUI is, the less the distal PAS is used, the shorter the 3′UTRs and *vice versa.* After calculating the PDUI values of all samples, DaPars quantifies the differential poly (A) site usage by the absolute mean difference ($\Delta PDUI$) of PDUIs of two samples.

$$|\Delta PDUI| = \left|PDUI_{sample\ 1} - PDUI_{sample\ 2}\right| \tag{3}$$

To further detect the most significant poly (A) site switching events (APA events), DaPars used the following criteria: first, given the expression levels of transcripts with distal and proximal poly (A) sites ($W_L^{\,i*}$ and $W_S^{\,i*}$), the Benjamini-Hochberg corrected Fisher's exact test p-value (FDR) of PDUI difference between two samples is controlled at 5%; second, the absolute mean difference of PDUIs ($\Delta PDUI$) is no less than 0.2.

$$\begin{cases} FDR \le 0.05 \\ |\Delta PDUI| = \left|PDUI_{sample\ 1} - PDUI_{sample\ 2}\right| \ge 0.2 \end{cases} \tag{4}$$

Even though multiple similar tools have been developed after DaPars (*e.g.*, APAtrap (Ye, Long, Ji, Li, & Wu, 2018), TAPAS (Arefeen, Liu, Xiao, & Jiang, 2018)), by far, DaPars is still the most popular tool for APA analysis in RNA-seq data (cited 266 times), primarily because it is a powerful and precise bioinformatics algorithm to identify poly (A) sites and quantify differential poly (A) site usage, but also partly because it is an easy-to-use bioinformatics software. Fig. 2 represents the bioinformatics pipeline for running DaPars using any arbitrary RNA-seq datasets. Users start by aligning downloaded RNA-seq data using any RNA-seq alignment tools such as STAR (Dobin et al., 2013) or TopHat (Trapnell, Pachter, & Salzberg, 2009), and the result of which will be stored in the bedgraph format for the downstream DaPars analysis. DaPars also requires additional gene model file in bed format downloaded from UCSC genome browser and the mapping of transcripts to gene symbols file as input. The DaPars results will be generated by two separate python scripts which identify both distal and proximal poly (A) sites and quantify differential poly (A) site usage among different samples. The output of DaPars will be a PDUI matrix with rows representing transcripts and columns representing samples. The minimum requirement for running DaPars is 8 GB RAM (16 GB recommended) and 500 GB free disk space.

3.2 Application of DaPars

Systematic APA analysis in The Cancer Genome Atlas (TCGA) by DaPars has reveal how pervasive and recurrent APA events are in large clinical cohorts across distinct cancer types (Feng, Li, Wagner, & Li, 2018; Xia et al., 2014). Global-scale coordinated APA events are commonly observed in cancers, yet, even though TCGA has devoted significant efforts to characterize numerous genomic, epigenomic, and transcriptomic features in thousands of tumors, they lack APA-level analysis due to the lack of Poly (A)-seq platform and the lack of computational tools to analyze existing RNA-seq data (Xia et al., 2014). DaPars retrospective analysis of existing TCGA RNA-seq, which were originally sequenced for gene expression, identified 244–744 genes with statistically significant and recurrent (occurrence rate > 20%) differential poly (A) site usage during tumorigenesis, adding up to a total of 1346 non-redundant APA genes across 7 tumor types (Xia et al., 2014). What's more, by combing APA information with clinical covariates (tumor stage, age, gender, and smoking status) into the Cox proportional hazards model, much more significant *p*-values are derived compared to use clinical covariates alone (Xia et al., 2014), indicating that APA induced 3′UTR shortening or lengthening is associated with poor survival of cancer patients.

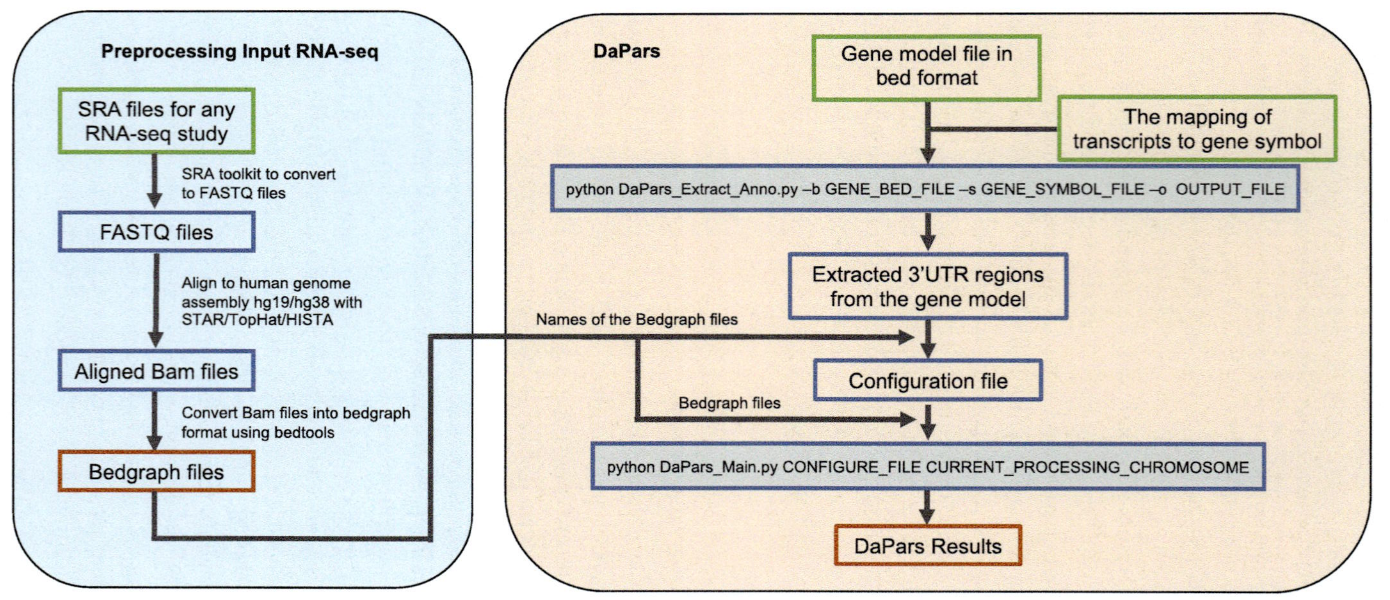

Fig. 2 Bioinformatics pipeline for running DaPars using RNA-seq data. The left panel shows RNA-seq preprocessing steps which generate the bedgraph files required for DaPars analysis. The right panel shows the steps for running DaPars.

In addition, DaPars constructed an atlas of tissue-specific, human APA events using 8277 RNA-seq datasets from the Genotype-Tissue Expression (GTEx) project, and further linked APA-associated genetic variants to the genetic risk of multiple diseases (Li, Gao, Peng, Wagner, & Li, 2019). For example, one single-nucleotide polymorphism (SNP) (rs10954213) in IRF5 has been associated with the alternative 3′UTR change in the same gene and affect mRNA stability, which further contribute to the susceptibility of systemic lupus erythematosus (SLE) (Li et al., 2019). In total, approximately 0.4 million common APA-associated genetic variants (3′aQTLs) were identified across 46 tissues of 467 individuals, which were able to explain up to 16.1% of human traits and diseases (*i.e.*, 16.1% trait-associated variants co-localized with 3′aQTLs in at least on tissue) (Li et al., 2019).

4. APA analysis in single cells

Single-cell RNA sequencing (scRNA-seq) has become one of the most widely used technologies in biomedical research by providing an unprecedented opportunity to quantify the abundance of diverse transcript isoforms among individual cells (Saliba, Westermann, Gorski, & Vogel, 2014; Shapiro, Biezuner, & Linnarsson, 2013). However, methods to quantify relative APA usage across single cells remain underdeveloped. Recently, Velten et al. (2015) developed an experimental protocol BATseq to quantify various 3′UTR-isoforms at the single-cell resolution. By integrating the standard scRNA-seq protocol and the 3′ enriched bulk RNA-seq protocol, Velten et al. found that cell types can be well separated based exclusively on their 3′UTR isoform usages, indicating that APA is a molecular feature intrinsic to cell states (Velten et al., 2015). While a compelling method, BATseq is hampered by its low sensitivity (~5%) and high procedural complexity, thereby not being widely adopted in practice. In contrast, standard scRNA-seq data is widely available, yet most of the scRNA-seq data has not been analyzed in a manner that would systematically identify and quantify APA events. Since scRNA-seq only captures a small fraction (typically 5–15%) of the total mRNAs in each cell (Stegle, Teichmann, & Marioni, 2015), it can falsely quantify genes, especially lowly expressed ones, as unexpressed; this phenomenon is termed as "dropout." Existing bulk RNA-seq based APA methods such as DaPars cannot overcome this vexing challenge when applied directly to scRNA-seq data, as they would lead to a high degree of sparsity in the resulting APA profiles.

4.1 The scDaPars algorithm

To fill the above knowledge gap, scDaPars (Dynamic analysis of Alternative PolyAdenylation from scRNA-Seq) was recently developed for quantifying and recovering APA usage at the single-cell and single-gene resolution using both 3′end and full-length single-cell RNA-seq (scRNA-seq) data (Li, Gao, Li, & Amos, 2020). scDaPars employs a regression model that enables sharing of APA information across related cells to tackle the sparsity, achieving considerable robustness when applied to noisy scRNA-seq data. Given a scRNA-seq dataset, scDaPars first uses DaPars to calculate raw relative APA usage, measured by the percentage of distal poly (A) site usage index (PDUI) (Fig. 3). Of note, the raw PDUI values can only be estimated for genes with sufficient read coverages (default coverage of 5 reads per base), which automatically separates genes into robust genes (genes unaffected by dropout events) and dropout genes for further analysis. Due to the intrinsically low coverage of scRNA-seq data, the resulting PDUI matrix from last step will be overly sparse with widespread missing data. To further recover the complete PDUI matrix independent of gene expression, we develop a new imputation method by sharing APA information across different cells (Fig. 3). The R package scDaPars has a minimum system requirement of 8 GB RAM for computing.

scDaPars begins by determining which cells are from the same cell subpopulation and therefore are neighboring cells. In order to quantify APA usage independent of gene expression, scDaPars uses raw APA usage for this task. However, due to the technical limitation of scRNA-seq data, it is unlikely to completely cluster cells into true subpopulations based on the sparse PDUI matrix generated in last step. Instead, the goal of this step is to determine a set of potential neighboring cells which scDaPars will fine-tune in the following steps. To increase the robustness and reliability of the clustering results and to find more plausible neighboring cells, scDaPars applies principle component analysis (PCA) to the raw PDUI matrix, so that principal components (PCs) that can together explain at least 40% of the variance in the data are used for clustering. The cells are then clustered into subpopulations using graph-based community detection algorithm *walkstrap* (Pons & Latapy, 2005). The single cells are the vertices in the graph, and community detection in graphs will identify groups of vertices with high probability of being connected to each other than to members of other groups. Suppose scDaPars divides cells into K subpopulations in this step, for each cell m, its potential neighboring cells N_m are the other cells in the same cell subpopulation k.

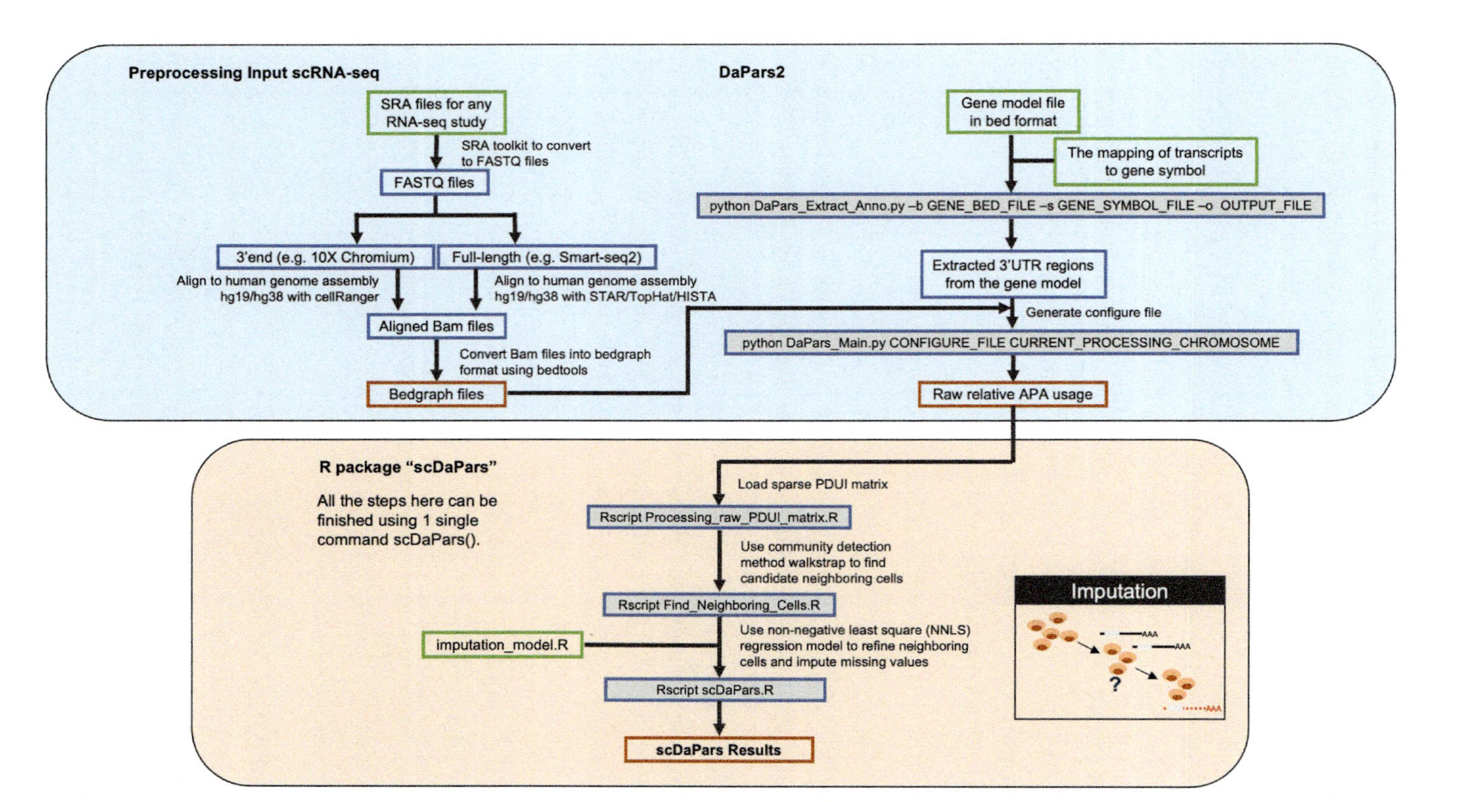

Fig. 3 Bioinformatics pipeline for running scDaPars using any scRNA-seq data. The upper panel shows RNA-seq preprocessing and DaPars steps which generate the raw relative APA usage for downstream imputation. The lower panel shows the steps for running scDaPars imputation steps.

$$N_m = \{i \in k, i \neq m\} \tag{5}$$

After potential neighboring cells N_m for each cell are determined, scDaPars imputes missing APA usage cell by cell. Recall that PDUIs can only be estimated for genes with sufficient read coverage, scDaPars thereby automatically separates genes into robust genes and dropout genes when calculating the PDUI matrix. Here, denote the set of robust genes for cell m as R_m and the set of dropout genes that will be imputed in this step as D_m. scDaPars then learns the cells' similarities through the robust gene set $G_{Robust,\ m}$ and impute the APA usage of D_m by borrowing information from the same gene's APA usage in other neighboring cells learned from R_m. To fine-tune the grouping of neighboring cells from N_m, scDaPars uses non-negative least squares (NNLS) regression:

$$\overline{\theta_m} = argmin_{\theta_m} \| PDUI_{R_m,m} - PDUI_{R_m,N_m}\theta_m \|_2^2, \theta_m > 0 \tag{6}$$

where N_m represents the indices of cells that are potential neighboring cells of cell m, $PDUI_{Gene_{robust},\ m}$ is a vector of response variables representing R_m rows in the m-th column (cell m) of the original PDUI matrix, $PDUI_{R_m,\ Nm}$ is a sub-matrix of the original PDUI matrix with dimensions $|R_m| \times |N_m|$. The goal is to find the optimal coefficients $\overline{\theta_m}$ of length $|N_m|$ that can minimize the deviation between APA usage of R_m in cell m and those in potential neighboring cells. The advantage of using NNLS is that it has the property of leading to a sparse estimate of θ_m, whose components may have exact zeros, so that true neighboring cells of cell m are conveniently selected from N_m. Once $\overline{\theta_m}$ is computed, scDaPars use this coefficient $\overline{\theta_m}$ estimated from the set R_m to impute the APA usage of genes in the set D_m in cell m.

$$\overline{PDUI_{g,m}} = \begin{cases} PDUI_{g,m}, \text{if } g \in R_m \\ PDUI_{g,N_m} \bullet \overline{\theta_m}, \text{if } g \in D_m \end{cases} \tag{7}$$

4.2 Application of scDaPars

scDaPars enables identification of novel cell subpopulations invisible to conventional gene expression analysis alone during endoderm differentiation. scDaPars analysis in a time-course Smart-seq2 scRNA-seq dataset (Chu et al., 2016) containing 758 cells sequenced at 0, 12, 24, 36, 72 and 96h of differentiation during human definitive endoderm (DE) emergence revealed clear and compact cell clusters from each time point along the differentiation process. In addition, scDaPars help delineate novel and potential meaning cell subpopulations in cells at 96h of differentiation by integrating APA profile with imputed gene expression using similarity network fusion (SNF) (Wang et al., 2014). The identified novel cell subpopulation exhibited more differentiated phenotype evidenced by upregulated marker genes (*e.g.*, *SOX17*, *GATA6*, *EOMES*) and enriched differentiation associated pathways (Li et al., 2020).

5. Summary

In this chapter, we introduced four categories of bioinformatics tools to identify poly (A) sites and quantify differential poly (A) site usage in RNA-seq data, primarily focused on the most popular APA analysis tool DaPars. DaPars not only performs powerful and precise APA analysis in RNA-seq, it is also easy to use and applicable to any large-scale RNA-seq datasets including TCGA and GTEx. Furthermore, with the ever-growing number of single-cell studies, we introduced scDaPars, which builds on DaPars, to systematically identify and quantify APA events in both 3′end and full-length scRNA-seq data.

References

An, J. J., Gharami, K., Liao, G.-Y., Woo, N. H., Lau, A. G., Vanevski, F., et al. (2008). Distinct role of long 3′ UTR BDNF mRNA in spine morphology and synaptic plasticity in hippocampal neurons. *Cell*, *134*(1), 175–187.

Arefeen, A., Liu, J., Xiao, X., & Jiang, T. (2018). TAPAS: Tool for alternative polyadenylation site analysis. *Bioinformatics*, *34*(15), 2521–2529.

Behr, J., Kahles, A., Zhong, Y., Sreedharan, V. T., Drewe, P., & Rätsch, G. (2013). MITIE: Simultaneous RNA-Seq-based transcript identification and quantification in multiple samples. *Bioinformatics*, *29*(20), 2529–2538.

Chen, M., Ji, G., Fu, H., Lin, Q., Ye, C., Ye, W., et al. (2020). A survey on identification and quantification of alternative polyadenylation sites from RNA-seq data. *Briefings in Bioinformatics*, *21*(4), 1261–1276.

Chu, L.-F., Leng, N., Zhang, J., Hou, Z., Mamott, D., Vereide, D. T., et al. (2016). Single-cell RNA-seq reveals novel regulators of human embryonic stem cell differentiation to definitive endoderm. *Genome Biology*, *17*(1), 1–20.

Derti, A., Garrett-Engele, P., MacIsaac, K. D., Stevens, R. C., Sriram, S., Chen, R., et al. (2012). A quantitative atlas of polyadenylation in five mammals. *Genome Research*, *22*(6), 1173–1183.

Dobin, A., Davis, C. A., Schlesinger, F., Drenkow, J., Zaleski, C., Jha, S., et al. (2013). STAR: Ultrafast universal RNA-seq aligner. *Bioinformatics*, *29*(1), 15–21.

Edwalds-Gilbert, G., Veraldi, K. L., & Milcarek, C. (1997). Alternative poly (A) site selection in complex transcription units: Means to an end? *Nucleic Acids Research*, *25*(13), 2547–2561.

Feng, X., Li, L., Wagner, E. J., & Li, W. (2018). TC3A: The cancer 3′ UTR atlas. *Nucleic Acids Research*, *46*(D1), D1027–D1030.

Flavell, S. W., Kim, T.-K., Gray, J. M., Harmin, D. A., Hemberg, M., Hong, E. J., et al. (2008). Genome-wide analysis of MEF2 transcriptional program reveals synaptic target genes and neuronal activity-dependent polyadenylation site selection. *Neuron*, *60*(6), 1022–1038.

Garneau, N. L., Wilusz, J., & Wilusz, C. J. (2007). The highways and byways of mRNA decay. *Nature Reviews Molecular Cell Biology*, *8*(2), 113–126.

Grassi, E., Mariella, E., Lembo, A., Molineris, I., & Provero, P. (2016). Roar: Detecting alternative polyadenylation with standard mRNA sequencing libraries. *BMC Bioinformatics*, *17*(1), 1–9.

Gruber, A. J., Schmidt, R., Ghosh, S., Martin, G., Gruber, A. R., van Nimwegen, E., et al. (2018). Discovery of physiological and cancer-related regulators of 3′ UTR processing with KAPAC. *Genome Biology*, *19*(1), 1–17.

Harrison, P. F., Powell, D. R., Clancy, J. L., Preiss, T., Boag, P. R., Traven, A., et al. (2015). PAT-seq: A method to study the integration of 3′-UTR dynamics with gene expression in the eukaryotic transcriptome. *RNA*, *21*(8), 1502–1510.

Hoffman, Y., Bublik, D. R., Ugalde, A. P., Elkon, R., Biniashvili, T., Agami, R., et al. (2016). 3'UTR shortening potentiates microRNA-based repression of pro-differentiation genes in proliferating human cells. *PLoS Genetics*, *12*(2), e1005879.

Hong, W., Ruan, H., Zhang, Z., Ye, Y., Liu, Y., Li, S., et al. (2020). APAatlas: Decoding alternative polyadenylation across human tissues. *Nucleic Acids Research*, *48*(D1), D34–D39.

Hoque, M., Ji, Z., Zheng, D., Luo, W., Li, W., You, B., et al. (2013). Analysis of alternative cleavage and polyadenylation by 3′ region extraction and deep sequencing. *Nature Methods*, *10*(2), 133–139.

Jan, C. H., Friedman, R. C., Ruby, J. G., & Bartel, D. P. (2011). Formation, regulation and evolution of Caenorhabditis elegans 3′ UTRs. *Nature*, *469*(7328), 97–101.

Ji, Z., Lee, J. Y., Pan, Z., Jiang, B., & Tian, B. (2009). Progressive lengthening of 3′ untranslated regions of mRNAs by alternative polyadenylation during mouse embryonic development. *Proceedings of the National Academy of Sciences*, *106*(17), 7028–7033.

Katz, Y., Wang, E. T., Airoldi, E. M., & Burge, C. B. (2010). Analysis and design of RNA sequencing experiments for identifying isoform regulation. *Nature Methods*, 7(12), 1009–1015.

Li, W., Gao, Y., Li, L., & Amos, C. I. (2020). *Dynamic analysis of alternative polyadenylation from single-cell RNA-Seq (scDaPars) reveals cell subpopulations invisible to gene expression analysis*. bioRxiv.

Li, L., Gao, Y., Peng, F., Wagner, E. J., & Li, W. (2019). Genetic basis of alternative polyadenylation is an emerging molecular phenotype for human traits and diseases. *SSRN Electronic Journal*. Available at SSRN 3351825.

Müller, S., Rycak, L., Afonso-Grunz, F., Winter, P., Zawada, A. M., Damrath, E., et al. (2014). APADB: A database for alternative polyadenylation and microRNA regulation events. *Database*, *2014*, bau076.

Ozsolak, F., Kapranov, P., Foissac, S., Kim, S. W., Fishilevich, E., Monaghan, A. P., et al. (2010). Comprehensive polyadenylation site maps in yeast and human reveal pervasive alternative polyadenylation. *Cell*, *143*(6), 1018–1029.

Pons, P., & Latapy, M. (2005). *International symposium on computer and information sciences*. Springer.

Saliba, A.-E., Westermann, A. J., Gorski, S. A., & Vogel, J. (2014). Single-cell RNA-seq: Advances and future challenges. *Nucleic Acids Research*, *42*(14), 8845–8860.

Sandberg, R., Neilson, J. R., Sarma, A., Sharp, P. A., & Burge, C. B. (2008). Proliferating cells express mRNAs with shortened 3'untranslated regions and fewer microRNA target sites. *Science*, *320*(5883), 1643–1647.

Shapiro, E., Biezuner, T., & Linnarsson, S. (2013). Single-cell sequencing-based technologies will revolutionize whole-organism science. *Nature Reviews Genetics*, *14*(9), 618–630.

Sherstnev, A., Duc, C., Cole, C., Zacharaki, V., Hornyik, C., Ozsolak, F., et al. (2012). Direct sequencing of Arabidopsis thaliana RNA reveals patterns of cleavage and polyadenylation. *Nature Structural & Molecular Biology*, *19*(8), 845.

Shi, Y. (2012). Alternative polyadenylation: New insights from global analyses. *RNA*, *18*(12), 2105–2117.

Stegle, O., Teichmann, S. A., & Marioni, J. C. (2015). Computational and analytical challenges in single-cell transcriptomics. *Nature Reviews Genetics*, *16*(3), 133–145.

Tian, B., Hu, J., Zhang, H., & Lutz, C. S. (2005). A large-scale analysis of mRNA polyadenylation of human and mouse genes. *Nucleic Acids Research*, *33*(1), 201–212.

Tian, B., & Manley, J. L. (2017). Alternative polyadenylation of mRNA precursors. *Nature Reviews Molecular Cell Biology*, *18*(1), 18–30.

Trapnell, C., Pachter, L., & Salzberg, S. L. (2009). TopHat: Discovering splice junctions with RNA-Seq. *Bioinformatics*, *25*(9), 1105–1111.

Trapnell, C., Williams, B. A., Pertea, G., Mortazavi, A., Kwan, G., Van Baren, M. J., et al. (2010). Transcript assembly and quantification by RNA-Seq reveals unannotated transcripts and isoform switching during cell differentiation. *Nature Biotechnology*, *28*(5), 511–515.

Ulitsky, I., Shkumatava, A., Jan, C. H., Subtelny, A. O., Koppstein, D., Bell, G. W., et al. (2012). Extensive alternative polyadenylation during zebrafish development. *Genome Research*, *22*(10), 2054–2066.

Velten, L., Anders, S., Pekowska, A., Järvelin, A. I., Huber, W., Pelechano, V., et al. (2015). Single-cell polyadenylation site mapping reveals 3′ isoform choice variability. *Molecular Systems Biology*, *11*(6), 812.

Wang, B., Mezlini, A. M., Demir, F., Fiume, M., Tu, Z., Brudno, M., et al. (2014). Similarity network fusion for aggregating data types on a genomic scale. *Nature Methods*, *11*(3), 333.

Wang, R., Nambiar, R., Zheng, D., & Tian, B. (2018). PolyA_DB 3 catalogs cleavage and polyadenylation sites identified by deep sequencing in multiple genomes. *Nucleic Acids Research*, *46*(D1), D315–D319.

Xia, Z., Donehower, L. A., Cooper, T. A., Neilson, J. R., Wheeler, D. A., Wagner, E. J., et al. (2014). Dynamic analyses of alternative polyadenylation from RNA-seq reveal a 3′-UTR landscape across seven tumour types. *Nature Communications*, *5*(1), 1–13.

Ye, C., Long, Y., Ji, G., Li, Q. Q., & Wu, X. (2018). APAtrap: Identification and quantification of alternative polyadenylation sites from RNA-seq data. *Bioinformatics*, *34*(11), 1841–1849.

You, L., Wu, J., Feng, Y., Fu, Y., Guo, Y., Long, L., et al. (2015). APASdb: A database describing alternative poly (A) sites and selection of heterogeneous cleavage sites downstream of poly (A) signals. *Nucleic Acids Research*, *43*(D1), D59–D67.

CHAPTER TWELVE

Quantifying alternative polyadenylation in RNAseq data with LABRAT

Austin E. Gillen[a], Raeann Goering[b,c], and J. Matthew Taliaferro[b,c,*]
[a]Division of Hematology, University of Colorado School of Medicine, Aurora, CO, United States
[b]Department of Biochemistry and Molecular Genetics, University of Colorado Anschutz Medical Campus, Aurora, CO, United States
[c]RNA Bioscience Initiative, University of Colorado Anschutz Medical Campus, Aurora, CO, United States
*Corresponding author: e-mail address: matthew.taliaferro@cuanschutz.edu

Contents

Abstract

Alternative polyadenylation (APA) generates transcript isoforms that differ in their 3′ UTR content and may therefore be subject to different regulatory fates. Although the existence of APA has been known for decades, quantification of APA isoforms from high-throughput RNA sequencing data has been difficult. To facilitate the study of

Methods in Enzymology, Volume 655
ISSN 0076-6879
https://doi.org/10.1016/bs.mie.2021.03.018

APA in large datasets, we developed an APA quantification technique called LABRAT (Lightweight Alignment-Based Reckoning of Alternative Three-prime ends). LABRAT leverages modern transcriptome quantification approaches to determine the relative abundances of APA isoforms. In this manuscript we describe how LABRAT produces its calculations, provide a step-by-step protocol for its use, and demonstrate its ability to quantify APA in single-cell RNAseq data.

1. Introduction

The 3′ ends of eukaryotic mRNAs are defined by the processes of cleavage and polyadenylation. In many genes, cleavage and polyadenylation can occur at more than one location, leading to the generation of transcript isoforms that differ in the compositions of their 3′ UTRs. Because the 3′ UTR content of an mRNA can modulate the stability, localization, and translational efficiency of the transcript, APA therefore contributes to the regulation of each of these processes (Cho et al., 2005; Mayr & Bartel, 2009; Sandberg, Neilson, Sarma, Sharp, & Burge, 2008; Taliaferro et al., 2016).

In general, APA events can be classified into two distinct structures. Alternative polyadenylation sites can be located within the same terminal exon, giving rise to a structure sometimes called "Tandem UTRs" (Fig. 1A, left). Conversely, APA sites can be located within different terminal exons, giving rise to an "alternative last exon" (ALE) structure (Fig. 1A, right). Although these two classes of APA have historically been considered separately, recent evidence has demonstrated that they are tightly coregulated, suggesting that they may be regulated by similar mechanisms (Goering et al., 2020; Taliaferro et al., 2016).

Given the effect that APA can have on RNA metabolism, the quantification of the relative abundance of APA isoforms for a given gene has been of interest for many years. The advent of high-throughput RNA sequencing opened up the possibility of performing this quantification for many genes at once and of identifying genes whose APA status changes across conditions. A variety of computational tools have been developed for this purpose (Grassi, Mariella, Lembo, Molineris, & Provero, 2016; Ha, Blencowe, & Morris, 2018; Xia et al., 2014).

Early tools (Grassi et al., 2016; Xia et al., 2014) relied on computed alignments between RNAseq reads and a supplied genome or transcriptome. Following this alignment, the abundances of APA isoforms could be calculated by counting the numbers of reads consistent with each isoform.

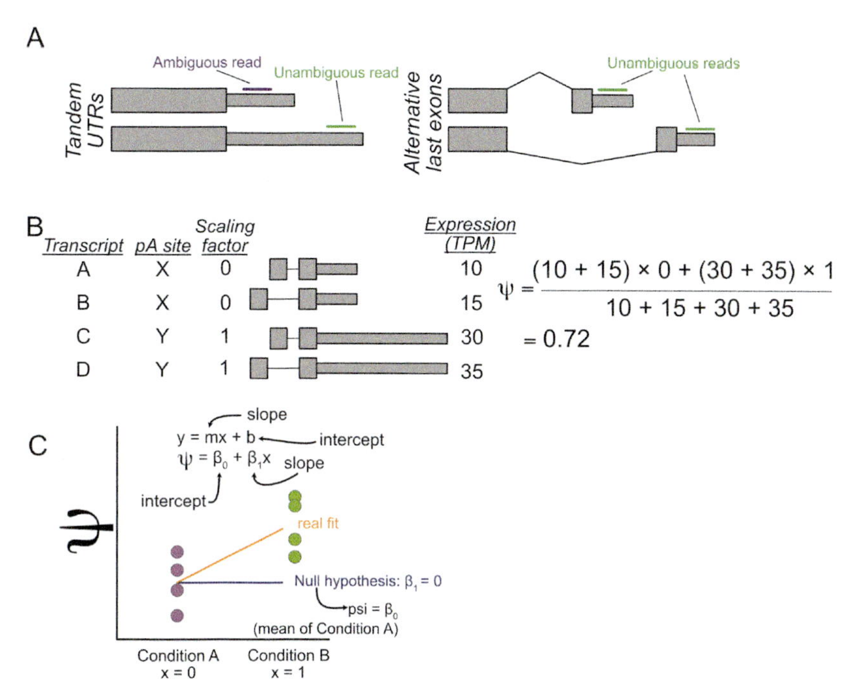

Fig. 1 Overview of LABRAT approach. (A) Tandem UTR (left) and alternative last exon (right) gene structures. (B) Visualization of procedure for calculating ψ. (C) Linear model used by LABRAT for identification of genes with significant changes in ψ across conditions.

However, these tools suffer from a disadvantage inherent to APA isoforms and tandem UTR structures in particular. In tandem UTR structures, there is often a large amount of shared sequence between the two isoforms. This makes reads that align to this common region less powerful in their ability to discriminate between the isoforms, since they could have arisen from either isoform. This reduces the overall statistical power and accuracy of the analysis.

Alternatively, later approaches, including LABRAT (Goering et al., 2020; Ha et al., 2018), take advantage of modern transcriptome quantification methods like Salmon and Kallisto (Bray, Pimentel, Melsted, & Pachter, 2016; Patro, Duggal, Love, Irizarry, & Kingsford, 2017). These quantification methods can fractionally and probabilistically assign multimapping "ambiguous" reads to individual transcripts, improving the accuracy and power of subsequent quantifications. LABRAT leverages the quantifications produced by Salmon of individual transcripts to compute quantifications of

APA site usage. It then compares relative APA site usage across conditions to identify genes whose relative abundance of APA isoforms changes between samples.

2. How LABRAT works

In this section, we will provide an overview of how LABRAT performs its computations. Generally, LABRAT takes a set of transcript-level quantifications produced by Salmon and aggregates them into APA site-level quantifications. These APA site-level quantifications are then used to create an overall value that represents the APA status of the gene called ψ. Values of ψ can range between 0 and 1 with a ψ value of 0 representing exclusive usage of the most upstream (gene-proximal) APA site and a ψ value of 1 representing exclusive usage of the most downstream (gene-distal) APA site. LABRAT then compares ψ values across conditions to identify genes whose APA is differentially regulated between conditions.

2.1 Filtering transcripts

LABRAT takes in a genome annotation in gff format. From this annotation it derives the 3′ ends of transcripts to be quantified. However, it does not consider *every* transcript. In many annotations, there are dubious transcripts that result from incomplete transcript assemblies, old idiosyncratic ESTs, RNAs that haven't yet been fully processed, and other error prone sources. Because these may negatively impact the accuracy of APA quantification, LABRAT uses a set of filters to remove these transcripts.

Some of these filters utilize specific transcript tags found in the supplied annotation. These tags may not be found in every annotation, but are always found in Gencode gff annotations. Because Gencode annotations are only offered for human and mouse genomes, this restricts the species compatible to analysis with LABRAT. To ameliorate this limitation, we wrote specific versions of LABRAT that are compatible with Ensembl annotations for rat and *Drosophila* genomes.

The first filter used ensures that the transcript is protein coding. Although APA may regulate noncoding transcripts including lncRNAs, a large fraction of the undesired, spurious transcripts are not protein coding. To filter these, LABRAT selects transcripts that have the "protein_coding" attribute.

Transcripts whose 3′ end is not well defined have the potential to induce artifacts in APA quantification. These transcripts often arise from degraded or partial transcripts, yet still end up in many genome annotations. To remove these transcripts from the analysis, LABRAT filters out transcripts that contain the attribute "mRNA_end_NF."

2.2 Truncating transcripts

Transcript models derived from genome annotations necessarily contain all the exons, both constitutive and alternative, that exist within the model. This directly and inflexibly links the inclusion of all exons within that transcript model with cleavage and polyadenylation at the site defined by the model's 3′ end. However, this may not be biologically meaningful. Although links between alternative exon inclusion and alternative polyadenylation have been observed (Pai et al., 2016), it is not necessarily true that the inclusion or exclusion of an alternative exon upstream in a transcript is always followed by the usage of a given APA site.

To decouple such potentially spurious links, LABRAT truncates each transcript to its final two exons prior to quantification. These transcript terminal fragments are supplied to Salmon for quantification, and the TPM (transcripts per million) values produced are used in downstream steps.

2.3 Calculating expression of APA sites

Salmon quantifies the abundances of individual transcripts. RNA expression at the gene level can be derived from these transcript quantifications by summing expression values across all transcripts within the gene. This strategy is used by the popular RNAseq analysis tool tximport (Soneson, Love, & Robinson, 2015) for usage with differential gene expression analysis tools. LABRAT employs a similar strategy. However, instead of summing expression values across all transcripts that belong to a *gene*, LABRAT sums expression values across all transcript fragments that belong to a given *APA site*.

Following quantification, transcript fragments that share 3′ ends are grouped together. By default, transcripts whose 3′ ends are within 25 nt of each other are grouped together, although this parameter is tunable. This allows for some microheterogeneity in APA site location which may arise either for biological reasons or to small inaccuracies in the genome annotation. The expression values (TPMs) of all transcript fragments that belong to a 3′ end are summed, generating an expression quantification *for the APA site*. Following this process, each gene will be associated with one expression quantification per APA site.

Genes that do not meet a *total* expression threshold (i.e., the sum across all of the APA sites for the gene) are filtered and removed from further analysis. By default, this threshold is set at 5 TPM, although this parameter is also tunable.

2.4 Calculating ψ values

Following APA site quantification, LABRAT summarizes the APA status of the gene using a metric called ψ. Each gene is therefore assigned one ψ value. A ψ value of 0 represents a gene where the most upstream (gene-proximal) APA site is exclusively used, and a ψ value of 1 represents a gene where the most downstream (gene-distal) APA site is exclusively used.

LABRAT begins by ordering the APA sites for a gene from most upstream to most downstream. For each gene, the number of distinct APA sites, n, is recorded, and each APA site is assigned a value, m, that is equal to its rank order from most upstream to most downstream. LABRAT then calculates a scaling factor for each APA site that is equal to $(m-1)/(n-1)$.

The scaled expression value of each APA site is then calculated by multiplying the expression of the APA site by its scaling factor. The scaled expression values are then summed across all APA sites. Unscaled expression values are similarly summed across all APA sites. ψ values are then calculated as the ratio between the scaled and unscaled expression values.

Consider a gene with four transcript fragments that belong to two distinct APA sites (Fig. 1B). Expression values for each APA site are calculated by summing expression values across all transcript fragments that belong to the site. The transcripts that belong to the upstream APA site will have a scaling factor of 0 while the transcripts that belong to the downstream APA site will have a scaling factor of 1. A ψ value can then be calculated by taking the ratio of scaled and unscaled expression values (Fig. 1B).

Importantly, this process scales to genes with more than two APA sites. For example, the scaling factors for a gene with three APA sites would be 0, 0.5 and 1. Each gene, regardless of the number of APA sites it contains, is assigned a single ψ value. For genes with more than two APA sites, ψ therefore gives an overall sense of the relative usage of upstream and downstream APA sites. This approach can be contrasted with the alternative of pairwise comparisons between all possible pairs of APA sites, which, if the gene contains many APA sites, can quickly become unwieldy.

2.5 Comparing ψ values across conditions

After computing ψ values for each gene in each sample, LABRAT compares ψ values across conditions to identify genes whose relative APA usage has changed. This is done using a linear mixed effects model (Fig. 1C). LABRAT fits a model to the ψ values across conditions and then asks if that model is a better fit than a null model that assumes no change in ψ across conditions using a log likelihood test. The raw p value from the log likelihood test is then corrected for multiple hypothesis testing using the Benjamini-Hochberg method (Benjamini & Hochberg, 1995). Importantly, this model can incorporate covariates (e.g., batch, library design).

2.6 RNAseq library design

Although the majority of publicly available RNAseq data is derived from RNAseq libraries that cover the whole transcript, these are perhaps not the best library design for quantifying APA. Newer approaches that specifically profile the 3′ ends of reads (Zheng, Liu, & Tian, 2016) offer high sensitivity and accuracy for APA quantification. To deal with these designs, LABRAT includes the –librarytype parameter. If this value is set to "RNAseq" then the quantification of ψ values proceeds as described in Sections 2.1–2.5. However, if this value is set to "3pseq," then some minor deviations are employed.

First, instead of quantifying the last two exons of every transcript, the last 300 nt of every transcript are used. This value is used because the insert sizes in many 3′ end sequencing libraries are 100–300 nt long. Second, the transcript quantification of these libraries uses Salmon's count output instead of the length-normalized TPM value. This is because length normalization of 3′ end data is not necessary nor desirable.

3. Quantifying alternative polyadenylation with LABRAT

3.1 Installing LABRAT

Although this depends on the size of the transcriptome being analyzed, LABRAT generally uses between 5 and 20 Gb of memory and takes between 30 min and 4 h to complete. Multithreading is supported during the quantification of APA isoform abundance with salmon (Patro et al., 2017). See the README at https://github.com/TaliaferroLab/LABRAT for detailed installation instructions, requirements and detailed documentation. The above GitHub repository contains a file (labratenv.yaml) that contains the prerequisites necessary to run LABRAT. The code in Fig. 2 uses this file and the *conda* package manager (available at https://docs.conda.io/projects/conda/en/latest/user-guide/install/) to install everything needed for running LABRAT.

```
conda env create -f labratenv.yml
source activate labrat
```

Fig. 2 Example code to install python environments compatible with LABRAT.

3.2 Generating transcript ends for quantification

LABRAT is run in three steps. The first filters transcripts in a genome annotation for those with high confidence 3′ ends. The second quantifies the abundances of those ends. The third calculates the relative usage of APA sites in each gene as ψ values and identifies genes whose ψ value changes significantly across conditions.

LABRAT creates and utilizes a database constructed from a gff genome annotation file to relate transcripts and genes. To create this database, LABRAT requires these genome annotations in uncompressed gff3 format. It is recommended to use annotation files from GENCODE (https://www.gencodegenes.org) as they contain specific flags used by LABRAT. Once created, this database will be stored in the same directory as the gff file from which it was made. Database generation only needs to be performed once but can take a few hours. If interrupted, the partial database created should be deleted before attempting again. Once this database is created, LABRAT uses it to make a fasta file of transcript ends to be quantified.

The relevant options for the creation of this fasta file are as follows:

- –mode: This argument defines which of the three steps in APA quantification LABRAT will be performing. For the generation of the transcript end fasta file, it should be set to "makeTFfasta."
- –gff: This is the path to the gff annotation to be used.
- –genomefasta: This is the path to the sequence of the genome in fasta format.
- –lasttwoexons: This flag tells LABRAT if the fasta file it creates should contain the entire sequence of a transcript, or just it's 3′ end. If included, the 3′ end is generated. If omitted, the entire sequence is generated. Including this flag can lead to higher accuracy of APA quantifications. This is because it removes the contribution of upstream alternative exons, which are rigidly associated with specific 3′ ends in the annotation, to the quantification of APA.
- –librarytype: The allowed values of this parameter are "RNAseq" and "3pseq." These correspond to the library design strategies used in generating the data (see Section 2.6).

The example in Fig. 3 will generate a fasta file containing the last two exons of filtered transcripts from a human genome annotation file in preparation for quantification with RNAseq data. This fasta file will be named TFseqs.fa.

3.3 Quantification of transcript fragments

After generation of the transcript fragments, their relative abundance in the supplied high-throughput sequencing data will be calculated. This step relies on the transcriptome quantification tool Salmon (Patro et al., 2017).

This step should be run in an empty directory. Compressed or uncompressed fastq or fasta read files that contain either single or paired end reads can be used. The code in Fig. 4 outputs a directory for each sample containing salmon quantification files.

```
#create human database and terminal fragment fasta
python .../PyScripts/LABRAT.py --mode makeTFfasta --gff
.../Annotations/hg38/gencode.v32.annotation.gff3 --genomefasta
.../Annotations/hg38/GRCh38.p13.genome.fa.gz --lasttwoexons --librarytype
RNAseq
#rename TFseqs fasta
mv TFseqs.fasta hsTFseqs.fasta
```

Fig. 3 Example code for creating both the terminal fragment fasta file and a genome annotation database. Gencode's genome annotation gff file and genome fasta are required inputs. TFseqs.fasta is created in the current directory while the database file is generated in the same directory as the gff.

```
#create salmon quantification files
python .../PyScripts/LABRAT.py --mode runSalmon --librarytype RNAseq
--txfasta .../LABRAT/hsTFseqs.fasta --reads1
.../fastq/BrainM1_1.fastq.gz,...fastq/BrainF1_1.fastq.gz,.../fastq/LiverM1_
1.fastq.gz,.../fastq/LiverF1_1.fastq.gz,.../fastq/LiverF2_1.fastq.gz
--reads2
.../fastq/BrainM1_2.fastq.gz,.../fastq/BrainF1_2.fastq.gz,.../fastq/LiverM1
_2.fastq.gz,.../fastq/LiverF1_2.fastq.gz,.../fastq/LiverF2_2.fastq.gz
--samplename BrainM1,BrainF2,LiverM1,LiverF1,LiverF2 --threads 4
```

Fig. 4 Example code for running LABRAT's runSalmon function. RNAseq forward reads (reads1), reverse reads (reads2) and sample names are required inputs. Three prime end sequencing reads can also be used however the librarytype option should reflect the type of library provided. This code must be run in an empty directory as it outputs quantifications in new salmon directories for each sample.

The relevant options for the quantification of these transcript fragments are as follows:

- –mode: This argument defines which of the three steps in APA quantification LABRAT will be performing. For the quantification of transcript abundances, it should be set to "runSalmon."
- –txfasta: This is the path to the fasta file of transcript fragments to be quantified that was generated in Section 3.2.
- –reads1: A comma separated list of files containing the forward sequencing reads.
- –reads2: A comma separated list of files containing the reverse sequencing reads. This can be omitted if single end data is being used. Importantly, the order of this list must be consistent with the order of the samples in –reads1.
- –samplename: A comma separated list of names for the samples being quantified. Importantly, the order of this list must be consistent with the order of the samples in –reads1.

3.4 Calculating ψ values

Transcript abundances are then used to calculate ψ values for every gene with at least two alternative polyadenylation sites. Gene-level ψ values are then compared across conditions to identify genes whose ψ value has significantly changed.

This requires knowledge of which samples belong to which conditions. This is supplied to LABRAT using a tab-delimited file. Minimally, this file contains two columns with the headers "sample" and "condition." Optionally additional columns can be added that specify covariates. It is required that the header for any covariate column contain the string "covariate." Sample names must match those given to LABRAT during the Salmon quantification, and the condition column must contain exactly two factors. An example of this file in tabular form is shown in Table 1.

Following quantification, ψ values for all genes in all conditions as well as raw and Benjamini-Hochberg corrected *p*-values are reported in a file named "LABRAT.psis.pval." Differences in mean ψ values across the conditions are also reported. Additionally the APA structure for each gene is reported. The possible values for this structure are "TUTR" (tandem UTR), "ALE" (alternative last exon), or if both structures are present within the gene, "mixed." A second file called "numberofposfactors.txt" is also generated. This file contains information of how each transcript was assigned

Table 1 Example of sampconds text file in tabulated format.

Sample	Condition	Covariate1
BrainM1	Brain	M
BrainF1	Brain	F
LiverM1	Liver	M
LiverF1	Liver	F
LiverF2	Liver	F

Sample and condition columns are required with additional "covariate" containing columns being optional. This file defines the conditions that psi values are calculated across.

```
#calculate psi values
python .../PyScripts/LABRAT.py --mode calculatepsi --gff
.../Annotations/hg38/gencode.v32.annotation.gff3 --salmondir .../salmon
--sampconds .../LABRAT/sampconds.txt --conditionA Brain --conditionB Liver
#rename output file
mv LABRAT.psis.pval LABRAT.psis.pval.BrainLiver
```

Fig. 5 Example code for running LABRAT's calculatepsi function. Gencode's genome annotation gff, the directories produced by runSalmon, a tab-delimited sampconds text file and defined conditions are required inputs. This code produces several output files within the current directory.

to an APA site. The code in Fig. 5 uses LABRAT to quantify ψ values in human brain and liver samples.

The relevant options for the quantification of the ψ values are as follows:

- –mode: This argument defines which of the three steps in APA quantification LABRAT will be performing. For the calculation of ψ values, it should be set to "calculatepsi."
- –gff: This is the path to the gff annotation to be used. It should be the same annotation used in earlier steps.
- –salmondir: A directory containing salmon quantification subdirectories with one for each sample. The names of these subdirectories are the sample names supplied during the runSalmon step.
- –sampconds: A tab delimited text file relating samples, conditions, and optionally, covariates. The names of the samples should match those found on the subdirectories in the salmondir directory.
- –conditionA and –conditionB: In order to define a difference in ψ across conditions (Δψ), the direction of comparison must be defined. Δψ for

each gene is defined as the mean ψ value in condition B minus the mean ψ value in condition A. Both conditionA and conditionB must be found in the condition column of the sampconds file.

3.5 Directory architecture

LABRAT expects and will create a defined directory structure in regard to the Salmon quantifications. An example of this directory structure is shown in Fig. 6.

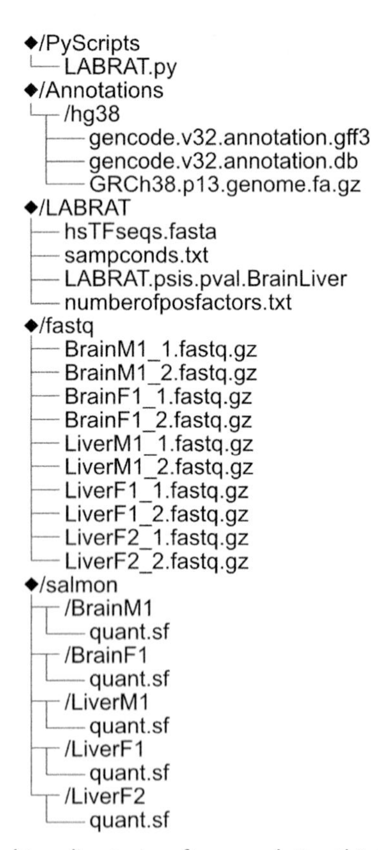

Fig. 6 Schematic of resulting directories after completing this LABRAT quickstart guide. While not explicitly required, similar directory organization for LABRAT projects is best practice.

4. Quantification of APA in single cell RNAseq data

LABRAT was originally designed to quantify APA from bulk RNAseq data. However, the growing plethora of available single cell RNAseq datasets presented an opportunity to look at the heterogeneity of APA regulation with cell populations. To take advantage of this, a companion script was written, LABRATsc (LABRAT single cell). LABRATsc uses the same approach that LABRAT does by using available tools to quantify transcript abundances, grouping transcripts that share polyadenylation sites together, and reporting APA status using ψ values that range between 0 and 1. In this section, we briefly detail how LABRATsc deals with single cell RNAseq data, special considerations to keep in mind, and present an example workflow and outputs.

4.1 How LABRATsc quantifies APA in single cell RNAseq data

LABRAT relies on Salmon to quantify transcripts from bulk RNAseq data. LABRATsc relies on an analogous tool for single cell RNAseq quantification, alevin (Srivastava, Malik, Smith, Sudbery, & Patro, 2019). While LABRAT compares ψ values between two conditions (e.g., treatment and control), LABRATsc compares ψ values between predefined groups or clusters of cells. These clusters can be defined using standard approaches such as tSNE and UMAP. Following quantification with alevin, transcripts are filtered to keep only those with confidently defined 3′ ends exactly as they are in LABRAT-based quantification. Quantification then proceeds upon one of two paths as indicated by the –mode parameter.

If the –mode parameter is set to "cellbycell" then a ψ value is calculated for every gene in every cell. In practice, the coverage for most genes in most cells is extremely low or nonexistent. Genes that do not pass a read coverage threshold (indicated by the –readcount in a given cell) have ψ values of NA in that cell. For each gene, ψ values are then compared across cell clusters using the ψ value in each individual cells as an independent observation.

If the –mode parameter is set to "subsampleClusters," then read counts for each transcript are first summed across all of the cells within a cluster. This has the advantage of raising the number of reads associated with each gene, but single cell resolution is lost. Statistical tests to identify genes with regulated APA across cell clusters are performed by creating a distribution of ψ values for each gene in each cluster through bootstrapping resampling.

4.2 Important considerations

Droplet-based single-cell RNA-sequencing libraries (10 × Genomics, Drop-seq, etc.) are particularly well suited to APA analysis as their construction is virtually identical to that of bulk PAS-seq libraries that capture 3′ ends directly using anchored oligo-d(T) primers ("TVN" priming) (Yao & Shi, 2014). However, several important caveats must be considered when using LABRATsc with these libraries.

First, low read depth, relative to bulk RNA-seq, and the so-called drop out effect limit the reliable detection of APA events in individual cells to relatively highly expressed (and robustly captured) genes. These inherent limitations make the selection of appropriate minimum read thresholds (set using the –readcountfilter argument) critical to identifying robust APA events with LABRATsc. We suggest thresholds of ≥ 100 counts per gene for cluster-level ψ value calculation (–mode subsampleClusters) and ≥ 5 counts for per-cell ψ value calculation (–mode cellbycell) as reasonable starting points, but these thresholds—particularly in the per-cell case—vary considerably between experiments due to differences in sequencing depth, per-cell RNA content, and cell type-specific gene expression patterns, among other factors. It is important to note that the accuracy of per-cell ψ value calculations with low read thresholds will be less accurate for genes with large numbers of 3′ end isoforms, but this concern is mitigated somewhat by the fact that 94% of the human GENCODE genes considered by LABRATsc have five or fewer isoforms.

Second, while these libraries putatively capture 3′ ends of mRNAs, they also contain substantial internal priming artifacts derived from TVN priming on genomically encoded poly(A) tracts. This is not a rare event—we have observed that internal priming accounts for up to 40% of molecules captured in 10 × Genomics 3′ libraries, and RNA velocity methods rely on this internal priming in introns to quantify pre-mRNAs (La Manno et al., 2018). If these internal priming events occur in close proximity to bonafide polyadenylation sites, they may skew raw ψ values substantially by "double counting" some fragments. However, while this may impact the accuracy of raw ψ values for some genes, the relative ψ values between cells are unaffected as the rate of internal priming is largely consistent across cells and we are thus still able to reliable identify APA events.

4.3 Generating input matrices for LABRATsc with alevin

Prior to running LABRATsc, cell-by-isoform count matrices must be produced. This step relies on the single-cell transcriptome quantification tool alevin, which is distributed with Salmon. For general information about alevin, please see https://salmon.readthedocs.io/en/latest/alevin.html and Srivastava et al. (2019).

The following arguments must be passed to alevin for use with LABRATsc, ideally in this order:

- –l: Library type. For most single-cell libraries, this will be "ISR."
- –1: A list of files containing the forward sequencing reads (also supports shell expansion of wildcards).
- –2: A list of files containing the forward sequencing reads (also supports shell expansion of wildcards) in the same order as –1.
- –dropseq/–chromium/–chromiumV3: One of these arguments must be provided depending on the sequencing platform used.
- -i: A salmon index, generated with LABRAT using the –librarytype 3pseq argument (as in Section 3.2).
- -p: Number of threads used by alevin (default is all available threads).
- -o: Output path for count matrix and metadata.
- –tgMap: A transcript-to-gene map file, which consists of each transcript ID in the salmon index listed twice per-line, separated by a tab.
- –fldMean 250: Expected mean fragment length (250 for consistency with LABRAT's execution of salmon).
- –fldSD 20: Expected standard deviation of mean fragment length (20 for consistency with LABRAT's execution of salmon).
- –validateMappings: Enables selective alignment of the sequencing reads.
- –whitelist: A whitelist of cell barcodes from a previous analysis to restrict quantitation to previously identified valid barcodes [Optional].

The example in Fig. 7 will generate a count matrix in the folder "sample1" from a hypothetical 10 × Genomics 3′ v3 single-cell RNA-seq library (lists of FASTQs indicated with placeholders [read1 FASTQs] and [read2 FASTQs]) against a pre-build salmon reference (hsTFseqs3pseq.fasta) with a transcript-to-gene map (hsTFseqs3pseq.fasta.tgMap) and optional whitelist of cell barcodes (sample1_barcodes.txt).

```
#create alevin quantification files
salmon alevin \
       -l ISR -1 [read1 FASTQs] -2 [read2 FASTQs]  --chromiumV3 \
       -i hsTFseqs3pseq.fasta -p 12 -o sample1 \
       --tgMap hsTFseqs3pseq.fasta.tgMap --fldMean 250 --fldSD 20 \
       --validateMappings --whitelist sample1_barcodes.txt
```

Fig. 7 Example code showing the use of alevin to generate input matrices for LABRATsc.

4.4 Calculating ψ values with LABRATsc

Transcript counts from one or more single-cell libraries are next used to calculate ψ values for every gene with at least two alternative polyadenylation sites. As when running LABRAT on bulk RNA-seq, gene-level ψ values are then compared across conditions to identify genes with significant ψ value changes. As described in Section 4.1, LABRATsc provides two different approaches for ψ calculation and significance testing: per-cell or using subsampled clusters.

The relevant options for the quantification of the ψ values with LABRATsc are as follows:

- –mode: cellbycell or subsampleClusters, as described in Section 4.1.
- –gff: This is the path to the gff annotation to be used. It should be the same annotation used to generate the salmon index provided to alevin.
- –alevindir: A directory containing alevin quantification subdirectories with one for each sample. The names of these subdirectories will be appended to the cell names in each sample matrix to form a "sample_barcode" cell id for each cell.
- –conditions: A tab delimited text file with column names "sample" and "condition." The first column contains cell ids and the second column contains cell condition or cluster. The cell ids in the sample column must follow the "sample_barcode" structure described above. Note that unlike LABRAT, LABRATsc does not currently support covariates.
- –readcountfilter: Minimum read count necessary for calculation of ψ values. Genes that do not pass this filter will have reported ψ values of NA. For "cellbycell" mode, this is the number of reads mapping to a gene in that single cell, while in "subsampleClusters" mode, this is the summed number of reads mapping to a gene across all cells in a predefined cluster.
- –conditionA and –conditionB: In order to define a difference in ψ across conditions (Δψ), the direction of comparison must be defined. Δψ for each gene is defined as the mean ψ value in condition B minus the mean ψ value in condition A. Both conditionA and conditionB must be found in the condition column of the sampconds file.

The code in Fig. 8 uses LABRATsc to quantify ψ values using both available modes in example data provided in the LABRAT github repo (paths are relative to the root directory of the repository).

Following quantification, ψ values for all genes in all conditions as well as raw and Benjamini-Hochberg corrected *p*-values are reported in files named

```
#calculate psi values cell-by-cell with a 5 read minimum
python LABRATsc.py \
       --mode cellbycell \
       --gff gencode.v32.annotation.gff3 \
       --alevindir testdata/alevin_example/alevin_out \
       --readcountfilter 5 \
       --conditions testdata/alevin_example/conditions.tsv \
       --conditionA Diagnosis --conditionB Relapse
#calculate psi values by subsampling clusters with a 100 read minimum
python LABRATsc.py \
       --mode subsampleClusters \
       --gff gencode.v32.annotation.gff3 \
       --alevindir testdata/alevin_example/alevin_out \
       --readcountfilter 100 \
       --conditions testdata/alevin_example/conditions.tsv \
       --conditionA Diagnosis --conditionB Relapse
```

Fig. 8 Example code showing the use of LABRATsc to calculate psi and delta psi values in both cellbycell and subsampleClusters modes.

"results.subsampleclusters.txt" (subsampleClusters mode) or "results.cellbycell.txt" (cellbycell mode). Differences in mean ψ values across the conditions are also reported. In cellbycell mode, the results file additionally includes the number of cells in each condition passing read depth filters for each gene. Finally, per-cell psi values are reported for each gene when run in cellbycell mode in a file named "psis.cellbycell.txt.gz." These results can be used to annotate existing single-cell analyses, as demonstrated by the example in Fig. 9.

Fig. 9 shows the significant alternative polyadenylation event at the *SAT1* gene in bone marrow mononuclear cells from a published acute myeloid leukemia dataset (Pei et al., 2020) (GEO accession: GSE143363). Fig. 9A shows a UMAP projection of the cells labeled with cluster names and colored by sample (Diagnosis or Relapse). Fig. 9B shows the per-cell calculated ψ values for the *SAT1* gene on the same projection, with higher ψ values observed in the "DX monocytic" cluster when compared to the "DX primitive" and "RL monocytic" clusters. This pattern is also evident when plotting distributions of ψ values in the three major clusters (Fig. 9C). Notably, this event is detectable using both the cellbycell and subsampleClusters modes, which report virtually identical ψ value changes (Fig. 9D). These results demonstrate that LABRATsc is able to identify significant APA events between different cell types ("DX primitive" vs "DX monocytic") and in closely related cells after in vivo therapy ("DX monocytic" vs "RL monocytic").

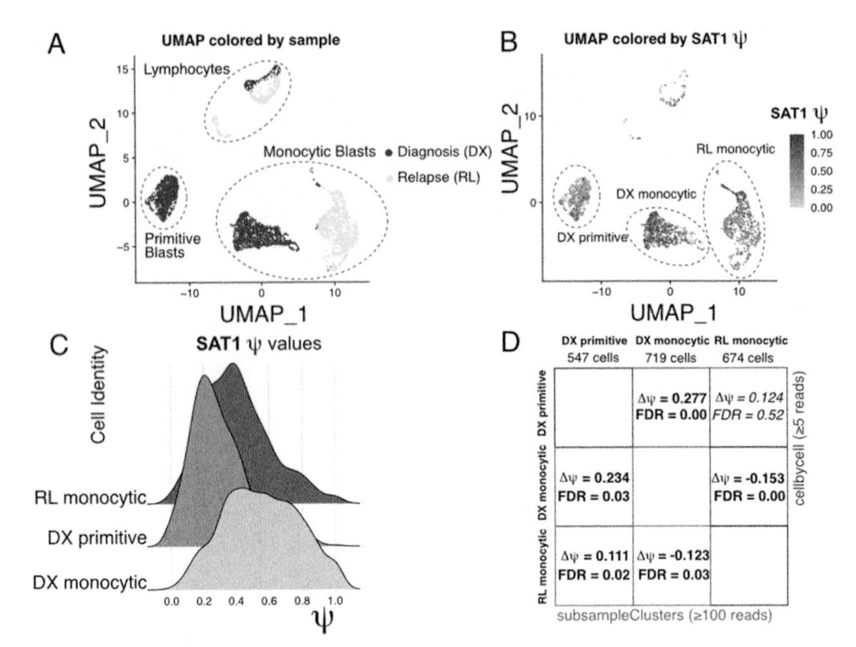

Fig. 9 Alternative polyadenylation of SAT1 in acute myeloid leukemia. (A) UMAP projection showing the diagnosis and relapse samples from GSE143363. Major cell types are indicated with dashed circles. (B) UMAP projection from (A), with cells colored by *SAT1* ψ value. Important clusters are indicated with dashed circles. (C) Ridge plot showing distributions of SAT1 ψ values in the clusters highlighted in (C). (D) Table comparing *SAT1* pairwise delta-ψ tests between the clusters highlighted in (C) using "--mode subsampleClusters" and "--mode cellbycell."

Acknowledgments

We thank Krysta Engel for helpful comments and suggestions. This work was funded by the National Institutes of Health (R35-GM133885) (JMT), the Boettcher Foundation (Webb-Waring Early Career Investigator Award AWD-182937), a Predoctoral Training Grant in Molecular Biology (NIH-T32-GM008730) (RG) and the RNA Bioscience Initiative at the University of Colorado Anschutz Medical Campus (RG and JMT).

References

Benjamini, Y., & Hochberg, Y. (1995). Controlling the false discovery rate: A practical and powerful approach to multiple testing. *Journal of the Royal Statistical Society. Series B, Statistical Methodology*, *57*(1), 289–300.

Bray, N. L., Pimentel, H., Melsted, P., & Pachter, L. (2016). Near-optimal probabilistic RNA-seq quantification. *Nature Biotechnology*, *34*(5), 525–527.

Cho, P. F., Poulin, F., Cho-Park, Y. A., Cho-Park, I. B., Chicoine, J. D., Lasko, P., et al. (2005). A new paradigm for translational control: Inhibition via 5′-3' mRNA tethering by Bicoid and the eIF4E cognate 4EHP. *Cell*, *121*(3), 411–423.

Goering, R., Engel, K. L., Gillen, A. E., Fong, N., Bentley, D. L., & Matthew Taliaferro, J. (2020). *LABRAT reveals association of alternative polyadenylation with transcript localization, RNA binding protein expression, transcription speed, and cancer survival*. Cold Spring Harbor Laboratory. 2020.10.05.326702 https://doi.org/10.1101/2020.10.05.326702.

Grassi, E., Mariella, E., Lembo, A., Molineris, I., & Provero, P. (2016). Roar: Detecting alternative polyadenylation with standard mRNA sequencing libraries. *BMC Bioinformatics, 17*(1), 423.

Ha, K. C. H., Blencowe, B. J., & Morris, Q. (2018). QAPA: A new method for the systematic analysis of alternative polyadenylation from RNA-seq data. *Genome Biology, 19*(1), 45.

La Manno, G., Soldatov, R., Zeisel, A., Braun, E., Hochgerner, H., Petukhov, V., et al. (2018). RNA velocity of single cells. *Nature, 560*(7719), 494–498.

Mayr, C., & Bartel, D. P. (2009). Widespread shortening of 3'UTRs by alternative cleavage and polyadenylation activates oncogenes in cancer cells. *Cell, 138*(4), 673–684.

Pai, A. A., Baharian, G., Pagé Sabourin, A., Brinkworth, J. F., Nédélec, Y., Foley, J. W., et al. (2016). Widespread shortening of 3' untranslated regions and increased exon inclusion are evolutionarily conserved features of innate immune responses to infection. *PLoS Genetics, 12*(9), e1006338.

Patro, R., Duggal, G., Love, M. I., Irizarry, R. A., & Kingsford, C. (2017). Salmon provides fast and bias-aware quantification of transcript expression. *Nature Methods, 14*(4), 417–419.

Pei, S., Pollyea, D. A., Gustafson, A., Stevens, B. M., Minhajuddin, M., Fu, R., et al. (2020). Monocytic subclones confer resistance to venetoclax-based therapy in patients with acute myeloid leukemia. *Cancer Discovery, 10*(4), 536–551.

Sandberg, R., Neilson, J. R., Sarma, A., Sharp, P. A., & Burge, C. B. (2008). Proliferating cells express mRNAs with shortened 3′ untranslated regions and fewer microRNA target sites. *Science, 320*(5883), 1643–1647.

Soneson, C., Love, M. I., & Robinson, M. D. (2015). Differential analyses for RNA-seq: Transcript-level estimates improve gene-level inferences. *F1000Research, 4*, 1521.

Srivastava, A., Malik, L., Smith, T., Sudbery, I., & Patro, R. (2019). Alevin efficiently estimates accurate gene abundances from dscRNA-seq data. *Genome Biology, 20*(1), 65.

Taliaferro, J. M., Vidaki, M., Oliveira, R., Olson, S., Zhan, L., Saxena, T., et al. (2016). Distal alternative last exons localize mRNAs to neural projections. *Molecular Cell, 61*(6), 821–833.

Xia, Z., Donehower, L. A., Cooper, T. A., Neilson, J. R., Wheeler, D. A., Wagner, E. J., et al. (2014). Dynamic analyses of alternative polyadenylation from RNA-seq reveal a 3′-UTR landscape across seven tumour types. *Nature Communications, 5*, ncomms6274.

Yao, C., & Shi, Y. (2014). Global and quantitative profiling of polyadenylated RNAs using PAS-seq. In J. Rorbach, & A. J. Bobrowicz (Eds.), *Polyadenylation: Methods and protocols* (pp. 179–185). Humana Press.

Zheng, D., Liu, X., & Tian, B. (2016). 3'READS+, a sensitive and accurate method for 3′ end sequencing of polyadenylated RNA. *RNA, 22*(10), 1631–1639.

CHAPTER THIRTEEN

Poly(A) tail dynamics: Measuring polyadenylation, deadenylation and poly(A) tail length

Michael Robert Murphy, Ahmet Doymaz, and Frida Esther Kleiman*
Department of Chemistry, Hunter College, City University of New York, New York, NY, United States
*Corresponding author: e-mail address: fkleiman@hunter.cuny.edu

Contents

Methods in Enzymology, Volume 655
ISSN 0076-6879
https://doi.org/10.1016/bs.mie.2021.04.005

Abstract

Transcription of mRNAs culminates in RNA cleavage and a coordinated polyadenylation event at the 3′ end. In its journey to be translated, the resulting transcript is under constant regulation by cap-binding proteins, miRNAs, and RNA binding proteins, including poly(A) binding proteins (PABPs). The interplay between all these factors determines whether nuclear or cytoplasmic exoribonucleases will gain access to and remove the poly(A) tail, which is so critical to the stability and translation capacity of the mRNA. In this chapter, we present an overview of two of the key features of the mRNA life-cycle: cleavage/polyadenylation and deadenylation, and describe biochemical assays that have been generated to study the activity of each of these enzymatic reactions. Finally, we also provide protocols to investigate mRNA's poly(A) length. The importance of these assays is highlighted by the dynamic and essential role the poly(A) tail length plays in controlling gene expression.

1. Introduction

Regulation of 3′ end processing is a crucial, dynamic post-transcriptional mechanism that alters transcript levels and/or translational potential in response to steady state as well as alterations in cellular homeostasis. Nearly all protein-coding mRNAs and long non-coding RNAs transcribed by RNA polymerase II (RNAP II) undergo a coupled co-transcriptional cleavage and polyadenylation (CpA) reaction, resulting in 200–300 adenosines synthesized on the 3′ end (Bentley & Coupling, 2014; Sheets & Wickens, 1989). More specifically, the newly transcribed precursor, which is complementary to the template DNA strand from which it was blueprinted, is catalytically cleaved in the phosphate backbone leaving a reactive 3′ OH group. Once endonucleolytic cleavage has occurred, a long string of non-templated adenosine residues is attached by a family of nucleotidyltransferases to the 3′ end of the transcript. For a more comprehensive overview of the process and history, the reader is directed to Proudfoot (2011). This two-part reaction is triggered by recognition of the poly(A) signal (PAS), the hexanucleotide sequence AAUAAA or one of its derivatives (Tian & Graber, 2012), which is required for both reactions to take place. The cleavage reaction takes place ~30 nucleotides downstream of the PAS and involves the recognition of less conserved sequence elements flanking either side of the poly(A) site. The poly(A) tail plays multiple roles in mRNA stability, translation regulation, nuclear export in an mRNA-specific fashion (Murphy & Kleiman, 2020a; Zhang, Virtanen, & Kleiman, 2010).

The CpA reaction is evolutionary conserved; while prokaryotes and archaea possess either oligonucleotide adenylate tracks or a CCA motif, multicellular eukaryotes have adopted polyadenosine (polyA) tails into the hundreds of bases. The universality of this two-step coupled process highlights its importance in normal cellular homeostasis. A myriad of factors from many different canonical pathways (Shi et al., 2009) are important for the efficacy of the CpA reaction, with a sizable repertoire of over 80 catalytic, structural, activating, repressing, upstream and downstream components in mammalian cells. A well-studied example of overlapping pathways is the role of transcription elongation/termination by RNAP II in CpA. RNAP II has a long C-terminal domain (CTD) of heptad repeats, which the presence and phosphorylation of are both key CpA requirements determined by structural, genome-wide and biochemical assays (Bentley & Coupling, 2014; Darnell, 2013; Ryan, Murthy, Kaneko, & Manley, 2002). An interesting early observation was that components of the cleavage and polyadenylation specificity factor (CPSF) and cleavage stimulation factor (CstF) complexes are deposited at the promoters of RNAP II-transcribed genes (Glover-Cutter, Kim, Espinosa, & Bentley, 2008; Hirose & Manley, 1998), allowing for co-transcriptional scanning of PAS. A consequence of such scanning and the selection of functional poly(A) hexamers is alternative polyadenylation within the same gene (Tian & Manley, 2017), a mechanism regulated by the core spliceosome component U1 snRNA genome-wide in a concentration dependent manner (Berg et al., 2012).

Once a transcript has been capped, polyadenylated, and released from the transcription machinery, several fail-safe methods have evolved to sense changes in cellular environment by changing the length of the poly(A) tail and controlling gene expression. Deadenylation is the physiological enzymatic removal of non-templated adenosine monophosphates primarily in mRNA and lncRNA, and this is accomplished by the action of deadenylase to trigger a cascade of events leading to regulation of mRNA transport, translation and degradation (reviewed in Goldstrohm & Wickens, 2008; Zhang et al., 2010). Deadenylases are categorized into two superfamilies based on their catalytic domains: DEDD-exonucleases, and exonuclease-endonuclease-phosphate (EEP)-nucleases. As there is a delay between deadenylation and mRNA decay (Decker & Parker, 1993) (Fig. 1A), deadenylation is commonly described as the initial and rate-limiting activity in mRNA turnover (Goldstrohm & Wickens, 2008; Zhang, Kleiman, & Devany, 2014). In fact, modeling predicts that deadenylation is by far the largest determinant of mRNA degradation, as

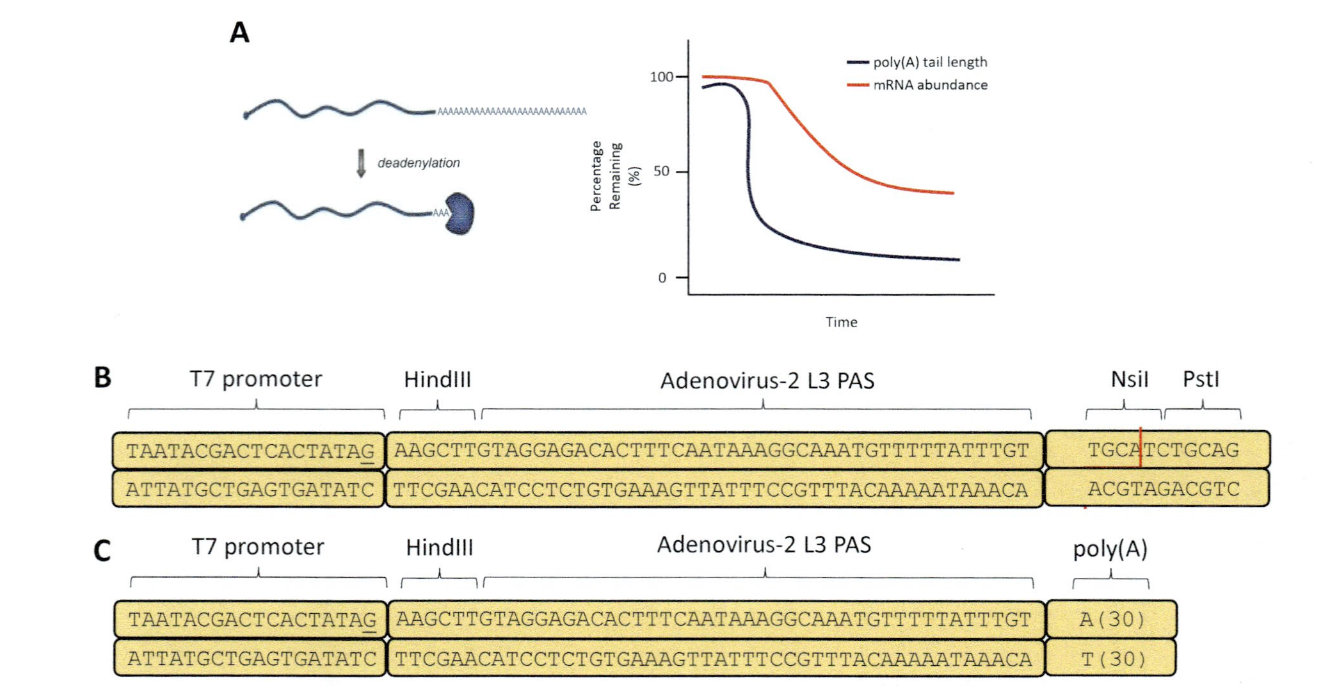

Fig. 1 (A) A simplified model of mRNA decay lagging behind deadenylation. Enzymatic deadenylation leads to a negligible level of substrate. Further enzymatic reactions of decapping and mRNA decay by exonucleases results in lag to qualify an mRNA as "decayed." (B-C) Constructs used to test cleavage and polyadenylation (A) and for deadenylation (B).

both 5′-3′ and 3′-5′ nuclease activities are dependent upon poly(A) tail removal (Cao & Parker, 2001). Deadenylation allows fine-tuning of post-transcriptional regulation by having a number of decay intermediates with different length of poly(A) tails. The existence of a pathway that can control mRNA functions in response to stimuli allows the cell to adapt swiftly to a changing environment. For example, deadenylation-mediated mRNA decay not only prevents the expression of genes that are not necessary for stimulus response, but also allows the redirection of cellular resources to recover from the insult, such as during stress conditions (Zhang et al., 2014).

Several cis-acting sequence elements present in the 3′ untranslated region of target mRNAs have been shown to participate in the regulation of poly-adenylation/deadenylation of specific mRNAs. Such cis-acting sequences are recognized by microRNAs, AU-rich element binding proteins, poly-adenylation factors or other RNA binding factors. As the processing of the 3′ end of an mRNA can drastically affect gene expression, a number of genomic analysis have highlighted the relevance of PAS selection, cis-acting sequences and how the message is ended. However, direct analysis by biochemical approaches is necessary to have the mechanistic understanding required for the processing of specific genes.

2. Equipment

- Thermal cycler (Thermo Fisher)
- Heat block set to 65 °C for RNA denaturation
- Water bath set to 37 °C for RNase H reactions
- Northern blotting equipment (gel electrophoresis, transfer, visualization)
- Common molecular biology tools (pipettes, micro centrifuge tubes, etc.)
- Equipment for polyacrylamide gel electrophoresis (glasses, chamber, Owl, Thermo Fisher)
- Phosphorimager (FLA7000IP Typhoon)
- Storage phosphor screen and exposure cassette (Thermo Fisher)
- Douncer (Thomas Scientific)
- Nanodrop 2000 (Thermo Scientific)
- ImageJ software (http://rsb.info.nih.gov/ij/)

3. Chemicals

3.1 Buffer A

- 10 mM Tris pH 7.9 (Sigma Aldrich)
- 1.5 mM $MgCl_2$ (Sigma Aldrich)

- 10 mM KCl (Sigma Aldrich)
- 0.5 mM *dithiothreitol* (DTT, Sigma Aldrich)
- 0.5 mM phenylmethylsulfonyl fluoride (PMSF, Sigma Aldrich)
- 100 × Protease Inhibitor Cocktail (PIC, MilliporeSigma)

3.2 Buffer C

- 20 mM Tris pH 7.9
- 25% glycerol (Sigma Aldrich)
- 1.5 mM $MgCl_2$
- 0.45 M NaCl
- 0.5 mM DTT
- 0.5 mM PMSF
- 100 × PIC

3.3 RNA-labeling mix

- Linearized pG3SVL-A and pG3L3-A (1 μL of 1 μg/μL)
- SP6 RNA Polymerase (1 μL of 10 U/μL, Promega)
- [α-^{32}P]-GTP (0.6 μL of 10 μCi/μL, PerkinElmer)
- 5′Me (Shi et al., 2009) G(5′)ppp(5′)G-Cap analog (3 μL of 5 mM, Promega)
- 5 × transcription buffer (3 μL)
  - 200 mM Tris·Cl, pH 7.9
  - 30 mM $MgCl_2$
  - 50 mM DTT
  - 50 mM NaCl
  - 10 mM spermidine (Sigma Aldrich)
- RNasin, 0.6 μL of 40 U/μL (Promega)
- DTT, 1.5 μL of 0.1 M
- rNTP mix (1.5 μL of 5 mM of ATP, CTP, UTP/1 mM GTP, Promega)
- RNase free water to 15 μL (Millipore)

3.4 RNA purification

- TRIzol (Thermo Scientific)
- Chloroform (Sigma Aldrich)
- Isopropanol (Sigma Aldrich)
- 75% Ethanol (Sigma Aldrich)

- Phosphate-buffered saline (PBS):
  - 137 mM NaCl (Sigma Aldrich)
  - 2.7 mM KCl
  - 10 mM Na_2HPO_4
  - 1.8 mM KH_2PO4

3.5 Urea-PAGE gels

- 5% Polyacrylamide (Sigma Aldrich)
- 8.3 M Urea (Sigma Aldrich)

3.6 Cleavage mix

- 0.2–0.5 ng of labeled RNA
- 250 ng of tRNA (Sigma Aldrich)
- 0.25 unit of RNasin
- 9.6 mM HEPES-NaOH pH 7.9 (Sigma Aldrich)
- 9.6% Glycerol (Sigma Aldrich)
- 24 mM $(NH_4)_2SO_4$ (Sigma Aldrich)
- 20 mM creatine phosphate (Sigma Aldrich)

3.7 Deadenylation mix

- 25 mM HEPES pH 7.0
- 100 mM NaCl
- 0.1 mM EDTA
- 1.5 mM $MgCl_2$
- 0.5 mM DTT
- 2.5% polyvinyl alcohol (Sigma Aldrich)
- 10% glycerol
- 0.25U RNasin
- 10 nM ^{7}MeGpppG capped in vitro transcribed L3(A_{30}) RNA substrate

3.8 Oligo(dT)/RNase H-Northern blot analysis

- Oligo(dT)$_{12–18}$ primer (Invitrogen)
- RNase H (Invitrogen) and 10 × RNase H reaction buffer (Invitrogen)
- RNA isolated from cells (at least 10 μg total RNA per condition)
- 0.25 U RNasin

3.9 PAT assay

- Oligo(dT)-anchor primer (5′-GGGGATCCGCGGTTTTTTTTTT-3′)
- Gene-specific primer (located 200–400 bp upstream the poly(A) site)
- GoScript Reverse Transcriptase (Promega)
- PCR Nucleotide Mix, 10 mM of each dNTP (Promega)
- 25 mM $MgCl_2$
- 5 × GoScript Reaction Buffer (Promega)
- GoTaq PCR mix (Promega)
- 1 μg per condition RNA isolated from cells
- T4 Polynucleotide Kinase (New England Biolabs)
- T4 PNK 10 × Reaction Buffer (New England Biolabs)
- γ-^{32}P ATP (3000 Ci/mmol, 5 mCi/mL, PerkinElmer)
- DNA ladder (New England Biolabs)

3.10 Tris-EDTA (TE) buffer

- 15.759 g of Tris-Cl pH 8
- 2.92 g of EDTA pH 8
- Add distilled water to 1 L

3.11 2 × RNA loading buffer

50% Formamide (Sigma Aldrich)
20% Glycerol
6.5% Formaldehyde (sigma Aldrich)
1 × MOPS buffer pH 7
  41.86 g of MOPS free acid (Sigma Aldrich)
  4.1 g of sodium acetate (Sigma Aldrich)
  3.72 g of Na_2 EDTA (Sigma Aldrich)
  Adjust pH using NaOH
  Add dH_2O until volume is 1 L
0.05% Bromophenol Blue (Fisher Scientific)
0.05% xylene cyanol (Fisher scientific)

3.12 Saline-sodium citrate (SSC) buffer

175.3 g of NaCl
77.4 g of sodium citrate (Sigma Aldrich)
Adjust to pH 7 with 14 N HCl
Add dH_2O until volume is 1 L

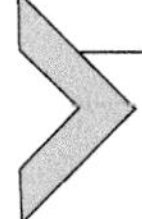

4. Cleavage and polyadenylation

4.1 Introduction

Most of our understanding of CpA has come from biochemical characterization of the process in mammals. Fractionation of HeLa cell nuclear extract resulted in the purification of the CPSF and CstF protein complexes, which recognize a functional PAS to promote RNA cleavage in the phosphate backbone leaving a reactive 3′ OH group. While the subunit CPSF-160 directly binds to the RNA at the PAS, subunit CPSF-73 contains the enzymatic activity for endonucleolytic cleavage at the poly(A) site, several nucleotides downstream of the PAS. Helping to coordinate the architecture of CpA machinery is recognition of a GU-rich sequence downstream of the poly(A) site by CstF complex. Similarly, poly(A) polymerase (PAP) was also found to be recruited to this 3′ processing complex and shown to add a long string of non-templated adenosine residues on the 3′ to denote the end of the transcript and affect stability on the RNA. The CTD of RNAP II is also necessary for efficient CpA, partly through recruitment of CPSF and CstF components. It was found that CPSF is required to induce polymerase pausing once it encounters a functional PAS, suggesting that CstF and CPSF travel along with RNAP II during nascent gene transcription in search of the PAS. Other major components of the CpA machinery include cleavage factor (*CF*) required for the cleavage reaction, *CF* IIm involved in the binding to RNAP II CTD and other CpA factors (reviewed in Proudfoot, 2011).

The universality of this two-step coupled process highlights its importance in normal cellular homeostasis. Biochemical characterization of CpA was further achieved using genetic screens in yeast, with some factors homologous to their mammalian counterparts and others apparently unique to yeast. CpA has been shown to be a vital part in controlling gene expression by affecting the RNA's lifespan, regulating its stability, translational efficiency, and export from the nucleus. As the presence/absence of a poly(A) regulates these processes, the length of a poly(A) tail can itself be regulated for the same purposes.

While the CpA reaction can be reconstituted in vitro using purified fractions from different cell nuclear extracts, it cannot be done using recombinant proteins due to the complexity of the factors needed in the CpA reaction and the extensive post-translational modifications they undergo. These modifications include: arginine methylation, lysine sumoylation, lysine acetylation, and the phosphorylation of serine, threonine and tyrosine residues. As these modifications occur in different cellular conditions, such as stress, proliferation, differentiation and development, it is important to study CpA in small-scale nuclear extracts from adherent cells. The use of these small-scale nuclear extracts allows for functional studies to be performed by depleting factors involved in the reaction. While CpA can be tested with endogenous substrates or tagged systems carrying functional PAS, purified fractions are generally tested with radiolabeled in vitro transcribed RNA carrying a PAS, such as simian virus 40 (SV40) late pre-mRNA and human adenovirus-2 L3 pre-RNA.

4.2 Nuclear extract preparation from adherent cells for CpA or deadenylation

1. Culture cells to ~80% confluency in a 10cm dish in Dulbecco's Modified Eagle Medium with 10% fetal bovine serum and 1% penicillin-streptomycin mix. Adherent cells such as human colon cancer HCT116 and cervical cancer HeLa cells have been used.
2. Remove media, wash once with *phosphate-buffered saline* (PBS; 137 mM NaCl, 2.7 mM KCl, 10 mM Na_2HPO_4, 1.8 mM KH_2PO_4.) and aspirate.
3. Treat cells with stressor or drug. For example, for UVC treatment, place plate in a Stratalinker 1800 (Stratagene) and deliver 2 pulses of $20 J/m^2$, replace media and incubate for 2h at 37 °C.
4. Harvest cells at ~80% confluency ($\sim 3 \times 10^7$ cells per 10cm plate).
5. Wash 2× with 10mL cold PBS, add 5mL cold PBS and scrape cells into 15ml falcon tube.
6. Spin 1000g for 5min at 4 °C.
7. Resuspend the pellet in 4mL Buffer A (10 mM Tris pH7.9, 1.5 mM $MgCl_2$, 10 mM KCl, 0.5 mM DTT, 0.5 mM PMSF and 1× PIC). Incubate for 10 min on ice.
8. Transfer resuspended cells into a 7 ml glass dounce tissue homogenizer, and dounce 20 times with a "tight" pestle (type B). While the hypotonic conditions will soften the cell membrane, douncing serves to break the cell membrane leaving the nucleus intact

   Tips:

   *Keep the douncer on ice as often as possible. Try to keep the douncer submerged the entire time, or be very gentle in removing the pestle from the liquid to avoid the introduction of bubbles.

*Place the douncer on a hard surface at a slight angle. Press the pestle into the liquid at a slightly higher angle to create friction against the side of the tube.

*When the bottom of the tube is reached, bring the pestle parallel with the tube and retract it slowly. Repeat 20 times.

*Between samples treated in different conditions, rinse the douncer and pestle with a small amount of detergent and milliQ water.

9. Transfer the sample to a fresh falcon tube and spin at 4000 rpm for 10 min at 4 °C to pellet the nuclei. Non-nuclear fraction is in the supernatant
10. Resuspend the pellet in 200 μL Buffer C (20 mM Tris PH 7.9, 25% glycerol, 1.5 mM $MgCl_2$, 0.3 M NaCl, 0.5 mM DTT, 0.5 mM PMSF, and 1 × PIC) and transfer to a microcentrifuge tube, keeping the sample on ice. (The high salt conditions will rupture the nuclear membrane, while the glycerol aids in freezing and storing aliquots).
11. Incubate the sample for 30 min at 4 °C in a rotator.
12. Centrifuge the samples at 13,000 rpm for 15 min at 4 °C. Aliquot the extracts into chilled Eppendorf tubes (~50 μL per aliquot) and store at −80 °C for future use. The extracts are good to use for CpA and deadenylation reactions for 2 weeks.
13. Dialysis of the extracts is optional and might be needed for certain samples. Dialysis in buffer D (20 mM Tris pH 7.9, 20% (v/v) glycerol, 0.2 mM EDTA, 0.5 mM DTT, 50 mM ammonium sulfate) might be required.

Tips:

*The resulting supernatant will contain the soluble nuclear fraction.

*Measure the concentration of soluble nuclear fraction with bicinchoninic acid (BCA) protein assay or equivalent.

*Fractionation of subcellular components can be tested by Western blot against Lamin A/C or topoisomerase II for the nuclear fraction, and GAPDH or actin for the cytoplasmic fraction.

4.3 RNA substrate for CpA or deadenylation

1. First, plasmids have to be linearized (Fig. 1B–C). For in vitro CpA reactions use either pG3SVL-A or pG3L3-A, which contain the SV40 late PAS and adenovirus 2 L3 PAS, respectively. pG3SVL-A and pG3L3-A plasmids are linearized at the *Dra*I and BamHI sites, respectively. For deadenylation, digest plasmid pT3L3(A_{30}), which contains L3 RNA

body followed by 30 residues of adenosine, with *NsiI*. To further investigate the specificity of the deadenylation activity, RNA substrates from ML43(G_{30}), ML40(C_{32}) and ML54(U_{30}) plasmids can be used.

2. Linearized pG3SVL-A and pG3L3-A are transcribed in vitro by SP6 RNA polymerase. Linearized pT3L3(A_{30}) and control plasmids are transcribed using T3 RNA polymerase.
3. The conditions for in vitro transcription and radioactive labeling are the following: in a 15 μL total volume reaction, substrate RNA is uniformly labeled by including [α-^{32}P]-GTP (0.6 μL of 10 μCi/μL) and 5′ capped during transcription using the 5′Me (Shi et al., 2009) G(5′)-ppp(5′)G-Cap analog (3 μL of 5 mM). Add appropriate RNA polymerase (1 μL of 10 U/μL) and 5 × transcription buffer (3 μL), (RNasin, 0.6 μL of 40 U/μL), linearized DNA (1 μL of 1 μg/μL), DTT (1.5 μL of 0.1 M), and rNTP mix (1.5 μL of 5 mM of ATP, CTP, UTP/ 1 mM GTP). This rNTP mix favors the incorporation of radiolabeled GTP into the RNA. Add RNase free water to 15 μL. Make sure to use Perspex shield at all times when working with radioactive substrate.
4. Incubate at 37 °C for 1 h. (Take 1 μL to assess transcription and RNA purification efficiency during step 18).
5. Add 1 volume of 2 × RNA Loading Dye to final concentration (47.5% formamide, 0.01% SDS, 0.01% bromophenol blue (BFB), 0.005% xylene cyanol and 0.5 mM EDTA). This ensures that the RNA in your sample does not reform secondary structures after the heat denaturation step.
6. Heat at 70 °C for 5 min to denature RNA and immediately transfer to ice. The use of a denaturing system after heating your sample maintains the denatured state of your RNA.
7. The transcript is then purified on a 0.4 mm denaturing polyacrylamide gel (6% polyacrylamide, 8.3 M urea) in 1 × Tris/Borate/EDTA (TBE) buffer. Use a 4-well comb. While heating the samples, setup the gel box and flush urea out of the wells with running buffer using a large tip.
8. Load samples and run gel at over 400 V for ~1.5 h until bromophenol blue reaches 1 cm to the end of the gel.
9. Carefully open the gel glasses using a thin spatula and cover the gel with plastic wrap. Place the covered gel on a tray and expose on X-ray film for 2 min in a dark room. Remember to align the film and the gel well to allow the extraction of the labeled RNA.

10. Locate labeled RNA by the shadow on film and cut the region. Place the selected portion of the gel in an Eppendorf tube using a spatula. Carefully discard the rest of the gel into radioactive waste.
11. To gel purify the labeled RNA, add 500 μL of 0.3 M sodium acetate and 0.2% SDS. Elute by rotating for 3 h at room temperature.
12. Centrifuge for 3 min at 3000 rpm. Take the solution and carefully discard the rest of the radioactive gel.
13. Add to the eluted RNA 1 μL of glycogen (20 mM) to facilitate the following precipitation.
14. Perform a phenol-chloroform extraction. Mix gently for 2–3 min and centrifuge at 10,000 rpm at room temperature for 10 min. Under acidic conditions, total RNA remains in the upper aqueous phase, while most of DNA and proteins remain either in the interphase or in the lower organic phase.
15. Labeled RNA is then recovered by precipitation with 1 volume of isopropanol at −80 °C for 1 h.
16. Centrifuge the sample at 13,000 rpm for 15 min at 4 °C. Wash the pellet with ethanol 70%. Let dry the pellet and then resuspend in 30 μL of nuclease-free water. Note: if the pellets dry out too much, the RNA crystallizes and is very difficult to solubilize. If it does not dry well, the RNA does not solubilize into solution. (Take 1 μL to assess transcription and RNA purification efficiency during step 18). Aliquot in tubes of 5 μL each and keep at −20 °C in a beta-box for emitting radioactive isotope samples.
17. To assess transcription and RNA purification efficiency, measure radioactivity from samples taken in step 4 and 17 (Fig. 2A). To the removed aliquots (1 μL), add 1 μL of glycogen (20 mM) and 100 μL of Tris-EDTA buffer. Add this mixture to an aqueous fluor solution and count in a scintillation counter to determine the total amount of radioactive nucleotide in the sample before and after purification of labeled RNA. The ratio of cpm of the sample after gel purification (step 17) to total cpm (step 4) is the fraction of labeled nucleotide incorporated into RNA. This value should generally be at least 30%. For more information on how to calculate RNA specific activity and yield please check https://www.thermofisher.com/us/en/home/references/ambion-tech-support/nuclease-enzymes/tech-notes/determining-rna-probe-specific-activity-and-yield.html

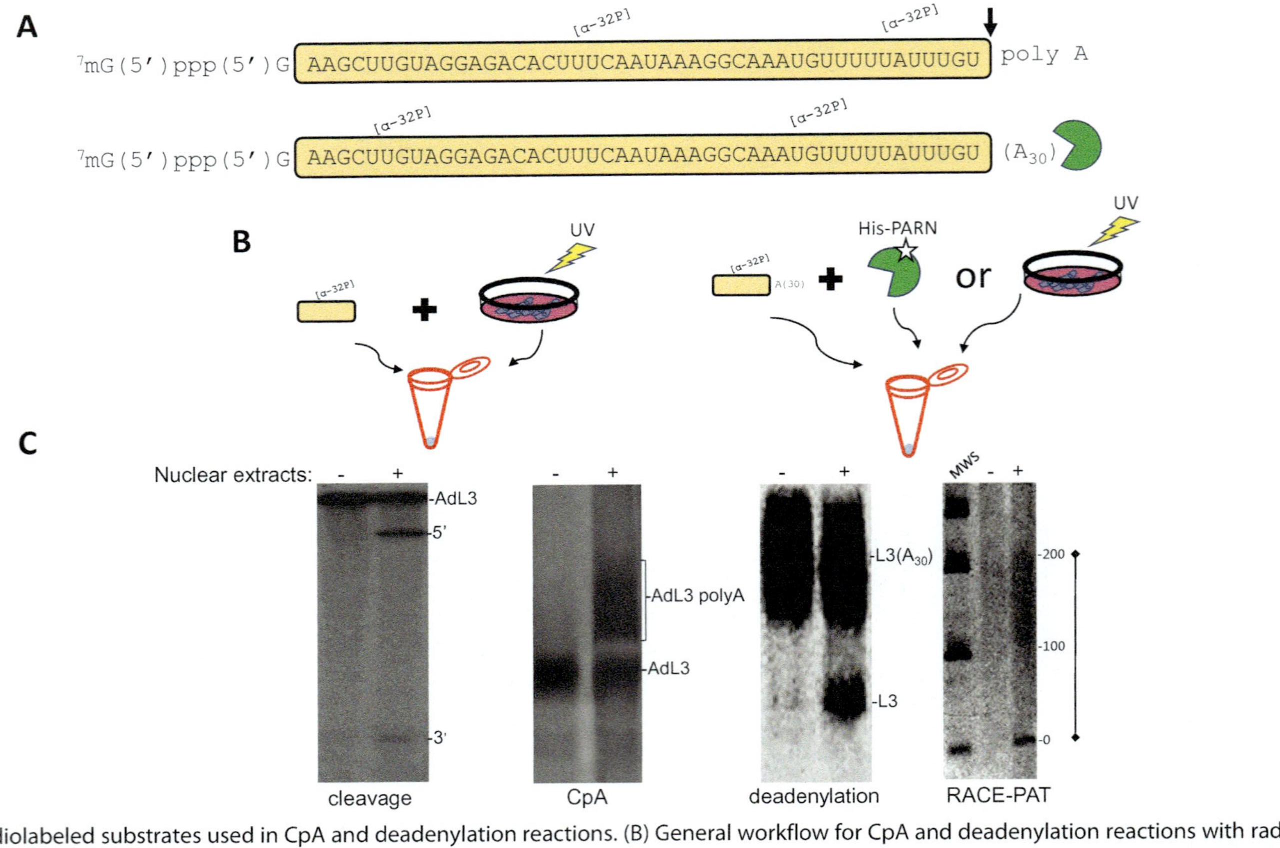

Fig. 2 (A) Capped radiolabeled substrates used in CpA and deadenylation reactions. (B) General workflow for CpA and deadenylation reactions with radiolabeled RNA and recombinant protein or cell lysates. (C) Representative cleavage, CpA, deadenylation and RACE-poly(A) test reactions. For CpA and deadenylation reactions, different capped radiolabeled substrates are incubated with nuclear extracts purified from human colorectal carcinoma HCT116 cells. For in vitro CpA reactions pG3L3-A, which contains adenovirus 2 L3 PAS, is used. For deadenylation, plasmid pT3L3 (A_{30}), which contain L3 RNA body followed by 30 residues of adenosine, is used. Samples were then analyzed in denaturing urea gels followed by autoradiography. For RACE-PAT assays, nuclear RNA from HCT116 cells are reverse transcribed using oligo(dT)-anchor primer and amplified using an oligonucleotide that hybridizes TP53 transcript. The products were separated on a non-denaturing PAGE and detected by ethidium bromide staining. Molecular weight standard (MWS, 100 base pair ladder) is also included.

4.4 In vitro 3′ cleavage reaction (Fig. 2B–C)

1. Master mix for a single reaction of total volume 12.5 μL contains 7.5 μL of the following: 2.5 μL of 12.5% polyvinyl alcohol (PVA), 1 μL of 0.25 M creatinine phosphate (CP), 1 μL of 0.5 mg/ml tRNA, 1 μL of 6.25 mM 3′ dATP (cordycepin), 0.25 μL of 20 U/μL nuclease-inhibitor, 1 μL of 25 mM EDTA pH 8, and 0.75 μL RNase-free water. As the nuclear extracts and the purified fractions might have PAP, 3′ dATP is added to facilitate observation of cleavage products without interfering with poly(A) addition. However, it has been shown that 3′ dATP can inhibit cleavage of some substrates, such as SV40 RNA. Extensive studies have shown that neither ATP nor hydrolysis of CP are essential for 3′ mRNA cleavage, suggesting that CP can function as a necessary cofactor or mimicking a phosphorylated one (Hirose & Manley, 1998; Khleborodova, Pan, Nagre, & Ryan, 2016). It is important to highlight that different RNA substrates might need slightly different conditions to reach maximum efficiency. For example, it has been shown that the requirements for magnesium concentrations can vary for different substrates (0–2.5 mM Khleborodova, Pan, Nagre, & Ryan, 2016).
2. Add to the 7.5 μL of master mix no more than 5 μL/reaction of nuclear extracts or purified fractions at concentrations no higher than 2 mg/mL. Sometimes dilution with buffer D is needed to compare nuclear extracts or purified fractions obtained from cells treated in different conditions.
3. Heat the labeled RNA to 80 °C for 2 min. Place on ice before adding 1 μL of it to the reaction mix in a concentration no higher than 0.5 ng/μL. Alternatively, the RNA may be included in the master mix.
4. After addition of the RNA, the reaction mix is incubated at 30 °C for 1.5 h.
5. The reaction is stopped with the addition of 112.5 μL of 0.4 mg/mL of proteinase K, 20 mM Tris pH 7.9, 10 mM EDTA, 1% SDS, 100 mM NaCl. Incubate the sample (total volume of 125 μL) at 30 °C for 15 min.
6. Perform acid phenol-chloroform extraction. Mix gently for 2–3 min and centrifuge at 10,000 rpm at room temperature for 10 min.
7. To the upper aqueous phase, add 1 μL of glycogen (20 mM), 0.1 volume of 3 M sodium acetate and 2.5 volumes of ice cold 100% ethanol. Mix thoroughly and precipitate at −80 °C for 1 h.
8. Centrifuge the sample at 13,000 rpm for 15 min at 4 °C. Wash the pellet with ethanol 70%. Let the pellet dry and then resuspend in 10 μL of 90%

formamide, 10 mM EDTA and BFB. Keep samples at −20 °C in a beta-box until analysis by gel electrophoresis.

9. Use 6% polyacrylamide/8.3 M urea gel in 1× TBE buffer with a 32-well comb. Before loading, heat the sample to 60 °C for 2 min then place on ice. Remember to flush urea out of the wells with running buffer using a large tip before loading. Load the gel with 3 μL of each sample.

10. Run the gel at over 400 V for ~1.5 h until BFB reaches 1 cm to the end of the gel.

11. Carefully open the gel glasses using a thin spatula and cover the gel with a piece of filter paper. Press the paper firmly onto the gel and carefully peel the paper, which should have the gel attached to it. Remember to label on the paper the directionality of the loaded samples. Cover the exposed gel side with plastic wrap.

12. Dry on a gel dryer with the paper side facing the vacuum. Dry for 15 min or until the gel is completely dry.

13. Expose to X-ray film in a light-tight cassette with intensifying screen at −80 °C. Alternatively, the gel can be visualized and quantitated using a Molecular Dynamics Storm Phosphorimager. Exposure time depends on strength of labeled RNA and efficiency of the reaction.

14. Determine the ratio between cleaved and un-cleaved RNA in each sample to quantitate cleavage efficiency. Relative cleavage is calculated as [cleaved fragment/(cleaved fragment + un-cleaved RNA substrate)] × 100. Quantifications can be done with ImageJ software (http://rsb.info.nih.gov/ij/).

4.5 In vitro polyadenylation reaction (Fig. 2B–C)

1. Master mix is similar to the one for cleavage reaction, with the replacement of 3′ dATP for ATP to allow the CpA reaction to proceed (2.5% PVA, 20 mM CP, 40 ng/mL tRNA, 0.25 mM ATP, 0.4 U/μL nuclease-inhibitor, 2 mM EDTA pH 8, and RNase-free water).

2. Samples are analyzed as in Section 4.4, steps 2–14.

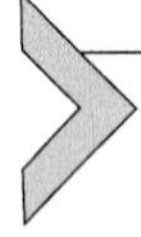

5. Deadenylation

5.1 Introduction

Early experiments into the role of the mRNA poly(A) tail in translation utilized 3′ OH exonucleases isolated from nuclei to remove the tail. Later studies identified that both poly(A) specific endo- and exonucleases can remove the poly(A) tail from target mRNAs (Goldstrohm & Wickens, 2008; Webster, Stowell, Tang, & Passmore, 2017; Zhang et al., 2014). Deadenylation can lead to either the removal of the cap structure at the 5′ end, exposing the transcripts to digestion by a 5′-to-3′ exonuclease, or transcript degradation in the 3′-to-5′ direction. Thus, mechanisms involving poly(A) tail length control and degradation represent an additional checkpoint for regulation of gene expression (reviewed in Goldstrohm & Wickens, 2008; Zhang et al., 2014). One of the most characterized deadenylases so far is the CCR4-POP2-NOT complex, which belongs to the family of EEP nucleases. The most studied deadenylases in the DEDD-exonucleases family are poly(A)-specific ribonuclease (PARN), which is a major deadenylase in mammals, and poly(A) nuclease (PAN), which is involved in early steps of poly(A) tail metabolism. Deadenylases localize both in the nucleus and the cytoplasm, and are involved in different cellular processes such as DNA damage response, non-sense mediated decay, cell differentiation, etc. Deadenylation results in either a decrease in mRNA stability followed by degradation by the exosome, or the transient modulation of the length of the poly(A) tail allowing a new round of poly(A) tail extension and translation.

As opposed to the CpA reaction, deadenylation can be reconstituted in vitro by both using recombinant proteins and purified fractions from different cell nuclear extracts. The in vitro reaction can be tested using an artificial radiolabeled substrate that is 5′ capped and includes sequences preceding the adenovirus-2 L3 PAS followed by 30 residues of adenosine. A construct containing a longer poly(A) is suitable for analyzing poly(A) intermediate regulation and the reader is directed to an in-depth protocol to that end (Webster et al., 2017). RNA substrates from ML43(G_{30}), ML40(C_{32}) and ML54(U_{30}) constructs can be used to further investigate the specificity of the deadenylating activity. Functional studies on deadenylases have been performed using reporter genes with modified cis-acting sequence elements or RNA structures or depleting factors involved in the reaction, allowing to test the effect of ribonuclease activities on mRNA stability of individual genes and on a global scale, respectively.

5.2 Deadenylation reaction (Fig. 2B–C)

1. The preparation of nuclear extracts from adherent cells and of RNA substrate is described in Sections 4.2 and 4.3.

2. Conditions for in vitro deadenylation in reaction volume of 15 μL: 1.5 mM $MgCl_2$, 2.5% PVA (molecular weight, 10,000), 100 mM KCl, 0.15 U ribonuclease inhibitor, 10 mM HEPES-KOH, pH 7, 0.1 mM EDTA, 0.25 mM DTT, and 10% glycerol. Heat the MeGpppG capped in vitro transcribed/radiolabeled L3(A_{30}) RNA substrate to 80 °C for 2 min, place on ice before adding 1 μL of it to the reaction mix in a concentration no higher than 0.5 ng/μL. To test the specificity of the deadenylating activity for poly(A), RNA substrates from ML43(G_{30}), ML40(C_{32}) and ML54(U_{30}) plasmids can be used.
3. Add no more than 5 μL/reaction of nuclear extracts to the master mix, purified fractions or recombinant factors at concentrations no higher than 2 mg/mL.

   Tips:

   *Sometimes dilution with buffer D is needed to compare different samples. When testing the influence of factors on recombinant ribonuclease activity, the in vitro reconstituted deadenylation reactions have to be done using a limiting amount of deadenylase and in the absence or presence of increasing amounts of the regulatory factor.
4. After addition of the RNA, the reaction mix is incubated at 30 °C for 1 h.
5. Reactions are terminated and analyzed as in Section 4.4, steps 5–13.
6. Determine the ratio between deadenylated and non-deadenylated RNA in each sample to quantitate deadenylation efficiency. Relative deadenylation is calculated as [L3 fragment/(L3 fragment + L3(A_{30}) substrate)] × 100. Quantifications can be done with ImageJ software (http://rsb.info.nih.gov/ij/).
7. Alternatively, the products formed by the deadenylation reaction of L3(A_{30}), ML43(G_{30}), ML40(C_{32}), or ML54(U_{30}) RNA substrates can be analyzed by one-dimensional thin-layer chromatography (TLC). Release of 5′ AMP mononucleotides is expected only if the ribonuclease activity is specific for poly(A) and is a 3′ exonuclease and not endonuclease. Deadenylation reactions are analyzed in TLC using 0.75 M KH_2PO_4, pH 3.5 (H_3PO_4), as the solvent and polyethylenimine-cellulose F plates. The liberated [^{32}P]AMP product is detected after TLC plates are dried and scanned by a 400 S Phosphorimager (Molecular Dynamics). The amount of released AMP is calculated through the specific activity of [^{32}P]AMP in the RNA substrate.

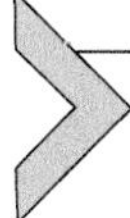

6. Poly(A) tail length measurements

6.1 Introduction

The length of the poly(A) tail has been implicated in mRNA transcripts transport to cytoplasm, stability , as well as translational efficiency (Garneau, Wilusz, & Wilusz, 2007; Schoenberg & Maquat, 2012). Several methods have been developed in order to measure poly(A) tail length, including the oligo(dT)/RNase H-Northern blot analysis (Murray & Schoenberg, 2008), the PCR poly(A) test (PAT) (Sallés & Strickland, 1999), and more recently, genome-wide sequencing techniques such as TAIL-seq (Chang, Lim, Ha, & Kim, 2014) and FLAM-seq (Legnini et al., 2019).

The oligo(dT)/RNase H-Northern blot assay allows for the direct, comparative analysis of poly(A) tail length. In brief, RNA with putative poly(A) tails are incubated with oligo(dT) primers and digested with RNase H, an endoribonuclease that cleaves phosphodiester bonds of RNA bound to DNA. Both the original RNA and the oligo(dT)-incorporated RNA are visualized using Northern blot analysis and their sizes are compared. A change in migration between samples treated with oligo(dT) and those not treated indicates the presence of a poly(A)tail. When visualized alongside a molecular weight marker, the distance of the shift indicates the length of the poly(A) tail. Several protocols have been developed for this technique (Murray & Schoenberg, 2008), as we describe it briefly in this chapter. While this technique allows for quick and direct analysis of poly(A) tail length, there are several drawbacks. One issue is the utilization of oligo(dT) primers and the possibility of short tracts of A-residues existing in the middle of transcript, thereby preventing the exclusive study of poly(A) tails. Another concern is the large starting material requirements of the assay. If the amount of starting RNA is low, it would be better to perform the PAT assay.

The PAT assay allows for the measurement of poly(A) tail length of specific mRNA targets in a population through real-time (RT)-PCR amplification (Sallés & Strickland, 1999). Variations of the PAT assay each include reverse transcription of a pool of RNA using oligo(dT)-anchor primer, followed by PCR amplification of the cDNA using a gene-specific primer and oligo(dT)-anchor primer again (to prevent shortening of the poly(A) sequence after each cycle of PCR). A "smear" of products indicates the heterogeneous pool of poly(A) tail lengths of a target RNA (Fig. 2C). This assay has several variations, outlined in Rio, Ares, Hannon, and Nilsen (2018), of which we describe the RACE-PAT protocol here.

While a myriad of genomic analysis has highlighted the relevance of poly(A) length, this chapter focuses on direct analysis by biochemical approaches.

6.2 RNA purification from adherent cells

1. Aspirate media from a cell culture in 10 cm plate (~10^6 cells) and wash once with ice cold PBS.
2. Aspirate the PBS and add 1 mL TRIzol, scrape the plate briefly and transfer TRIzol/cell lysate into a 1.5 mL Eppendorf tube. (TRIzol followed by chloroform extraction can be used to isolate a large amount of starting material while maintaining RNA purity).
3. Leave the cell lysates at room temperature for 5 min.
4. Add 250 μL chloroform and shake the tube vigorously. Leave at room temperature for 5 min.
5. Centrifuge at 10,000 rpm for 5 min. Remove the aqueous phase, avoiding the organic phase.

   Tips:

   *It may be easier to avoid contamination by tilting the Eppendorf tube, pipetting along the side of the tube, and by leaving a small amount of aqueous phase behind.
6. Add 550 μL isopropanol to the aqueous phase and mix gently. Leave at room temperature for 5 min.
7. Centrifuge at 14,000 rpm for 30 min at 4 °C.
8. Place samples on ice. Wash the pellet with 1 mL 75% ethanol. Mix gently.
9. Centrifuge at 9500 rpm for 5 min at 4 °C.
10. Remove supernatant and let the pellets air-dry. Centrifuge the tubes briefly to force remaining fluid on the side of the tube to the bottom, then pipette off as much of the ethanol as is feasible.
11. Add approximately 15–25 μL (depending on yield) of either Tris-EDTA (TE) buffer or water to the RNA pellet. Determine the concentration of a 1/40 dilution of the purified RNA by measuring absorbance at 260 and 280 nm. The 260/280 ratio should be greater than 1.8. If the ration is <1.8, the RNA is likely partially degraded or there is DNA contamination. The equivalent of the OD at 260 nm (in μg/μL) is the concentration.

6.3 Oligo(dT)/RNase H-northern blot analysis

1. RNA is purified as described in Section 6.2. Importantly, the oligo(dT)/RNase H assay requires a large amount of starting material (>10 μg).

   Tips:

   *It is also important to make sure any ribonuclease inhibitors added during or after extraction do not interfere with RNase H activity.
2. Set up the reaction by combining 10 μg of denatured RNA, 6 μL of oligo(dT)$_{12-18}$ primer (0.5 μg/μL), and 2 μL of 10 × RNase H buffer (Invitrogen) for a total volume of 18 μL. Set up another reaction without oligo(dT) primer for comparison (adding an equal amount of RNase-free water instead). If the RNA transcript being studied is greater than 2 kb and/or it will be difficult to visualize a shift in size between the oligo(dT) ± lanes, 3 μg of a gene-specific primer antisense to a 3′ sequence near the poly(A) site can be included. This will produce a truncated RNA transcript and allow for easier visualization of a transcript otherwise too large to discern differences when its tail is digested.
3. Denature at 65 °C for 5–10 min.
4. Add 1 U of RNase H (Invitrogen), and 0.2 U of ribonuclease inhibitor (RNasin) into each tube. Incubate reaction for 1 h at 37 °C.
5. Inactivate by heating at 65 °C for 10 min, or by adding 1 μL of 0.5 M EDTA.
6. Digested RNA at this stage can be extracted through a variety of different purification protocols. Ethanol precipitation is a quick and efficient technique to purify RNA, and suitable at this point (Rio et al., 2018). Add 1/10 volume of 3 M sodium acetate to the reaction and include 2.5 volumes of cold 100% EtOH. Mix briefly, and keep at −80 °C for 1 h. Recover precipitated RNA by centrifugation at 12,000 × *g* for 10 min at 4 °C, and decant supernatant. Wash the pellet twice, by adding cold 70% ethanol, centrifuging at 12,000 × *g* for 10 min at 4 °C, and removing the supernatant. Let the pellet air dry for 5 min, and dissolve RNA pellet in RNase-free water
7. Continue to visualize the two reactions (oligo (dT) ±) alongside starting RNA material through Northern blot analysis. In brief, RNAs are separated using gel electrophoresis in either agarose or polyacrylamide in the presence of formaldehyde. Mix 15 μg RNA sample with equal volume of 2 × RNA loading buffer (50% formamide, 20% glycerol, 6.5% formaldehyde, 1 × MOPS buffer, 0.05% BFB, 0.05% xylene cyanol). Incubate at 65 °C for 12–15 min and put samples on ice immediately. The RNA is transferred by overnight capillary transfer onto a nylon

membrane (HYBOND N+, GE Healthcare). All other manipulations are essentially as described in Sambrook (1989).

Tips:

*After the blotting, mark the gel wells on the nylon membrane with a needle or blade.

8. Prepare radiolabeled RNA probes complementary to RNA transcript of interest using the protocol described in Section 4.3, steps 3–18. Use either T7, T3, or SP6 RNA polymerases accordingly.
9. Wash the membrane with 0.1% SDS, 0.1 × SSC. Then pre-hybridize for 1 h at 65 °C with ULTRAHyb ultrasensitive hybridization buffer (Thermo Fisher Scientific).

   Tips:

   *Pre-warm ULTRAHyb to 68 °C until the SDS fully dissolves.
10. The labeled probes are then hybridized to the nylon membrane overnight in ULTRAHyb ultrasensitive hybridization buffer (Thermo Fisher Scientific) at 65 °C overnight.
11. Nonspecifically bound probes are washed away after hybridization first by rinsing with 6 × SSC, then with 6 × SSC at 65 °C for at least 20 min, followed by two washes with 0.1 × SSC 0.1% SDS at 65 °C for 20 min.
12. Pull the membrane out of the hybridization bottle, leave semidry, wrap the membrane in Saran Wrap, and expose the membrane to a phosphor screen in a cassette for different times

    Tips:

    *Use Perspex shield while handling the probe and the membrane a substantial amount of radioactivity is used.

    *The phosphor screen should be blanked just before use.

6.4 PAT assay (Fig. 2C)

1. RNA is purified as described in Section 6.2. cDNAs are synthesized from purified RNA using oligo(dT)-anchor primers (5′-GGGGATCCGCG GTTTTTTTTTT-3′), which can anneal to poly(A) transcripts and then extended in a cDNA synthesis reaction. cDNAs with a uniform size are obtained from RNAs with short poly(A) tails as they have limited number of sites for oligo(dT) binding. cDNA products of heterogeneous sizes are obtained from RNAs with long poly(A) tails that have multiple sites for binding the oligo(dT) primer/adapter.
2. Incubate 1 μg (or more) of RNA with 0.5 μg oligo(dT)-anchor primer at 80 °C for 5 min (in a total reaction volume of 5 μL), quick-spin centrifugation and then keep at 4 °C for 10 min.

3. Combine the RNA/oligo(dT)-anchor primer mixture (5 μL) with 15 μL of master mix. For a single reaction of 15 μL total volume add 3 μL of 5 × GoScript reaction buffer, 2 μL of 25 mM $MgCl_2$, 2 μL 10 mM dNTPs, 1 μL of GoScript reverse transcriptase, 20 U RNase inhibitor and 7 μL of RNase-free water.
4. Reactions are left at 25 °C (anneal) for 5 min, 42 °C (extend) for 1.5 h, and lastly at 70 °C for 15 min (inactivation of reverse transcriptase). The cDNA may be used immediately as the template for PCR or stored at −20 °C.
5. The PCR reaction is performed using 1 μL of cDNA, 1.25 μL of oligo(dT)-anchor primer and 1.25 μL of gene-specific primer located 200–400 bp upstream the poly(A) site. Using this strategy, the PCR products will be between 100 and 300 nucleotides, a size easily resolved on a 6% polyacrylamide gel. Combine the cDNA/primers mixture (3.5 μL) with PCR mix: 2 μL 10 × PCR buffer, 3 μL 50 mM $MgCl_2$, 3 μL 1.25 mM dNTPs, 8 μL of RNase-free water and 0.5 μL Taq DNA polymerase. The reactions are incubated in a thermal cycler first at 95 °C for 2 min, and then for 30 cycles, which comprises of a denaturation step at 95 °C for 30 s, an annealing step at a temperature 5 °C below the melting temperature of the gene-specific primer for 45 s, and an extension step at 72 °C for 1.5 min. After the cycles, a final extension at 72 °C is performed for 5 min

   Tips:

   *PCR steps may be optimized by adjusting the annealing temperature and length of time and/or adjusting primer/cDNA concentrations.
6. Reactions are kept at 4 °C until gel electrophoresis analysis or stored at −20 °C.
7. For better detection of PCR products, alternatively, the PAT assay can be performed using the upstream gene-specific primer radiolabeled at the 5′ end with [γ-^{32}P]-ATP. It is convenient to also radiolabel a size marker. For a reaction volume of 20 μL add 250 ng primer or 1 μg of DNA ladder, 2 μL 10 × polynucleotide kinase buffer (supplied with the enzyme), 0.5 μL of 10 μCi/μL of [γ-^{32}P]-ATP, 1 μL of T4 polynucleotide kinase, and RNase-free water to 20 μL.
8. Incubate reaction at 37 °C for 30 min, and then inactivate the T4 polynucleotide kinase by keeping reaction at 65 °C for 5 min.
9. Purification to eliminate any unincorporated radioactivity is not required as it does not interfere with the analysis of PCR products in the gel. The labeled oligo may be purified on a silica spin-column for nucleotide removal. The labeled oligo may be used immediately for PCR or stored at −80 °C.

10. Repeat PCR as described in step 5 using 1 μL of cDNA, 1.25 μL of oligo(dT)-anchor primers and 1.25 μL of radiolabeled gene-specific primer.
11. PCR products are then resolved through gel electrophoresis. Combine 3 μL of each PCR reaction with 3 μL formamide loading buffer (95% formamide, 0.025% xylene cyanol and BFB, 18 mM EDTA and 0.025% SDS), heat to 95 °C for 5 min and resolve on a 6% polyacrylamide gel or 1% agarose gel. To differentiate PCR products of different length the gel has to be long enough to allow the resolution. Run the gel in TBE buffer at 60 W until the lower dye front is at the bottom of the gel. Remove one glass plate and treat as in Section 4.4, steps 11–13. While mRNAs with a short poly(A) tail will appear as a discrete peak, mRNAs with long tails will appear as a range of PCR products of different sizes that appear as a smear of bands. Sometimes peaks are observed at 25 nucleotides intervals, which corresponds in size to the distribution of PABP.

7. Summary

The intricate network of factors impacting poly(A) tail metabolism highlights the relevance of these pathways in cellular homeostasis. In recent years, numerous studies have described genome-wide changes in different cellular conditions, such as stress, proliferation, differentiation and development. Direct connections between the mRNA 3′ end processing machinery and factors involved in those cellular pathways provide an obvious link to different diseases, such as cancer and other clinical conditions (Murphy & Kleiman, 2020b). In this chapter, we provide an overview of two key steps in the mRNA life-cycle, CpA and deadenylation, and biochemical assays generated to study each of these enzymatic reactions and their effect on mRNA's poly(A) length. While bioinformatic studies facilitate our understanding of the dynamics of these reactions and their effect on cellular functions, this research will benefit from a more mechanistic approach that offers details on the factors and sequences involved in this complex process of gene expression.

Acknowledgment

This work was supported by National Cancer Institute, National Institutes of Health (NIH) 1U54CA221704-01A (to F.E.K).

References

Bentley, D., & Coupling, L. (2014). mRNA processing with transcription in time and space. *Nature Reviews. Genetics*, *15*, 163–175.

Berg, M. G., et al. (2012). U1 snRNP determines mRNA length and regulates isoform expression. *Cell*, *150*, 53–64.

Cao, D., & Parker, R. (2001). Computational modeling of eukaryotic mRNA turnover. *RNA*, 7, 1192–1212.

Chang, H., Lim, J., Ha, M., & Kim, V. N. (2014). TAIL-seq: Genome-wide determination of poly(A) tail length and 3′ end modification. *Molecular Cell*, *53*, 1044–1052.

Darnell, J. E., Jr. (2013). Reflections on the history of pre-mRNA processing and highlights of current knowledge: A unified picture. *RNA*, *19*, 443–460.

Decker, C. J., & Parker, R. (1993). A turnover pathway for both stable and unstable mRNAs in yeast: Evidence for a requirement for deadenylation. *Genes & Development*, 7, 1632–1643.

Garneau, N. L., Wilusz, J., & Wilusz, C. J. (2007). The highways and byways of mRNA decay. *Nature Reviews. Molecular Cell Biology*, *8*, 113–126.

Glover-Cutter, K., Kim, S., Espinosa, J., & Bentley, D. L. (2008). RNA polymerase II pauses and associates with pre-mRNA processing factors at both ends of genes. *Nature Structural & Molecular Biology*, *15*, 71–78.

Goldstrohm, A. C., & Wickens, M. (2008). Multifunctional deadenylase complexes diversify mRNA control. *Nature Reviews Molecular Cell Biology*, *9*, 337–344.

Hirose, Y., & Manley, J. L. (1998). RNA polymerase II is an essential mRNA polyadenylation factor. *Nature*, *395*, 93–96.

Khleborodova, A., Pan, X., Nagre, N. N., & Ryan, K. (2016). An investigation into the role of ATP in the mammalian pre-mRNA 3′ cleavage reaction. *Biochimie*, *125*, 213–222.

Legnini, I., et al. (2019). FLAM-seq: Full-length mRNA sequencing reveals principles of poly(A) tail length control. *Nature Methods*, *16*, 879–886.

Murphy, M. R., & Kleiman, F. E. (2020a). Connections between 3′ end processing and DNA damage response: Ten years later. *Wiley Interdisciplinary Reviews: RNA*, *11*, e1571.

Murphy, M. R., & Kleiman, F. E. (2020b). Connections between 3′-end processing and DNA damage response. *Wiley Interdisciplinary Reviews: RNA*, *11*, e1571.

Murray, E. L., & Schoenberg, D. R. (2008). Assays for determining poly(A) tail length and the polarity of mRNA decay in mammalian cells. *Methods in Enzymology*, *448*, 483–504.

Proudfoot, N. J. (2011). Ending the message: Poly(A) signals then and now. *Genes & Development*, *25*, 1770–1782.

Rio, D. C., Jr., Ares, M., Hannon, G. J., & Nilsen, T. W. (2018). Ethanol precipitation of RNA and the use of carriers. *Cold Spring Harbor Protocols*, *2010*, 1–5.

Ryan, K., Murthy, K. G. K., Kaneko, S., & Manley, J. L. (2002). Requirements of the RNA polymerase II C-terminal domain for reconstituting pre-mRNA 3′ cleavage. *Molecular and Cellular Biology*, *22*, 1684–1692.

Sallés, F. J., & Strickland, S. (1999). Analysis of poly(A) tail lengths by PCR: The PAT assay. *Methods in Molecular Biology*, *118*, 441–448.

Sambrook, H. C. (1989). *Molecular cloning: A laboratory manual*. New York: Cold Spring Harbor.

Schoenberg, D. R., & Maquat, L. E. (2012). Regulation of cytoplasmic mRNA decay. *Nature Reviews Genetics*, *13*, 246–259.

Sheets, M. D., & Wickens, M. (1989). Two phases in the addition of a poly(A) tail. *Genes & Development*, *3*, 1401–1412.

Shi, Y., et al. (2009). Molecular architecture of the human pre-mRNA 3′ processing complex. *Molecular Cell*, *33*, 365–376.

Tian, B., & Graber, J. H. (2012). Signals for pre-mRNA cleavage and polyadenylation. *Wiley Interdisciplinary Reviews: RNA*, *3*, 385–396.

Tian, B., & Manley, J. L. (2017). Alternative polyadenylation of mRNA precursors. *Nature Reviews Molecular Cell Biology*, *18*, 18–30.
Webster, M. W., Stowell, J. A. W., Tang, T. T. L., & Passmore, L. A. (2017). Analysis of mRNA deadenylation by multi-protein complexes. *Methods*, *126*, 95–104.
Zhang, X., Kleiman, F. E., & Devany, E. (2014). Deadenylation and its regulation in eukaryotic cells. *Methods in Molecular Biology*, *1125*, 289–296.
Zhang, X., Virtanen, A., & Kleiman, F. E. (2010). To polyadenylate or to deadenylate: That is the question. *Cell Cycle*, *9*, 4437–4449.

CHAPTER FOURTEEN

Reconstitution and biochemical assays of an active human histone pre-mRNA 3′-end processing machinery

Yadong Sun[a,†], Wei Shen Aik[a,‡], Xiao-Cui Yang[b], William F. Marzluff[b,c], Zbigniew Dominski[b,c,*], and Liang Tong[a,*]

[a]Department of Biological Sciences, Columbia University, New York, NY, United States
[b]Integrative Program for Biological and Genome Sciences, University of North Carolina at Chapel Hill, Chapel Hill, NC, United States
[c]Department of Biochemistry and Biophysics, University of North Carolina at Chapel Hill, Chapel Hill, NC, United States
*Corresponding authors: e-mail address: zbigniew_dominski@med.unc.edu; ltong@columbia.edu

Contents

† Present address: School of Life Science and Technology, ShanghaiTech University, Shanghai, China.
‡ Present address: Department of Chemistry, Hong Kong Baptist University, Kowloon Tong, Kowloon, Hong Kong SAR, China.

Methods in Enzymology, Volume 655
ISSN 0076-6879
https://doi.org/10.1016/bs.mie.2021.03.021

Abstract

In animal cells, replication-dependent histone pre-mRNAs are processed at the 3′-end by an endonucleolytic cleavage carried out by the U7 snRNP, a machinery that contains the U7 snRNA and many protein subunits. Studies on the composition of this machinery and understanding of its role in 3′-end processing were greatly facilitated by the development of an *in vitro* system utilizing nuclear extracts from mammalian cells 35 years ago and later from *Drosophila* cells. Most recently, recombinant expression and purification of the components of the machinery have enabled the full reconstitution of an active machinery and its complex with a model pre-mRNA substrate, using 13 proteins and 2 RNAs, and the determination of the structure of this active machinery. This chapter presents protocols for preparing nuclear extracts containing endogenous processing machinery, for assembling semi-recombinant and fully reconstituted machineries, and for histone pre-mRNA 3′-end processing assays with these samples.

1. Introduction

Eukaryotic messenger RNA precursors (pre-mRNAs) must undergo processing at the 3′-end before they can be exported to the cytoplasm as mature mRNAs for translation into proteins. The processing events typically consist of cleavage of a pre-mRNA at a specific location followed by the addition of a poly(A) tail. A large machinery with many protein subunits is required for this canonical processing (Shi & Manley, 2015; Sun, Hamilton, & Tong, 2020; Zhao, Hyman, & Moore, 1999). The mammalian cleavage factor (mCF) (Chan et al., 2014; Schonemann et al., 2014), with CPSF73, CPSF100 and symplekin as its subunits, catalyzes the cleavage reaction, with CPSF73 as the endonuclease (Mandel et al., 2006).

In animal cells, replication-dependent histone pre-mRNAs are processed through a different mechanism, as they are cleaved at the 3′-end but not polyadenylated (Dominski & Marzluff, 2007; Romeo & Schumperli, 2016). Histone pre-mRNAs contain two distinct sequence elements encompassed within an approximately 60-nucleotide region located less than

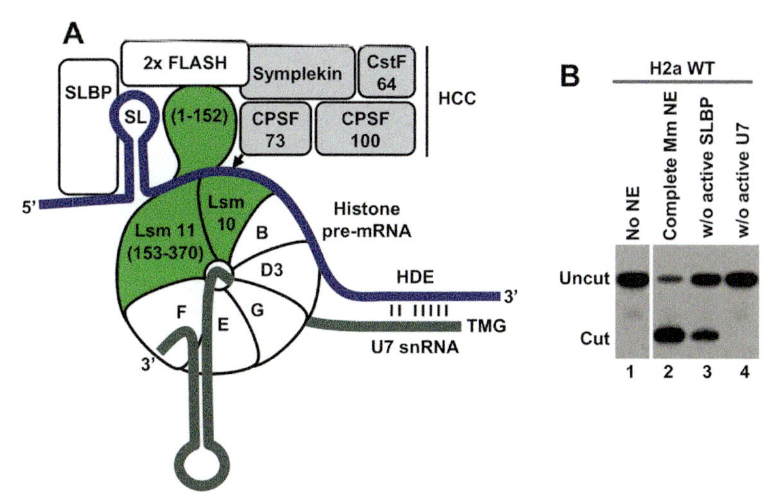

Fig. 1 The human histone pre-mRNA 3′-end processing machinery. (A) Schematic representation the human of U7 processing machinery determined by biochemical studies using nuclear extracts from mammalian cells. Base pairing between the HDE of histone pre-mRNA (dark blue line) and the 5′ region of U7 snRNA (dark green line) is illustrated by vertical lines. Components of the HCC are indicated in gray and the two U7-specific subunits of the Sm ring are indicated in dark green. (B) Processing of wild-type mouse histone pre-mRNA in a complete mouse nuclear extracts or in modified extracts where SLBP or U7 snRNA were pre-bound and hence inactivated by appropriate competitors.

100 nucleotides (nts) 3′ of the stop codon (Fig. 1A). The upstream sequence element folds into a conserved stem-loop structure (Table 1), which is recognized by the stem-loop binding protein (SLBP) (Tan, Marzluff, Dominski, & Tong, 2013; Zhang et al., 2014). Residues 125–200 of human SLBP constitute its RNA binding domain (RBD), which together with residues 201–223 are essential for processing (Dominski, Zheng, Sanchez, & Marzluff, 1999). The cleavage site is located between the stem-loop and the second sequence element, the histone downstream element (HDE) (Table 1), which is recognized through base pairing with the 5′-end of U7 snRNA, a component of the U7 snRNP (Bond, Yario, & Steitz, 1991; Mowry & Steitz, 1987; Schaufele, Gilmartin, Bannwarth, & Birnstiel, 1986) (Fig. 1A).

U7 snRNP is a minor snRNP and exists at low concentrations in most mammalian cells, not exceeding 5×10^3 particles per cell. Of the seven proteins in its Sm ring, five are shared with the spliceosomal snRNPs: SmB, SmD3, SmE, SmF and SmG. The two remaining proteins, Lsm10 and Lsm11, are specific for the U7 snRNP, and they replace SmD1 and

Table 1 Sequences of RNAs used for structural studies and 3′-end processing assays.

RNA	Sequence[a]
U7 snRNA (1-63)	CA**GUGUUACAGCUCUUU**UAGAAUUUGUCUAGUAG GCUUUCUGGCUUUUUACCGGAAAGCCCCU
U7 Mut snRNA	CA**GUGUUACAG**gagaaaUAGAAUUUGUCUAGUAG GCUUUCUGGCUUUUUACCGGAAAGCCCCU
H2a* (52-mer), for structural studies	CCAAAGGCUCUUUUCAGAGCCACCCA↓CUGAAUCAG AU**AAAGAGCUGUAACAC**
H2a* (60-mer), doubly fluorescently labeled	TAMRA-CCAAAGGCUCUUUUCAGAGCCACCCA↓CU GAAUCAGAU**AAAGAGCUGUAACAC**GGUAGCCA-FAM
H2a* (64-mer), radio-labeled	^{32}P-CAAAAGGCUCUUUUCAGAGCCACCCA↓CUGAAU CAGAU**AAAGAGCUGUGACAC**GGUAGCCGGUCU
H2a* Mut (64-mer), radio-labeled	^{32}P-CAAAAGGCUCUUUUCAGAGCCACCCA↓CUGAAU CAGAUuuucuc**CUGUGACAC**GGUAGCCGGUCU
Downstream cleavage product (26-mer)	CUGAAUCAGAU**AAAGAGCUGUAACAC**-FAM

[a]The cleavage sites in the H2a* substrates are indicated with the downward arrow, and the stem-loop is underlined. The HDE in H2a* (bold) can form 15 consecutive Watson-Crick base pairs with the 5′ region of U7 snRNA (bold). In U7 Mut snRNA, six nucleotides of this region were substituted with complementary nucleotides (lower case letters). This mutation reduces the length of duplex with H2a* to 9 base pairs and restores 15-base-pair duplex with H2a* Mut containing a compensatory mutation in the HDE (lower case letters). For the 52-mer H2a*, the cleavage reaction produces two 26-mers; hence the 60-mer fluorescently labeled H2a* is used for 3′-end processing assays in order to distinguish the two products.

SmD2 found in the same positions of the spliceosomal ring. The complex of U7 snRNA with the Sm ring is referred to as the U7 Sm core. Lsm11 contains a long N-terminal extension that interacts with an N-terminal segment of the protein FLASH (Yang, Burch, Yan, Marzluff, & Dominski, 2009; Yang et al., 2011), which forms a coiled-coil dimer (Aik et al., 2017). The two interacting proteins recruit the histone pre-mRNA cleavage complex (HCC) to the U7 snRNP histone pre-mRNA 3′-end processing machinery, also referred to as the U7 machinery in short here. Remarkably, HCC is equivalent to mCF, with CPSF73, CPSF100, symplekin and CstF64 as its subunits. CPSF73 is the endonuclease for the cleavage reaction of histone pre-mRNA 3′-end processing as well (Dominski, 2010; Dominski, Yang, & Marzluff, 2005), indicating that the cleavage module is shared between the canonical and U7 machineries.

Histone pre-mRNA 3′-end processing has been studied extensively over the past 35 years using nuclear extracts (Gick, Kramer, Keller, & Birnstiel, 1986). Recently, components of the machinery have been expressed recombinantly, purified and studied individually to gain structural information, including SLBP (Tan et al., 2013), FLASH (Aik et al., 2017), and HCC (Zhang, Sun, Shi, Walz, & Tong, 2020). In addition, the U7 Sm core was reconstituted from all seven recombinant Sm/Lsm proteins and synthetic U7 snRNA and shown to support accurate processing of histone pre-mRNA in nuclear extracts by binding endogenous HCC and forming a semi-recombinant machinery (Bucholc et al., 2020). Most importantly, a fully reconstituted machinery, containing human U7 Sm core, FLASH and HCC, has been found to be active in cleaving model histone pre-mRNA substrates *in vitro* (Sun, Zhang, et al., 2020; Yang et al., 2020). Such a reconstituted machinery was also crucial in enabling its structure to be determined in complex with a model substrate and SLBP (13 proteins and 2 RNAs), providing the first molecular insights into an active pre-mRNA 3′-end processing machinery (Sun, Zhang, et al., 2020). This chapter will present protocols for preparing nuclear extracts, semi-recombinant as well as fully reconstituted histone machineries, and for histone pre-mRNA 3′-end processing assays with these samples.

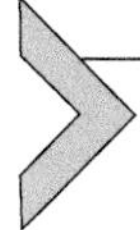

2. Preparation of nuclear extracts for histone pre-mRNA 3′-end processing

Nuclear extracts active in 3′-end processing of replication-dependent histone pre-mRNAs can be prepared using a number of mammalian cell lines, including HeLa cells (Gick et al., 1986), K21 mouse mastocytoma cells (Stauber, Soldati, Lüscher, & Schümperli, 1990), U2OS cells (ZD, unpublished results), CHO cells (ZD, unpublished results) and 66-2 mouse myeloma cells (Dominski, Sumerel, Hanson, & Marzluff, 1995; Skrajna, Yang, Dadlez, Marzluff, & Dominski, 2018). Nuclear extracts from HeLa cells contain relatively small amounts of U7 Sm core and FLASH and typically process histone pre-mRNAs inefficiently. This low processing activity can be enhanced by the addition of bacterially expressed N-terminal segment of FLASH, which interacts with endogenous Lsm11 and promotes the recruitment of the HCC that is abundant in HeLa nuclear extracts, converting an inactive U7 Sm core into an active U7 machinery (Yang et al., 2011). Besides mammalian cells, nuclear extracts active in 3′-end processing of histone pre-mRNAs were also obtained from two *Drosophila* cells lines: S2 (Dominski et al., 2002) and Kc (Dominski, Yang, Purdy, & Marzluff, 2005).

Mouse myeloma cells may contain three- to fivefold more U7 snRNP than HeLa cells, reflecting natural differences between cell lines and species (Smith et al., 1991). They yield nuclear extracts that are highly active in 3′-end processing, converting as much as 90% of ^{32}P-labeled pre-mRNA substrate into mature mRNA during 1 h incubation (Fig. 1B). Extracts from mouse myeloma cells were instrumental in identifying CPSF73 as the catalytic component of the 3′-end processing machinery (Dominski, Yang, & Marzluff, 2005; Yang, Sullivan, Marzluff, & Dominski, 2009) and were successfully used for one-step purification of U7 machinery and determining its composition by mass spectrometry (Skrajna, Yang, & Dominski, 2019).

2.1 Equipment

1. Shaker (Eppendorf New Brunswick Innova 44) for bacterial and insect cell cultivation
2. Fluorescence microscope (for insect cell culture)
3. Hemocytometer to count cells
4. Sonicator (Misonix S-4000)
5. Centrifuge with different rotors: Beckman TY.JS4.2 rotor for harvesting cells; F15-8 × 50c carbon fiber rotor for clearing lysates
6. Beckman GS-6R benchtop refrigerated centrifuge and GH 3.8 Swing Bucket Rotor for collecting cells and nuclei
7. Sorvall RC-5B refrigerated centrifuge with rotor SS-34 for high-speed spinning

8. Eppendorf 5424 benchtop mini-centrifuge
9. Geiger counter
10. Corning stirring plate
11. Tube rotator
12. Water bath
13. Power supply
14. Vertical gel electrophoresis apparatus V16-2 for RNA separation
15. Protein purification system: ÄKTA Pure (Cytiva)
16. HiTrap™ Heparin HP 5 mL column (Cytiva)
17. HiTrap™ Q HP 5 mL column (Cytiva)
18. Size-exclusion chromatography column (Sephacryl S-300; Cytiva)
19. Superose 6 10/300 GL column (Cytiva)
20. Biological safety cabinet (Thermo)
21. Freezer (−80 °C) (Thermo)
22. Nanodrop spectrophotometer (Thermo)
23. Protein or RNA electrophoresis equipment (Bio-Rad)
24. ChemiDoc MP imaging system (Bio-Rad)

2.2 Reagents or resources

1. Cellfectin™ II reagent (Gibco)
2. Transfection medium (Expression Systems)
3. Sf9 cells and Hi5 cells (Expression Systems)
4. ESF 921 medium (Expression Systems)
5. Protease inhibitor cocktail tablet (Sigma), PMSF
6. Ni-NTA resin (Qiagen)
7. HEPES, Tris
8. Isopropyl-β-D-thiogalactopyranoside (IPTG)
9. Diethyl pyrocarbonate (DEPC)
10. Filtered buffer stocks: 1 M Tris–HCl (pH 8.0), 5 M NaCl, 1 M Imidazole, 1 M DTT, 2 M KCl, 0.5 M EDTA (pH 8.0)
11. *E. coli* DH10MultiBac cells
12. *E. coli* BL21 (DE3) Star cells
13. X-gal
14. RNasin plus ribonuclease inhibitor (Promega-Fisher)
15. Urea
16. Kanamycin, Ampicillin, Gentamycin
17. Ethanol, isopropanol
18. β-mercaptoethanol, DTT

19. Antimycotic solution for insect cell media
20. LB powder, LB agar tablets
21. Plasmid miniprep purification kit (Thermo)
22. Amicon filter concentrators
23. Purified TEV protease
24. NaCl, KCl
25. Glycerol
26. Spermine, spermidine (Sigma)
27. Streptavidin-agarose (Sigma)
28. Membrane tubing MW cutoff 6000–8000 (23 mm) for dialysis of nuclear extracts (Spectrapor)
29. Adenosine 5′-triphosphate (γ-^{32}P) 3000 Ci/mmol (PerkinElmer)
30. T4 polynucleotide kinase (New England Biolabs)

2.3 Growing and handling mouse myeloma cells

In mammalian cells, SLBP starts accumulating in cells at the G1/S phase transition and reaches the highest level in S phase, concomitant with DNA replication, when histone mRNAs are generated, and it is rapidly degraded by the proteasome pathway during the G2 phase (Zhang et al., 2014). Similar profile of cell cycle-regulated expression was also observed for FLASH, suggesting that the active form of U7 snRNP, i.e., containing the HCC, may be present at the highest concentrations during S phase (Barcaroli et al., 2006). It is therefore critical that cell cultures used for preparation of nuclear extracts active in processing of histone pre-mRNAs are maintained in the exponential growth phase and contain a high percentage of dividing cells.

Large amounts of mouse myeloma cells (e.g., 20 L culture and more) can be ordered from Cell Culture Company (C3, Minneapolis, MN). The cells are grown at the company's production site at 37 °C in the presence of 5% CO_2 in Joklik's Minimal Essential Medium (MEM) containing 10% heat inactivated horse serum to a density of 1.2–1.4 × 10^6 cells/mL. Mouse myeloma cells are exceptionally fragile and their inappropriate handling during harvesting and extract preparation can lead to uncontrolled nuclear lysis and significant reduction of both the volume and protein concentration of the resultant extract. Immediately prior to shipment, the cells are gently harvested in separate 250 mL conical tubes using Beckman GH 3.8 swinging bucket rotor under gentle conditions of centrifugation (5 min at 2000 rpm or 650 × *g*), avoiding multiple spinning of the same cell pellet. Cell pellets are softened by tube swirling or tapping, and combined by gentle pipetting using a small amount of growth media. Tubes containing cell suspension are shipped on wet ice. Upon arrival the following day, the cells are used immediately for preparation of nuclear extract. Smaller amounts of mouse myeloma cells (1–5 L) can be grown at 37 °C in the laboratory using spinner flasks of appropriate size and 5% CO_2 incubator.

2.4 Preparation of nuclear extracts from mouse myeloma cells

Mouse myeloma cells are processed using a modified method of Dignam et al. (Abmayr, Yao, Parmely, & Workman, 2006; Dignam, Lebovitz, & Roeder, 1983; Dignam, Martin, Shastry, & Roeder, 1983), with most modifications incorporated into the method serving to stabilize the integrity of the nuclear envelope and to limit the extent of nuclear lysis during high salt extraction (Dominski et al., 1995).

Nuclear extracts with the highest activity in supporting 3′-end processing of histone pre-mRNAs are routinely obtained by adjusting the final KCl concentration during extraction to 0.25 M, significantly lower than the 0.42 M required to yield extracts active in transcription (Dignam, Lebovitz, et al., 1983; Dignam, Martin, et al., 1983) or splicing (Krainer, Maniatis, Ruskin, & Green, 1984). This concentration is well tolerated by nuclei isolated from mouse myeloma cells and is sufficient to extract all essential components of the processing machinery. Higher salt concentrations bring a risk of nuclear breakage during extraction and yield extracts with lower processing activity, possibly as a result of extracting uncharacterized inhibitors or irreversibly disrupting U7 processing machinery.

The protocol given below describes how to prepare active extracts from mouse myeloma cells. All steps during the preparation of nuclear extracts are conducted on ice or in a cold room, using pre-cooled solutions, centrifuge tubes and glassware. Note that this protocol is a list of specific recommendations and suggestions, and some critical steps may need to be optimized experimentally to adjust for differences among cell lines, growth conditions and available equipment.

1. Collect cells shipped as a high concentration suspension in growth medium by spinning 5 min at 650 × *g* and gently aspirate the supernatant using a vacuum pump, making sure that no residual medium is left over the cell pellet.
2. Estimate the volume of the pellet and loosen it by gently tapping the bottom of the tube. This step is important as it eliminates the need for excessive pipetting in the following step.

3. Resuspend cells in 5 volumes of hypotonic buffer (10 mM HEPES-KOH pH 7.9, 1.5 mM $MgCl_2$, 20 mM KCl, 0.5 mM DTT, 0.75 mM spermidine, 0.15 mM spermine) by slowly pipetting up and down the suspension and occasionally swirling the tube.

   Tip: DTT, spermidine and spermine are added to buffers just before use from the following stock solutions kept frozen or freshly prepared: 1 M DTT (2000 ×), 75 mM spermidine (100 ×), and 15 mM spermine (100 ×). KCl in all buffers can be replace by NaCl without noticeable loss of activity in 3′-end processing of histone pre-mRNAs.
4. Transfer cell suspension to a Dounce homogenizer of appropriate size and allow cells to swell on ice for 10 min.

   Tip: In contrast to the extremely fragile mouse myeloma cells, HeLa cells and some other cell types, including *Drosophila* cells, typically require one wash with hypotonic buffer prior to the cell swelling step. The cells are gently centrifuged immediately after adding 5 volumes of the hypotonic buffer, and resuspended in 2 volumes of the same buffer for the subsequent 10 min swelling step, as in the original protocol (Abmayr et al., 2006; Dignam, Lebovitz, et al., 1983; Dignam, Martin, et al., 1983). This extra washing step removes residual amounts of the media trapped in the pellet and enhances cell swelling. Since it is followed by adding only two volumes of hypotonic buffer for cell lysis, it also ultimately results in a more concentrated cytoplasmic fraction that can be used for other purposes. If required, this protocol can be used for preparation of nuclear and cytoplasmic extracts from mouse myeloma cells, although it may compromise the integrity of nuclei during the extraction step.
5. Place a tight pestle in the homogenizer and move it up and down to lyse the cells, avoiding fast and forceful strokes. After 10 initial up and down strokes, check the ratio of unbroken cells to free nuclei under a phase-contrast microscope by placing ~50 μL of cell suspension on a glass slide and covering the drop with a cover slip. Use five additional strokes if the percentage of unbroken cells is larger than 20% but do not continue if no additional lysis occurs. Note that this is one of the most critical steps in extract preparation and it should be carefully monitored as different cell batches and different cell lines require different time for swelling and different number of strokes for lysis. As a rule of thumb, use minimal number of strokes to get sufficient percentage of free nuclei in the suspension.

6. Add 1/10 volume of restore buffer (67.5% sucrose in hypotonic buffer) and spin 5 min at 1500 × *g*. This step is a modification of the original protocol of Dignam, Lebovitz, et al. (1983) and Dignam, Martin, et al. (1983) and is intended to stabilize the nuclei and prevent their lysis during subsequent steps of the protocol.
7. Carefully remove the supernatant containing the cytosolic fraction and mitochondria, and estimate the volume of the darker and more compact nuclear pellet at the bottom of the tube.
8. Soften the nuclear pellet by gently tapping the bottom of the tube with a finger, add ¼ pellet volume of low salt extraction buffer (20 mM HEPES-KOH pH 7.9, 1.5 mM $MgCl_2$, 20 mM KCl, 0.5 mM DTT, 0.2 mM EDTA, 25% glycerol, 0.75 mM spermidine, 0.15 mM spermine) and resuspend the nuclei by gently swirling the tube.
9. Use a pipette of appropriate size to complete the process of resuspending the nuclei by slowly pipetting the suspension up and down. Transfer the resuspended nuclei to a glass beaker of an appropriate size placed on ice.
10. Use an additional ¼ volume of low salt buffer to rinse the nuclei remaining in the centrifuge tube and in the pipette and combine the recovered volume with the nuclei already placed in the beaker.

    Tip: Make sure to carefully record the volume of the suspension that is being collected in the beaker. For smaller scale preparations, this could be facilitated by using a graduated conical centrifuge tube. Note that the suspension is relatively thick and partially viscous, leaving a sizeable fraction inside the pipette, potentially leading to overestimation of the transferred volume. Note also that the overall volume of low salt extraction buffer added to resuspend nuclei equals 50% of the nuclear pellet. This amount could be reduced to 30–40%, yielding more concentrated and potentially more active nuclear extracts.
11. Insert a magnetic bar of an appropriate size into the suspension of the nuclei, place the beaker on ice over a stirring plate and set up the speed of stirring to achieve visible mixing.
12. Use the following formula to calculate the volume (X) of high salt extraction buffer required to give final salt concentration of 0.25 M: $X = 0.25 \times Y/0.95$, where Y is the volume of nuclear suspension in low salt extraction buffer.

    Tip: The final salt concentration during the extraction step will depend on the accuracy of measuring the volume of the transferred nuclear suspension. Overestimation of the volume may result in the actual salt concentration during the extraction being much higher than anticipated, causing excessive lysis of nuclei during the extraction step, and hence reducing protein concentration, processing activity and total volume of the extract. Nuclei of mouse myeloma cells, when properly handled, should survive extraction in 0.25 M salt with only limited nuclear lysis.

13. While constantly stirring the suspension, drop-wise add high salt extraction buffer (20 mM HEPES-KOH pH 7.9, 1.5 mM $MgCl_2$, 1.2 M KCl, 0.5 mM DTT, 0.2 mM EDTA pH 8.0, 25% glycerol, 0.75 mM spermidine, 0.15 mM spermine). Note that high salt buffer has the same composition as low salt buffer with the exception that 20 mM KCl is replaced with 1.2 M KCl. Avoid adding larger amounts of the buffer in short time as this can create local high salt concentration and accelerate lysis of nuclei.
14. Reduce the intensity of stirring and extract nuclei for 45–60 min. Limited lysis may occur during this time, as indicated by the presence of a viscous clump of released chromatin in the center of the beaker.
15. Spin down the extracted nuclei by centrifugation at 12,000 × *g* for 30 min and carefully collect the supernatant.
16. Transfer the supernatant to dialysis tube (12–14 kDa MWCO) that was boiled in deionized water for 15 min prior to using or prepared as recommended by the manufacturer. Leave a small air bubble in the tube that will facilitate its flow in a vertical orientation during dialysis. Make sure that the two open ends are securely sealed with dialysis clips by squeezing the liquid in the tube and checking for potential leaks.
17. Place the dialysis tube in 100 × volume of ice-cold dialysis buffer (20 mM HEPES-KOH pH 7.9, 100 mM KCl, 0.5 mM DTT, 0.2 mM EDTA, 20% glycerol) in a beaker or cylinder containing a magnetic bar and dialyze on a stirring plate in a cold room for 1.5–2 h, making sure that the tube slowly spins in the buffer. Transfer the tube to a fresh change of dialysis buffer and continue dialyzing for additional 1.5–2 h.
18. Transfer the dialyzed nuclear extract to a centrifuge tube and spin at 12,000 × *g* for 20 min to remove visible precipitates.
19. Aliquot the nuclear extract, snap freeze in dry ice and store at −80 °C. The extracts can be stored at this temperature for decades and survive multiple thawing-freezing cycles without losing activity in 3′-end processing of histone pre-mRNAs.

Typically, 1 L myeloma cell culture cultured in suspension to a density of 1×10^6 cells/mL yields 1 mL of a highly active nuclear extract with a protein concentration approaching 8–10 mg/mL. The yield of the extract and the reproducibility of the protocol can be substantially improved by scaling up the cell culture to 10–20 L or more. The same protocol can be used to prepare nuclear extracts from other mammalian cell lines and was also successfully used to generate highly active nuclear extracts from two *Drosophila* cell lines grown in suspension: Schneider's line S2 (S2) and Kc (Dominski, Yang, Purdy, & Marzluff, 2003; Dominski, Yang, Purdy, et al., 2005; Dominski et al., 2002; Sabath et al., 2013). As indicated above, both HeLa and *Drosophila* cells are relatively sturdy and their efficient breakage into the nuclear and cytoplasmic fractions typically requires one wash with the hypotonic buffer (see above) after the cells are harvested. Spermine and spermidine, the two related polyamines added to buffers to prevent nuclear breakdown and chromatin leakage, are omitted when working with HeLa and *Drosophila* cells.

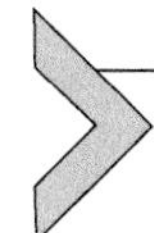

3. Reconstitution of an active human histone pre-mRNA 3′-end processing machinery

For the U7 Sm core, the expression of SmB, SmD3, SmE, SmF and SmG in *E. coli* as SmB-SmD3 and SmE-SmF-SmG sub-complexes, and the purification conditions, have been established in previous reports on spliceosomal snRNPs (Grimm et al., 2013; Kambach et al., 1999; Leung et al., 2010; Leung, Nagai, & Li, 2011; Pomeranz Krummel, Oubridge, Leung, Li, & Nagai, 2009; Raker, Hartmuth, Kastner, & Lührmann, 1999). FLASH N-terminal segment can be readily expressed in *E. coli* as well (Aik et al., 2017). However, SLBP produced from *E. coli* is not stable on its own and readily precipitates when concentrated. HCC and Lsm10-Lsm11 heterodimer are uninduced or insoluble when expressed in *E. coli*. Therefore, expression in insect cells using MultiBac technology (Sari et al., 2016) (Geneva Biotech) is used for SLBP, HCC and Lsm10-Lsm11. For HCC, symplekin must be co-expressed with CPSF73, CPSF100 and CstF64, but the protein yield is quite low. To improve the yield, adding a SUMO tag at the N terminus of symplekin is necessary. Similarly, an MBP tag is added at the N terminus of Lsm10 to increase solubility. After cell lysis, the proteins are purified by nickel affinity followed by either ion exchange or size-exclusion chromatography. A detailed list of the constructs, expression and purification conditions is given in Table 2. The first step of U7 machinery reconstitution is the formation of U7 Sm core (Bucholc et al., 2020), which is then combined with the other components (Sun, Zhang, et al., 2020). The active fully reconstituted machinery contains 13 recombinant proteins and 2 RNAs.

3.1 Expression and purification from bacteria (SmB-SmD3, SmE-SmF-SmG and FLASH)

1. Inoculate 100 mL LB containing appropriate antibiotic(s) with a colony of bacteria harboring construct(s) of interest and grow overnight with vigorous shaking at 37 °C, as starter culture.
2. Inoculate 6 L LB containing appropriate antibiotic(s) with the starter culture (10 mL/L) and incubate with vigorous shaking at 37 °C until OD_{600} reaches 0.8–1.
3. Induce protein expression by adding isopropyl β-D-1-thiogalactopyranoside (IPTG) to a final concentration of 0.4 mM and grow at 20 °C for 18 h.
4. Harvest cells by centrifugation at 4000 rpm in a Beckman TY.JS4.2 rotor for 15 min at 4 °C and discard the supernatant.
5. Resuspend cell pellet in 100 mL lysis buffer containing 20 mM Tris (pH 7.5), 500 mM NaCl, 10 mM imidazole, 5% (v/v) glycerol, 17.8 μg/mL PMSF, 10 mM β-mercaptoethanol.
6. Sonicate the cell suspension at 70% output for 6 min with 1 s on and 2 s off (Misonix S-4000).
7. Centrifuge at 13,000 rpm in a F15-8 × 50c carbon fiber rotor at 4 °C for 30 min.
8. Equilibrate 250 μL Ni-NTA resin in each of two 50 mL Falcon tubes with 4 volumes of Ni wash buffer containing 20 mM Tris (pH 7.5), 500 mM NaCl, 40 mM imidazole, and 10 mM β-mercaptoethanol.
9. Transfer 50 mL of the supernatant to equilibrated Ni-NTA resin in each tube and slowly rotate for 1 h in a cold room.
10. Transfer the mixture to a 14 cm high, gravity flow column (Bio-Rad).
11. Wash resin with 3 × 15 mL Ni wash buffer.
12. Elute protein with 4 mL Ni elution buffer containing 20 mM Tris (pH 7.5), 500 mM NaCl, 500 mM imidazole, 5% (v/v) glycerol, and 10 mM β-mercaptoethanol.

    Tip: For SmB-SmD3 complex, the elution buffer needs 600 mM NaCl.

Table 2 List of constructs, expression and purification conditions.

Protein(s)	Residue ranges	Tags[a]	Vectors	Expression conditions	Purification conditions
Lsm10-Lsm11	Lsm11 1-360 (Δ211-322) Lsm10 1-123	N-6 × His on Lsm11 N-MBP on Lsm10	pFL	Hi5 cell	Nickel affinity HiTrap™ Heparin HP column
SmB-SmD3	SmB 1-95 SmD3 1-126	N-6 × His on SmD3	SmB in pCDF Duet SmD3 in pET-28a	Co-express in *E. coli* BL21 Star(DE3) strain	Nickel affinity HiTrap™ Heparin HP column
SmE-SmF-SmG	SmE 1-92 SmF 1-86 SmG 1-76	C-6 × His on SmG	SmG in pET-26b SmE and SmF in pCDF Duet	Co-express in *E. coli* BL21 Star (DE3) strain	Nickel affinity HiTrap™ Heparin HP column
FLASH	51-137 C54S/C83A mutant	C-6 × His	pET-26b	*E. coli* BL21 Star (DE3) strain	Nickel affinity Size-exclusion chromatography
HCC	Symplekin 30-1101 CPSF73 1-684 CPSF100 1-782 CstF64 1-597	N-6 × His-SUMO on symplekin	Symplekin in pFL CstF64 in pSPL fused with pFL containing CPSF73 and CPSF100	Co-infect Hi5 cell	Nickel affinity HiTrap™ Q HP column
SLBP	125-223	N-6 × His	pFL	Hi5 cell	Nickel affinity HiTrap™ Q HP column

[a]N is short for N-terminal; C is short for C-terminal.

13. Dilute with 6 mL buffer A (20 mM HEPES (pH 7.5) and 5 mM DTT) to ~200 mM NaCl.
14. Load onto HiTrap™ Heparin HP 5 mL column (Cytiva), and elute the protein complex with a salt gradient starting with 80% buffer A and 20% buffer B (20 mM HEPES (pH 7.5), 1 M NaCl, 5 mM DTT) to 100% buffer B.

    Tip: For SmB-SmD3 complex, load onto Heparin column without dilution, since dilution will cause protein precipitation. The complex is eluted with 40–100% buffer B.
15. For FLASH, skip steps 13 and 14. Instead, load onto size-exclusion chromatography column (Sephacryl S-300; Cytiva) with a buffer containing 20 mM Tris (pH 8.5), 250 mM NaCl, and 5 mM DTT.
16. Run SDS-PAGE gel to analyze the peak fractions.
17. Concentrate the factions of interest to 5–10 mg/mL. Divide into aliquots, snap freeze in liquid nitrogen and store at −80 °C.

3.2 Expression and purification from insect cells (SLBP, Lsm10-Lsm11 and HCC)

1. Bacmid preparation: Transform the final plasmids into competent *E. coli* DH10MultiBac cells and grow on plates at 37 °C for 48 h with appropriate antibiotics, IPTG and X-gal for color selection. Select positive/white clones for bacmid extraction. For more detailed information refer to the MultiBac protocol (Sari et al., 2016).
2. P1 baculovirus production: Pre-incubate 1.5 μg bacmid DNA and 8 μL Cellfectin™ II reagent (Gibco) in 200 μL Transfection medium (Expression Systems) at room temperature for 30 min, then supplement with Transfection medium to 1 mL. For every bacmid DNA, seed $0.8–1 \times 10^6$ freshly diluted Sf9 cells (Expression Systems) in a six-well tissue culture plate. Use the 1 mL Cellfectin-DNA suspension to replace the supernatant from seeded cell. Incubate overnight (~16 h) at 27 °C. Change the supernatant to 2 mL ESF 921 medium (Expression Systems). Incubate at 27 °C for 4–5 days. Collect supernatant (P1 virus) by centrifugation for 5 min at 5000 rpm and store in a 2 mL sterile Eppendorf tube at 4 °C.

    Tip: The standard protocol only requires incubating the 1 mL Cellfectin-DNA suspension and the seeded cell for 5 h at 27 °C. However, we have found that in many cases incubation overnight can give viruses that produce more recombinant protein.

3. P2 baculovirus production: Culture 50 mL Sf9 cells to a density of 1.5–2 × 10^6 cells/mL and add all P1 virus to the cells, then incubate at 27 °C at 120 rpm. After 3 days, harvest the cells by centrifugation for 10 min at 2000 rpm, then transfer the supernatant (P2 virus) into a 50 mL sterile Falcon tube and store at 4 °C.

   Tip: The quality of the viruses, in terms of how much recombinant protein they can produce, will decay over time. P1 virus can be stored at 4 °C for less than 3–4 weeks or at −80 °C for 1 year. P2 virus is usually stored at 4 °C for 1–2 months, but the virus for HCC can be stored for much shorter time (~2 weeks) before it loses the ability to produce recombinant protein. It is best to immediately use the P1 virus for amplification and P2 virus for protein production. The cell pellets can be stored at −80 °C for a long time.
4. Protein production in Hi5 cells: Infect or co-infect 1 L Hi5 cells (1.8 × 10^6 cells/mL, expression systems) with 25 mL corresponding P2 virus(es) (2.5% (v/v)) in a 3 L flask, then incubate at 27 °C for 48 h, shaking at 120 rpm.

   Tip: Steps 2–4 should be performed in a sterile hood. P1 and P2 viruses need to be protected from light. Florescent proteins had been co-integrated into the MultiBac baculoviral genome (Sari et al., 2016), so its expression can be monitored by fluorescence spectroscopy and become the marker for the expression analysis of the respective proteins. The strength of P2 viruses can vary between different batches. Therefore, the amount of virus for infection can vary from 1% to 5% (v/v).
5. Harvest cells by centrifugation at 2000 rpm in a Beckman TY.JS4.2 rotor for 13 min at 4 °C, discard the supernatant, flash freeze in liquid nitrogen and store at −80 °C.
6. Resuspend cell pellet in 100 mL lysis buffer containing 20 mM Tris (pH 7.5), 500 mM NaCl, 10 mM imidazole, 5% (v/v) glycerol and one protease inhibitor cocktail tablet (Sigma). For HCC complex, the lysis buffer is 25 mM Tris (pH 8.0), 300 mM NaCl and one protease inhibitor cocktail tablet.
7. Sonicate the cell suspension at 70% output for 6 min with 1 s on and 2 s off.
8. Centrifuge at 13,000 rpm in a F15-8 × 50c carbon fiber rotor at 4 °C for 45 min.
9. Equilibrate 250 μL Ni-NTA resin in each of two 50 mL Falcon tubes with 4 volumes of Ni wash buffer.

10. Transfer 50 mL of the supernatant to equilibrated Ni-NTA resin in each tube and slowly rotate for 1 h in a cold room.
11. Transfer the mixture to a 14 cm high, gravity flow column (Bio-Rad).
12. Wash resin with 3 × 15 mL Ni wash buffer containing 20 mM Tris (pH 7.5), 500 mM NaCl and 40 mM imidazole. For HCC complex, the Ni wash buffer is 25 mM Tris (pH 8.0), 150 mM NaCl and 20 mM imidazole.
13. Elute protein with 4 mL Ni elution buffer containing 20 mM Tris (pH 7.5), 500 mM NaCl, 500 mM imidazole, and 5% (v/v) glycerol. For HCC complex, the Ni elution buffer is 25 mM Tris (pH 8.0), 150 mM NaCl and 250 mM imidazole. 10 mM β-mercaptoethanol can also be included in the wash and elution buffers.
14. Dilute with 6 mL buffer A (20 mM HEPES (pH 7.5) and 5 mM DTT) to ~200 mM NaCl.

    Tip: It is necessary for Lsm10-Lsm11 to be eluted in a buffer with high concentration of salt first, then slowly diluted a little bit for ion exchange. Lsm10-Lsm11 tends to precipitate if eluted directly in a buffer with ~200 mM NaCl.
15. Load onto HiTrap™ Heparin HP 5 mL column (Cytiva), and elute the protein with a salt gradient starting with 80% buffer A and 20% buffer B (20 mM HEPES (pH 7.5), 1 M NaCl, 5 mM DTT) to 100% buffer B. For SLBP, load onto HiTrap™ Q HP column. For HCC, skip step 14. Instead, load onto HiTrap™ Q HP column and elute with a gradient starting with 15–60% buffer B (25 mM Tris pH 8.0, 1 M NaCl, 5 mM DTT).
16. Concentrate the factions of interest. Aliquot, snap freeze and store at −80 °C.

3.3 Reconstitution of U7 Sm core in complex with FLASH, SLBP and histone pre-mRNA

1. Anneal an equimolar mixture of human U7 snRNA and modified mouse H2a pre-mRNA (H2a*) in 100 μL reconstitution buffer A containing 20 mM HEPES (pH 7.5), 500 mM NaCl, 5 mM EDTA, and 5 mM DTT by heating to >90 °C for 5 min. The sequences of the RNAs are given in Table 1.
2. Snap cool on ice for 10 min.
3. Add equimolar Lsm10-Lsm11, SmE-SmF-SmG, SmB-SmD3, the annealed RNAs, and 2 molar equivalents of FLASH into appropriate volume of reconstitution buffer A. The total volume should be under 900 μL.

Tip: In most assembly experiments, N-terminal FLASH segment contained two point mutations, C54S/C83A, to prevent potential formation of disulfide bridges between two FLASH molecules in its dimer (Aik et al., 2017). However, wild-type FLASH segment is equally suitable.

4. Incubate the mixture at 30 °C for 30 min, followed by 37 °C for 15 min (Leung et al., 2010).
5. Cool on ice for 10 min.
6. Add equimolar SLBP to the mixture and incubate on ice for 5 min.
7. Add appropriate volume of 2.5 mg/mL TEV protease (at a ratio of 1:7 (w/w)) to remove the MBP tag on Lsm10. Incubate overnight at 4 °C. The total volume should be under 1 mL. The SUMO tag on symplekin is not removed.
8. Purify U7 Sm core-FLASH-SLBP-H2a* complex by gel filtration using a Superose 6 10/300 GL column (Cytiva), in reconstitution buffer B containing 20 mM HEPES (pH 7.5), 100 mM NaCl, 10 mM EDTA, and 5 mM DTT.

   Tip: Some components of U7 Sm core are unstable and prone to precipitation in low salt buffer, e.g., Lsm10-Lsm11 and SmB-SmD3. Therefore, to reconstitute U7 Sm core, the buffer must have high concentration of salt, otherwise the proteins will precipitate or aggregate. The high salt condition also promotes the spontaneous assembly of the three Sm sub-complexes on the U7 snRNA to form the heptameric ring. However, the high salt concentration will prevent interaction with HCC. After U7 Sm core is formed, the complex can tolerate low salt buffer, allowing a change to the low salt buffer during SP6 purification.
9. Fractions of interest are concentrated and stored at −80 °C. The presence of RNA in the complex is confirmed with an A_{260}/A_{280} ratio of ~1.7.

3.4 Reconstitution of the active processing machinery in complex with pre-mRNA substrate and SLBP

The two processing signals required for 3′-end processing of histone pre-mRNAs, the stem-loop and the HDE, are encompassed within an approximately 60-nucleotide RNA segment (Fig. 1A). This model pre-mRNA substrate can be chemically synthesized by commercial sources (for example Horizon Discovery/Dharmacon, Lafayette, CO or Integrated DNA Technologies, Inc., Coralville, IA) or prepared by *in vitro* transcription. Multiple assembly experiments demonstrated that chemical synthesis and T7/SP6-mediated transcription are equally suitable methods of generating U7 snRNA for U7 snRNP assembly (Bucholc et al., 2020; Sun, Zhang, et al., 2020; Yang et al., 2020).

1. Mix purified HCC and U7 Sm core-FLASH-SLBP-H2a* complex at a molar ratio of 1:1.3, in reconstitution buffer B.

   Tip: It is important that all manipulations with RNA, its handling and storage are carried out in an RNase-free environment, paying particular attention to avoid contaminations with the powerful RNase A routinely used for DNA isolation from bacterial cells and other general laboratory techniques.
2. Incubate on ice for 1 h.
3. Purify the histone pre-mRNA 3′-end processing machinery substrate complex using a Superose 6 10/300 GL column, in reconstitution buffer B.

This purified complex is used primarily for structural studies. For cleavage assays, the active U7 machinery can be assembled *in situ*, by combining its various components in the reaction buffer (see Section 4).

3.5 An alternative protocol for assembling U7 Sm core-FLASH complex

This protocol differs in some details from the protocol described above and was successfully used to reconstitute active processing machinery for functional studies with radio-labeled substrate.

1. Mix 75 μL of the assembly buffer (600 mM KCl, 15 mM HEPES pH 7.9, 15% glycerol, 0.1 μg/μL yeast tRNA and 20 mM EDTA) with 2000 pmol each of Lsm10-Lsm11, SmE-SmF-SmG, and U7 snRNA, and incubate 90 min at 32 °C.
2. Add 2000 pmol of SmB-SmD3 and, if necessary, bring up the volume to 100 μL with the assembly buffer. Incubate at 32 °C for an additional 90 min.
3. Add 4000 pmol (twofold molar excess compared to the other subunits) of bacterially expressed FLASH (amino acid residues 51–137) and incubate overnight at 4 °C.
4. Separate U7 Sm core-FLASH complex from the unbound components by size-exclusion chromatography using Superose™ 6 Increase 3.2/300 (Cytiva) or other suitable column. To equilibrate the column, use buffer compatible with *in vitro* 3′-end processing reaction: 75 mM KCl, 15 mM HEPES-KOH pH 7.9, 5% glycerol and 20 mM EDTA pH 8. Note that compared to processing buffer, the concentration of glycerol was reduced from 15% to 5% to improve liquid flow during purification.

5. Combine fractions containing the assembled complex, evaluate its concentration and analyze a fraction by SDS-PAGE to confirm the presence of all subunits. Store at −80 °C.

4. Histone pre-mRNA 3′-end processing assays using radio-labeled substrate

4.1 Labeling of synthetic histone pre-mRNA at the 5′-end with ^{32}P

1. Set up a 5′-end labeling reaction in a final volume of 30 μL by mixing the following components: 3 pmol of synthetic histone pre-mRNA substrate containing 5′ OH group (Table 1), 3 μL of 10× buffer (New England Biolabs, NEB, Ipswich, MA), 1.5 μL of T4 polynucleotide kinase (NEB), 25 μCi of γ-^{32}P-ATP, and water to total volume of 30 μL.

   Tip: Chemically synthesized RNAs contain a 5′-end hydroxyl group and are ready for 5′-end labeling without prior treatment with calf intestinal phosphatase.
2. Incubate 1 h at 37 °C.
3. Run through G50 spin column (Cytiva) to remove unincorporated radioactive ATP, following manufacturer's protocol.
4. Use Geiger counter to measure the amount of radiation (counts per minute, cpm) in 1 μL of the purified probe. Depending on the efficiency of labeling, the probe can be used for as long as 4–6 weeks.

   Tip: An RNA segment containing the two processing signals can be alternatively generated by *in vitro* transcription using a bacteriophage RNA polymerase (SP6 or T7) and an appropriate DNA template (a PCR-generated segment or a linearized plasmid). This method is recommended for generating longer pre-mRNA substrates that exceed the maximum length limit suitable for chemical synthesis (typically over 120 nucleotides). The generated RNA transcript is next treated with calf intestinal phosphatase (NEB) to remove the 5′ triphosphate and labeled at the 5′-end with ^{32}P using T4 polynucleotide kinase. The quality of labeled RNAs generated by these procedures should be checked at each step by electrophoresis in a denaturing gel.

4.2 Assays using nuclear extract

In vitro processing of histone pre-mRNAs is not inhibited by the presence EDTA, with a concentration as high as 50 mM having no major effect on the efficiency of the reaction (Kolev, Yario, Benson, & Steitz, 2008) (ZD, unpublished results). Cleavage of histone pre-mRNAs is catalyzed by CPSF73, a member of the metallo-β-lactamase family of nucleases (Callebaut, Moshous, Mornon, & de Villartay, 2002; Dominski, 2007) and contains two zinc ions required for catalysis (Mandel et al., 2006). Structural and biochemical studies indicate that the zinc ions are tightly coordinated by a number of amino acids of the catalytic center and are not accessible for binding EDTA (Kolev et al., 2008). Routinely, *in vitro* processing reactions are carried out in the presence of 20 mM EDTA (Gick et al., 1986), a concentration that effectively inhibits the activity of metal-dependent nonspecific nucleases and phosphatases.

1. Set up a processing reaction on ice by combining 7.5 μL of nuclear extract and 2.5 μL of 80 mM EDTA (pH 8) containing pre-mRNA substrate labeled at the 5′-end with ^{32}P. A typical 10 μL reaction contains 5000 counts per minute of the radioactive substrate, which equals, depending on the efficiency of 5′ labeling, 0.01–0.025 pmol (1–2.5 nM) or 0.2–0.5 ng of single-stranded RNA (60 nucleotides of length and 20,000 kDa of molecular weight).

   Tip: If desired, the reaction can be scaled up, or lower amounts of nuclear extracts can be tested, with the remaining volume made up using dialysis buffer. To achieve high ratio of final product relative to input substrate, use the smallest amount of the substrate possible. For older probes that have lost most of their radioactive signal due to short half-life of ^{32}P isotope (~14 days), keep the amount of substrate at constant level, extending instead the time of subsequent exposure by autoradiography.
2. Mix by gentle vortexing and incubate at 32 °C for 60 min.

   Tip: Carry out a time course reaction to establish the shortest time of incubation necessary to achieve maximum processing efficiency (Dominski et al., 1995; Gick et al., 1986). Cleavage of histone pre-mRNAs proceeds rapidly at 32 °C without a significant lag time and in some extracts most of the cleavage product is generated within 20–30 min after the start of incubation. Once the reaction reaches the plateau phase, further extension of this time does not significantly increase the amount of the final product and may instead cause its non-specific degradation, in particular in samples where EDTA was omitted or used at low concentration.
3. Stop the reaction by adding 2 μL of proteinase K solution (5 μg/μL) in 50 mM Tris pH 7.5 and 2.5% SDS, and incubate 1 h at 37 °C.

Tip: Alternatively, each sample can be processed by adding 200 μL of 0.3 M sodium acetate, followed by extraction with the same volume of phenol followed by precipitation of the aqueous phase with 500 μL ethanol. RNA precipitated at −20 °C is recovered by centrifugation, rinsed with ethanol, air dried and dissolved in 10–15 μL of loading dye (8 M urea, 0.01% bromophenol blue and 0.01% xylene cyanol). This approach takes longer time, may result in partial loss of RNA and is relatively laborious when multiple samples are processed, but it produces samples of much higher purity and is recommended for longer RNA substrates (>150 nucleotides) that may run during electrophoresis as a smear rather than a sharp band if proteinase K is used for their purification.

4. Add 40 μL of loading dye (8 M urea, 0.01% bromophenol blue and 0.01% xylene cyanol), and boil 5 min to denature RNA.
5. Cast an 8% polyacrylamide/7 M urea denaturing gel (30:1 acrylamide to bisacrylamide) using 20 × 20 cm gel plates and 0.75 mm spacer and pre-run the gel for at least 15–20 min to increase its temperature and remove contaminating salt ions.
6. Rinse the wells from excess of urea using a syringe and load 5 μL of each sample.
7. Apply appropriate voltage to keep the gel at 45–50 °C (for a gel 0.75 mm thick this will typically require 400–500 V) and run until desired separation of the labeled input pre-mRNA and the 5′ cleavage product is achieved. For a 60 nucleotide RNA and its product of 45 nucleotides (Fig. 1A), this takes between 20 and 25 min.
8. Manually transfer the gel to Whatman paper, place in a cassette with an X-ray film and intensifying screen and expose at −80 °C.

The 5′-labeled substrate is cleaved in the extract yielding two products. The upstream product that terminates with the stem-loop retains the label at the 5′-end and is visible upon autoradiography. The downstream product lacks radioactive label and its fate cannot be followed with this substrate. Studies with uniformly or site-specifically labeled substrates demonstrate that this product is degraded in a U7-dependent manner by a 5′-3′ exonuclease activity of CPSF73 (Dominski, Yang, & Marzluff, 2005; Yang, Sullivan, et al., 2009; Yang et al., 2020).

4.3 Assays using semi-recombinant machinery

Semi-recombinant U7 machinery is assembled by adding purified recombinant U7 Sm core-FLASH N-terminal segment complex to a nuclear extract.

Nuclear extracts contain relatively large amounts of free HCC (Yang et al., 2013) that spontaneously assembles with U7 Sm core-FLASH, giving rise to semi-recombinant U7 machinery (Bucholc et al., 2020). Studying processing activity of semi-recombinant U7 machinery requires that its activity is not obscured by the activity of the endogenous machinery from the extract. One way of achieving this objective is to assemble U7 Sm core on a mutant U7 snRNA (U7 Mut) that contains an extensive mutation within its 5′-end region and analyze the activity of semi-recombinant U7 machinery together with a mutant histone pre-mRNA containing a compensatory mutation within the HDE (H2a* Mut) that restores the base pairing interaction with the mutant U7 snRNA (Table 1), as previously described (Bond et al., 1991; Schaufele et al., 1986). This mutant substrate is not recognized by the endogenous machinery, with the entire processing detected in the extract being contributed by the semi-recombinant machinery.

1. Add 1 pmol of recombinant U7 Sm core that was assembled on U7 Mut snRNA and bound to N-terminal FLASH to a 10 μL processing reaction containing 7.5 μL of nuclear extract, 20 mM EDTA and 0.025 pmol (2.5 nM) of radio-labeled H2a* Mut pre-mRNA substrate.

   Tip: Titration experiments demonstrated that this amount of nuclear extract contains sufficient amount of free HCC to assemble 1 pmol of U7 Sm core into an active semi-recombinant U7 machinery. Note that using too much of U7 Sm core-FLASH complex relative to the amount of the HCC present in the extract may result in a significant fraction of the machinery lacking HCC. This form of the U7 machinery is capable of binding histone pre-mRNA substrate but is inactive, inhibiting processing through a dominant negative effect.

2. Incubate 60 min at 32 °C.

   Tip: To assure that the mutant histone pre-mRNA is not processed by the endogenous machinery, set up a control reaction containing nuclear extract but lacking recombinant Mut U7 Sm core-FLASH complex.

3. Process samples and separate RNA by electrophoresis, as described in Section 4.2 (steps 3–8).

U7 Sm core assembled from recombinant Sm/Lsm subunits and U7 snRNA interacts with the HCC from nuclear extract and supports accurate processing of histone pre-mRNAs (Bucholc et al., 2020). One important conclusion from these studies is that the ring proteins and U7 snRNA do not require any essential modifications to function in processing. No inhibitory effect on processing activity of semi-recombinant U7 machinery was observed when synthetic U7 snRNA was replaced with T7-generated RNA that contained additional nucleotides at both the 5′ and 3′-ends. Thus, *in vitro* transcription with bacteriophage RNA polymerases provides a cost-effective alternative to chemical RNA synthesis, allowing for generation of multiple U7 snRNA mutants of choice in a short time. For example, these studies demonstrated that the 3′ terminal stem-loop in U7 snRNA can be deleted without causing any adverse effects on the reaction, and that the spliceosomal type Sm binding site in U7 snRNA supports formation of the U7-specific ring but the resultant semi-recombinant U7 machinery displays reduced accuracy and efficiency of processing (Bucholc et al., 2020). The MBP tag on Lsm10 had no negative effect on processing activity of semi-recombinant U7 machinery and therefore it was left in functional assays. By altering the recombinant components of the ring, studies with semi-recombinant U7 machinery demonstrated that the long unstructured C-terminal extension present in SmB and the large loop separating motifs Sm1 and Sm2 in Lsm11 are dispensable for *in vitro* processing. With this simple approach, the importance of other regions in the ring proteins and U7 snRNA for 3′-end processing can be readily assessed.

4.4 Single-step purification of semi-recombinant U7 machinery *via* a photo-cleavable linker in the U7 snRNA

In this method recombinant U7 Sm core is assembled on chemically synthesized U7 snRNA containing biotin and a photo-cleavable linker at the 5′-end, which is then bound to N-terminal segment of FLASH and the complex immobilized on streptavidin beads and thoroughly washed to remove unincorporated subunits. In the next step, the beads are incubated with a nuclear extract to recruit endogenous HCC and the assembled semi-recombinant U7 machinery is eluted from the beads by exposure to long-wavelength UV (Skrajna et al., 2018, 2019).

1. Assemble U7 Sm core bound to N-terminal FLASH, as described in Section 3.5, using chemically synthesized U7 snRNA containing biotin and photocleavable linker at the 5′-end (commercially available from Horizon Discovery/Dharmacon or Integrated DNA Technologies, Inc.).

   Tip: Synthesis of RNA of this length (60 nucleotides) and containing the two modifications may present a challenge, resulting in a relatively low yield of the full-length product. It is therefore recommended to order a larger scale synthesis and request gel purification to remove shorter products.

2. Add 100 pmol of the U7 Sm core-FLASH complex to 1 mL of processing buffer (75 mM KCl, 15 mM HEPES-KOH pH 7.9, 15% glycerol and 20 mM EDTA pH 8), vortex and incubate on ice 5 min.
3. Spin down potential precipitates in a microcentrifuge (10 min at 10,000 × *g*).
4. Leaving a small amount at the bottom, transfer the supernatant over 30–40 μL of streptavidin beads that have been washed twice with processing buffer prior to use.
5. Rotate 1 h at 4 °C to immobilize recombinant U7 Sm core-FLASH complex.
6. Gently spin down the beads in a microcentrifuge (3 min at 30 × *g*) and aspirate the supernatant.
7. Briefly rinse the pellet of beads three times with processing buffer using the same spinning conditions and rotate 1 h with 1 mL of the same buffer.
8. Remove the supernatant and place the tube on ice.
9. Add 250 μL of 80 mM EDTA to 750 μL of a nuclear extract (to final EDTA concentration of 20 mM) and spin down 10 min in cold microcentrifuge at 10,000 × *g* to remove potential precipitates.
10. Transfer pre-cleared nuclear extract to the tube containing pellet of streptavidin beads and immobilized complex of U7 Sm core-FLASH (see step 8).
11. Rotate 1 h at 4 °C to reconstitute semi-recombinant U7 machinery by binding endogenous HCC.
12. Gently spin down the beads in a microcentrifuge (3 min at 30 × *g*) and aspirate the supernatant.
13. Apply three brief rinses of the pellet using each time 1 mL of processing buffer, as described above (step 7).
14. Rotate the beads with 1 mL of processing buffer for 1 h at 4 °C.
15. Spin as above (3 min at 30 × *g*), aspirate the supernatant and add 75 μL of processing buffer.
16. Transfer the suspension to a 500 μL tube, place on ice 0.5 in. away from the surface of a lamp emitting high intensity long-wavelength UV.
17. Irradiate 30 min, frequently vortexing and inverting the tube to avoid overheating and ensure equal exposure, as described in detail (Skrajna et al., 2018, 2019).

Tip: For most efficient elution, pre-warm the lamp prior to exposing the samples until the full brightness and intensity of UV irradiation are achieved (typically 5 min).

18. Spin 5 min at 60 × *g* and collect the supernatant.

19. Re-spin to remove residual beads and collect the supernatant, which contains purified semi-recombinant U7 machinery.

20. Analyze a fraction of the UV eluted material by SDS-PAGE and silver staining to confirm the presence of the four subunits of endogenous HCC: symplekin, CPSF100, CPSF73 and CstF64. If necessary, confirm the identity of visualized proteins by Western blotting and/or mass spectrometry.

21. Test processing activity of the UV-eluted semi-recombinant U7 machinery by mixing together 5 μL of processing mix (see above) containing 0.1 μg/μL yeast tRNA (Invitrogen), 0.5–2.5 μL of the eluted solution, SLBP (1–5 pmol), 0.025 pmol (0.5 ng) of 5′-labeled pre-mRNA substrate. Bring up the volume to 10 μL using processing/tRNA mix.

22. Incubate 60 min at 32 °C, inhibit the reaction by using proteinase K solution and separate radioactive RNA, as described in Section 4.2 (steps 3–8).

This is the first example of an *in vitro* reconstituted processing reaction. This reaction, while using some components purified from nuclear extract, is essentially free of endogenous U7 snRNP so wild-type histone pre-mRNA can be used as a substrate. In addition, with this system it is possible to make various mutations within U7 Sm core, FLASH and SLBP and analyze their effects on the efficiency of processing.

4.5 Assays using reconstituted machinery

One disadvantage of assembling semi-recombinant U7 machinery is that it requires endogenous HCC and therefore it can be isolated only in limited quantities, and even when purified using photo-cleavable RNA, it contains small amount of other polyadenylation factors (CPSF160, WDR33, Fip1 and CPSF30) and nuclear proteins that tend to associate with the HCC and may play unknown role in processing. In addition, semi-recombinant U7 machinery is relatively heterogeneous in size due to partial proteolysis of individual HCC components. The assembly of fully recombinant U7 machinery became possible following successful expression of the four components of HCC (symplekin, CPSF100, CPSF73 and CstF64) as a soluble complex in insect cells (Zhang et al., 2020). Below we describe how to use recombinant components to reconstitute a U7 machinery that accurately cleaves a radio-labeled histone pre-mRNA. All components of the processing reaction are mixed directly prior to the start of the incubation, resulting in rapid assembly of catalytically active U7 machinery and accurate cleavage of the histone pre-mRNA.

1. To 7.5 μL of processing mix containing yeast tRNA at 0.1 μg/μL add the following recombinant components of the processing reaction: 2.5 pmol of U7 Sm core-FLASH (amino acid residues 51–137) complex, 2.5 pmol of the HCC containing either near full-length or N-terminally truncated symplekin, 2.5 pmol SLBP (amino acid residues 125–223) and 0.025 pmol (0.5 ng, 2.5 nM) of 5′-labeled pre-mRNA substrate. Bring up the volume to 10 μL with processing mix containing yeast tRNA at 0.1 μg/μL.
2. Incubate 1 h at 32 °C.
3. Process the samples and electrophoretically separate radioactive RNA, as described for 3′-end processing with nuclear extracts (Section 4.2, steps 3–8).

Functional studies based on this approach demonstrated that recombinant U7 machinery accurately cleaves histone pre-mRNAs and yielded a number of unexpected findings that explained the molecular mechanism of cleavage by U7 snRNP (Sun, Zhang, et al., 2020; Yang et al., 2020). With the successful generation of fully recombinant and active U7 machinery, various mutations were incorporated within the HCC components and tested for their effects on processing. Using this approach, CstF64 was shown to be dispensable *in vitro*, suggesting that its association with the machinery may reflect a role in coupling transcription with processing *in vivo*. Most surprisingly, functional studies with the reconstituted U7 machinery demonstrated that processing of histone pre-mRNAs critically depends on the N-terminal domain (NTD) of symplekin (amino acids 30-360) (Kennedy et al., 2009; Xiang et al., 2010). This NTD was shown to support processing not only as a part of full-length symplekin and an integral component of the HCC but also as a separate polypeptide added to the processing reaction in *trans* even though it is unable to form a stable complex with the HCC (Sun, Zhang, et al., 2020). Among the most important findings, the D75N/H76A mutation within the active site of CPSF73 abolished processing activity, providing ultimate evidence that CPSF73 is the catalytic component of the U7 machinery (Sun, Zhang, et al., 2020; Yang et al., 2020), supporting previous, more circumstantial conclusions from UV-cross linking studies (Dominski, Yang, & Marzluff, 2005). The mutant version of U7 machinery was also shown to be unable to degrade the downstream cleavage product in the 5′-3′ direction, supporting the notion that CPSF73 is both an endonuclease and a 5′-3′ exonuclease, at least in the U7 machinery (Yang et al., 2020).

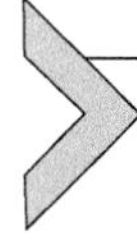

5. Histone pre-mRNA 3′-end processing assays using fluorescently labeled substrate

Fluorescence-based reporter assay is a non-radioactive and rapid technique to monitor RNA processing and degradation. For this assay, a 60-mer modified mouse histone H2a pre-mRNA (H2a*) substrate (Sun, Zhang, et al., 2020) is labeled with two fluorophores: 5′-end TAMRA and 3′-end FAM (Table 1). Two products are generated by the endonuclease activity of CPSF73 in the U7 machinery, an upstream product with TAMRA label and a downstream product with FAM label. While the upstream product is stable, the downstream product is degraded to mononucleotides, providing direct evidence for the 5′-3′ exonuclease activity of CPSF73. This exonuclease activity can also be demonstrated with an RNA substrate corresponding to the downstream cleavage product, labeled at the 3′-end with FAM (Table 1).

1. Set up a cleavage reaction on ice by mixing 1 μM U7 Sm core-FLASH, SLBP, HCC protein mixture and 0.1 μM labeled substrate in a 10 μL reaction containing 15 mM HEPES (pH 8.0), 75 mM KCl, 15% (v/v) glycerol, 20 mM EDTA (pH 8.0) and RNasin plus ribonuclease inhibitor (1 U/reaction, Promega-Fisher).

   Tip: Protease inhibitors are often added to protein purification buffers. Some of the commercial protease inhibitor cocktails contain the metallo-protease inhibitor phosphoramidon, which appears to also inhibit CPSF73 at high concentrations. The chemical structure of this compound bears some similarity to that of activated JTE-607, a known inhibitor of CPSF73 (Kakegawa, Sakane, Suzuki, & Yoshida, 2019; Ross et al., 2020). Therefore, care should be taken with the presence of protease inhibitor cocktails in cleavage assays.
2. Incubate at 30 °C for 1 h.
3. Add 10 μL 2× urea denaturing RNA loading buffer (8 M urea, 1× TBE) to each reaction and boil the sample for 10 min.

4. Load the boiled samples directly to 15% denaturing urea polyacrylamide gel (8 M urea).
5. Run the gel at 280 V for 25 min.
6. Image the gel by the ChemiDoc MP imaging system (Bio-Rad) or other systems. Typical exposure time is 10 s.

The assay clearly demonstrates the accumulation of the upstream cleavage product (Fig. 2). On the other hand, the downstream cleavage product does not accumulate, and is readily degraded to mononucleotides, with apparently no indication of intermediates, suggesting a processive 5′-3′ exonuclease activity for CPSF73 (Yang et al., 2020). A 3′-end FAM-labeled downstream cleavage product RNA can also be degraded. However, the reaction is much slower on ice, and a mutation in the active site of CPSF73 abolishes the activity with both substrates.

The fluorescence assay is not as sensitive as the radio-labeled assay, requiring 50–100 nM of fluorescently labeled substrate as compared to 2 nM radio-labeled substrate. However, the two labels on the substrate confer a

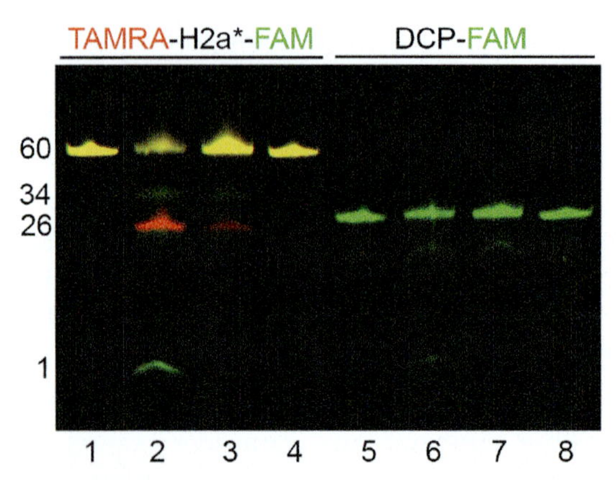

Fig. 2 Histone pre-mRNA 3′-end processing assays with fluorescently labeled substrates. The substrate for lanes 1–4 is a 60-mer RNA (H2a*) labeled at the 5′-end with TAMRA (red color) and 3′-end with FAM (green), and that for lanes 5–8 is a 26-mer downstream cleavage product (DCP) labeled at the 3′-end with FAM. Lanes 1 and 5: RNA alone. Lanes 2 and 6: RNA incubated with the reconstituted machinery at 30 °C for 1 h. For the H2a* RNA, the 5′ cleavage product contains 26 nts and the 3′ product contains 34 nts, which is degraded to mononucleotides by the 5′-3′ exonuclease activity of CPSF73. Lanes 3 and 7: RNA incubated with the reconstituted machinery on ice for 1 h. The cleavage activity is much weaker. Lanes 4 and 8: RNA incubated with the reconstituted machinery containing the catalytically inactive D75N/H76A mutant CPSF73 at 30 °C for 1 h. The symplekin N-terminal domain was provided in *trans* for these two reactions (Sun, Zhang, et al., 2020).

great advantage by readily allowing both products to be monitored in the assay and providing direct evidence for the processive degradation of the downstream cleavage product.

6. Summary

The successful reconstitution of an active human histone pre-mRNA 3′-end processing machinery has already led to the determination of its structure, as well as some detailed characterizations of the functional roles of its various components. For example, various mutations were incorporated in the HCC components and tested for their effects on processing. Many additional experiments should be possible, for example assessing the functional importance of interaction interfaces among the subunits observed in the structure as well as the functional roles (if any) of many other components that have been identified to be associated with the machinery.

Acknowledgments

This research is supported by NIH grants R35GM118093 (to LT) and R01GM029832 (to WFM and ZD). WSA was also supported by a fellowship from the Raymond and Beverley Sackler Center for Research at Convergence of Disciplines at Columbia University Medical Center.

References

Abmayr, S. M., Yao, T., Parmely, T., & Workman, J. L. (2006). Preparation of nuclear and cytoplasmic extracts from mammalian cells. *Current Protocols in Molecular Biology, Chapter 12*, Unit 12.11.

Aik, W. S., Lin, M. H., Tan, D., Tripathy, A., Marzluff, W. F., Dominski, Z., et al. (2017). The N-terminal domains of FLASH and Lsm11 form a 2:1 heterotrimer for histone pre-mRNA 3′-end processing. *PLoS One, 12*, e0186034.

Barcaroli, D., Bongiorno-Borbone, L., Terrinoni, A., Hofmann, T. G., Rossi, M., Knight, R. A., et al. (2006). FLASH is required for histone transcription and S-phase progression. *Proceedings of the National Academy of Sciences of the United States of America, 103*, 14808–14812.

Bond, U. M., Yario, T. A., & Steitz, J. A. (1991). Multiple processing-defective mutations in a mammalian histone pre-mRNA are suppressed by compensatory changes in U7 RNA both in vivo and in vitro. *Genes & Development, 5*, 1709–1722.

Bucholc, K., Aik, W. S., Yang, X.-C., Wang, K., Zhou, Z. H., Dadlez, M., et al. (2020). Composition and processing activity of a semi-recombinant holo U7 snRNP. *Nucleic Acids Research, 48*, 1508–1530.

Callebaut, I., Moshous, D., Mornon, J.-P., & de Villartay, J.-P. (2002). Metallo-b-lactamase fold within nucleic acids processing enzymes: The b-CASP family. *Nucleic Acids Research, 30*, 3592–3601.

Chan, S. L., Huppertz, I., Yao, C., Weng, L., Moresco, J. J., Yates, J. R., III, et al. (2014). CPSF30 and Wdr33 directly bind to AAUAAA in mammalian mRNA 3′ processing. *Genes and Development, 28*, 2370–2380.

Dignam, J. D., Lebovitz, R. M., & Roeder, R. G. (1983). Accurate transcription initiation by RNA polymerase II in a soluble extract from isolated mammalian nuclei. *Nucleic Acids Research*, *11*, 1475–1489.

Dignam, J. D., Martin, P. L., Shastry, B. S., & Roeder, R. G. (1983). Eukaryotic gene transcription with purified components. *Methods in Enzymology*, *101*, 582–598.

Dominski, Z. (2007). Nucleases of the metallo-b-lactamase family and their role in DNA and RNA metabolism. *Critical Reviews in Biochemistry and Molecular Biology*, *42*, 67–93.

Dominski, Z. (2010). The hunt for the 3′ endonuclease. *Wiley Interdisciplinary Reviews. RNA*, *1*, 325–340.

Dominski, Z., & Marzluff, W. F. (2007). Formation of the 3′ end of histone mRNA: Getting closer to the end. *Gene*, *396*, 373–390.

Dominski, Z., Sumerel, J., Hanson, R. J., & Marzluff, W. F. (1995). The polyribosomal protein bound to the 3′ end of histone mRNA can function in histone pre-mRNA processing. *RNA*, *1*, 915–923.

Dominski, Z., Yang, X.-C., & Marzluff, W. F. (2005). The polyadenylation factor CPSF-73 is involved in histone-pre-mRNA processing. *Cell*, *123*, 37–48.

Dominski, Z., Yang, X. C., Purdy, M., & Marzluff, W. F. (2003). Cloning and characterization of the Drosophila U7 small nuclear RNA. *Proceedings of the National Academy of Sciences of the United States of America*, *100*, 9422–9427.

Dominski, Z., Yang, X. C., Purdy, M., & Marzluff, W. F. (2005). Differences and similarities between Drosophila and mammalian 3′ end processing of histone pre-mRNAs. *RNA*, *11*, 1835–1847.

Dominski, Z., Yang, X. C., Raska, C. S., Santiago, C., Borchers, C. H., Duronio, R. J., et al. (2002). 3′ end processing of Drosophila melanogaster histone pre-mRNAs: Requirement for phosphorylated Drosophila stem-loop binding protein and coevolution of the histone pre-mRNA processing system. *Molecular and Cellular Biology*, *22*, 6648–6660.

Dominski, Z., Zheng, L. X., Sanchez, R., & Marzluff, W. F. (1999). Stem-loop binding protein facilitates 3′-end formation by stabilizing U7 snRNP binding to histone pre-mRNA. *Molecular and Cellular Biology*, *19*, 3561–3570.

Gick, O., Kramer, A., Keller, W., & Birnstiel, M. L. (1986). Generation of histone mRNA 3′ ends by endonucleolytic cleavage of the pre-mRNA in a snRNP-dependent in vitro reaction. *The EMBO Journal*, *5*, 1319–1326.

Grimm, C., Chari, A., Pelz, J. P., Kuper, J., Kisker, C., Diederichs, K., et al. (2013). Structural basis of assembly chaperone-mediated snRNP formation. *Molecular Cell*, *49*, 692–703.

Kakegawa, J., Sakane, N., Suzuki, K., & Yoshida, T. (2019). JTE-607, a multiple cytokine production inhibitor, targets CPSF3 and inhibits pre-mRNA processing. *Biochemical and Biophysical Research Communications*, *518*, 32–37.

Kambach, C., Walke, S., Young, R., Avis, J. M., de la Fortelle, E., Raker, V. A., et al. (1999). Crystal structures of two Sm protein complexes and their implications for the assembly of the spliceosomal snRNPs. *Cell*, *96*, 375–387.

Kennedy, S. A., Frazier, M. L., Steiniger, M., Mast, A. M., Marzluff, W. F., & Redinbo, M. R. (2009). Crystal structure of the HEAT domain from the pre-mRNA processing factor symplekin. *Journal of Molecular Biology*, *392*, 115–128.

Kolev, N. G., Yario, T. A., Benson, E., & Steitz, J. A. (2008). Conserved motifs in both CPSF73 and CPSF100 are required to assemble the active endonuclease for histone mRNA 3′-end maturation. *EMBO Reports*, *9*, 1013–1018.

Krainer, A. R., Maniatis, T., Ruskin, B., & Green, M. R. (1984). Normal and mutant human beta-globin pre-mRNAs are faithfully and efficiently spliced in vitro. *Cell*, *36*, 993–1005.

Leung, A. K. W., Kambach, C., Kondo, Y., Kampmann, M., Jinek, M., & Nagai, K. (2010). Use of RNA tertiary interaction modules for the crystallisation of the spliceosomal snRNP core domain. *Journal of Molecular Biology, 402*, 154–164.

Leung, A. K. W., Nagai, K., & Li, J. (2011). Structure of the spliceosomal U4 snRNP core domain and its implication for snRNP biogenesis. *Nature, 473*, 536–539.

Mandel, C. R., Kaneko, S., Zhang, H., Gebauer, D., Vethantham, V., Manley, J. L., et al. (2006). Polyadenylation factor CPSF-73 is the pre-mRNA 3′-end-processing endonuclease. *Nature, 444*, 953–956.

Mowry, K. L., & Steitz, J. A. (1987). Identification of the human U7 snRNP as one of several factors involved in the 3′ end maturation of histone premessenger RNA's. *Science, 238*, 1682–1687.

Pomeranz Krummel, D. A., Oubridge, C., Leung, A. K., Li, J., & Nagai, K. (2009). Crystal structure of human spliceosomal U1 snRNP at 5.5 Å resolution. *Nature, 458*, 475–480.

Raker, V. A., Hartmuth, K., Kastner, B., & Lührmann, R. (1999). Spliceosomal U snRNP core assembly: Sm proteins assemble onto an Sm site RNA nonanucleotide in a specific and thermodynamically stable manner. *Molecular and Cellular Biology, 19*, 6554–6565.

Romeo, V., & Schumperli, D. (2016). Cycling in the nucleus: Regulation of RNA 3′ processing and nuclear organization of replication-dependent histone genes. *Current Opinion in Cell Biology, 40*, 23–31.

Ross, N. T., Lohmann, F., Carbonneau, S., Fazal, A., Weihofen, W. A., Gleim, S., et al. (2020). CPSF3-dependent pre-mRNA processing as a druggable node in AML and Ewing's sarcoma. *Nature Chemical Biology, 16*, 50–59.

Sabath, I., Skrajna, A., Yang, X.-C., Dadlez, M., Marzluff, W. F., & Dominski, Z. (2013). 3′-end processing of histone pre-mRNAs in Drosophila: U7 snRNP is associated with FLASH and polyadenylation factors. *RNA, 19*, 1726–1744.

Sari, D., Gupta, K., Thimiri Govinda Raj, D. B., Aubert, A., Drncova, P., Garzoni, F., et al. (2016). The MultiBac baculovirus/insect cell expression vector system for producing complex protein biologics. *Advances in Experimental Medicine and Biology, 896*, 199–215.

Schaufele, F., Gilmartin, G. M., Bannwarth, W., & Birnstiel, M. L. (1986). Compensatory mutations suggest that base-pairing with a small nuclear RNA is required to form the 3′ end of H3 messenger RNA. *Nature, 323*, 777–781.

Schonemann, L., Kuhn, U., Martin, G., Schafer, P., Gruber, A. R., Keller, W., et al. (2014). Reconstitution of CPSF active in polyadenylation: Recognition of the polyadenylation signal by WDR33. *Genes and Development, 28*, 2381–2393.

Shi, Y., & Manley, J. L. (2015). The end of the message: Multiple protein-RNA interactions define the mRNA polyadenylation site. *Genes and Development, 29*, 889–897.

Skrajna, A., Yang, X.-C., Dadlez, M., Marzluff, W. F., & Dominski, Z. (2018). Protein composition of catalytically active U7-dependent processing complexes assembled on histone pre-mRNA containing biotin and a photo-cleavable linker. *Nucleic Acids Research, 46*, 4752–4770.

Skrajna, A., Yang, X. C., & Dominski, Z. (2019). Single-step purification of macromolecular complexes using RNA attached to biotin and a photo-cleavable linker. *Journal of Visualized Experiments*. https://doi.org/10.3791/58697.

Smith, H. O., Tabiti, K., Schaffner, G., Soldati, D., Albrecht, U., & Birnstiel, M. L. (1991). Two-step affinity purification of U7 small nuclear ribonucleoprotein particles using complementary biotinylated 2'-O-methyl oligoribonucleotides. *Proceedings of the National Academy of Sciences of the United States of America, 88*, 9784–9788.

Stauber, C., Soldati, D., Lüscher, B., & Schümperli, D. (1990). Histone-specific RNA 3′ processing in nuclear extracts from mammalian cells. *Methods in Enzymology, 181*, 74–89.

Sun, Y., Hamilton, K., & Tong, L. (2020). Recent molecular insights into canonical pre-mRNA 3′-end processing. *Transcription, 11*, 83–96.

Sun, Y., Zhang, Y., Aik, W. S., Yang, X. C., Marzluff, W. F., Walz, T., et al. (2020). Structure of an active human histone pre-mRNA 3′-end processing machinery. *Science*, *367*, 700–703.

Tan, D., Marzluff, W. F., Dominski, Z., & Tong, L. (2013). Structure of histone mRNA stem-loop, human stem-loop binding protein, and 3'hExo ternary complex. *Science*, *339*, 318–321.

Xiang, K., Nagaike, T., Xiang, S., Kilic, T., Beh, M. M., Manley, J. L., et al. (2010). Crystal structure of the human symplekin-Ssu72-CTD phosphopeptide complex. *Nature*, *467*, 729–733.

Yang, X.-C., Burch, B. D., Yan, Y., Marzluff, W. F., & Dominski, Z. (2009). FLASH, a proapoptotic protein involved in activation of caspase-8, is essential for 3′ end processing of histone pre-mRNAs. *Molecular Cell*, *36*, 267–278.

Yang, X.-C., Sabath, I., Debski, J., Kaus-Drobek, M., Dadlez, M., Marzluff, W. F., et al. (2013). A complex containing the CPSF73 endonuclease and other poyadenylation factors associates with U7 snRNP and is recruited to histone pre-mRNA for 3′-end processing. *Molecular and Cellular Biology*, *33*, 28–37.

Yang, X.-C., Sullivan, K. D., Marzluff, W. F., & Dominski, Z. (2009). Studies of the 5′ exonuclease and endonuclease activities of CPSF-73 in histone pre-mRNA processing. *Molecular and Cellular Biology*, *29*, 31–42.

Yang, X. C., Sun, Y., Aik, W. S., Marzluff, W. F., Tong, L., & Dominski, Z. (2020). Studies with recombinant U7 snRNP demonstrate that CPSF73 is both an endonuclease and a 5′-3′ exonuclease. *RNA*, *26*, 1345–1359.

Yang, X.-C., Xu, B., Sabath, I., Kunduru, L., Burch, B. D., Marzluff, W. F., et al. (2011). FLASH is required for the endonucleolytic cleavage of histone pre-mRNAs but is dispensable for the 5′ exonucleolytic degradation of the downstream cleavage product. *Molecular and Cellular Biology*, *31*, 1492–1502.

Zhang, Y., Sun, Y., Shi, Y., Walz, T., & Tong, L. (2020). Structural insights into the human pre-mRNA 3′-end processing machinery. *Molecular Cell*, *77*, 800–809.

Zhang, J., Tan, D., DeRose, E. F., Perera, L., Dominski, Z., Marzluff, W. F., et al. (2014). Molecular mechanisms for the regulation of histone mRNA stem-loop-binding protein by phosphorylation. *Proceedings of the National Academy of Sciences of the United States of America*, *111*, E2937–E2946.

Zhao, J., Hyman, L., & Moore, C. L. (1999). Formation of mRNA 3′ ends in eukaryotes: Mechanism, regulation, and interrelationships with other steps in mRNA synthesis. *Microbiology and Molecular Biology Reviews*, *63*, 405–445.

CHAPTER FIFTEEN

Comprehensive RNP profiling in cells identifies U1 snRNP complexes with cleavage and polyadenylation factors active in telescripting

Zhiqiang Cai, Byung Ran So, and Gideon Dreyfuss*

Department of Biochemistry and Biophysics, School of Medicine, Howard Hughes Medical Institute, University of Pennsylvania, Philadelphia, PA, United States

*Corresponding author: e-mail address: gdreyfuss@hhmi.upenn.edu

Contents

Methods in Enzymology, Volume 655
ISSN 0076-6879
https://doi.org/10.1016/bs.mie.2021.04.017

Abstract

Full-length transcription in the majority of protein-coding and other genes transcribed by RNA polymerase II in complex eukaryotes requires U1 snRNP (U1) to co-transcriptionally suppress transcription-terminating premature 3′-end cleavage and polyadenylation (PCPA) from cryptic polyadenylation signals (PASs). This U1 activity, termed telescripting, requires U1 to base-pair with the nascent RNA and inhibit usage of a downstream PAS. Here we describe experimental methods to determine the mechanism of U1 telescripting, involving mapping of U1 and CPA factors (CPAFs) binding locations in relation to PCPA sites, and identify U1 and CPAFs interactomes. The methods which utilizes rapid reversible protein-RNA and protein-protein chemical crosslinking, immunoprecipitations (XLIPs) of components of interest, and RNA-seq and quantitative proteomic mass spectrometry, captured U1–CPAFs complexes in cells, providing important insights into telescripting mechanism. XLIP profiling can be used for comprehensive molecular definition of diverse RNPs.

1. Introduction

Recent studies have revealed an unexpected and essential role for U1 snRNP (U1), an abundant small nuclear RNA-protein particle, in RNA polymerase II (Pol II) transcription of protein-coding genes and long non-coding RNAs (lncRNAs) genes. In this role, U1 is a suppressor of transcription-terminating premature cleavage and polyadenylation (PCPA) from cryptic polyadenylation signals (PASs) in introns and 3′-untranslated regions (3′UTRs) (Berg et al., 2012; Kaida et al., 2010). This activity, termed telescripting, is required for full-length Pol II transcription of mRNA precursors (pre-mRNAs). Telescripting is an additional and separable U1 activity from its well-established role in pre-mRNA splicing. Long introns are particularly susceptible to PCPA and U1 telescripting dependent compared to short introns, due to more cryptic PASs (Lepennetier & Catania, 2017). The differential sensitivity enables gene expression regulation according to intron length (the strongest determinant of gene length) by changes in the relative amounts of transcription output to U1 level. Modulation of available U1 levels, using antisense oligonucleotides as described below, illustrate the diverse and far-reaching roles of U1 telescripting, including gene size-function regulation (Oh et al., 2017), mRNA 3′-untranslated regions (3′UTR) length (Berg et al., 2012), and regulation of upstream antisense transcription from bidirectional promoters (Almada, Wu, Kriz, Burge, & Sharp, 2013; Ntini et al., 2013). U1 has been extensively characterized for its role in 5′-splice site (5′ss) recognition, an initiating step in spliceosome assembly (Mount, Pettersson, Hinterberger, Karmas, & Steitz, 1983), however, the mechanism of U1 telescripting was unknown.

U1 is comprised of 11 components: U1 snRNA (164 nucleotides (nt) in vertebrates), seven Sm proteins (SmB, SmD1, SmD2, SmD3, SmE, SmF, SmG) organized as a ring (Sm core) surrounding a short single stranded Sm site, and three U1-specific proteins, U1-70K, U1A and U1C. The atomic structure of U1 has been determined (Pomeranz Krummel, Oubridge, Leung, Li, & Nagai, 2009; Weber, Trowitzsch, Kastner, Lührmann, & Wahl, 2010), although substantial portions of the U1-specific proteins are not present or visible in these structures, including domains known to interact with other spliceosome proteins. U1 binds 5′ss by base-pairing of U1 snRNA 9-nucleotide (nt) in its 5′-sequence. This interaction is highly degenerate and generally insufficient by itself to distinguish functional 5′ss from cryptic 5′ss or other RNA sequences (Roca, Krainer, & Eperon, 2013). Thus, additional U1 interactions with pre-mRNAs and its associated RBPs (hnRNP complexes), and snRNPs, particularly U2, are important for U1 binding. For example, interactions with U2, which binds the branch site near the 3′ss, enhance U1 recruitment to internal 5′ss, and interactions with the 5′-cap binding complex (CBC) enhance U1 binding to the first 5′ss (Lewis, Gunderson, & Mattaj, 1995). Importantly, the multitude of potential binding sites and factors that influence U1 binding, have made it difficult to predict actual U1 binding locations, requiring experimental methods to map actual U1 binding sites. Nevertheless. U1 base-pairing is necessary both for splicing and telescripting, as complementary antisense oligonucleotides to U1 snRNA 5′-sequence inhibit splicing and elicits PCPA (Berg et al., 2012; Kaida et al., 2010).

PASs consist of three RNA sequence elements: a PAS hexamer, generally AAUAAA and variants thereof (Tian & Graber, 2012); a UGUA sequence upstream element (USE); and a downstream G/U-rich element (DSE) (Fig. 4). Each of these recruits a distinct subset of the CPAFs organized into three main subunits (Eckmann, Rammelt, & Wahle, 2011; Shi & Manley, 2015; Tian & Manley, 2017). The CPSF subunit, comprising CPSF160/CPSF1–WDR33–CPSF30/CPSF4–Fip1, binds to the PAS hexamer (Casañal et al., 2017; Chan et al., 2014; Clerici, Faini, Aebersold, & Jinek, 2017; Schönemann et al., 2014; Sun et al., 2018). This CPSF, together with CPSF100/CPSF2, the endonuclease CPSF73/CPSF3, and the poly(A) polymerase (PAP), are necessary and sufficient for the CPA reaction in vitro. The USE-binding CFIm subunit, is a tetrameric complex comprising CFIm25/CPSF5/NUDT21 dimer and CFIm59/CPSF7 and/or CFIm68/CPSF6. CFIm59 and CFIm68 and have serine and arginine (SR) repeats domains, a common protein-protein interaction domain in many RBPs involved in RNA processing (Rüegsegger, Blank, & Keller, 1998). The DSE-binding subunit, CstF, is a trimer-dimer of CstF64/CSTF2, CstF77/CSTF3, and CstF50/CSTF1 (Takagaki, Manley, MacDonald, Wilusz, & Shenk, 1990). Several additional components are also involved, including the poly(A)-stimulating factor, PABPN1 (Preker, Lingner, Minvielle-Sebastia, & Keller, 1995), Symplekin/SYMPK, and CFIIm (Pcf11, Clp1). Multiple interactions among CPAFs and with Pol II C-terminal domain (CTD) contribute to assembly of CPAFs and regulate CPA (Shi & Manley, 2015).

Here, we describe experimental methods to determine where on RNAs U1 and CPAFs bind, their compositions and protein interactions in cells. The focus of the studies is to identify potential interactions between U1 and CPAFs in order to advance understanding of U1 activities in telescripting and splicing, and regulation of PAS-dependent transcription termination.

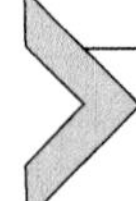

2. Methods

2.1 Overview of a strategy for comprehensive definition of RNPs in cells

As interactions of RNA binding proteins (RBPs) with RNAs and other proteins that make up RNPs can be disrupted and adventitious associations can occur during cell lysis and sample processing, we designed an experimental strategy to preserve native RNPs with formaldehyde, a "zero distance" protein-protein, protein-RNA, and RNA-RNA crosslinker that readily penetrates cells (Fig. 1). In contrast, UV light, which induces RNA-protein crosslinks by photo-activating uridines, does not induce protein-protein or RNA-RNA crosslinks (e.g., U1 snRNA base-paired to pre-mRNA), and is not reversible. The utility of formaldehyde for protein-RNA crosslinking in cells and IP had been demonstrated previously with antibodies to hepatitis delta virus (HDVg) and HDV RNA detection by RT-PCR (Niranjanakumari, Lasda, Brazas, & Garcia-Blanco, 2002). We have optimized formaldehyde in-cell crosslinking (XL), cell lysis, RNA fragmentation and stringent immunopurification (IP) conditions for multiple protein components and RNA selection in parallel and combined them with RNA-seq and quantitative mass spectrometry to map RNA binding sites and protein composition and stoichiometry. Use of multiple antibodies or other specific probes for components of RNPs of interest from the same samples increases

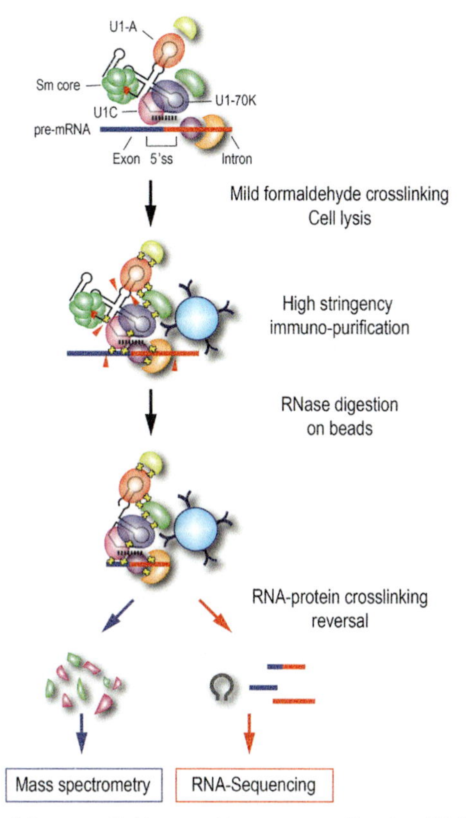

Fig. 1 A schematic of the crosslinking and immunopurification (XLIP) strategy for capturing native RNPs in cells and their comprehensive characterization by RNA-seq and mass spectrometry.

confidence in calling specific interactions and provides a large matrix of information. This strategy, XLIPs, provides high definition of RNPs, including interactions of individual components within them and the RNP's sphere of interactions. We have applied this strategy to the SMN complex (Yong, Kasim, Bachorik, Wan, & Dreyfuss, 2010), and more recently with further enhancements to U1 and CPAFs (So et al., 2019). We describe detailed experimental protocols and data analysis, followed by examples illustrating applications to studies of U1 and CPAFs.

2.2 Formaldehyde crosslinking in cells, cell transfection and sample preparation

Standard formaldehyde crosslinking, typically 3% (v/v) or higher for >30 min, is widely used to fix cells and preserve morphological features for applications such as immunohistochemistry. We sought instead to minimize crosslinking to ensure that only direct interactors are crosslinked. Testing a series of milder conditions, including lower formaldehyde concentrations and shorter incubation times on cells in tissue culture for XLIPs of several complexes showed that 0.2% formaldehyde for 10 min was sufficient to co-IP known interactors with little or no non-specific components. Higher formaldehyde concentrations and incubation times tended to increase non-specific background and decrease IP efficiency, likely due to decreased solubility and inaccessibility of the IP target. Based on this we used the following protocol for human HeLa cells and other cell types grown in tissue culture, attached or in suspension.

Typically, experiments started with 10 million sub-confluent cells grown in DMEM supplemented with 10% FBS, L-glutamine, penicillin and streptomycin, the medium was first aspirated and the cells were washed three times with PBS at room temperature. Crosslinking was performed in 1 mL PBS containing freshly prepared 0.2% formaldehyde (Sigma-Aldrich Cat. F8775) for 10 min at room temperature. PBS is suitable because it maintains physiological pH and because it does not contain amines, which could react with formaldehyde, decreasing its effective concentration. The crosslinking was stopped by addition of glycine, to 150 mM, which reacts with excess formaldehyde. After 10 min, the glycine was removed and the cells were washed twice with ice cold PBS. Care is taken to avoid the cells from drying in between steps. After the last step, the cell pellet can be processed for IP immediately or flash frozen in liquid nitrogen and stored for later use.

Studies concerning U1 made use of antisense morpholino oligonucleotide (AMO) complementary to U1 snRNA's 5′-end sequence (U1 AMO) which interferes with its base-pairing with pre-mRNAs and thereby elicits PCPA and splicing inhibition. U1 AMO and control, non-targeting AMO (cAMO), were transfected by electroporation using 50 nmol AMO per 10 million cells (Berg et al., 2012; Kaida et al., 2010; Oh et al., 2017, 2020) at 6–8 h prior to crosslinking. This U1 AMO dose inhibits most of U1 (>90%). The number of cells can be changed depending on availability and on how many IPs are planned, and the amounts of reagents for the various steps can be adjusted accordingly.

For IPs, pellets of 10 million cells were suspended in 750 μL of RSB300 (10 mM Tris-HCl, pH 7.8, 300 mM NaCl, and 2.5 mM $MgCl_2$) containing 1% Empigen BB and 0.5% TritonX-100 (Empigen buffer) and sonicated 3 times for 10 s at 4 W output. The lysate was then clarified by centrifugation at

10,000 rpm for 10 min at 4 °C and the supernatant was collected for immunoprecipitation. The zwitterionic detergent Empigen BB is highly effective for immunoprecipitations, as it solubilizes cells and protein complexes without denaturing proteins, unlike SDS. For example, a monoclonal antibody (4F4) to the hnRNP C1/C2 proteins (hnRNPC), which are abundant, nuclear, avid pre-mRNA binding proteins, was used for immunopurification of native hnRNPs under high salt (300 mM NaCl) and Empigen BB (1%) to minimize adventitious association (Choi & Dreyfuss, 1984).

2.3 Parallel immunoprecipitations with antibody-conjugation to magnetic beads

To avoid large amounts of antibodies from eluting with the IPed proteins and increasing background in downstream analyses, we covalently coupled antibodies to epoxy derivatized magnetic beads (Dynabeads M270 epoxy; Invitrogen) according to the manufacturer's directions. This procedure couples sulfhydryl groups in antibodies to the immobilized epoxy groups. It works well for many antibodies, but not others, and needs to be determined in each case. Antibodies that do not couple well to epoxy beads could be coupled to other derivatized beads, or deployed as in standard IP procedures by attaching them to protein A or protein G coated beads.

The use of magnetic beads allowed multiple IPs to be performed in parallel from the same cell lysate, which improves accuracy between replicates and data comparisons between different antibodies. For this, cell lysate (2–2.5 mg in 500 μL; total protein was determined using Bio-Rad Protein Assay Kit) was incubated with 50–60 μL of antibodies-crosslinked beads for 1.5 h in 96-well plates (5–6 wells per immunoprecipitation). The beads were washed four times with RSB300 (200 μL) and once with RSB150 containing 0.02% TritonX-100 using a Kingfisher 96 magnetic particle processor (ThermoFischer Scientific). Alternatively, for a smaller number of samples, handheld magnetic manifolds also work well.

2.4 RNP pulldowns from crosslinked cells with biotin-conjugated oligonucleotides

The same XLIP procedures can also be used for pulldowns with antisense oligonucleotides instead of antibodies. For U1, we have successfully used biotin-coupled U1 AMO and cAMO control transfected the same as the unmodified AMOs (Oh et al., 2020; So et al., 2019). RNA-seq confirmed that biotin-U1 AMO elicited the same PCPA as U1 AMO. After cell lysis using the Empigen buffer, the cell lysates (2–2.5 mg in 500 μL) were incubated with 25 μg Dynabeads MyOne Streptavidin C1 (Invitrogen) beads at 4 °C for 1 h. The beads were washed with lysis buffer (200 μL) and captured RNPs were digested with RNAse, eluted and processed for RNA-seq and mass spectrometry as described below.

2.5 Purification of RNA fragments and RNA-seq

To develop genome wide maps of U1 and CPAFs, the crosslinked IPs (XLIPs) were digested with RNase and the RNA fragments and proteins that remained were eluted and identified by high throughput RNA-sequencing (RNA-seq) and mass spectrometry, respectively. The RNP-bound beads were treated with RNAse T1 (0.1 unit/μL; ThermoFisher) for 6 min to moderately fragment long RNAs or with RNase T1 (1 unit/μL) plus RNase A (0.4 mg/mL) to yield RNA fragments <150 nt, followed by five washes in the lysis buffer. The RNase digestion is aimed to generate fragments small enough for better mapping of RNA binding sites and minimize inclusion of proteins that bind RNAs far from and unrelated to the IP target.

For RNA-sequencing, washed beads were incubated in buffer containing 20 mM Tris-HCl/pH 7.8, 150 mM NaCl, 1 mM EDTA and 5 mM DTT at 70 °C for 16 h for crosslinking reversal and then treated with 1 mg/mL protease K (Sigma-Aldrich) for 30 min at room temperature. The beads were discarded; the RNA in solution was purified by phenol-chloroform, treated with TURBO DNase (0.1 unit/μL, Ambion), and ethanol precipitated. The RNA fragment size distribution prior to library preparation was between 100 and 500 bp as analyzed on a Bioanalyzer, with peaks around 150–200 bp. cDNA synthesis and RNA-seq libraries were prepared as described above, excluding any further fragmentation with high heat (65–94 °C) and $MgCl_2$.

2.6 Processing RNA-seq reads

RNA-seq technologies and data processing tools continue to developed at rapid pace. For the studies describe here RNA-seq was performed on Illumina HiSeq 2500, and we have used the following data analysis pipeline. Paired-end RNA-seq reads were trimmed of any adaptor sequences with the FASTX-Toolkit (version 0.0.14). The two paired reads were merged into one single fragment using PEAR (version 0.9.8), and then fragments larger than 150 nt were filtered out. The remaining reads were aligned to the GRCh38/hg38 reference genome using STAR (version 2.5.3a) with the following parameters: twopassMode Basic—alignSJoverhangMin 5—

alignSJDBoverhangMin 5—outSAMmapqUnique 255—outFilterMultimap Nmax 1—outSJfilterReads Unique. Reads per exon were grouped, from which RPKM (reads per kilobase per million mapped reads) values were calculated using SAMtools (version 0.1.19).

In order to directly compare samples that have a different number of mapped reads, the read coverage for each sample was normalized to the total number of mapped reads per million (RPM). This normalized value was also used to scale the samples for visualization on the UCSC Genome Browser. External CLIP-seq and Pol II ChIP-seq datasets were downloaded in raw format from the Gene Expression Omnibus (https://www.ncbi.nlm.nih.gov/geo/) and then processed and aligned as described above. All RNA-seq datasets mentioned in this article were deposited at the same publicly available database.

2.7 Metagene analysis of RNA-seq data

To remove potential background binding, mapped reads in Biotin-U1-AMO XLPD or XLIPs RNA-seq were normalized using a log2 ratio to the corresponding input RNA or SP2/0 XLIP, respectively, using bamCompare from deepTools (v3.1.3). All profiles were generated using computeMatrix and plotProfile from the same package. Metagene plots of U1–CPAFs co-localization were carried out starting from EU polyA RNA peaks located in introns ($n = 1485$). These peaks were determined by searching for the local signal maxima of EU polyA RNA peaks using Piranha (version 1.2.0) with the following parameters: -s -b 50 -u 50 -p 0.05 -a 0.95 -v -d ZeroTruncatedNegativeBinomial. The middle-points of these peaks were defined as the peak center and used for centering the metagene plots with the aforementioned normalized reads of U1 and CPAFs binding from control and U1 AMO transfected XLIP-seq.

For metagene profiles around the TSS, first 5′ss, or last 3′ss, genes with either directly overlapping transcripts or transcripts within 2.5 kb of these points were excluded from the metagenes so as not to skew the plots with off-target signals. The resulting 15,091 genes were selected for metagene analysis. These metagene profiles were then plotted from the normalized read values across a 4 kb window.

2.8 Determination of protein composition and stoichiometry

The proteins eluted from the XLIPs can be probed directly for specific proteins by sodium dodecyl sulfate (SDS)-polyacrylamide gel electrophoresis (SDS-PAGE) and immune-blotting (Western blots). However, mass spectrometry provides much more and unbiased information about the composition and stoichiometry of the proteins that crosslinked to the target, and was applied routinely to all the samples.

For this, RNP complexes were eluted with 30 μL of 1 × lithium dodecyl sulfate (LDS) sample buffer (NuPAGE, Invitrogen) without DTT for 10 min at room temperature; after bead removal, were reversed by incubating samples with 2 mM DTT at 70 °C for 16 h. Samples (25 μL) eluted from each XLIPs with LDS buffer were run separately on a 4–12% Tris-Bis SDS-PAGE gel (Invitrogen) with a short path-length (~1 cm) and stained with Coomassie blue. The bands were cut and subjected to in-gel trypsin digestion and peptides were resolved by liquid chromatography and subjected to tandem mass spectrometry (LC-MS/MS) (Reyes et al., 2017).

2.9 Proteomic data analysis

Raw data were analyzed by MaxQuant using the UniProt Human Proteome (http://www.uniprot.org) and protein-protein interactions were determined using label-free quantification (Cox & Mann, 2008; Hubner & Mann, 2011; Schwanhäusser et al., 2011). IBAQ values for each IP target were adjusted to compare the relative stoichiometries. Known common contaminants were removed (e.g., keratins, immunoglobin heavy/light chains and trypsin/LysC-proteases). Protein functions were annotated based on the GeneCards database (www.genecards.org) and literature searches. The proteins shown in the figures and described in the text have been cross-verified by two or more biological repeats.

2.10 Applications to U1 and CPAFs

Applications of the methods to U1 and CPAFs illustrate the wealth of information they provide, practical considerations in their use, and their limitations. As U1 and CPAFs were central to the studies, XLIPs for U1 were performed with antibodies to the three U1 specific proteins, U1A, U1C, and U1-70K, and to several CPAFs from the main subunits in CPA, CFIm, CPSF, and CstF subunits, targeting CFIm25, Fip1, and CstF64, respectively. XLIPs were also performed for SF3B1, spliceosomal U2 snRNP protein, to determine binding U1 and CPAFs binding locations and compositions in relations to spliceosomes. Non-specific antibodies (SP2/0), total input, and IPs without crosslinking were included as references for the background. For most of the targets more than one antibody was available from commercial suppliers. Some antibodies do not work well for IP because the epitope is inaccessible in the native folded protein or in XLIPs if the epitope is irreversibly modified or crosslinked, although this was uncommon. For the main protein targets, we tested several antibodies and selected the most efficient for each that also couples to epoxy beads.

2.11 Constructing maps of RBPs and RNP RNA-binding sites from XLIPs-RNA-seq

Alignments of RNA-seq reads from several XLIPs (XLIPs-RNA-seq) to representative genes, presented in Fig. 2, show prominent peaks of U1 and CPAFs co-localized in introns. These peaks were notable in the longest intron, frequently the first or second intron in a gene, at or near the 5′ss and in many cases additionally tens of kilobases (kb) from the intron's 5′ss. The binding locations were frequently a series of peaks, which may represent one or more U1 binding site and a cluster of and CPAFs hundreds or more nucleotides downstream with looped-out pre-mRNA regions between them. This pattern is consistent with clusters of PASs silenced by U1 from the same 5′ss (Almada et al., 2013; Chiu et al., 2018). The U1 and CPAFs peaks were undetected or significantly less prominent in the input and in XLIPs with non-specific antibodies, supporting the specificity of the XLIPs (Fig. 2).

In addition, U1 XLIPs co-localized with SF3B1 in exons and splice sites, which represent spliceosome assembly sites. SF3B1 XLIPs were generally under-represented in intronic peaks of U1 and CPAFs, indicating that they are specific and distinct from spliceosomes. CPAFs XLIPs corresponded well with the positions of CPAFs determined by UV crosslinking and available in public databases (Martin, Gruber, Keller, & Zavolan, 2012; Schönemann et al., 2014), which cross-validates the methodologies. Importantly, the same peaks were detected with different antibodies to U1 and CPAFs, confirming the specificity of the procedure and greatly increasing confidence in the data. Determination of the amounts of snRNAs in the XLIPs from the RNA-seq of the same experiments also supported the specificity of the XLIPs-RNA-seq and the conclusions that can be drawn from them (So et al., 2019). U1A and U1-70K XLIPs contained U1 snRNA nearly exclusively. U1C XLIPs contained U1 snRNA preferentially and smaller amount of U2 snRNA (17% compared to U1 snRNA), consistent with pervasive U1–U2 associations. SF3B1 XLIPs contained U2 snRNA, preferentially, as well as significant amounts of U1 snRNA (about 30% compared to

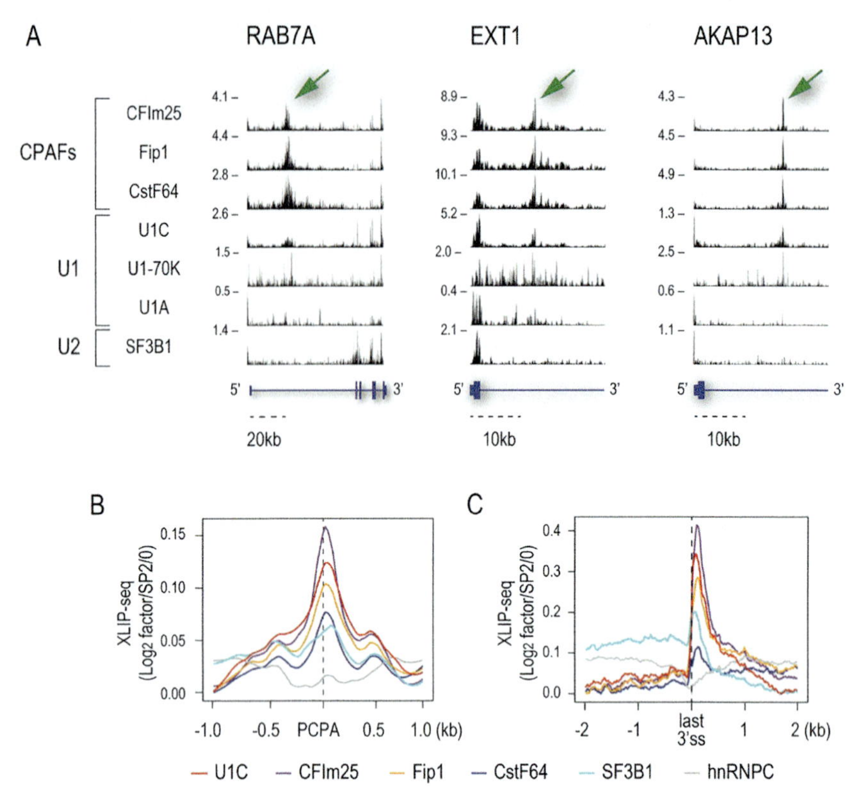

Fig. 2 U1 and CPAFs co-localize at PCPA locations in introns. (A) Genome browser views of XLIP-seq data for the indicated factors from HeLa cells transfected with U1 AMO or cAMO in select regions of representative genes (*RAB7A*, *EXT1*, and *AKAP13*). The *Y*-axis indicates reads per million (RPM) for the highest peak within the genome browser field for each sample. Annotated RefSeq gene structures are shown in blue with thin horizontal lines indicating introns and thicker blocks indicating exons. (B) Metagene plots for U1–CPAFs co-localization at PCPA sites ($n = 1485$). Normalized XLIP binding (log2 RPM in XLIPs over SP2/0) of the factors was rescaled using the lowest values as a baseline and were plotted around the PCPA sites within a 2 kb window. (C) Metagene plots of the normalized U1–CPAFs XLIP binding for PCPA genes as shown in (B) around the last 3′ss ($n = 1469$), within a 4 kb window. For better clarity, colored version of this figure is available online.

U2 snRNA). Importantly, CstF64, CFIm25 and Fip1 XLIPed large amounts of U1 snRNA, suggesting an association between CPAFs and U1 in cells.

Comparisons of U1 and CPAFs peaks with PCPA position determined from non-genomic 3′-poly(A) RNA sequences in nascent transcripts from cells transfected with U1 AMO compared with cAMO, showed that U1 and CPAFs peaks were frequently bound at or near PCPA positions, suggesting that U1 and CPAFs bind at or near actionable PASs that are normally suppressed (Fig. 1). Metaplots of the XLIPs-RNA-seq demonstrated the generality of these observations, showing co-localization of U1 and CPAFs at 5′ss, particularly the first 1′ss, and at PCPA locations in introns and 3′UTRs. These findings suggested that U1 binding did not inhibit binding of CPAFs to a nearby PAS. Instead, it raised the possibility that U1 and CPAFs interact with each other as U1 base-pairs to a 5′ss or other complimentary sequence and CPAFs bind to a downstream PAS(s). These conclusions were clear within the resolution of the methods. UV crosslinking can be used pinpoint crosslinked nucleotides to an RBP, however, it provides no information about the RBP's interactions with other proteins. The resolution of XLIPs-RNA-seq maps can be increased by more extensive RNase digestion, although that could result in loss of interacting proteins. It should be straightforward to add UV crosslinking as a step in the XLIPs procedure, if needed.

2.12 Constructing RBPs and RNP interactomes from XLIPs-MS: Discovery of a complex of U1 and CPAFs (U1–CPAFs)

Proteins captured in the XLIPs were released with LDS and analyzed by liquid chromatography-mass spectrometry (LC-MS/MS), which provided their composition and stoichiometry using label-free intensity based absolute quantification (IBAQ) (Schwanhäusser et al., 2011). Several calculations can be applied to determine enrichment or de-enrichment of a protein in a sample and between samples, including ratio to the sum of IBAQs in a sample in comparison to the ratio of the same in the total input, in IP with the same antibody without crosslinking, ratio of proteins to the IP target, and normalization of the sum of IBAQs between samples being compared. The choice of calculation method depends on the objective of the experiments. For the U1 and CPAFs studies, several calculations we tested yielded similar conclusions for the main proteins. The more variable part comes from user selected cutoff values (inclusion filter), as is the case for calculations of RNA-seq expression changes. Ideally, each IP should be done in triplicates to assess statistical significance of the magnitude of change. In experiments testing various conditions with multiple antibodies at the same time, this is sometimes impractical, for example, if one of the replicates is lost. In such cases, valuable information about the main components can still be obtained if it is supported by other IPs of the same target or complex. In most cases the target protein is by far the most abundant protein in the XLIP, consistent

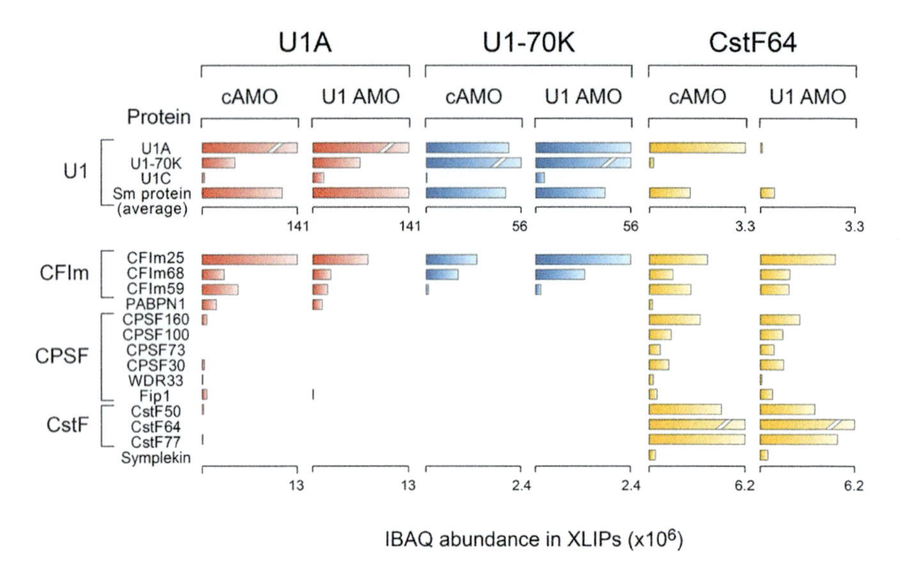

Fig. 3 Summary of U1 snRNP (U1) and 3′-processing cleavage and polyadenylation factors CPAFs enriched in XLIPs of U1 proteins U1A and U1-70K and the CPAF CstF64 from cells transfected with non-targeting control morpholino oligonucleotide (cAMO) or antisense to U1 snRNA 5′-sequence (U1 AMO). The bar graph indicates the IBAQ abundance determined from label-free mass spectrometry. The proteins were classified into functional groups according to protein (UniProt) and gene databases (GeneCards). The CPAFs are indicated according to subunits, CFlm, CPSF, and CstF. Note selective enrichment of CPAFs in U1 XLIPs and of U1 proteins in CstF64 XLIPs, as well as changes with U1 AMO, such as strong loss of U1A crosslinks in CstF64. A complete listing of proteins detected in these and additional XLIPs, total input and IPs without in-cell crosslinking can be found in So et al. (2019). They include spliceosomal components, including all snRNP; transcription and export complex (TREX) proteins; RBPs of the hnRNP proteins family and the SR domain subgroup; and transcription regulators. The scale of the IBAQs for each group is indicated under the bar graphs. Sm indicates the average IBAQ of the seven Sm proteins.

with the experimental design that aims to achieve specificity with minimal crosslinking rather than increase crosslinking and more background.

XLIPs-MS showed strong enrichment of known components of U1 and CPAFs in XLIPs, respectively (Fig. 3). The same IP procedure from the same cell cultures but without in-cell crosslinking (noXL), detected the target protein nearly exclusively, demonstrating the specificity of the antibodies and the stringency of the procedure. Relatively low IBAQs of proteins that are much more abundant in the input than the target protein, such as histones, cytoskeletal proteins, metabolic enzymes, and ribosomes, showed that the crosslinking was not excessive. As expected, U1 XLIPs contained

U1-specific and Sm proteins. U1C was consistently under-represented in U1A and U1-70K XLIPs, which may be explained by its limited interaction surface with U1-70K (Pomeranz Krummel et al., 2009; Weber et al., 2010), and its propensity to dissociate from U1 during the sample preparation (Hernández et al., 2009). Consistent with U1's role in splicing, U1 XLIPs were also highly enriched in spliceosomal proteins, including U2 (U2A′/SNRPA1, U2B″/SNRPB2, SF3A, and SF3B), U5 (U5-40K/SNRNP40, Brr2/SNRP200, and hSNU114/EFTUD2), the 19 complex (NTC; PRP19/PRPF19), and the exon-junction complex (EJC; eIF4A3 and Y14/RBM8A).

Importantly, U1 and CPAFs XLIPs-detected components of each other and of other complexes that both interacted with. U1A XLIPs contained CPAFs of the CFIm and CPSF subunits, with greater enrichment of CFIm (CFIm25 > CFIm59 > CFIm68 in an order of IBAQ values). U1-70K XLIPs were enriched in the CFIm subunit CFIm25 > CFIm68 and only small amounts of CFIm59 (Fig. 2). CFIm25, for example, was as highly represented as the U2 associated splicing factor U2AF65 in U1A XLIPs. CstF64 XLIPs contained CPAFs of all three CPA subunits, as expected for an assembled and functional CPA complex, as well as comparable amounts of U1A and Sm proteins with CFIm25. PAP and CPA-regulating Pcf11 and Clp1 (CFIIm) were undetected in the XLIPs, likely because they interact transiently, which makes crosslinking inefficient, and their low abundance. Additional components enriched in U1 and CstF64 XLIPs, include the transcription elongation and export complex (TREX), Pol II transcription regulators, mRNA degradation factors, chromatin remodeling proteins, and hnRNP and SR proteins, known for their roles in every aspect of pre-mRNA processing (Fig. 3). The XLIPs-MS therefore revealed complexes of U1 and CPAFs (U1–CPAFs), consistent with the co-localization of their components on nascent Pol II transcripts determined by XLIPs-RNA-seq of RNA fragments from the same complexes.

2.13 Changes in U1–CPAFs with U1 AMO reveal key mechanisms of telescripting

The binding of U1–CPAFs to full-length pre-mRNAs indicate that these complexes are active in telescripting. U1 AMO elicits PCPA, which makes it useful for probing U1 and CPAFs that are inactive in telescripting. XLIPs-RNA-seq showed that U1 and CPAFs were co-localized in the same positions in U1 AMO as in cAMO, suggesting that U1 was bound in the same locations without its normal base-pairing via U1 snRNA 5′-sequence. The

RNA-seq confirmed PCPA in the same U1 AMO transfected cells. XLIPs-MS indicated that that U1 AMO did not disrupt U1 itself, as XLIPs with U1 proteins still co-IPed each other and U1 snRNA. Consistent with the general splicing inhibition at the high U1 AMO dose used in these experiments (Berg et al., 2012; Kaida et al., 2010; Oh et al., 2017), U1 interactions with spliceosome components were reduced by 75–90% (Fig. 2) though U1 interactions with U2 were less affected. XLIPs with CstF64 showed that U1 AMO did not disrupt interactions among CPAFs either. Moreover, U1-70K XLIPs with CFIm25 and CFIm68 were also maintained. However, U1A XLIPs with CstF64 were strongly reduced (90%), indicating that U1 AMO altered U1A interactions with CPAFs. A notable change was an increase in the IBAQ ratio of CFIm68 compared to CFIm59 with U1 AMO, both in CstF64 XLIPs and U1-70K XLIPs. This suggested that U1 AMO remodels U1–CPAFs, which can form despite U1 AMO. Mass spectrometry and RNA-seq from pulldowns with biotin-U1 transfected into cells, confirmed and extended these observations, and showed that U1 with U1 AMO bound to it associated with CPAFs bound to pre-mRNAs.

Many additional details about U1–CPAFs revealed by these experiments are described in So et al. (2019). These include interactions of U1 with the cap binding complex, Pol II, transcription elongation and nuclear mRNA export factors, adaptors to exosomes with roles in nuclear RNA degradation, and the CTD kinases, CDK11A and CDK12, which are required for Pol II pause release (Bartkowiak et al., 2010). Thus, the XLIPs methods are useful for detecting changes in RNPs in cells. Key interaction in U1–CPAFs and its changes in relation to telescripting activity are summarized in a schematic model in Fig. 4.

3. Summary

The XLIP procedure captured U1–CPAFs and enabled comprehensive profiling of their composition and RNA binding locations. U1–CPAFs, are distinct from U1 complexes with spliceosomes, and regulate PAS-dependent 3′-end processing and thereby Pol II transcription elongation and termination. The observations, summarized in Fig. 4, suggest that U1 telescripting is mediated by direct interaction of U1 base-paired to nascent with CPAFs bound at PASs downstream. Other mechanisms, such as that U1 hinders binding of CPAFs to PASs, may also be possible in specific contexts if the U1 binding site is very close to the PAS. The XLIPs are consistent with several earlier studies, including interactions between U1-free U1A or U1-70K with individual CPAFs (Awasthi & Alwine, 2003; Lutz et al.,

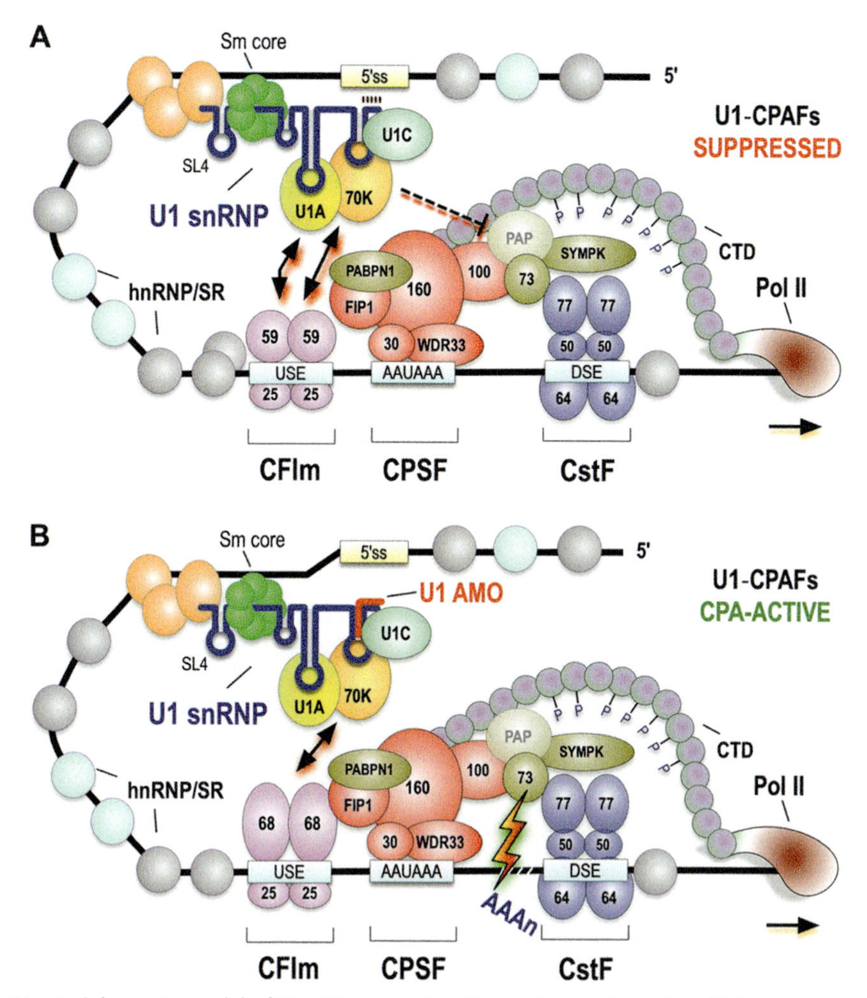

Fig. 4 Schematic model of U1–CPAFs and its role in telescripting. U1 snRNPs comprises U1 snRNA and proteins, U1-70K and U1A bind stem loop (SL) 1 and 2, respectively, while U1C associates with U1 through U1-70K. A heptameric Sm core on U1's Sm site between SL3 and SL4. The CPAFs proteins are shown organized into subunits CFIm, CPSF, CstF, which bind the marked PAS RNA sequence elements. (A) U1–CPAF active in telescripting bound to the first 5′ss by U1 snRNA 5′-sequence base-pairing and to a cryptic PAS in the first introns bound by CPAFs. U1–CPAFs suppresses premature transcription termination at the cryptic PAS. (B) U1–CPAFs with U1 AMO bound to U1 snRNA 5′-sequence, which interferes with U1 base-pairing with 5′ss and causes U1 loss of function in telescripting. In this state, U1 still binds nascent Pol II transcripts in the same positions and interacts with CPAFs, however, U1–CPAFs are remodeled and switched CPA-active states. Two notable changes are loss of likely inhibitory interactions between U1A and CPAFs, and increased association of CFIm68, a CFIm25 binding protein that stimulates CPA. Also shown are interactions of U1–CPAFs with other functional complexes and factors, including Pol II, the pre-mRNA cap binding complex (CPC), the transcription elongation and export complex (TREX), and adaptors to exosomes, likely poised to degrade PCPA products.

1996). Previous studies to map U1 binding (Engreitz et al., 2014) or identify the U1 interactome (Chi et al., 2018; Chu et al., 2015; Masuda et al., 2020) used much more extensive chemical crosslinking, and did not include multiple antibodies to U1 or U1 AMO. U1–CPAFs have been detected previously, likely because they readily dissociate during cell lysis unless they are first crosslinked in cells, and U1 is not essential for CPA. CPA can be reconstituted in cell-free systems without U1 on in vitro transcribed minimal substrates and uncoupled from transcription, which likely contributes to U1–CPAFs assembly.

XLIPs-detected changes in the composition and crosslinks in U1–CPAFs with U1 AMO (Fig. 4), suggest at least two potential mechanisms that contribute to telescripting. First, U1 prevents or limits association of the CPA-stimulating factors, CFIm68 and PABPN1 with the other CPAFs (Rüegsegger et al., 1998; Zhu et al., 2018). Second, U1A inhibition of CPAFs. U1A inhibits PAP in specific contexts, such as in vitro, in the last exon, and as U1-free protein (Boelens et al., 1993; Gunderson, Polycarpou-Schwarz, & Mattaj, 1998; Workman, Veith, & Battle, 2014), and could potentially inhibit CPAFs in general, in U1–CPAFs. This notion is suggested by the selective decrease in U1A-CPAFs crosslinks with U1 AMO.

The many interactions of U1 with the transcription, RNA processing, export and degradation machineries, reveal a central role of U1 in regulation of mRNA biogenesis, well beyond splicing. Prominent U1 XLIPs peaks at the first 5′ss suggest that U1 is important in controlling transcription elongation close to the transcription start site as well as throughout nascent transcripts. The XLIPs methodology illustrated here for U1 and CPAFs is a general approach that should facilitate comprehensive definitions of other types of RNPs, including changes in composition and interactions that give insights into their function.

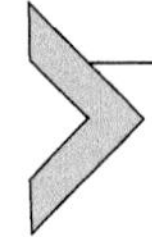

4. Key resources table

Reagent or resource	Source	Identifier
Antibodies		
U1-70K	Synaptic system	203,011
U1A	Abcam	ab55751
U1C	Sigma-Aldrich	SAB4200188

—cont'd

Reagent or resource	Source	Identifier
SF3B1	Bethyl laboratories	A300-996A
CFIm25	Proteintech	10,322-AP
CFIm68	Abcam	ab175237
Fip1	Bethyl laboratories	A301-462A
CstF64	Bethyl laboratories	A301-092A
hnRNPC	Choi and Dreyfuss (1984)	4F4
SP2/0	Choi and Dreyfuss (1984)	SP20
Chemicals, peptides, and recombinant proteins		
Formaldehyde	Sigma-Aldrich	F8775
RNAse T1	ThermoFisher	EN0541
RNAse A	ThermoFisher	EN0531
TURBO DNAse	ThermoFisher	AM2238
Ribo-Zero rRNA removal kit	Illumina	MRZH11124
Click-iT nascent RNA capture kit	Invitrogen	C10365
KAPA stranded RNA-Seq library preparation kit	Kapa Biosystems	KK8400
Oligotex kit	Qiagen	70,022
Dynabeads MyOne streptavidin C1	Invitrogen	65,001
Dynabeads antibody coupling kit	Invitrogen	14311D
NuPAGE LDS sample Buffer (4 ×)	Invitrogen	NP0008
Deposited data		
Raw RNA-sequencing data	This study	GEO: to be reported
PAR-CLIP RNA-seq	Martin et al. (2012)	GEO: GSE37398
Pol II ChIP RNA-seq	Oh et al. (2017)	GEO: GSE103252

Continued

—cont'd

Reagent or resource	Source	Identifier
Ensembl release 91		
Experimental models: Cell lines		
Human: HeLa cells		
Oligonucleotides		
Control antisense morpholino oligonucleotide	Kaida et al. (2010)	
U1 antisense morpholino oligonucleotide	Kaida et al. (2010)	
3′-biotinylated control antisense morpholino oligonucleotide	This study, GeneTools	
3′-biotinylated U1 antisense morpholino oligonucleotide	This study, GeneTools	
Software and algorithms		
FASTX-Toolkit (version 0.0.14)	http://hannonlab.cshl.edu/fastx_toolkit/	http://hannonlab.cshl.edu/fastx_toolkit/
PEAR (version 0.9.8)	Zhang, Kobert, Flouri, and Stamatakis (2014)	http://www.exelixis-lab.org/web/software/pear
STAR (version 2.5.3a)	Dobin et al. (2013)	https://github.com/alexdobin/STAR
SAMtools (version 0.1.19)	Li et al. (2009)	http://samtools.sourceforge.net/
Piranha (version 1.2.0)	Uren et al. (2012)	http://smithlabresearch.org/software/piranha/
deepTools (version 3.1.3)	Ramírez et al. (2016)	http://deeptools.readthedocs.io/en/latest/
Bedtools (version 2.15.0)	Quinlan and Hall (2010)	http://bedtools.readthedocs.io/en/latest/
Bowtie 2 (version 2.3.1)	Langmead and Salzberg (2012)	http://bowtie-bio.sourceforge.net/bowtie2/index.shtml

5. Lead contact for reagent and resource sharing

Further information and requests for reagents should be directed to the Lead contact, Gideon Dreyfuss (gdreyfuss@hhmi.upenn.edu).

Acknowledgments

We thank members of our laboratory for helpful discussions and comments on the manuscript. We thank Dr. Steven Seeholzer and Ms. Lynn Spruce of the Children's Hospital of Philadelphia Proteomics Core facility, and Dr. Benjamin Garcia and Dr. Xing-Jun Cao of the University of Pennsylvania School of Medicine Quantitative Proteomics Resource Core facility for expert help with mass spectrometry experiments. This work was supported by the US National Institutes of Health (R01GM112923 to G.D.). G.D. is an Investigator of the Howard Hughes Medical Institute.

References

Almada, A. E., Wu, X., Kriz, A. J., Burge, C. B., & Sharp, P. A. (2013). Promoter directionality is controlled by U1 snRNP and polyadenylation signals. *Nature*, *499*, 360–363.

Awasthi, S., & Alwine, J. C. (2003). Association of polyadenylation cleavage factor I with U1 snRNP. *RNA*, *9*, 1400–1409.

Bartkowiak, B., et al. (2010). CDK12 is a transcription elongation-associated CTD kinase, the metazoan ortholog of yeast Ctk1. *Genes & Development*, *24*, 2303–2316.

Berg, M. G., et al. (2012). U1 snRNP determines mRNA length and regulates isoform expression. *Cell*, *150*, 53–64.

Boelens, W. C., et al. (1993). The human U1 snRNP-specific U1A protein inhibits polyadenylation of its own pre-mRNA. *Cell*, *72*, 881–892.

Casañal, A., et al. (2017). Architecture of eukaryotic mRNA 3′-end processing machinery. *Science*, *358*, 1056–1059.

Chan, S. L., et al. (2014). CPSF30 and Wdr33 directly bind to AAUAAA in mammalian mRNA 3′ processing. *Genes & Development*, *28*, 2370–2380.

Chi, B., et al. (2018). Interactome analyses revealed that the U1 snRNP machinery overlaps extensively with the RNAP II machinery and contains multiple ALS/SMA-causative proteins. *Scientific Reports*, *8*, 8755.

Chiu, A. C., et al. (2018). Transcriptional pause sites delineate stable nucleosome-associated premature polyadenylation suppressed by U1 snRNP. *Molecular Cell*, *69*, 648–663.e7.

Choi, Y. D., & Dreyfuss, G. (1984). Monoclonal antibody characterization of the C proteins of heterogeneous nuclear ribonucleoprotein complexes in vertebrate cells. *The Journal of Cell Biology*, *99*, 1997–1204.

Chu, C., et al. (2015). Systematic discovery of Xist RNA binding proteins. *Cell*, *161*, 404–416.

Clerici, M., Faini, M., Aebersold, R., & Jinek, M. (2017). Structural insights into the assembly and polyA signal recognition mechanism of the human CPSF complex. *eLife*, *6*.

Cox, J., & Mann, M. (2008). MaxQuant enables high peptide identification rates, individualized p.p.b.-range mass accuracies and proteome-wide protein quantification. *Nature Biotechnology*, *26*, 1367–1372.

Dobin, A., et al. (2013). STAR: Ultrafast universal RNA-seq aligner. *Bioinformatics*, *29*, 15–21.

Eckmann, C. R., Rammelt, C., & Wahle, E. (2011). Control of poly(A) tail length. *Wiley Interdisciplinary Reviews: RNA*, *2*, 348–361.

Engreitz, J. M., et al. (2014). RNA-RNA interactions enable specific targeting of noncoding RNAs to nascent pre-mRNAs and chromatin sites. *Cell*, *159*, 188–199.

Gunderson, S. I., Polycarpou-Schwarz, M., & Mattaj, I. W. (1998). U1 snRNP inhibits pre-mRNA polyadenylation through a direct interaction between U1 70K and poly(A) polymerase. *Molecular Cell*, *1*, 255–264.

Hernández, H., et al. (2009). Isoforms of U1-70k control subunit dynamics in the human spliceosomal U1 snRNP. *PLoS One*, *4*, e7202.

Hubner, N. C., & Mann, M. (2011). Extracting gene function from protein-protein interactions using quantitative BAC InteraCtomics (QUBIC). *Methods*, *53*, 453–459.

Kaida, D., et al. (2010). U1 snRNP protects pre-mRNAs from premature cleavage and polyadenylation. *Nature*, *468*, 664–668.

Langmead, B., & Salzberg, S. L. (2012). Fast gapped-read alignment with bowtie 2. *Nature Methods*, *9*, 357–359.

Lepennetier, G., & Catania, F. (2017). Exploring the impact of cleavage and polyadenylation factors on pre-mRNA splicing across eukaryotes. *G3: Genes, Genomes, Genetics*, 7, 2107–2114.

Lewis, J. D., Gunderson, S. I., & Mattaj, I. W. (1995). The influence of 5′ and 3′ end structures on pre-mRNA metabolism. *Journal of Cell Science. Supplement*, *19*, 13–19.

Li, H., et al. (2009). The sequence alignment/map format and SAMtools. *Bioinformatics*, *25*, 2078–2079.

Lutz, C. S., et al. (1996). Interaction between the U1 snRNP-A protein and the 160-kD subunit of cleavage-polyadenylation specificity factor increases polyadenylation efficiency in vitro. *Genes & Development*, *10*, 325–337.

Martin, G., Gruber, A. R., Keller, W., & Zavolan, M. (2012). Genome-wide analysis of pre-mRNA 3′ end processing reveals a decisive role of human cleavage factor I in the regulation of 3' UTR length. *Cell Reports*, *1*, 753–763.

Masuda, A., et al. (2020). tRIP-seq reveals repression of premature polyadenylation by co-transcriptional FUS-U1 snRNP assembly. *EMBO Reports*, *21*, e49890.

Mount, S. M., Pettersson, I., Hinterberger, M., Karmas, A., & Steitz, J. A. (1983). The U1 small nuclear RNA-protein complex selectively binds a 5′ splice site in vitro. *Cell*, *33*, 509–518.

Niranjanakumari, S., Lasda, E., Brazas, R., & Garcia-Blanco, M. A. (2002). Reversible cross-linking combined with immunoprecipitation to study RNA-protein interactions in vivo. *Methods*, *26*, 182–190.

Ntini, E., et al. (2013). Polyadenylation site-induced decay of upstream transcripts enforces promoter directionality. *Nature Structural & Molecular Biology*, *20*, 923–928.

Oh, J.-M., et al. (2017). U1 snRNP telescripting regulates a size-function-stratified human genome. *Nature Structural & Molecular Biology*, *24*, 993–999.

Oh, J.-M., et al. (2020). U1 snRNP regulates cancer cell migration and invasion in vitro. *Nature Communications*, *11*, 1.

Pomeranz Krummel, D. A., Oubridge, C., Leung, A. K. W., Li, J., & Nagai, K. (2009). Crystal structure of human spliceosomal U1 snRNP at 5.5 A resolution. *Nature*, *458*, 475–480.

Preker, P. J., Lingner, J., Minvielle-Sebastia, L., & Keller, W. (1995). The FIP1 gene encodes a component of a yeast pre-mRNA polyadenylation factor that directly interacts with poly(A) polymerase. *Cell*, *81*, 379–389.

Quinlan, A. R., & Hall, I. M. (2010). BEDTools: A flexible suite of utilities for comparing genomic features. *Bioinformatics*, *26*, 841–842.

Ramírez, F., et al. (2016). deepTools2: A next generation web server for deep-sequencing data analysis. *Nucleic Acids Research*, *44*, W160–W165.

Reyes, E. D., et al. (2017). Identifying host factors associated with DNA replicated during virus infection. *Molecular & Cellular Proteomics, 16*, 2079–2097.

Roca, X., Krainer, A. R., & Eperon, I. C. (2013). Pick one, but be quick: 5′ splice sites and the problems of too many choices. *Genes & Development, 27*, 129–144.

Rüegsegger, U., Blank, D., & Keller, W. (1998). Human pre-mRNA cleavage factor Im is related to spliceosomal SR proteins and can be reconstituted in vitro from recombinant subunits. *Molecular Cell, 1*, 243–253.

Schönemann, L., et al. (2014). Reconstitution of CPSF active in polyadenylation: Recognition of the polyadenylation signal by WDR33. *Genes & Development, 28*, 2381–2393.

Schwanhäusser, B., et al. (2011). Global quantification of mammalian gene expression control. *Nature, 473*, 337–342.

Shi, Y., & Manley, J. L. (2015). The end of the message: Multiple protein-RNA interactions define the mRNA polyadenylation site. *Genes & Development, 29*, 889–897.

So, B. R., et al. (2019). A complex of U1 snRNP with cleavage and polyadenylation factors controls telescripting, regulating mRNA transcription in human cells. *Molecular Cell, 76*, 590–599.e4.

Sun, Y., et al. (2018). Molecular basis for the recognition of the human AAUAAA polyadenylation signal. *Proceedings of the National Academy of Sciences of the United States of America, 115*, E1419–E1428.

Takagaki, Y., Manley, J. L., MacDonald, C. C., Wilusz, J., & Shenk, T. (1990). A multisubunit factor, CstF, is required for polyadenylation of mammalian pre-mRNAs. *Genes & Development, 4*, 2112–2120.

Tian, B., & Graber, J. H. (2012). Signals for pre-mRNA cleavage and polyadenylation. *Wiley Interdisciplinary Reviews: RNA, 3*, 385–396.

Tian, B., & Manley, J. L. (2017). Alternative polyadenylation of mRNA precursors. *Nature Reviews. Molecular Cell Biology, 18*, 18–30.

Uren, P. J., et al. (2012). Site identification in high-throughput RNA-protein interaction data. *Bioinformatics, 28*, 3013–3020.

Weber, G., Trowitzsch, S., Kastner, B., Lührmann, R., & Wahl, M. C. (2010). Functional organization of the Sm core in the crystal structure of human U1 snRNP. *The EMBO Journal, 29*, 4172–4184.

Workman, E., Veith, A., & Battle, D. J. (2014). U1A regulates 3' processing of the survival motor neuron mRNA. *Journal of Biological Chemistry, 289*, 3703–3712.

Yong, J., Kasim, M., Bachorik, J. L., Wan, L., & Dreyfuss, G. (2010). Gemin5 delivers snRNA precursors to the SMN complex for snRNP biogenesis. *Molecular Cell, 38*, 551–562.

Zhang, J., Kobert, K., Flouri, T., & Stamatakis, A. (2014). PEAR: A fast and accurate Illumina paired-end reAd mergeR. *Bioinformatics, 30*, 614–620.

Zhu, Y., et al. (2018). Molecular mechanisms for CFIm-mediated regulation of mRNA alternative polyadenylation. *Molecular Cell, 69*, 62–74.e4.

CHAPTER SIXTEEN

Simultaneous studies of gene expression and alternative polyadenylation in primary human immune cells

Joana Wilton[a,b,c], Michael Tellier[d], Takayuki Nojima[d,e], Angela M. Costa[f,g], Maria Jose Oliveira[f,g,h], and Alexandra Moreira[b,c,i,*]

[a]Graduate Program in Areas of Basic and Applied Biology (GABBA) PhD Program, ICBAS-Instituto de Ciências Biomédicas Abel Salazar, Universidade do Porto, Porto, Portugal
[b]Gene Regulation, i3S—Instituto de Investigação e Inovação em Saúde, Universidade do Porto, Porto, Portugal
[c]IBMC-Instituto de Biologia Molecular e Celular, Porto, Portugal
[d]Sir William Dunn School of Pathology, University of Oxford, Oxford, United Kingdom
[e]Medical Institute of Bioregulation, Kyushu University, Fukuoka, Japan
[f]Tumor and Microenvironment Interactions Group—i3S—Instituto de Investigação e Inovação em Saude, Universidade do Porto, Porto, Portugal
[g]INEB-Instituto Nacional de Engenharia Biomédica, Porto, Portugal
[h]Faculdade de Medicina, Universidade do Porto, Porto, Portugal
[i]ICBAS-Instituto de Ciências Biomédicas Abel Salazar, Universidade do Porto, Porto, Portugal
*Corresponding author: e-mail address: alexandra.moreira@i3s.up.pt

Contents

Methods in Enzymology, Volume 655
ISSN 0076-6879
https://doi.org/10.1016/bs.mie.2021.04.004

Abstract

Transcription termination in eukaryotic cells involves the recognition of polyadenylation signals (PAS) that signal the site of pre-mRNA cleavage and polyadenylation. Most eukaryotic genes contain multiple PAS that are used by alternative polyadenylation (APA), a co-transcriptional process that increases transcriptomic diversity and modulates the fate of the mRNA and protein produced. However, current tools to pinpoint the relationship between mRNAs in different subcellular fractions and the gene expression outcome are lacking, particularly in primary human immune cells, which, due to their nature, are challenging to study. Here, we describe an integrative approach using subcellular fractionation and RNA isolation, chromatin-bound and nucleoplasmic RNA-Sequencing, 3′ RNA-Sequencing and bioinformatics, to identify accurate APA mRNA isoforms and to quantify gene expression in primary human macrophages. Our protocol includes macrophage differentiation and polarization, co-culture with cancer cells, and gene silencing by siRNA. This method allows the simultaneous identification of macrophage APA mRNA isoforms integrated with the characterization of nuclear APA events, the identification of the molecular mechanisms involved, as well as the gene expression alterations caused by the cancer-macrophage crosstalk. With this methodology we identified macrophage APA mRNA signatures driven by the cancer cells that alter the macrophage inflammatory and transcriptomic profiles, with consequences for macrophage physiology and tumor evasion.

1. Introduction

All human protein-coding genes with the exception of histones contain polyadenylation signals (PAS), pointing the 3′ end of the pre-mRNA and where a polyadenylation tail is added for correct mRNA 3′ end formation. Seventy-four percent of all human genes contain more than one PAS (Derti, Garrett-Engele, MacIsaac, et al., 2012; Lee et al., 2018; Shi, 2012; Tian, Hu, Zhang, & Lutz, 2005; Wang, Sandberg, Luo, et al., 2008) that may be used by alternative polyadenylation (APA) increasing transcriptomic diversity. APA produces several mRNA isoforms transcribed from the one single gene that may have different mRNA stabilities, result in diverse protein expression levels, distinct mRNA and protein subcellular locations. This is due to the regulatory properties of the alternative 3′ UTRs conferred by the differential binding of RNA Binding Proteins (RBPs) and miRNAs (Berkovits & Mayr, 2015; Ma & Mayr, 2018; Mitra, Johnson, & Coller, 2016; Wurth & Gebauer, 2015).

The total estimated number of highly conserved PAS in the human genome exceeds 569,000 (Herrmann et al., 2020), and recent advances in the transcriptomic field highlighted the contribution of APA to the establishment of gene- and cell-specific programs such as cell activation and differentiation (Castello, Fischer, Eichelbaum, et al., 2012; Mayr, 2017; Pai, Baharian, Pagé Sabourin, et al., 2016; Sandberg, Neilson, Sarma, Sharp, & Burge, 2008; Tian & Manley, 2016). In particular, there are genome-wide trends consisting of proximal PAS selection in cancer cells and activated immune cells (Jia, Yuan, Wang, et al., 2017; Mayr & Bartel, 2009; Pai et al., 2016; Sandberg et al., 2008), and distal PAS selection in more differentiated cells such as neurons and during organism development (Braz, Cruz, Lobo, et al., 2017; Ji, Lee, Pan, Jiang, & Tian, 2009). These APA profiles have been established by powerful transcriptomic methodologies, including dedicated RNA-sequencing and bioinformatic protocols.

Cleavage and polyadenylation are mostly co-transcriptional events (Dye & Proudfoot, 1999) occurring in the nucleus, but the expression of APA mRNA isoforms containing alternative 3′ UTRs is also modulated by microRNAs, which occurs in the cytoplasm. Thus, it is often difficult to identify APA mRNA isoforms correctly by simple RNA-Sequencing methodologies without the use of subcellular fractionation procedures or bioinformatic algorithms. Nevertheless, it is possible to accurately pinpoint APA mRNA isoforms as they are transcribed by RNAPII by dedicated RNA sequencing methodologies such as chromatin RNA-Seq (Nojima, Gomes, Carmo-Fonseca, & Proudfoot, 2016) combined with 3′ mRNA-Sequencing, allowing to quantify and distinguish mRNA isoform levels produced by APA or other post-transcriptional events.

The use of primary human immune cells in transcriptomic studies has been hampered by the plastic nature of those cells and the different protocols for differentiation and polarization in use that may result in diverse transcriptomic data. Macrophages are plastic immune cells that depending on their inflammatory profile, function as phagocytes to infectious pathogens (named M1, or pro-inflammatory) and in tissue healing (named M2, or anti-inflammatory). They are also abundant cells of the tumor microenvironment where they may either cooperate with or abrogate the progression of cancer (Locati, Mantovani, & Sica, 2013; Mantovani & Locati, 2013).

Here we describe an integrative approach using chromatin bound and nucleoplasm RNA-Seq, 3′ mRNA-Seq and bioinformatics, to understand how cancer cells affect the transcriptome of primary human macrophages. We describe a robust protocol for macrophage differentiation and polarization and how to co-culture M1 polarized macrophages with CRC cells in a transwell chamber. This model system allows a proper communication between the two different cell types while permits an efficient separation of the two cellular populations for the transcriptomic studies. Comparing the RNA profiles and data obtained by the different RNA-Seq methods using RNA isolated from the same donor allow us to ascertain whether changes in gene expression are due to APA, or if they are due to post-transcriptional regulation. We also describe a successful method for siRNA transfection in difficult to transfect primary human macrophages. Functional studies in physiologically-relevant macrophage and cancer cellular populations can then be easily be performed to explore the molecular mechanisms that regulate gene expression and APA.

2. Overview of the method

Here, we describe in detail how to perform transcriptomic studies in primary human macrophages in co-culture with CRC cells, which can be extended to other immune cells or organotypic cultures. An overview and workflow of the full protocol is presented on Figs. 1A, 2 and 3A. This strategy allows the analyses of steady-state, chromatin-bound (ChrRNA-Seq) and nucleoplasmic RNA (NpRNA-seq), as well as 3′ RNA-Seq by Quant-Seq, and the bioinformatic processing needed to obtain simultaneous differential gene and mRNA isoform expression data.

The cell culture protocol lasts for 9 days, starting with primary human monocyte isolation from healthy blood donors, and macrophage differentiation and polarization into M1 macrophages which are subsequently co-cultured with CRC cells in a transwell chamber (workflow in Fig. 1A). Subsequently, total RNA is extracted from total, nuclear or cytoplasmic fractions, libraries are constructed and chromatin-bound RNA-Seq and 3′ RNA-Seq is performed and the data is analyzed bioinformatically (Fig. 3).

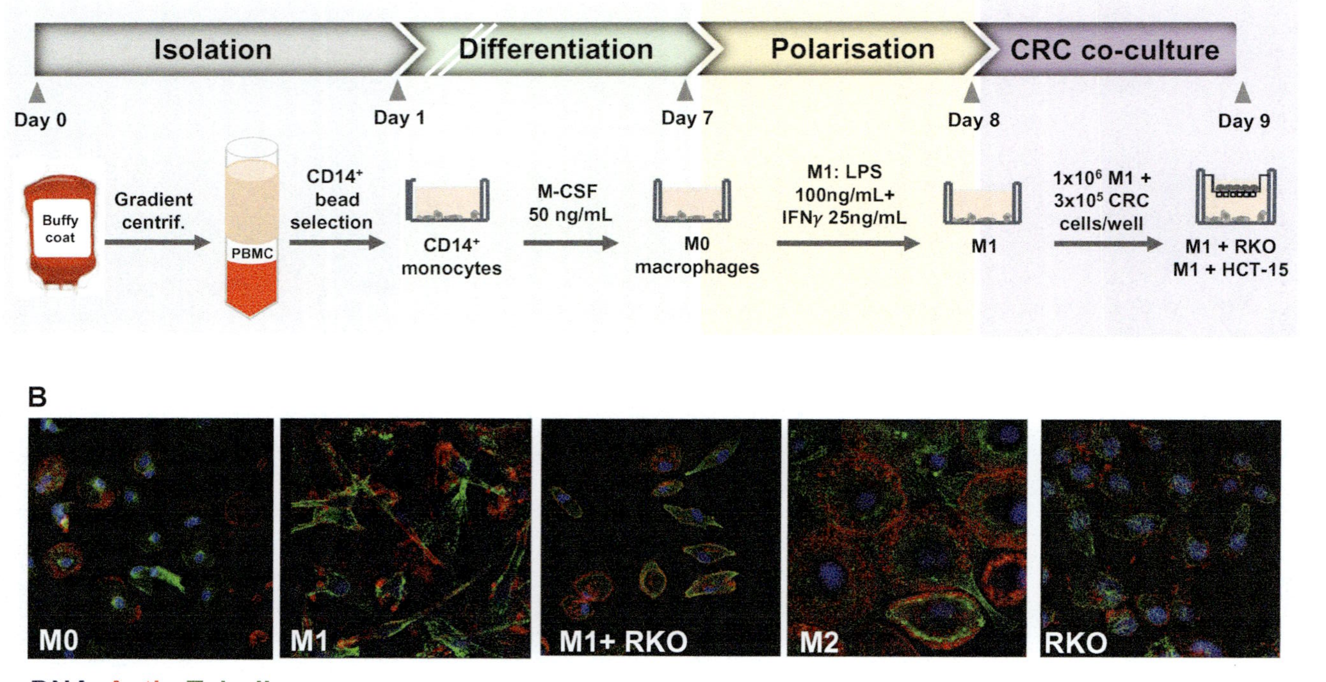

Fig. 1 (A) Workflow and experimental setup: buffy coats are subject to gradient centrifugation to obtain PBMCs, which are then exposed to $CD14^+$ beads to select $CD14^+$ monocytes. Monocytes are differentiated for 7 days in presence of M-CSF, obtaining M0 macrophages which are then polarized to M1 or M2 macrophages for 24h. M1 macrophages are put into contact with CRC cell lines RKO or HCT-15 for 24h. (B) Confocal microscopy images of unpolarized macrophages (M0) showing actin-tubulin cytoskeleton of pro-inflammatory macrophages (M1), M1 macrophages after RKO co-culture (M1 + RKO), anti-inflammatory macrophages (M2) and RKO cells for reference (RKO). Actin shown in red, tubulin shown in green, DNA shown in blue.

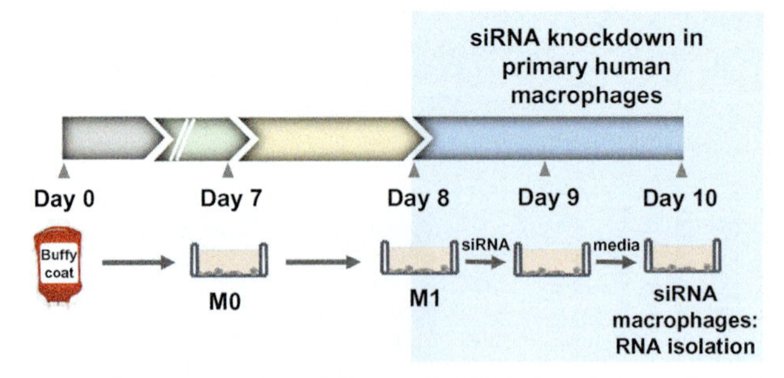

Fig. 2 Experimental setup and workflow of siRNA knockdown in primary human macrophages.

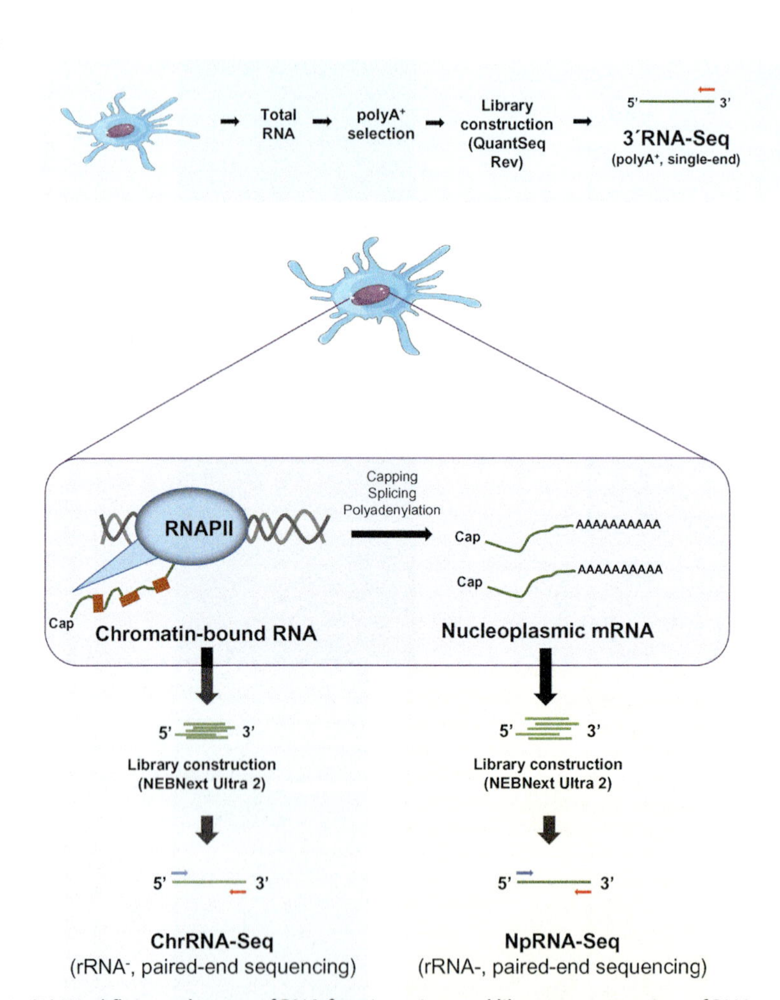

Fig. 3 (A) Workflow and setup of RNA fractionation and library construction of RNA samples from primary human macrophages. Top, conversion of total RNA into single-end 3′ mRNA-Seq libraries. Left, chromatin-bound RNA is converted into rRNA-depleted paired-end RNA-Seq libraries. Right, nucleoplasmic RNA is converted into rRNA-depleted paired-end RNA-Seq libraries. (B) Scheme of chromatin-bound and nucleoplasmic RNA fractionation. *Panel B adapted from Nojima T., Gomes T., Grosso A. R. F, et al. (2015) Mammalian NET-seq reveals genome-wide nascent transcription coupled to RNA processing.* Cell, 161*(3), 526–540. doi:10.1016/j.cell.2015.03.027.

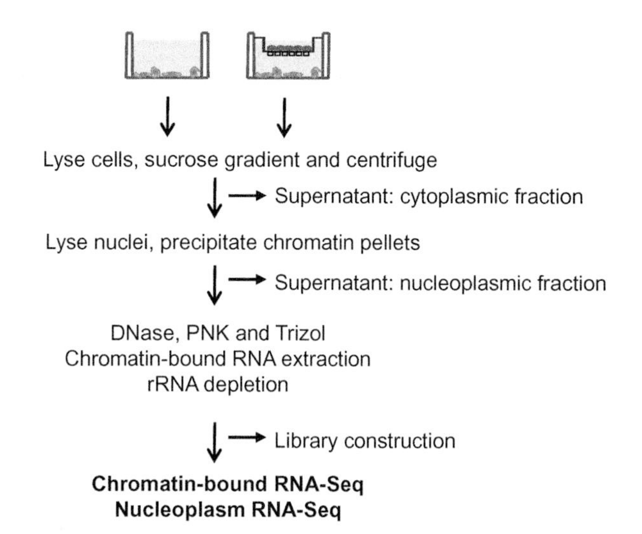

Fig.3—Cont'd

3. Detailed protocol

3.1 Before you begin

Timing: 4h

1. When manipulating RNA, work in an RNase-free workstation: clean pipettes, benches, tip boxes, centrifuge, thermocycler, thermomixer, etc., with RNase Xterminator. Preventive measures such as not sharing reagents to avoid RNA degradation and cross-contamination are mandatory
2. Keep macrophage cell number consistent between conditions and biological replicates to isolate similar amounts of RNA
3. Maintain colorectal cancer (CRC) cell lines and primary human macrophages in a humified atmosphere at 37°C and 5% CO_2, in complete RPMI medium with 10% FBS and 1× penicillin/streptomycin. When polarized to pro-inflammatory conditions, macrophages are kept for 24h in RPMI medium with 2% FBS.
4. Prepare cell culture solutions and buffers:

(a) CD14 beads buffer to isolate $CD14^+$ monocytes: PBS pH 7.2/5% FBS/2 mM EDTA. For each PBMC sample, a total of 15 mL are needed
(b) Red blood cell lysis buffer: 0.15 M NH_4Cl/0.01 M $NaHCO_3$/ 1.3 mM EDTA. For each PBMC sample, a total of 5 mL are needed
(c) RPMI Complete: RPMI/10% FBS/1 × penicillin/streptomycin
(d) RPMI polarization media: RPMI/2% FBS

5. Prepare subcellular fractionation solutions for chromatin-bound, nucleoplasm and cytoplasmic RNA isolation
Critical: Always use RNase-free water. For the buffers used in RNA isolation add protease inhibitors (cOmplete Mini), phosphatase winhibitors (PhosStop) and RNase inhibitors (Ribolock) just before use, according to manufacturer's instructions. For Ribolock, use a final concentration of 2 U/μL.
(a) HLB/NP40 lysis buffer: 10 mM Tris-HCl pH 7.5/10 mM NaCl/ 2.5 mM $MgCl_2$/0.5% (v/v) NP-40. For each macrophage sample, 4 mL of HLB/NP40 lysis buffer are used, therefore we recommend to prepare 30 mL, adding in the following order: 28.02 mL RNase-free H_2O + 300 μL 1 M Tris-HCl pH 7.5 + 30 μL 5 M NaCl + 75 μL 1 M $MgCl_2$ + 1.5 mL 10% NP-40
(b) HLB/NP40/Sucrose buffer: 10 mM Tris-HCl pH 7.5/10 mM NaCl/2.5 mM MgC12/0.5% (v/v) NP-40/10% (w/v) sucrose. For each macrophage sample, 1 mL of HLB/NP40/Sucrose buffer is used, therefore we recommend to prepare 10 mL, adding in the following order: 7.355 mL RNase-free H_2O + 100 μL 1 M Tris-HCl pH 7.5 + 20 μL 5 M NaCl + 25 μL 1 M $MgCl_2$ + 0.5 mL 10% NP-40 + 2 mL 50% (w/v) sucrose
(c) NUN1 buffer: 20 mM Tris-HCl pH 7.9/75 mM NaCl/0.5 mM EDTA/50% (v/v) glycerol. For each macrophage sample, 125 μL of NUN1 buffer are used, therefore we recommend to prepare 10 mL, adding in the following order: 4.625 mL RNase-free H_2O + 200 μL 1 M Tris-HCl pH 7.9 + 150 μL 5 M NaCl + 25 μL 0.2 M EDTA + 5 mL 100% (v/v) glycerol. Add protease and phosphatase inhibitors and Ribolock
(d) NUN2 buffer: 20 mM HEPES-KOH pH 7.6/300 mM NaCl/ 0.2 mM EDTA/7.5 mM $MgCl_2$/1% (v/v) NP-40/1 M urea. For each macrophage sample, 1.2 mL of NUN2 buffer is used, therefore we recommend to prepare 30 mL, adding in the following order: 24.345 mL RNase-free H_2O + 0.6 mL 1 M HEPES-KOH

pH 7.6 + 1.8 mL 5 M NaCl + 30 μL 0.2 M EDTA + 225 μL 1 M $MgCl_2$ + 3 mL 10% (v/v) NP-40 + 1.8 g urea. Add protease and phosphatase inhibitors and Ribolock

(e) HSB buffer: 10 mM Tris-HCl pH 7.5/500 mM NaCl/10 mM $MgCl_2$/0.25 U/μL TURBO DNase. For each macrophage sample, 50 μL of HSB buffer is used, therefore we recommend to prepare 500 μL, adding in the following order: 382.5 μL RNase-free H_2O + 5 μL 1 M Tris-HCl pH 7.5 + 50 μL 5 M NaCl + 62.5 μL 2 U/μL TURBO DNase

6. Before starting the library construction

(a) Always centrifuge microtubes before and after thermocycler incubation steps

(b) Prepare mastermixes of reagents for library construction (steps 50–55 and 63–68) and include an extra 10% of volume surplus per reaction

(c) Thermocycler lid must be maintained at 99 °C except where indicated otherwise

(d) Resuspend and equilibrate magnetic beads at RT 30 min before using. They must have a uniformly brown color, without clumps

4. Key resources table

Reagent or resource	Source	Identifier
Antibodies		
AlexaFluor488 conjugated secondary antibody	Life Technologies	Cat # A32723
DAPI-Vectashield	Vector Laboratories	Cat # H1200-10
Monoclonal anti-α-tubulin antibody	Sigma-Aldrich	Cat # T9026
AlexaFluor 647 Phalloidin	Life Technologies	Cat # A22287
Biological Samples		
Buffy coats from anonymized blood donors	Centro Hospitalar Universitário *São João, Porto, Portugal*	–

Continued

—cont'd

Reagent or resource	Source	Identifier
Chemicals, Peptides, and Recombinant Proteins		
CD14 Microbeads	Miltenyi Biotec	Cat # 130-050-201
Chloroform	Merck	Cat # R49352845737
CompleteMINI protease inhibitors	Sigma	Cat # 11836153001
Recombinant DNase I	Roche	Cat # EN0521
GenMute transfection reagent for Primary Human Macrophages	SignaGen	Cat # SL100568-PMG
Interferon-gamma (IFNγ)	Peprotech	Cat # 300-02
Interleukin-4 (rhIL-4)	Peprotech	Cat # 200-04
Interleukin-10 (rhIL-10)	Peprotech	Cat # 200-10
Lipopolysaccharide (LPS from *Escherichia coli* subtype O55:B5)	Sigma-Aldrich	Cat # L4005
Lymphoprep	Axis-Shield	Cat # 1114547
Macrophage Colony Stimulating Factor (M-CSF)	Peprotech	Cat # 300-25
NP-40 Igepal	Sigma-Aldrich	Cat # 18896
SuperScript IV reverse transcriptase	ThermoFisher Scientific	Cat # 18090050
Sybr Green I for qPCR detection	ThermoFisher	Cat # S7563
Trizol reagent	Ambion	Cat # 15596018
TURBO DNase 0.25 U/μL	ThermoFisher Scientific	Cat # AM2238
Critical Commercial Assays		
AMPure XP beads	Beckman Coulter	Cat # A63880
DynaBeads mRNA Purification kit	Invitrogen	Cat # 61006
High Sensitivity DNA kit	Agilent	Cat # 5067-4626, 5067-4627
KAPA Library Quantification kit	Roche	Cat # 07960140001

—cont'd

Reagent or resource	Source	Identifier
NextSeq High-Output kit for 75 cycles	Illumina	Cat # FC-404-2005
NEBNext Ultra II directional library prep kit for Illumina	New England Biolabs	Cat # E7760
Qubit RNA HS Assay Kit	ThermoFisher	Cat # Q32852
QuantSeq REV kit	Lexogen	Cat # 016.24
QuantSeq 3′ mRNA-Seq PCR Add-on kit for Illumina	Lexogen	Cat # 020.96
Ribo-Zero Gold rRNA removal kit for Human/Mouse/Rat	Illumina	Cat # MRZG12324
Experimental Models: Cell Lines		
HCT-15 cell line	ATCC	CCL-225
RKO cell line	ATCC	CRL-2577
Oligonucleotides		
Non-targeting siRNA control (ONTARGETplus SMARTpool siNTC)	Dharmacon	Cat # D-001810-10-05: 5′-UGGUUUACAUGUCG ACUAA-3′ 5′-UGGUUUACAUGU UGUGUGA-3′ 5′-UGGUUUACAUGUUU UCUGA 3′ 5′-UGGUUUACAUGU UUUCCUA-3′

Software and Algorithms

Name	Source
7500 v2.0.6 software	https://www.thermofisher.com/pt/en/home/technical-resources/software-downloads/applied-biosystems-7500-real-time-pcr-system.html
Bedtools	https://bedtools.readthedocs.io/en/latest/
Bowtie	http://bowtie-bio.sourceforge.net/bowtie2/index.shtml

Continued

—cont'd
Software and Algorithms

Name	Source
Cutadapt	https://cutadapt.readthedocs.org/en/stable
DESeq2	https://bioconductor.org/packages/release/bioc/html/DESeq2.html
DEXSeq	https://bioconductor.org/packages/devel/bioc/vignettes/DEXSeq/inst/doc/DEXSeq.html
FastQC	http://www.bioinformatics.babraham.ac.uk/projects/fastqc
FIJI package for ImageJ	https://fiji.sc/
Galaxy	https://usegalaxy.org/
Human genome release hg38	https://www.gencodegenes.org/human/
GraphPad Prism Software version 7	https://www.graphpad.com/scientific-software/prism/
htseq-count	https://htseq.readthedocs.io/en/master/
ImageLab Software (Bio-Rad)	https://www.bio-rad.com/en-pt/product/image-lab-software?ID=KRE6P5E8Z
Python v. 2.7	https://www.python.org/
RStudio	https://rstudio.com/products/rstudio/download/
SAMtools	http://samtools.sourceforge.net/
STAR aligner	https://github.com/alexdobin/STAR
TopHat	http://ccb.jhu.edu/software/tophat/index.shtml

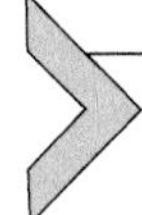

5. Materials, reagents and equipment

5.1 Materials

- 6 multi-well plates made of plasma-treated tissue culture polystyrene (TCPS) (Ref. 657,160, Greiner Bio-One)
- 6 multi-well plate permeable transwell insert (Corning, Ref 353,102) with 1.0 μm pore size

- 24 multi-well plates (Ref. 4,430,400, Frilabo) for immunofluorescence
- LS column (Ref. 130-042-401, Miltenyi Biotec)
- Nuclease-free microcentrifuge tubes, 1.5 mL (Ref. 616,201, Greiner bio-one)
- Nuclease-free PCR tubes 0.2 mL (72,991,002, Sarstedt)
- Nuclease-free 15 mL Falcon tubes (Ref. 430,055, Corning)
- Nuclease-free 50 mL Falcon tubes (Ref. 430,828, Corning)
- T-flasks for cell culture (Corning)

5.2 Alternative materials and reagents

- Commercial alternatives to RNase Xterminator may be used, or it may be substituted by 3% H_2O_2 and 0.5% SDS, followed by 70% ethanol
- Equivalent equipment and consumables may be used, except for TCPS plastic cell culture plates for macrophage cultures and co-cultures with CRC cells

5.3 Equipment

- 48-well Thermocycler (Biometra)
- 7500 Fast Real-Time PCR System (Applied Biosystems)
- Bioanalyzer 2100 (G2939BA, Agilent)
- Cell culture CO_2 incubator (CO_2 at 5%, humidified, 37 °C, Binder)
- Cell culture laminar flow hood (Thermo MSC ADV 1.2)
- Centrifuges (Eppendorf 5415R and 5810R)
- DynaMag2 magnetic rack for microcentrifuge tubes (Ref. 12321D, ThermoFisher)
- Inverted light microscope (Zeiss Axiovert 25)
- Minispinner (Elmi DOS-20L)
- Nanodrop ND-1000 (Thermo Scientific)
- Qubit fluorometer (ThermoFisher, cat. no. Q33216)
- TCS-SP5 laser confocal microscope (Leica Microsystems, Germany)
- Thermomixer (Eppendorf)
- Vortex (ZX3, VELP Scientifica)

6. Step-by-step method details

6.1 Preparation of primary human macrophages for co-cultures/transfection

See Fig. 1A for workflow.

6.1.1 Isolation of PBMCs from human peripheral blood and $CD14^{+}$ enrichment (adapted from Ohradanova-Repic, Machacek, Fischer, & Stockinger, 2016)

Timing: 8 h (day 1).

Critical

- As this method is designed for primary human macrophages, special attention should be taken to design your experiment to use PBMCs from as many donors (3–4) as possible in each experiment
- Work as quickly as possible at RT to avoid macrophage cell death
- Cell culture plastic materials as well as all the reagents should be selected from the Key Resources Table and the Materials and Methods section. In particular, we recommend TCPS 6-multiwell plates to increase macrophage adhesion and prevent macrophage polarization (Lerman, Lembong, Muramoto, Gillen, & Fisher, 2018; Rostam, Singh, Salazar, et al., 2016)
- All reagents must be sterile; on steps 1–2, reagents must be at room temperature (RT).

1. Separate the PBMC ring from the other buffy coat sections by density gradient centrifugation using Lymphoprep Ficoll reagent
2. Red Blood Cell (RBC) removal and PBMC counting:
   (a) Add 5 mL RBC lysis buffer to PBMCs. Incubate at 37 °C for 5 min, add 25 mL PBS and centrifuge at 300 g for 10 min at 4 °C
   (b) Resuspend pellet in RPMI medium (10% FBS, 1 × streptomycin/penicillin) and count cells, using the trypan blue exclusion test
   (c) Determine cell number and centrifuge the appropriate number of cells at 300 g for 10 min, at 4 °C

Critical

- From this point on, the protocol should be performed as quickly as possible until cells are plated, keeping the cells at 4 °C whenever possible to prevent cell mortality.
- PBMCs contain 10–20% macrophage content, therefore you should multiply the amount of macrophages you need by 5 to determine the amount of PBMCs to be centrifuged in this step.
   (a) $CD14^{+}$ monocytes were purified from PBMCs by magnetic-activated cell sorting using CD14 microbeads as follows: resuspend pellet in 47.5 μL $CD14^{+}$ beads buffer and 5 μL CD14 microbeads per each 10×10^{6} cells. Mix and incubate for 20 min at 4 °C, mixing every 5 min

   - **(b)** Wash cells in 2–3 mL of CD14 beads buffer and centrifuge at 300 g for 10 min at 4 °C
   - **(c)** Remove the supernatant completely and resuspend in 500 μL $CD14^+$ beads buffer, for a maximum of 100×10^6 PBMCs
     Critical: ensure that the PBMC suspension is clump-free.

3. $CD14^+$ positive selection (adapted from Ohradanova-Repic et al., 2016):
   - **(a)** In the flow hood, equilibrate each column by adding 500 μL $CD14^+$ beads buffer
   - **(b)** For each donor used, load PBMC suspension on the column. Add 500 μL in the Falcon tube containing the remaining PBMCs, resuspend and load in the column
   - **(c)** Wash the column three times with 3 mL $CD14^+$ beads buffer

4. $CD14^+$ cell elution:
   - **(a)** Load column with 5 mL $CD14^+$ beads buffer and immediately insert the plunger, eluting the $CD14^+$ cells into the Falcon tube
   - **(b)** Centrifuge cells at 300 g for 10 min at 4 °C, resuspend in complete RPMI
   - **(c)** Count cells using the trypan blue exclusion test

6.1.2 Macrophage differentiation and M1 polarization

Timing: 2 h + 6 days incubation (days 1–7)

Critical

- Plate cells in a cell density of 1×10^6/well (for co-culture with CRC cells) or 1.5×10^6/well (for transfection).
- If performing immunofluorescence, seed macrophages on 6-wells containing 4 glass coverslips each

1. $CD14^+$ cell plating:
   - **(a)** Resuspend $CD14^+$ cells in complete RPMI medium
   - **(b)** Add recombinant human macrophage colony stimulating factor (M-CSF) to a final concentration of 50 ng/mL, to improve differentiation and adhesion
   - **(c)** Seed cells into 6-well plates, 2 mL in each well
   - **(d)** Incubate in a CO_2 incubator at 37 °C for 7 days
   - **(e)** Add M-CSF-containing RPMI at day 4
     Critical: On day 7, check if macrophages have adhered to the well surface. If so, macrophages may be efficiently polarized. If not, wait until day 7.

Note: This differentiation protocol yields ~93% $CD14^+$ cells (Ohradanova-Repic et al., 2016).

2. Macrophage polarization:
   (a) Take out media and slowly add 2mL polarization medium (RPMI/2% FBS, no antibiotics) containing polarization agents: M1—lipopolysaccharide (LPS) 100ng/mL and interferon-gamma (IFNγ) 25ng/mL. Incubate for 24h at 37°C
   (b) Keep unpolarized macrophages as controls, adding the same volume of RPMI polarization media without polarization agents

Critical

- Add the polarization media drop by drop, to avoid macrophage detachment
- LPS and IFNγ concentrations are critical to proper macrophage polarization, as different concentration combinations will give rise to distinct transcriptional profiles
- The lack of antibiotics in the media is purposefully done to avoid immunomodulatory effects that could hinder proper polarization (Brooks, Hart, & Coleman, 2005)
- The decreased amount of FBS is used to increase attachment and phagocytosis, and lower local TGFβ concentrations, which diverts macrophage polarization toward anti-inflammatory conditions (Oida & Weiner, 2010; Suganuma, Fahey, Bryan, Healy, & Talalay, 2011; Zhang, Wang, Wang, et al., 2016)

Note: If desired, one can also polarize macrophages toward anti-inflammatory conditions (M2) through addition of RPMI 2% FBS without antibiotics containing 10ng/mL recombinant human IL-10 (Ref. 200-10, Peprotech) or 20ng/L IL-4 (Ref. 200-04, Peprotech).

6.1.3 Macrophage and colorectal cancer co-cultures

Timing: 3+24h incubation (days 7–8)

1. Prepare cell lines for co-culture:
   (a) Trypsinize RKO and HCT-15 cells, resuspend separately in 5mL RPMI and count cells via trypan blue exclusion test
   (b) Resuspend in a cell density of 3.75×10^5 cells/mL

Critical

- Ensure initial cell confluency is equal in both cell lines
- For a typical sequencing run, using 6×10^6 macrophages and 3 biological replicates, we recommend to use two T75 flasks with at least 80% confluency of CRC cell line, to obtain sufficient cells for co-culture

Note: We used RKO and HCT-15 CRC cell lines for co-culture with macrophages. Optimize co-culture conditions by titrating cell confluency in the transwells, using different timepoints and initial seeding cell densities.

2. Prepare macrophages for co-culture:
   a. Remove macrophage polarization media from wells
   b. Add 1.2 mL complete RPMI per 6-multiwell that will have co-cultured CRC cells, and 2 mL complete RPMI in control wells
      Note: As macrophage polarization has been successfully achieved, macrophages can be exposed again to antibiotic-containing media without detrimental polarization effects.
3. Perform co-cultures:
   a. Carefully add co-culture 1.0 μm inserts to each respective well
      Critical: The porosity level of the co-culture insert filter stops cancer cells from crossing to the lower compartment while still allowing soluble factor exchange between the upper and lower compartment cell populations.
   b. Add 800 μL RKO or HCT-15 cells in the insert, making sure that its surface is fully covered
   c. Incubate co-cultures for 24 h

6.1.4 Macrophage and CRC immunofluorescence

Timing: 6 h (staining) + ~2 h (imaging) (day 9)

1. Transfer glass coverslips containing macrophages to 24-well plates after mono- or co-culture
2. Prepare the samples at room temperature:
   a. Fix macrophages by adding 4% PFA for 10 min
      Pause point: at this point, PFA-fixed cells may be kept at 4 °C protected from light for a maximum of 3 weeks.
   b. Wash cells with PBS 3 × 5 min each, under gentle agitation
   c. Quench cells by adding 50 mM NH_4Cl for 10 min
   d. Repeat step (b).
   e. Permeabilize cells by adding 0.2% Triton X-100 for 10 min
   f. Repeat step (b).
   g. Block cells by adding 5% BSA for 1 h
3. Sample incubation, protected from light:
   a. First incubate with monoclonal alpha-tubulin antibody (1:4000 dilution) for 1 h
   b. Secondly, incubate with AlexaFluor 488 conjugated secondary antibody (1:1000 dilution) for 45 min

  c. Lastly, stain actin by incubating with AlexaFluor 647 Phalloidin (1:40 dilution) for 20 min

4. Mount coverslips using DAPI-Vectashield and sealing with polish
Pause point: coverslips may be kept at −20 °C for a maximum of 6 weeks, away from light.

5. Image cells on a confocal microscope (e.g., Leica TCS-SP5), using the 40 × oil objective

6. Perform image analysis on appropriate software (e.g., FIJI)

Fig. 1B shows macrophage and CRC cell images obtained through this protocol.

6.1.5 Macrophage siRNA transfection

Timing: 2 + 48 h incubation.

Fig. 2 shows the workflow for macrophage siRNA transfection.

1. Take out co-culture or polarization media and add 1 mL complete RPMI media 30 min to 1 h before transfection.

2. Prepare GenMute Transfection Buffer working solution.
Note: This 1 x working solution is shelf-stable for 24 months, but we recommend to prepare it fresh with RNase-free water before every use, keeping it at RT until needed.

3. For transfection of 6-wells, dilute 50 pmol siRNA in 100 µL GenMute Transfection Buffer working solution. Add 4 µL GenMute Transfection Reagent and incubate for 15 min at RT.
   - Never incubate the transfection complex containing the transfection reagent and the siRNA for longer than 30 min, as transfection complexes increase in size proportionally to the incubation time. If the transfection complex incubation is too long, transfection efficiency will decrease.

4. Add the transfection mix to the cells drop-by-drop, on the center of the well.

5. Gently rock the 6-multiwell plate back and forth, and incubate at 37 °C overnight.

Critical

- Macrophages must be plated at a cell density of 1.5×10^6/well on Day 1, to obtain at least 50% confluency on transfection day.
- Primary human macrophages are notoriously hard to transfect, due to their inherent ability to detect non-self molecules. Therefore, we recommend using GenMute.

6. Take out the transfection media and add complete RPMI to macrophages and incubate for 48 h.
7. Isolate RNA (Section 6.2).

Note: Knockdown levels can be assessed by RT-qPCR.

6.2 RNA isolation

RNA is extracted from macrophages after 24 h of co-culture and monoculture, and after transfection with siRNA.

Critical

- In order to obtain intact RNA samples, keep the bench, all materials and reagents RNase-free by wiping with RNase Xterminator or similar
- Similarly, avoid RNases by changing gloves frequently during the protocol
- Labcoats should also be frequently changed
- It is also useful to create a dedicated RNA-only work space to reduce cross-contamination
- We recommend to use commercially available RNase-free water, and to avoid DEPC-treated water due to its reactivity to extracted RNA and incompatibility with many of the solutions used for RNA subcellular fractionation (e.g., Tris).

6.2.1 Total RNA extraction

Timing: 4 + 16 h incubation at −80 °C.

Total RNA extraction for 3′-mRNA-Seq is shown on the upper panel of Fig. 3A.

1. Cell lysis
   - **(a)** Remove RPMI from the macrophage culture
   - **(b)** Wash with ice-cold PBS
   - **(c)** Add 250 μL Trizol per 6-multiwell and pipet up and down to homogenize

     Note: Trizol contains acidic phenol to separate RNA from the remaining fractions, and guanidinium thiocyanate to inactivate endogenous RNases.
   - **a.** Transfer to RNase-free microcentrifuge tube
   - **b.** Incubate samples at RT for 2 min to dissociate the nucleoprotein complex
   - **c.** Add 200 μL of chloroform per each mL Trizol reagent
   - **d.** Mix them thoroughly by inverting the tube
   - **e.** Incubate at RT for 1 min

f. Centrifuge at 4 °C for 15 min at 12000 g
g. Transfer the upper aqueous phase to a new 1.5 mL RNase-free microcentrifuge tube

Critical: avoid taking out the organic layer and lower phase containing contaminants (phenol from Trizol, DNA, and protein).

2. RNA precipitation
   h. Add an equal volume of isopropanol to the samples. Add 1 μL of glycogen to act as co-precipitant
   i. Vortex briefly
   j. Incubate at −80 °C for 16 h, to maximize RNA yield
   k. Centrifuge at 4 °C for 20 min at 12000 g
   l. Remove the supernatant carefully
   m. Wash with 1 mL of 70% Ethanol, flick the tube to release the pellet from the bottom of the tube
   n. Centrifuge at 4 °C for 5 min at 7500 g
   o. Discard the supernatant and air dry the RNA pellet

Critical
- After isopropanol precipitation, macrophage RNA needs to be incubated at −80 °C for a minimum of 4 h, but the optimum time is 16 h
- Avoid to dry the RNA completely to facilitate resuspension

Notes
- RNA can optionally be resuspended at 55 °C for 5 min before quantification
- RNase inhibitors (e.g., Ribolock) can be added at a 1:20 dilution to prevent RNA degradation

3. Quantify RNA concentration with a fluorescent dye-based method such as the Qubit RNA HS Assay Kit. Briefly, dilute 1 μL resuspended RNA in a solution containing RNA HS reagent in a 1:200 dilution with Qubit RNA HS Buffer, incubate at RT for 2 min and quantify the RNA in a Qubit fluorometer. Calculate the RNA concentration taking into account the previously made dilution.

Pause point: RNA can be kept at −80 °C for 2 weeks before proceeding with library construction; if samples are kept any further, we recommend to check RNA quality on a electrophoretic chip such as RNA 6000 Pico Kit (Agilent), or to run an RNase-free 0.8% agarose gel to confirm RNA integrity, visualizing the gel on a UV trans illuminator such as GelDoc or similar imaging system. Total RNA should appear as two sharp rRNA bands (28S and 18S) in a 2:1 ratio, while degraded RNA should appear as a smear.

6.2.2 RNA extraction from nuclei and cytoplasm fractions

Timing: 8 + 16 h precipitation time.

The protocol is an adaptation of Nojima, Gomes, Grosso, et al. (2015) (Fig. 3A and B).

Critical

- For one RNA-sequencing experiment, 6×10^6 macrophage cells were used as input material, at a cell density of 1×10^6/well
- Prepare HLB/NP40, HLB/NP40/sucrose, NUN1, NUN2, and HSB buffers beforehand, adding protease, phosphatase and RNase inhibitors just before use as shown in the initial Before you Begin section
- If possible, work in the cold room on the Nuclei and Cytoplasm Fractionation and Chromatin and Nucleoplasm Fractionation steps
- Continuously renew the ice inside the icebox to ensure RNA samples are kept at 4 °C at all times

1. Wash cells in 2 mL of ice-cold PBS twice, keeping the multiwell chambers on ice
   Critical: ensure PBS is added slowly as to avoid removing or disrupting macrophages.
2. Lyse cells in 4 mL of HLB/NP40 buffer and incubate on ice for 5 min
3. Transfer to a 15 mL Falcon tube
4. Underlay slowly 1 mL ice-cold HLB/NP40/sucrose buffer and centrifuge at 300 g for 5 min at 4 °C
5. Collect the pellet
   Note: The supernatant can be kept at −80 °C for up to a month if cytoplasmic fraction needs to be analyzed.

6.2.3 RNA extraction from chromatin and nucleoplasm fractions

1. Resuspend completely the isolated nuclei with 125 μL NUN1 buffer
2. Add ice-cold 1.2 mL NUN2 buffer
3. Mix thoroughly by vortex with max speed
4. Note: NUN2 buffer includes 1 M urea which improves chromatin isolation.
5. Incubate on ice for 15 min
6. Mix thoroughly by vortex with max speed every 5 min
7. Centrifuge at 15680 g for 10 min at 4 °C
8. Keep both fractions: the supernatant is nucleoplasmic RNA, the pellet is chromatin-bound RNA
9. Isolate chromatin-bound RNA:
   a. Resuspend the chromatin pellet in 50 μL HSB (containing DNase)

b. Incubate at 37 °C for 10 min, 1400 rpm on a Thermomixer
c. Add 50 μL Proteinase K/SDS solution
d. Incubate at 37 °C for 10 min, 1400 rpm on a Thermomixer
e. Extract RNA with acid phenol-chloroform to remove TURBO DNase:
  i. Add 100 μL acid phenol: chloroform: isoamyl alcohol and shake for 15 s
  ii. Centrifuge at 10000 g at 4 °C for 2 min
  iii. Transfer upper aqueous phase to a new tube
  iv. Add 0.1 volumes of 3 M sodium acetate, 2 μL glycogen and 250 μL of 100% ethanol. Vortex for 5 s
  v. Incubate at −80 °C overnight for increased yield and centrifuge at 16000 g for 20 min at 4 °C
  Critical: After isopropanol precipitation, macrophage RNA needs to be incubated at −80 °C for a minimum of 4 h, but the optimum time is 16 h.
  vi. Remove ethanol carefully and dry RNA pellet
  vii. Resuspend in 50 μL RNase-free water

10. Isolate RNA with Trizol:
a. Add 500 μL Trizol
b. Vortex until RNA pellet disappears
c. Incubate samples at RT for 2 min to dissociate the nucleoprotein complex
d. Add 100 μL chloroform
e. Mix them thoroughly by inverting the tube
f. Incubate at RT for 10 min
g. Centrifuge for 10 min at 12000 g at 4 °C
h. Transfer the upper aqueous phase (~100 μL) containing RNA to a new 1.5 mL RNase-free microcentrifuge tube

Critical: avoid taking out the organic layer and lower phase containing contaminants (phenol from Trizol, DNA, and proteins).

11. RNA precipitation:
a. Add an equal volume (~100 μL) of isopropanol to the samples
b. Add 1 μL of glycogen to act as co-precipitant
c. Mix them by vortex briefly and incubate at −80 °C for 16 h, to maximize RNA yield
d. Centrifuge at 12000 g for 20 min at 4 °C
e. Remove the supernatant. Keep the pellet
f. Wash the pellet with 500 μL of 70% Ethanol

g. Centrifuge at 7500 g for 5 min at 4 °C
h. Remove carefully the supernatant. Keep the pellet
i. Air dry the pellet
j. Critical: Avoid complete dry to facilitate resuspension.
k. Resuspend the pellet with 10 μL of nuclease-free water

Notes

- Resuspended RNA can optionally be heated at 55 °C for 5 min before quantification
- RNase inhibitors (e.g., Ribolock) may be added at a 1:20 dilution to prevent RNA degradation

12. Quantify RNA concentration with a fluorescent dye-based method such as the Qubit RNA HS Assay Kit

13. Before proceeding, analyze subcellular fractionated RNA by RT-qPCR: *MALAT1*, a highly expressed nuclear lncRNA (Zhang, Hamblin, & Yin, 2017) is recommended to check successful subcellular fractionation. *MALAT1* should only be detected in the chromatin-bound and nucleoplasmic fractions
Critical: This is a quality control step in the protocol, to ensure correct subcellular fractionation. An example of expected results is shown on Fig. 4A.

14. Prior to library construction, check for potential bacterial contamination
Critical: Cross-contamination of cell culture samples due to manipulation or donor nosocomial infection can ruin the experiment. Therefore, it is essential to rule out the presence of microbes before library construction.

(a) Convert 100 ng RNA to cDNA using a reverse transcriptase such as SuperScript IV (SSIV), according to manufacturer's instructions. Briefly, to 100 ng RNA, add RNase-free water up to 12 μL, 1 μL of 50 μM random hexamers and 1 μL of 10 mM dNTPs, to a final concentration of 2.5 μM and 0.5 mM respectively. Incubate at 65 °C for 5 min. 4 μL SSIV buffer, 1 μL DTT, 0.5 μL Ribolock, and 0.5 μL SSIV

(b) Perform the following PCR program:

i. 23 °C for 10 min
ii. 55 °C for 10 min
iii. 80 °C for 10 min

(c) Analyze the cDNA with PCR using primers for the bacterial 16S gene (27F: 5′-AGAGTTTGATCCTGGCTCAG-3′ and 1492R:

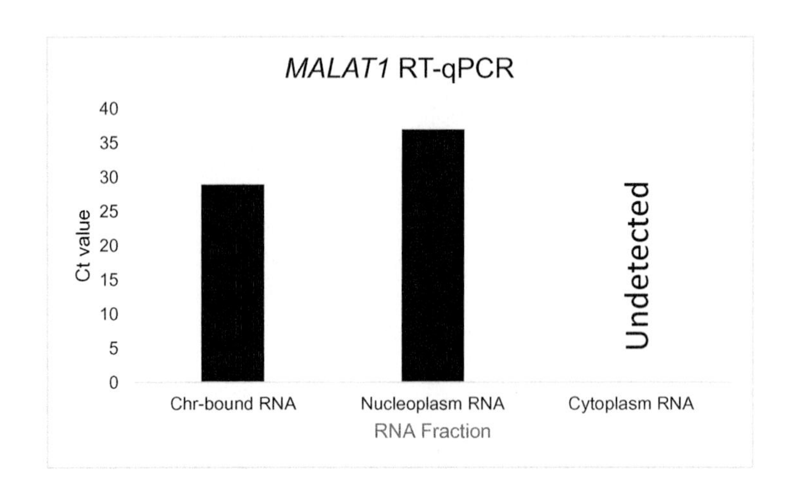

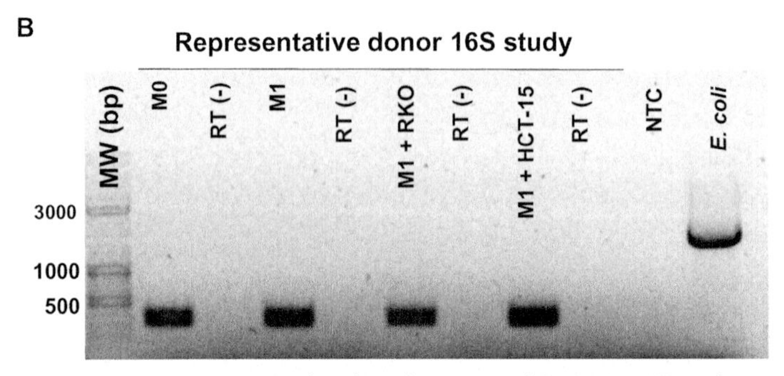

Fig. 4 (A) *MALAT1* expression levels in fractionated RNA. *MALAT1* is detected in ChrRNA and nucleoplasm RNA fraction but not detected in the cytoplasmic RNA fraction. (B) Assessment of bacterial contamination in macrophage samples by PCR of 16S (0.8% agarose gel, GeneRuler DNA Ladder mix (Invitrogen) as molecular weight (MW) marker). RT—no reverse transcriptase added; NTC—non-target control. Bands in macrophage lanes were subject to Sanger sequencing and correspond to human precursor 45S pre-rRNA.

5′-CGGTTACCTTGTTACGACTT-3′). Use bacterial genomic DNA (e.g., *E. coli* DH5α) as a positive control. Perform the following PCR program:

i. 95 °C for 5 min

ii. 30 cycles of 30 s at 94 °C, 90 s at 48 °C, and 2 min at 72 °C

iii. final extension for 10 min at 72 °C

Subsequently, run samples on a 0.8% agarose gel. A representative example is shown in Fig. 4B.

6.3 Library construction, RNA sequencing and QC

Workflows on Figs. 3A and B, 5A.

Critical

- Use the compatible DynaMag2 magnetic stand for accurate and quick nucleic acid separation

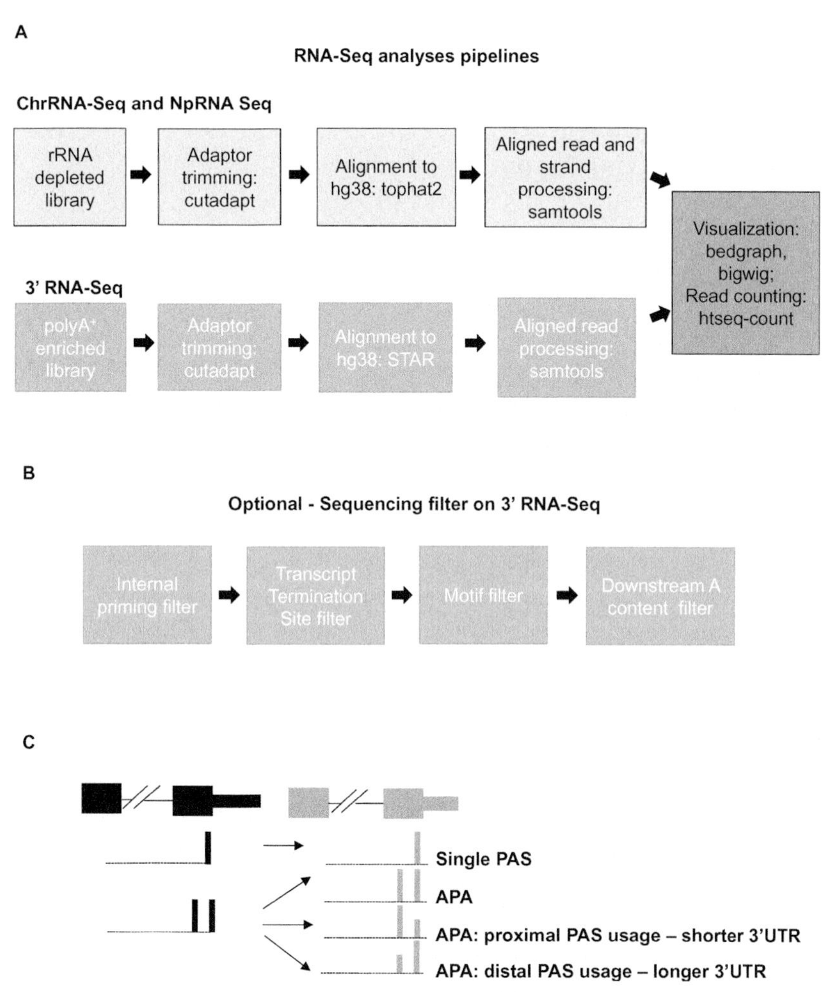

Fig. 5 (A) Pipelines of the bioinformatic RNA-Seq analyses. Top, pipeline used on the chromatin-bound (ChrRNA-Seq) and nucleoplasmic RNA-Seq (NpRNA-Seq) data. Bottom, pipeline used on the total 3′ mRNA-Seq data. (B) Scheme showing the consecutive sequencing filters on 3′ mRNA-Seq for identification of true PAS. (C) Scheme showing possible outcomes on PAS choice after stimuli on cells.

- Do not change position of the tube containing beads, or take it out while it sits on the magnetic stand, since it might disrupt both beads and RNA

6.3.1 Total RNA

Timing: 5.5 + 2 h qPCR.

Total RNA libraries for 3′ mRNA-Seq are prepared by first performing mRNA enrichment using oligo (dT)25 Dynabeads. Libraries are prepared using the QuantSeq REV kit, according to manufacturer's instructions. Both protocols are briefly described below.

6.3.2 mRNA enrichment

1. Prepare total RNA for mRNA enrichment:
   a. Use up to 1 μg total RNA as a starting material in 100 μL
      Critical: This kit has a maximum of 75 μg starting total RNA. While the recommended cell quantities will not exceed this amount, it is important to consider it as excess RNA quantities will inhibit final mRNA yield.
   b. Incubate at 65 °C for 2 min
   c. Place the tube immediately on ice
2. Prepare oligo (dT)25 Dynabeads:
   a. Completely resuspend the Dynabeads
   b. Transfer 200 μL of resuspended oligo (dT)25 Dynabeads to a 1.5 mL RNase-free microcentrifuge tube
   c. Place the tube in the magnetic stand for 30 s
   d. Wash beads once with 100 μL of the binding buffer
   e. Keep the tube on the magnetic stand
3. Incubate RNA with oligo (dT)25 Dynabeads:
   a. Add immediately 100 μL of the binding buffer to the oligo (dT)25 Dynabeads
   b. Add denatured RNA to the oligo (dT)25 Dynabead suspension
   c. Mix by pipetting
   d. Rotate at RT for 5 min to allow for annealing of RNA to the oligo (dT) on the Dynabeads
4. Wash and elute pA^+ mRNA:
   a. Place the tube on the magnetic stand
   b. Wait for 30 s until the solution becomes clear
   c. Remove the supernatant
   d. Remove the tube from the magnetic strand
   e. Wash the Dynabeads with 200 μL of washing buffer

f. Place the tube on the magnetic stand
g. Repeat the steps (b–f) once more
Critical: Remove all the supernatant before adding the next buffer (either wash or elution). If needed, tilt the whole system (magnetic stand + tube) at an angle to remove all the liquid phase.
h. Add 10 μL of 10 mM Tris-HCl pH 7.5 to each tube
Critical: mRNA must be eluted through addition of a low-salt buffer.
i. Incubate at 65 °C for 2 min
j. Place the tube on the magnetic stand until the solution becomes clear
k. Transfer the eluted mRNA to a new RNase-free tube
l. Quantify polyA$^+$ mRNA using the Qubit RNA HS kit

Pause point: We recommend not to pause once mRNA is enriched, as the first pause point of the library construction is fairly soon after (see below). As freeze-thaw cycles damage RNA, it is best to pause after RNA conversion to cDNA during the library construction.

6.3.3 Library construction—Enriched mRNA from total RNA using the QuantSeq REV kit from Lexogen

5. First Strand (FS) Synthesis:
a. Add 10–500 ng RNA with 5 μL of FS1 buffer in 0.2 mL microcentrifuge tube. If needed, add RNase-free water up to 10 μL
b. Mix by pipetting

Critical

- Use as much RNA as you can as input (maximum 500 ng), provided the amounts do not vary considerably between samples—e.g., do not synthesize RNA from 10 ng from one sample and 100 ng from the other
- If sample input is 10 ng, consult the manufacturer's instructions to adapt the protocol for low input
- All master mixes should be performed with a 10% surplus to account for pipetting errors

Note: Input RNA must be in a volume lower than 5 μL.

c. Pre-warm a mixture containing 9.5 μL of FS2 and 0.5 μL of E1 at 42 °C
Note: This mixture should be pre-warmed in a thermomixer.
d. Incubate RNA sample tube (step b) at 85 °C for 3 min on the thermocycler
e. Cool down to 42 °C
f. Centrifuge the tube briefly
g. Place the tube on the thermocycler

   h. Add 10 μL of pre-warmed FS2 + E1 mixture (step c) into the sample tube (step g).
   i. Mix by pipetting
   Critical: It is imperative that the temperature of the sample does not lower from 42 °C at this stage, to prevent mishybridization. Therefore, both RNA + FS1 and FS2 + E1 mixes have to be kept at 42 °C at all times.
   j. Centrifuge the tube briefly
   k. Place the tube on the thermocycler
   l. Incubate at 42 °C for 15 min

6. RNA Removal:
   a. Centrifuge the tube briefly
   b. Add 5 μL of RS solution into the sample tube (step a).
   c. Mix well with a pipette set for 15 μL
   d. Centrifuge the tube briefly
   e. Incubate at 95 °C for 10 min
   f. Cool down to 25 °C
   g. Centrifuge the tube briefly

7. Second Strand (SS) Synthesis:
   Note: The second strand synthesis occurs through binding of Illumina-compatible random primers at the 5′ end of the cDNA sequence.
   a. If not pausing, place the beads and purification solution at RT to re-equilibrate, to use at step 8
   b. Add 10 μL of SS1 solution to the sample
   c. Mix with a pipette set for 30 μL
   d. Incubate at 98 °C for 1 min
   e. Cool down slowly to 25 °C, setting ramp speed to 0.5 °C/second
   Critical: Test your thermocycler for the ramp speed, using a timer if necessary to ensure correct ramp speed at this stage.
   f. Incubate at 25 °C for 30 min
   g. Prepare a pool containing 4 μL of SS2 and 1 μL of E2 per sample
   h. Add 5 μL of the pool to each tube
   i. Mix with a pipette set for 30 μL
   j. Incubate at 25 °C for 15 min
   k. Transfer to a 1.5 mL microcentrifuge tube
   Pause point: Unpurified, unamplified dsDNA libraries may be kept at −20 °C overnight or up to a week.

8. First Purification

Critical

- Do not open 1.5 mL microcentrifuge tubes while on the magnetic stand to avoid accidental spillage
- Place a paper tissue over the tubes whenever they are on the magnetic stand, to prevent aerosol contamination
- Prepare 80% EtOH and use it for up to a week to prevent evaporation bringing changes to the protocol

a. Ensure all components are at RT
b. Resuspend purification beads (PB)
c. Spin down sample PCR tube
d. Add 16 μL of resuspended PB to each sample
e. Mix by pipetting
f. Incubate at RT for 5 min
g. Place tubes on the magnetic stand until the solution becomes completely clear
Critical: The manufacturer's protocol mentions this step as taking 2–5 min, but as time depends on the strength of your magnetic stand, this period may vary.
h. Remove and discard supernatant, not disturbing the bead suspension
i. Add 40 μL of EB buffer
j. Remove the tube from the magnetic stand
k. Resuspend beads
l. Incubate at RT for 2 min
m. Add 56 μL of PS solution to the beads and EB mix to reprecipitate library
n. Mix by pipetting
o. Incubate at RT for 5 min
p. Repeat steps (g) and (h).
q. Add 120 μL of freshly made 80% EtOH, and incubate for 30 s to wash
r. Remove and discard supernatant
Critical: Leave tubes in contact with the magnetic stand to avoid resuspending beads, which would lower library yield.
s. Repeat step (q).
t. Remove the supernatant completely
Critical: EtOH traces can inhibit subsequent PCR reactions.
u. Let beads dry at RT until all EtOH is evaporated

Critical: Environmental conditions of your laboratory affect the drying time of beads. While the manufacturer's instructions mention 5–10 min, if your laboratory is particularly dry, this time may be shortened.

v. Add 20 μL of EB buffer into the tube
w. Remove the tube from magnetic stand
x. Resuspend beads
y. Incubate at RT for 2 min
z. Place tube on the magnetic stand
aa. Leave the tube until the solution becomes clear
bb. Transfer 17 μL of clear solution into a new 0.2 mL PCR tube

Pause point: store libraries at −20 °C overnight or up to a week.

9. Calculate exact PCR cycle number

Critical: Differences in cell types, sample conditions and RNA integrity may cause different outcomes. Therefore, optimizing the precise endpoint PCR number of cycles for each sample is mandatory for library construction. For this optimization, the QuantSeq 3′ mRNA-Seq PCR Add-on kit for Illumina kit in the Key Resources Table is required.

a. Add 2 μL EB buffer to the previous 17 μL purified library

b. Dilute Sybr Green I 1:4000 using DMSO. This will be the 2.5 × stock concentration

Critical: Sigma is the only supplier for Sybr Green I recommended by the manufacturer.

c. For each sample, mix 1.7 μL of diluted library + 7 μL PCR + 5 μL 7000 primer + 1 E + 1.2 μL Sybr Green I 2.5 × + 14.1 μL EB buffer. All these reagents, except for Sybr Green I, are included in the PCR Add-on Kit. Add one extra reaction tube for non-template control, substituting the diluted library for EB buffer

d. Perform the following qPCR program:

i. 98 °C for 30 s
ii. 35 cycles of 98 °C for 10 s, 65 °C for 20 min, 72 °C for 30 s
iii. Final extension 72 °C for 1 min
iv. Hold at 10 °C

Critical: Adjust the emission maximum at 520 nm (Sybr Green I) on the RT-qPCR apparatus if needed.

e. Calculate the number of cycles for the subsequent endpoint PCR, according to the Annex E on manufacturer's protocol (Lexogen QuantSeq Rev). Briefly, obtain the Ct at half of the maximum fluorescence. From the Ct value obtained, subtract 3 Ct values, as one-tenth of the library has been used. The final Ct value will be the number of cycles needed to amplify each library sample by endpoint PCR (step 10 below).

10. Library amplification:

In this section, libraries are amplified by endpoint PCR to generate enough material for sequencing, after adding complete adapters required for cluster generation and unique indices to each sample to enable multiplexing.

Note: This section will only describe single-indexed libraries (i7). To perform amplification of dual-indexed libraries (i5 and i7), please consult the manufacturer's instructions.

a. Place the beads and purification solution at RT to re-equilibrate, to use at step 11

b. Pool 7 μL of PCR and 1 μL of E3 per sample

c. Add 8 μL of the pool to each of the purified library

d. Spin down the i7 index plate before opening. Pierce the sealing foil of each desired index with a needle. Add 5 μL of each respective index to each sample. Mix well by pipetting. Seal the opened wells of each i7 index with two layers of parafilm

Critical: Change needles every time, discarding safely to avoid cross-contamination.

e. Spin down the tubes. Perform x number of PCR cycles (11 to 22, as determined by step 54), using the following PCR program:

i. 98 °C for 30 s

ii. 35 cycles of 98 °C for 10 s, 65 °C for 20 min, 72 °C for 30 s

iii. Final extension 72 °C for 1 min

iv. Hold at 10 °C

11. Second Purification: repeat step 8, using 30 μL instead of the recommended volumes of EB and PS, but maintain 20 μL as the final EB addition

6.3.4 Library quality control

After construction, libraries must go through a quality control step. This can be performed at the sequencing facility, but we recommend it is done beforehand, in order not to send unsuitable, under- or overamplified libraries.

We recommend the use of two independent quality control methods: KAPA quantification (via qPCR), to infer library concentration, and a DNA quality chip (e.g., TapeStation), to infer library size.

Libraries are subsequently sequenced with NextSeq500 using Single-End 75-nucleotide reads, a read length needed to precisely map the mRNA 3′ ends. In order to obtain enough reads per sample to identify statistically significant differences in mRNA isoforms between conditions, special attention should be taken not to overmultiplex each library sequencing lane. It is recommended not to go beyond 6 samples of human macrophage RNA-Seq per lane, which allows to achieve at least 42 M reads per sample.

Critical: It is extremely important not to overamplify the libraries, to avoid sequence bias. Therefore, researchers should keep to the number of cycles quantified in the add-on protocol.

6.3.5 Fractionated RNA (nucleoplasmic and chromatin-bound RNA)

Timing: 6 + 2 h at −80 °C.

Fractionated RNA are rRNA-depleted using the Ribo-Zero Gold rRNA removal kit for Human/Mouse/Rat. Sequencing libraries are constructed using the NEBNext Ultra II directional library prep kit for Illumina, according to manufacturer's instructions. Both protocols are briefly described below.

6.3.6 rRNA removal

Ribosomal RNA consists of ~80% of total RNA (Westermann, Gorski, & Vogel, 2012). If left on the RNA sample to be sequenced, most of the sequencing reads will reflect this portion to the detriment of other RNAs.

Critical: Pre-treat your RNA with DNase, according to the manufacturer's protocol.

1. Wash magnetic beads
    - **a.** Leave all required components (beads, RNase-free water, bead resuspension solution) at RT for 30 min
    - **b.** Add 225 μL of beads in a 1.5 mL microcentrifuge tube
    - **c.** Place the tube on magnetic stand and leave for 1 min or until the solution becomes clear

      Critical: As in 3′ mRNA-Seq, the strength of your magnetic stand may alter the incubation times.
    - **d.** Discard the supernatant
    - **e.** Remove the tube from magnetic stand
    - **f.** Add the beads with 225 μL of RNase-free water to wash the beads
    - **g.** Resuspend by vortex

   - **h.** Place the tube on magnetic stand and leave for 1 min or until the solution becomes clear
   - **i.** Repeat steps (d–h) once
   - **j.** Discard the supernatant
   - **k.** Remove tube from magnetic stand
   - **l.** Add 65 μL of bead resuspension solution
   - **m.** Resuspend by vortex
   - **n.** Optional step: Add 1 μL of Ribolock RNase inhibitor and mix by pipetting
2. Hybridize probes:
   - **a.** Thaw all necessary ingredients on ice
   - **b.** Mix up to 2.5 μg RNA with 4 μL of RB buffer, 8 μL of RS solution, and RNase-free water up to 40 μL in total
   - **c.** Incubate at 68 °C for 10 min
   - **d.** Centrifuge the tube briefly
   - **e.** Incubate the reaction mixture at RT for 5 min
3. rRNA removal:
   - **f.** Add sample to the washed magnetic beads (65 μL, prepared in 14).
   - **g.** Mix thoroughly by vortex for 10 s
   - **h.** Incubate the mixture at RT for 5 min
   - **i.** Transfer the tube to another heat block and incubate at 50 °C for 5 min
   - **j.** Place the tube on the magnetic stand
   - **k.** Leave the tube for 1 min or until the solution becomes clear
   - **l.** Transfer 90 μL of the supernatant (rRNA-depleted RNA fraction) to a new 1.5 mL microcentrifuge tube

   Pause point: samples may be kept overnight at −20 °C or at −80 °C up to a month.
4. rRNA Purification:
   - **m.** Add 90 μL of RNase-free water to each sample
   - **n.** Add 18 μL $NaCH_3COO$ 3 M and 2 μL glycogen
   - **o.** Mix thoroughly by vortex
   - **p.** Incubate for 2 h at −80 °C
   - **q.** Centrifuge at 4 °C for 20 min at 10000 g
   - **r.** Discard the supernatant
   - **s.** Wash the pellet with 200 μL of freshly made 70% EtOH
   - **t.** Centrifuge the tube at 4 °C for 5 min at 10000 g
   - **u.** Discard the supernatant

v. Repeat steps (g)–(i) once
w. Discard the supernatant
x. Air dry at RT for 5 min
y. Dissolve the pellet in 10 μL
z. Quantify the rRNA-depleted RNA using the Qubit RNA HS kit

Pause point: samples may be kept overnight at −80 °C or up to a month.

6.3.7 Library construction—rRNA-depleted fractionated RNA—Using the NEB Next Ultra 2 directional library prep for Illumina

1. RNA Fragmentation and Priming
   Note: Maximum input for this kit is 100 ng rRNA-depleted RNA in a maximum of 5 μL.
   a. Add 5 μL of rRNA-depleted RNA (>100 ng) to 4 μL of FSSB and 1 μL of RP
   b. Mix by pipetting
   c. Spin down the mixture briefly
   d. Incubate for at 94 °C 15 min
   e. Transfer the tube on ice
2. cDNA Synthesis
   a. Add the first strand synthesis mixture, 8 μL of SSR and 2 μL of FSSE
   b. Mix by pipetting
   c. Spin down the reaction mixture briefly
   d. Incubate for 10 min at 25 °C, 15 min at 42 °C and 15 min at 70 °C, hold at 4 °C. Keep lid temperature at 99 °C
   e. Transfer the tube to ice
   f. Add the second strand synthesis mixture, 48 μL of RNase-free water, 8 μL of SSSB, and 4 μL of SSSE
   g. Mix by pipetting
   h. Keep the tube on ice
      Critical: SSSB contains dUTP, which confers strand-specificity to this protocol.
   i. Centrifuge the tube briefly
   j. Incubate for 1 h at 16 °C
3. First purification
   a. Ensure all components of the purification module are completely thawed before start the following steps
   b. Resuspend magnetic beads by vortex

c. Transfer libraries to 1.5 mL microcentrifuge tubes and add 144 μL of the magnetic beads
d. Mix them thoroughly by vortex
e. Incubate for 5 min at RT
f. Centrifuge the tube briefly
g. Place tube on the magnetic stand
h. Incubate for 1 min or until the solution becomes clear
i. Carefully discard the supernatant without disturbing the beads
j. Add 200 μL of freshly made 80% EtOH at RT
k. Leave for 30 s
l. Discard supernatant
Critical: always leave tubes in contact with the magnet to avoid resuspending beads, which would lower library yield.
m. Repeat steps (j)–(l) once
n. Air dry beads at RT for 5 min on the magnetic stand
o. Remove tube from magnetic stand
p. Elute DNA by adding 53 μL of RNase-free water
q. Mix them by vortex
r. Centrifuge the tube briefly
s. Incubate for 2 min at RT
t. Place the tube on magnetic stand and leave until the solution becomes clear
u. Transfer 50 μL of the supernatant to a new 0.2 mL microcentrifuge tube

4. End prep and adaptor ligation:
a. Keep RNA sample on ice
b. Add 7 μL of EPB and 3 μL of EPE to the sample
c. Mix them with a pipette set to 50 μL
d. Centrifuge the tube briefly
e. Incubate for 30 min at 20 °C and 30 min at 65 °C, hold at 4 °C. Keep lid temperature at 99 °C on the thermal cycler
f. Dilute sequencing adaptor in ice-cold dilution buffer (5-fold dilution for high input, 25-fold for 1–10 ng initial RNA). Keep on ice
g. Add 2.5 μL diluted adaptor, 1 μL of LE and 30 μL of LMM
h. Mix them with a pipette set to 80 μL
i. Centrifuge the tube briefly
j. Incubate at 20 °C for 15 min
k. Add 3 μL of USER enzyme to the ligation mixture
l. Mix them with a pipette set to 80 μL

m. Incubate the reaction mixture at 37 °C for 15 min. Keep lid temperature at 99 °C on the thermal cycler

5. Second library purification: repeat step 20. Use 87 μL of magnetic beads and incubate for 10 min. Elute RNA sample with 32 μL of RNase-free water. Transfer 30 μL of eluted sample to a new 0.2 mL microcentrifuge tube for the following PCR amplification

6. Library PCR amplification:
Note: as with 3′ mRNA-Seq, we will describe single-indexed library prep.

a. Divide the 30 μL library into two fractions. Keep the remaining half at −20 °C in case of mishandling such as over amplification

b. Mix 15 μL of the RNA sample, 25 μL of Q5 MM, 5 μL of universal i5 primer, and 5 μL index i7 primer

c. Mix them by pipetting

d. Perform the following PCR program, keeping lid temperature set to 105 °C on the thermal cycler:

1. °C for 30 s (Initial denaturation)

2. 7 cycles (for 100 ng input RNA) of 98 °C for 10 s, 65 °C for 75 s

3. 65 °C for 5 min (Final extension)

4. Hold at 4 °C

7. Third library purification: repeat step 20. Use 45 μL of magnetic beads. Elute libraries with 23 μL of RNase-free water. Transfer 20 μL of the PCR sample to a new 0.2 mL microcentrifuge tube

6.3.8 Library quality control

Libraries *should* be subject to a quality control step before proceeding. This can be performed at the sequencing facility, but we recommend it is done beforehand in order not to send unsuitable, under- or overamplified libraries.

We recommend the use of two independent quality control methods: KAPA quantification (via qPCR), to infer library concentration, and a DNA quality chip (e.g., TapeStation and Bioanalyzer), to infer library size.

Sequence the libraries in a HiSeq2000 using Paired-End 42 base pairs, a read length needed to determine sense and antisense RNA from fractionated samples. In order to obtain enough reads per sample to identify statistically significant differences in gene expression in subcellular fractionated RNA in several macrophage conditions, special attention should be taken not to overmultiplex each library sequencing lane. It is recommended not to go beyond 6 samples of human macrophage RNA-Seq per lane, which allows for obtaining at least 66 M paired-end reads per sample.

Critical: It is extremely important not to overamplify the libraries, to avoid sequence bias. Therefore, keep to the number of cycles recommended on the manufacturer's instructions (NEBNext Ultra 2 directional kit for Illumina). If libraries are overamplified, you need to repeat the protocol starting from step 23.

It is crucial to mention that changes to the polarization stimuli of macrophages may give rise to dramatically different RNA-Seq results.

6.4 Bioinformatics pipeline

Timing: 28 h for preliminary analysis, but variable depending on number of samples and computational power.

Pipelines are shown on Fig. 5A and B.

As in library construction, bioinformatics pipeline for sequencing data analyses differs between total and fractionated RNA samples. However, some steps are common and hence will only be described once. This pipeline is mostly performed on the command line, but it is also possible to perform the same steps in platforms such as Galaxy of Bioconductor. For ease of explanation, code excerpts will be used to exemplify some steps.

Critical

- You must use the same human genome version and release as reference for the whole sequencing project
- Any changes made to the following pipelines or to software versions may generate distinct data

6.4.1 Chromatin-bound and nucleoplasm RNA-Seq (paired-end)

1. Sequencing QC was performed using FastQC
   Note: Pay attention to the per base sequence quality and adapter dimer contamination. High levels of the former (score >30) and low levels of the latter (adaptor content <2%) are indicative of high-quality reads.
2. Remaining adapters were trimmed with Cutadapt, removing low-quality reads under 10 bases, using the following code:
   ```
   cutadapt -minimum-length 10 -q 15,10 -a adaptersequence_Fwd -A adaptersequence_Rev -o rawfile_read1_trimmed.fastq.gz -p rawfile_read2_trimmed.fastq.gz rawfile_read1.fastq.gz rawfile_read2.fastq.gz
   ```
3. Trimmed data were aligned to the reference human genome using TopHat2, with Bowtie2 indexes

Notes

- In this step, we recommend to allow for read pairs to be separated by up to 3 kb (chromatin-bound) and up to 200 bp (nucleoplasm) and only one alignment to the reference genome for each read
- Aligned read mapping rate should be high at this stage (~90%).

```
tophat2 -r 3000 -g 1 -a 5 -library-type fr-firststrand --no-coverage-search -o/path/to/index rawfile_read1_trimmed.fastq.gz rawfile_read2_trimmed.fastq.gz
```

(for Chromatin-bound)

```
tophat2 -r 200 -g 1 -a 5 -library-type fr-firststrand --no-coverage-search -o/path/to/index rawfile_read1_trimmed.fastq.gz rawfile_read2_trimmed.fastq.gz
```

(for Nucleoplasm)

4. Process aligned read pairs with SAMtools (Li, Handsaker, Wysoker, et al., 2009). Index the bam files with SAMtools index. Call the total number of mapped reads using SAMtools idxstats. Every sample was normalized by dividing by 100×10^6 reads
   Critical: Samples were mapped in proper orientation and strand-specific.
5. Sort samples with SAMtools sort and create bedgraph using Bedtools with the following code:

```
bedtools genomecov -ibam mappedfile_read1_strand+.sorted.bam -bg -scale #readsby100M -split > mappedfile_read1_strand+.bedGraph
```

Bigwig files can be created from the bedgraph files with the bedGraphToBigWig tool and a file containing the chromosome sizes of the genome version used (both accessible from the UCSC genome browser):

```
bedGraphToBigWig mappedfile_read1_strand+.bedGraph hg3b.chrom.sizes mappedfile_strand+.bw
```

6. Count reads with htseq-count using the following code:

```
htseq-count -f bam -s yes -r name -t exon mappedfile_read1_strand+.sorted.bam Homo_sapiens.GRCh38.90.chr+.gtf > mappedfile_read1_strand+_ht.txt
```

After obtaining gene counts, all sample replicates must be compared through a correlation matrix, to ensure reproducibility. A representative example is shown on Fig. 6A and B. Alternatively, a Principal Component Analysis (PCA) graph may instead be produced. Subsequently, gene expression studies are performed in the fractionated samples (see Section 6.5.1).

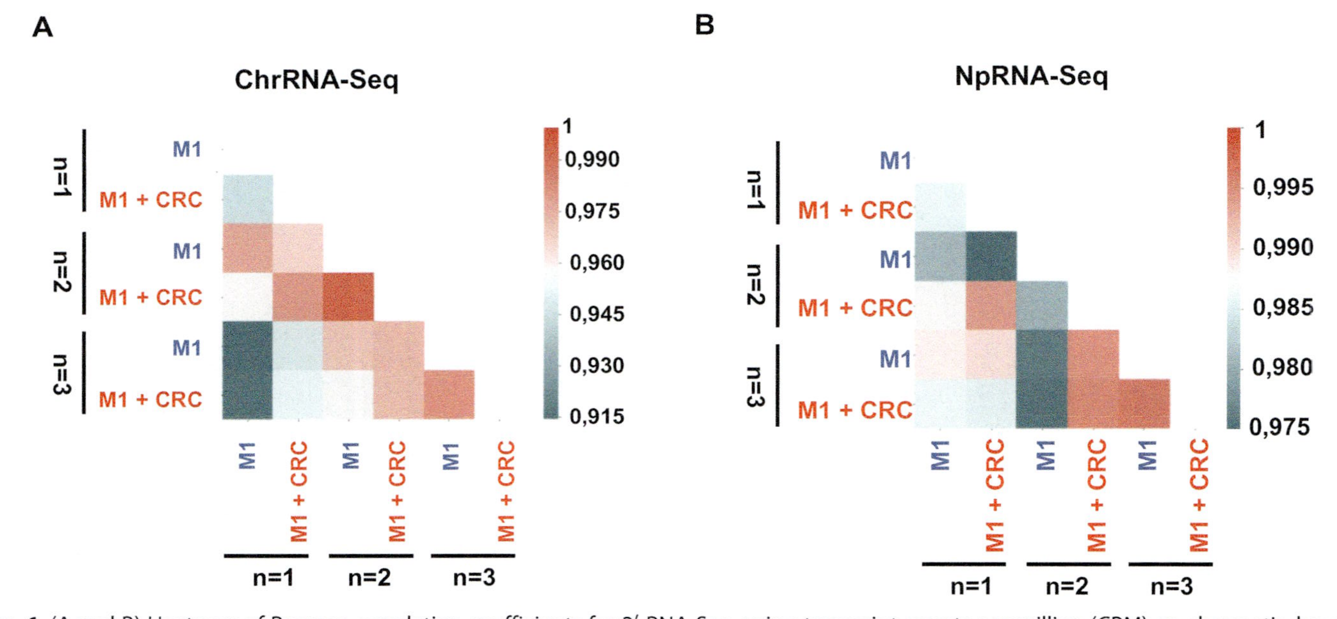

Fig. 6 (A and B) Heatmap of Pearson correlation coefficients for 3′ RNA-Seq, using transcript counts per million (CPM) on chromatin-bound RNA-Seq data (A) or nucleoplasmic RNA-Seq data (B) performed with three different blood donors; 0 means total lack of correlation, while 1 means total correlation.

6.4.2 3′ mRNA-Seq (single read)

The first optional step to analyze 3′ mRNA-Seq data is to identify the true used PAS, which is performed sequentially. Due to possible leakiness in PAS detection, 3′ mRNA-Seq results were filtered by the sequencing company.

Briefly, the input fastq file was mapped to the reference and 3 filters were applied in sequence in the pipeline shown on Fig. 5B:

(1) a Transcript Termination Site (TTS) filter, taking out transcripts not lying within ~10 nt of an annotated transcript end site;

(2) a Motif filter removed all sequencing peaks whose read start contained a subset of hexamer sequences upstream (Livak & Schmittgen, 2001);

(3) a downstream A content filter checked the amount of As in a window downstream of the read start. If a given peak passed the validated threshold, the read was considered internal priming and was removed.

All reads passing the three consecutive filters were retained as non-internal priming and put into subsequent differential expression and alternative mRNA isoform analyses.

7. The sequencing QC is performed similarly to the paired-end data, as is cutadapt, with the following adaptations due to the single read format:

```
cutadapt -minimum-length 10 -q 10 -a adaptersequence -o rawfile_trimmed.fastq.gz rawfile.fastq.gz
```

8. Trimmed data were aligned using STAR aligner using the same genome build as for chromatin and nucleoplasm RNA-Seq. Aligned read mapping rate is lower due to the library used, but should exceed 75%.

```
STAR -runThreadN 8 -genomeDir/path/to/index -readFilesIn/path/to/sample/rawfile_trimmed.fastq.gz -readFilesCommand zcat -outFileNamePrefix/path/to/sample/mappedfile.bam
```

9. Aligned reads were also processed using SAMtools, and reads were counted with htseq-count:

```
htseq-count -f bam -r name -s reverse -r mappedfile_read1_strand.sorted.bam Homo_sapiens.chr.gtf > mappedfile_read1_strand_ht.txt
```

Subsequently, gene expression and mRNA APA isoform analyses are performed.

6.5 Gene expression analyses of the RNA-Seq results

6.5.1 Differential gene expression

Differences in gene expression were calculated using DESeq2. To detect true gene expression, differentially expressed genes were followed up if their

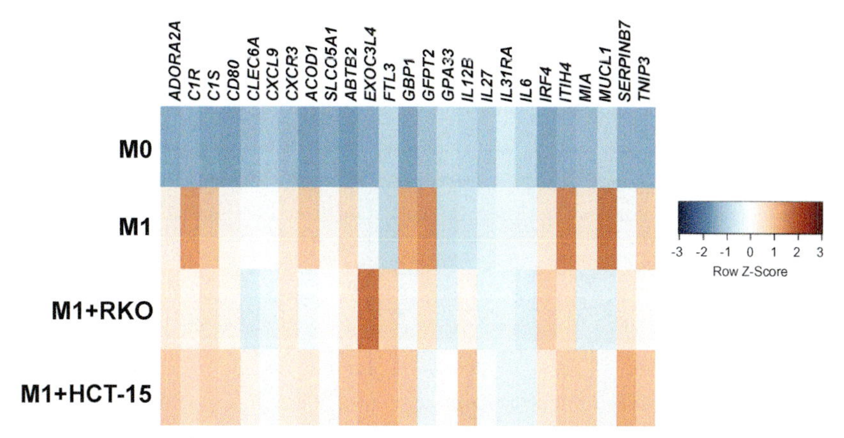

Fig. 7 Heatmap of a representative donor, showing differentially expressed genes (DEGs) in M0, M1 and M1 macrophages co-cultured with RKO or HCT-15 cell lines.

transcripts per million (TPM) are >5 mean TPM, Log2Fold Changes greater than 1 or smaller than −1, and adjusted *p*-value <0,05, calculated by DESeq2 (Love, Huber, & Anders, 2014).

A representative example of results is represented in Fig. 7, showing a heatmap of statistically significant differentially expressed genes in M0, M1, and M1 macrophages co-cultured with two different CRC cell lines (RKO and HCT-15).

6.5.2 Differential expression of APA mRNA isoforms

Differences in the expression of mRNA isoforms due to APA were calculated using DEXSeq (Love et al., 2014) in the 3′ mRNA-Seq data, with the Galaxy pipeline. Firstly, sorted bam files were subjected to DEXSeq-count, to count the sequencing reads associated with each PAS—and hence each mRNA isoform. Afterwards, DEXSeq-Count files were compared between each condition, and the resulting table was organized by corrected p-value, to identify genes in which at least one mRNA isoform varied with statistical significance.

Alternative polyadenylation was considered to be present when two or more alternative 3′ UTRs (depicted in Fig. 5C) were expressed, and when the second highest mRNA isoform level was at least 10% of the highest isoform in at least two replicates.

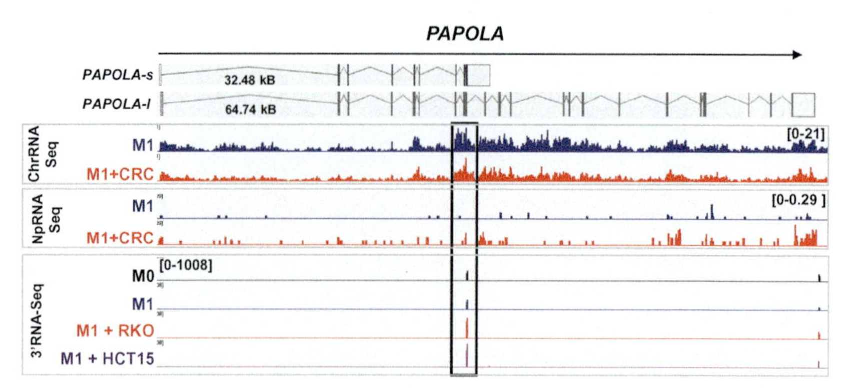

Fig. 8 IGV gene profiles and transcript counts of *PAPOLA*. Data obtained through ChrRNA-Seq (top), NpRNA-Seq (middle) and 3′ mRNA-Seq (bottom) of macrophages, M1 macrophages cultured alone and co-cultured with the CRC cell lines RKO or HCT-15.

A representative example of results is represented in Fig. 8, showing *PAPOLA* (Poly(A) Polymerase gene) gene counts and mRNA isoform differences between macrophage conditions.

6.5.3 RT-qPCR validation of the RNA sequencing data

Total RNA was digested with DNase I (Ref. EN0521, Roche) for 25 min at 37 °C, inactivating the enzyme for 10 min at 80 °C. RNA was reverse transcribed using SuperScript IV (Ref. 18,090,010, ThermoFisher) and random hexamers, according to manufacturer's instructions and above. RT-qPCR reactions were performed in triplicate using SYBR Select Master Mix (Ref. 4,472,908, Applied Biosystems) and 0.125 μM of primers, using the following protocol:

(a) Hold 20 s at 20 °C

(b) Incubate 10 min at 95 °C

(c) 40 cycles of 15 s at 95 °C and 1 min at 60 °C

(d) Perform melt curve by incubating for 15 s at 95 °C, 1 min at 60 °C and until 95 °C with a slope of 1%, 20 s each

A representative example of results is represented in Fig. 9, showing *RAC1* (Ras-related C3 botulinum toxin substrate 1 gene) gene counts and mRNA isoforms (Fig. 9A) and validation by RT-qPCR (Fig. 9B and C) in M1 macrophages.

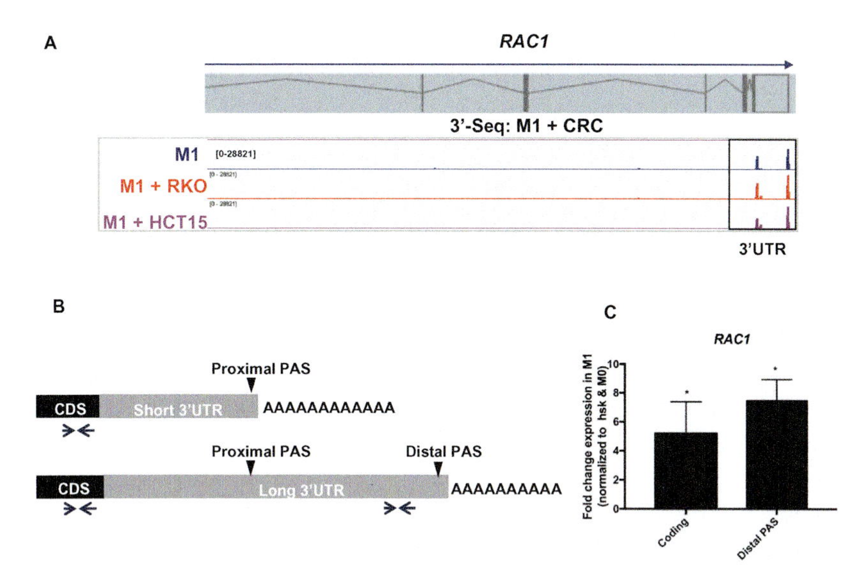

Fig. 9 (A) IGV gene profiles and transcript counts of *RAC1*. Data obtained through ChrRNA-Seq (top), NpRNA-Seq (middle) and 3′ mRNA-Seq (bottom) of M0 macrophages, M1 macrophages cultured alone and co-cultured with the CRC cell lines RKO or HCT-15. (B) Schematic representation of the primers used in RT-qPCR validation of 3′ RNA-Seq data. (C) Fold change of the coding region and the long 3′ UTR isoform (distal PAS) of the *RAC1* gene in M1 vs. M0 macrophages.

7. Concluding remarks

Studying the transcriptome of macrophages interacting with cancer cells and other components of the tumor microenvironment is crucial to better understand immune evasion. RNA-Seq data on primary human macrophages co-cultured with CRC thus allows for the development of more efficient diagnostic and therapeutic tools. Isolating macrophages from healthy donors, exposing them to inflammatory stimuli conditions and co-culturing them indirectly with cancer cells mimics the macrophage-cancer cell crosstalk that occurs in the colon. This model system is ideal for transcriptomic studies, bypassing the difficulty of isolating primary immune cells from the stroma of colorectal tumors. Importantly, our primary human macrophages closely mimic the microenvironment to which the macrophages are subject, which would not be the case if monocytic cell lines such as THP-1 were used, which do not share embryonic progenitors nor transcriptomic profiles (Tedesco, De Majo, Kim, et al., 2018).

With the co-cultures with cancer cells and human primary macrophages homogenously polarized to M1 inflammatory populations, and the use of Chromatin-bound RNA-Seq, Nucleoplasm RNA-Seq and 3′ mRNA-Seq methodologies and dedicated bioinformatics, we are able to characterize precisely APA mRNA isoform profiles and quantify gene expression changes in a highly plastic immune cell population from the same human donor. By quantifying the statistical significance of the mRNA isoform changes between conditions, we observed that different genes display different APA profiles and that CRC cells modulate macrophage APA.

The described protocol is a low-cost, big data approach for transcriptomic studies adapted to the need and budget of research laboratories, and particularly in the macrophage—CRC model system. This protocol requires low quantities of primary macrophages to obtain simultaneously nuclear and total RNA for sequencing and analyse differential gene expression, as well as identify novel APA mRNA isoforms, in Illumina-compatible systems. Using this methodology a thorough, unbiased transcriptomic characterization of plastic immune cells in a relevant pathophysiologic context is achieved. Furthermore, with this method we can identify genome-wide APA mRNA signatures with important roles in immune responses that may serve as potential diagnostic and therapeutic targets, using relatively few macrophage cells. With subsequent siRNA depletion of genes with functions in gene expression regulation may lead to the identification of the molecular mechanisms involved.

In conclusion, our method provides a basis for a comprehensive transcriptomic analyses using co-cultures of primary human macrophages with cancer cells (Supplementary Fig. S1 in the online version at https://doi.org/10.1016/bs.mie.2021.04.004).

8. Safety considerations

Chloroform, HCl, Isopropanol, PFA, and TRIzol reagent are dangerous. Manipulate them in the chemical hood using gloves and eyeglasses. If exposed, thoroughly wash skin in running water.

9. Expected outcomes

This protocol presents an integrated approach to study gene expression and alternative polyadenylation events. It accurately determines true APA isoforms, due to comparison of subcellular fractionated RNA-Seq (chromatin-bound RNA and nucleoplasm RNA) and whole-cell 3′ mRNA-Seq. Therefore, researchers are able to distinguish accurate APA mRNA isoforms produced in the nucleus, from the cytoplasmic mRNA isoforms, which may be subjected to modulation by miRNAs and RNA binding proteins.

Raw sequencing data by 3′ mRNA-Seq, chromatin-bound RNA-Seq and nucleoplasm RNA-Seq can be converted into differential gene expression in macrophage populations. Chromatin-bound and nucleoplasm RNA-Seq approaches further inform researchers of regulatory RNAs such as long non-coding RNA and antisense transcripts genome-wide.

Globally, the RNA sequencing data obtained using this protocol reveals transcriptomic profiles relevant for the innate immune system and for anti-tumoral therapy studies, as it pinpoints gene expression alterations occurring in primary human macrophages.

Timing: full protocol

Isolation of PBMCs from human peripheral blood and $CD14^+$ enrichment: 8h.

Differentiation of macrophages from $CD14^+$ cells and inflammatory polarization: 2h + 6 days incubation.

Macrophage and Colorectal cancer co-cultures: 3 + 24h incubation.

Macrophage and CRC Immunofluorescence: 6h (staining) + ~2h (imaging).

Macrophage siRNA transfection: 2 + 48h incubation.

Total RNA isolation: 4 + 16h incubation at −80 °C.

Fractionated RNA isolation: 8 + 16h precipitation time.

Total RNA library construction: 5.5 + 2h qPCR.

Fractionated RNA library construction: 6 + 2h at −80 °C.

Bioinformatics: 28h.

10. Quantification and statistical analysis

Statistical analyses for bioinformatics were performed using the following tests:

- DESeq2: *p*-value calculated with the Wald test, adjusted using Benjamini-Hochberg False Discovery Rate (FDR), selecting for padj <0.05
- DEXSeq: *p*-value calculated with the LRT test, adjusted using Benjamini-Hochberg False Discovery Rate (FDR), selecting for padj <0.05

Differentially expressed genes selected for validation by RT-qPCR should have a large differential gene expression fold differences (>2 Fold Change (FC)) between mono and co-culture in all donors. This criteria is applied to identify non-random transcriptomic changes with potential physiological consequences.

For APA analysis, mRNA isoforms with differential expression between conditions were validated by RT-qPCR using two sets of primer pairs as described, one pair for the coding region and another for the distal PAS. Relative expression was calculated using the reference gene *18S* (Fwd: 5′-GCAGAATCCACGCCAGTACAAGA-3′; Rev.: 5′-CCCTCTAT GGGCCCGAATCTT-3′) and results were analyzed applying the ΔΔCt method.

Statistical analyses for qPCR were performed using GraphPad Prism Software version 7. Presented data is shown as mean ± standard deviation (SD). Comparison between two independent groups was performed with the Student's *t*-test. *P*-values below 0.05 were considered statistically significant ($* = p < 0.05$).

11. Advantages

This approach allows for the simultaneous study of fractionated and total RNA of primary human macrophages. Moreover, true APA mRNA isoforms identification and differential gene expression results are obtained in the same human donor due to this integrative protocol.

Primary human macrophages are successfully polarized to the extremes of the inflammatory spectrum (M1, M2), and researchers will obtain homogenous populations able to survive ex-vivo for up to 10 days.

The establishment of the co-culture system allows for the sequencing of both cell populations—macrophages and cancer cells—if desired, allowing researchers to obtain reciprocal transcriptomic signatures and to further characterize their model system.

The use of RNA molecules present in different cellular compartments for each sequencing method (ChrRNA-Seq, NpRNA-Seq, cytoplasmic RNA-Seq) is achieved by subcellular fractionation, which is validated by independent techniques (*MALAT1* expression by RT-qPCR). Ribosomal RNA removal and mRNA enrichment ensures that the libraries are made with the RNA fractions of interest.

Additionally, the whole protocol is easily scaled-up to include more inflammatory conditions, increased input for sequencing, and includes crucial quality control stages at each step to ensure robust and high-quality data.

12. Limitations

This approach does not enable the identification of ribosomal RNAs from primary human macrophages. Therefore, transcriptomic studies to identify protein translation regulators, which are relevant for inflammation (Poddar et al., 2013), should follow a different approach when selecting macrophage rRNA for sequencing libraries.

This protocol recommends avoiding excessive multiplexing of samples. We favor increasing the number of sequencing runs if possible to avoid the risk of lowering sequencing read per sample. But if many macrophage replicates and conditions are needed, the cost of sequencing will increase.

We recommend the selection of blood donors below 40 years old, due to a decreased risk of hematological malignancies or chronic anti-inflammatory utilization by the donor, which may interfere with the results.

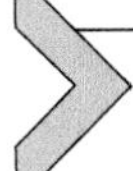

13. Optimization and troubleshooting

13.1 Problem and solution

$CD14^+$ cells are non-adherent—Use fresh M-CSF for macrophage isolation.

Heterogeneous macrophage polarization—control LPS, IFNγ, and IL-10 lots. Additionally, add these regents when macrophages are adherent and sufficiently differentiated with M-CSF, and keep the polarization RPMI media 2% FBS without antibiotics.

Low amounts of RNA are present by the end of RNA isolation—This may be due to RNA degradation or low amount of RNA due to a small number of macrophages. Therefore, use RNase-free reagents on the whole procedure or use more cells as starting material.

No PCR product is obtained at the end of library construction—PCR products are underamplified, hence we suggest to use higher PCR cycle numbers, or higher RNA input for the library.

PCR products are longer than expected (450 bp on QuantSeq Rev. 300 bp on NEB Next)—PCR products are overamplified, therefore we suggest to reduce PCR cycle number and to restart library construction.

Matrix correlation data or heatmap showing differences between replicates and/or gene expression results and APA mRNA isoforms are too variable—This indicates intrinsic donor variability. Therefore, the number of biological replicates should be increased.

Ethical statement

Buffy coats isolated in accordance with the Ethics Committee protocol of Centro Hospitalar Universitário São João (CHUJ) (90/19) consist of waste products of healthy anonymized blood donations containing high amounts of leukocytes. Donors filled informed consent prior to each donation, and their identity was archived exclusively at CHUJ.

Acknowledgments

We thank members of the Gene Regulation, Cell Activation and Gene Expression, and Tumor and Microenvironment Interactions laboratories at i3S, and Nick Proudfoot and Shona Murphy laboratories at the Sir William Dunn School of Pathology, University of Oxford, for critical and fruitful discussions of this work. This work was supported by the Fundação para a Ciência e Tecnologia (FCT) by funding to AM (FEDER—Fundo Europeu de Desenvolvimento Regional funds through the COMPETE 2020—POCI/Portugal 2020 and POCI-01-0145-FEDER-007274 projects), by Norte-01-0145-FEDER-000051—“Cancer Research on Therapy Resistance: From Basic Mechanisms to Novel Targets,” supported by Norte Portugal Regional Operational Programme (NORTE 2020), under the PORTUGAL 2020 Partnership Agreement, through the European Regional Development Fund (FEDER) and by PhasAge. This project has received funding from European Union’s Horizon 2020 research and innovation program under grant agreement No 952334. J.W. was supported by studentships from FCT (PD/BD/114168/2016) and (H2020-NMP-PILOTS-2015-683356-FOLSMART). The authors acknowledge Centro Hospitalar Universitário São João—Immunohemotherapy Department for their kind buffy coats donations, and the i3S Bioimaging (María Lázaro, PPBI-POCI-01-0145-FEDER-022122), and Scientific Genomics Platforms (Mafalda Rocha and Rob Mensink).

Author contributions

J.W., M.J.O., and A.M. designed the project; J.W., A.M.C., M.T., T.N., M.J.O. and A.M. wrote the paper.

References

Berkovits, B. D., & Mayr, C. (2015). Alternative 3′ UTRs act as scaffolds to regulate membrane protein localization. *Nature*, *522*(7556), 363–367. https://doi.org/10.1038/nature14321.

Braz, S. O., Cruz, A., Lobo, A., et al. (2017). Expression of Rac1 alternative 3′ UTRs is a cell specific mechanism with a function in dendrite outgrowth in cortical neurons. *Biochimica et Biophysica Acta, Gene Regulatory Mechanisms*, *1860*(6), 685–694. https://doi.org/10.1016/j.bbagrm.2017.03.002.

Brooks, B. M., Hart, C. A., & Coleman, J. W. (2005). Differential effects of β-lactams on human IFN-γ activity. *The Journal of Antimicrobial Chemotherapy*, *56*(6), 1122–1125. https://doi.org/10.1093/jac/dki373.

Castello, A., Fischer, B., Eichelbaum, K., et al. (2012). Insights into RNA biology from an atlas of mammalian mRNA-binding proteins. *Cell*, *149*(6), 1393–1406. https://doi.org/10.1016/j.cell.2012.04.031.

Derti, A., Garrett-Engele, P., MacIsaac, K. D., et al. (2012). A quantitative atlas of polyadenylation in five mammals. *Genome Research*, *22*(6), 1173–1183. https://doi.org/10.1101/gr.132563.111.

Dye, M. J., & Proudfoot, N. J. (1999). Terminal exon definition occurs cotranscriptionally and promotes termination of RNA polymerase II. *Molecular Cell*, *3*(3), 371–378. https://doi.org/10.1016/S1097-2765(00)80464-5.

Herrmann, C. J., Schmidt, R., Kanitz, A., Artimo, P., Gruber, A. J., & Zavolan, M. (2020). PolyASite 2.0: A consolidated atlas of polyadenylation sites from 3′ end sequencing. *Nucleic Acids Research*, *48*(D1), D174–D179. https://doi.org/10.1093/nar/gkz918.

Ji, Z., Lee, J. Y., Pan, Z., Jiang, B., & Tian, B. (2009). Progressive lengthening of 3′ untranslated regions of mRNAs by alternative polyadenylation during mouse embryonic development. *Proceedings of the National Academy of Sciences of the United States of America*, *106*(17), 7028. https://doi.org/10.1073/pnas.0905246106.

Jia, X., Yuan, S., Wang, Y., et al. (2017). The role of alternative polyadenylation in the antiviral innate immune response. *Nature Communications*, *8*, 14605. https://doi.org/10.1038/ncomms14605.

Lee, S. H., Singh, I., Tisdale, S., Abdel-Wahab, O., Leslie, C. S., & Mayr, C. (2018). Widespread intronic polyadenylation inactivates tumour suppressor genes in leukaemia. *Nature*, *561*(7721), 127–131. https://doi.org/10.1038/s41586-018-0465-8.

Lerman, M. J., Lembong, J., Muramoto, S., Gillen, G., & Fisher, J. P. (2018). The evolution of polystyrene as a cell culture material. *Tissue Engineering. Part B, Reviews*, *24*(5), 359–372. https://doi.org/10.1089/ten.teb.2018.0056.

Li, H., Handsaker, B., Wysoker, A., et al. (2009). The sequence alignment/map format and SAMtools. *Bioinformatics*, *25*(16), 2078–2079. https://doi.org/10.1093/bioinformatics/btp352.

Livak, K. J., & Schmittgen, T. D. (2001). Analysis of relative gene expression data using real-time quantitative PCR and the 2-ΔΔCT method. *Methods*, *25*(4), 402–408. https://doi.org/10.1006/meth.2001.1262.

Locati, M., Mantovani, A., & Sica, A. (2013). Macrophage activation and polarization as an adaptive component of innate immunity. *Advances in Immunology*, *120*, 163–184. https://doi.org/10.1016/B978-0-12-417028-5.00006-5.

Love, M. I., Huber, W., & Anders, S. (2014). Moderated estimation of fold change and dispersion for RNA-seq data with DESeq2. *Genome Biology*, *15*(12), 1–21. https://doi.org/10.1186/s13059-014-0550-8.

Ma, W., & Mayr, C. (2018). A membraneless organelle associated with the endoplasmic reticulum enables 3′UTR-mediated protein-protein interactions. *Cell*, *175*(6), 1492–1506. e19. https://doi.org/10.1016/j.cell.2018.10.007.

Mantovani, A., & Locati, M. (2013). Tumor-associated macrophages as a paradigm of macrophage plasticity, diversity, and polarization lessons and open questions. *Arteriosclerosis, Thrombosis, and Vascular Biology*, *33*(7), 1478–1483. https://doi.org/10.1161/ATVBAHA.113.300168.

Mayr, C. (2017). Regulation by 3′–untranslated regions. *Annual Review of Genetics*, *51*(1), 171–194. annurev-genet-120116-024704 https://doi.org/10.1146/annurev-genet-120116-024704.

Mayr, C., & Bartel, D. P. (2009). Widespread shortening of 3′UTRs by alternative cleavage and polyadenylation activates oncogenes in cancer cells. *Cell*, *138*(4), 673–684. https://doi.org/10.1016/j.cell.2009.06.016.

Mitra, M., Johnson, E. L., & Coller, H. A. (2016). Alternative polyadenylation can regulate post-translational membrane localization. *Trends in Cell & Molecular Biology*, *10*, 37–47.

Nojima, T., Gomes, T. T., Carmo-Fonseca, M., & Proudfoot, N. J. (2016). Mammalian NET-seq analysis defines nascent RNA profiles and associated RNA processing genome-wide. *Nature Protocols*, *11*(3), 413–428. https://doi.org/10.1038/nprot.2016.012.

Nojima, T., Gomes, T., Grosso, A. R. F., et al. (2015). Mammalian NET-seq reveals genome-wide nascent transcription coupled to RNA processing. *Cell*, *161*(3), 526–540. https://doi.org/10.1016/j.cell.2015.03.027.

Ohradanova-Repic, A., Machacek, C., Fischer, M. B., & Stockinger, H. (2016). Differentiation of human monocytes and derived subsets of macrophages and dendritic cells by the HLDA10 monoclonal antibody panel. *Clinical & Translational Immunology*, *5*(1). https://doi.org/10.1038/cti.2015.39, e55.

Oida, T., & Weiner, H. L. (2010). Depletion of TGF-β from fetal bovine serum. *Journal of Immunological Methods*, *362*(1–2), 195–198. https://doi.org/10.1016/j.jim.2010.09.008.

Pai, A. A., Baharian, G., Pagé Sabourin, A., et al. (2016). Widespread shortening of 3′ untranslated regions and increased exon inclusion are evolutionarily conserved features of innate immune responses to infection. *PLoS Genetics*, *12*(9), 1–24. https://doi.org/10.1371/journal.pgen.1006338.

Poddar, D., Basu, A., Baldwin, W. M., Kondratov, R. V., Barik, S., & Mazumder, B. (2013). An Extraribosomal function of ribosomal protein L13a in macrophages resolves inflammation. *Journal of Immunology*, *190*(7), 3600–3612. https://doi.org/10.4049/jimmunol.1201933.

Rostam, H. M., Singh, S., Salazar, F., et al. (2016). The impact of surface chemistry modification on macrophage polarisation. *Immunobiology*, *221*(11), 1237–1246. https://doi.org/10.1016/j.imbio.2016.06.010.

Sandberg, R., Neilson, J. R., Sarma, A., Sharp, P. A., & Burge, C. B. (2008). Proliferating cells express mRNAs with shortened 3′ UTRs and fewer microRNA target sites. *Science*, *320*(5883), 1643–1647. https://doi.org/10.1126/science.1155390.Proliferating.

Shi, Y. (2012). Alternative polyadenylation: New insights from global analyses. *RNA*, *18*(12), 2105–2117. https://doi.org/10.1261/rna.035899.112.

Suganuma, H., Fahey, J. W., Bryan, K. E., Healy, Z. R., & Talalay, P. (2011). Stimulation of phagocytosis by sulforaphane. *Biochemical and Biophysical Research Communications*, *405*(1), 146–151. https://doi.org/10.1016/j.bbrc.2011.01.025.

Tedesco, S., De Majo, F., Kim, J., et al. (2018). Convenience versus biological significance: Are PMA-differentiated THP-1 cells a reliable substitute for blood-derived macrophages when studying in vitro polarization? *Frontiers in Pharmacology*, *9*(FEB), 1–13. https://doi.org/10.3389/fphar.2018.00071.

Tian, B., Hu, J., Zhang, H., & Lutz, C. S. (2005). A large-scale analysis of mRNA polyadenylation of human and mouse genes. *Nucleic Acids Research*, *33*(1), 201–212. https://doi.org/10.1093/nar/gki158.

Tian, B., & Manley, J. L. (2016). Alternative polyadenylation of mRNA precursors. *Nature Reviews. Molecular Cell Biology*, *18*(1), 18–30. https://doi.org/10.1038/nrm.2016.116.

Wang, E. T., Sandberg, R., Luo, S., et al. (2008). Alternative isoform regulation in human tissue transcriptomes. *Nature*, *456*(7221), 470–476. https://doi.org/10.1038/nature07509.

Westermann, A. J., Gorski, S. A., & Vogel, J. (2012). Dual RNA-seq of pathogen and host. *Nature Reviews. Microbiology*, *10*(9), 618–630. https://doi.org/10.1038/nrmicro2852.

Wurth, L., & Gebauer, F. (2015). RNA-binding proteins, multifaceted translational regulators in cancer. *Biochimica et Biophysica Acta, Gene Regulatory Mechanisms*, *1849*(7), 881–886. https://doi.org/10.1016/j.bbagrm.2014.10.001.

Zhang, X., Hamblin, M. H., & Yin, K. J. (2017). The long noncoding RNA Malat1: Its physiological and pathophysiological functions. *RNA Biology*, *14*(12), 1705–1714. https://doi.org/10.1080/15476286.2017.1358347.

Zhang, F., Wang, H., Wang, X., et al. (2016). TGF-β induces M2-like macrophage polarization via SNAILmediated suppression of a pro-inflammatory phenotype. *Oncotarget*, 7(32), 52294–52306. https://doi.org/10.18632/oncotarget.10561.

CHAPTER SEVENTEEN

RIPiT-Seq: A tandem immunoprecipitation approach to reveal global binding landscape of multisubunit ribonucleoproteins

Zhongxia Yi and Guramrit Singh*
Department of Molecular Genetics, Center for RNA Biology, The Ohio State University, Columbus, OH, United States
*Corresponding author: e-mail address: singh.734@osu.edu

Contents

Abstract

RNA-binding proteins (RBPs) regulate all aspects of RNA metabolism. The ability to identify RNA targets bound by RBPs is critical for understanding RBP function. While powerful techniques are available to identify binding sites of individual RBPs at high resolution, it remains challenging to unravel binding sites of multicomponent ribonucleoproteins (RNPs) where multiple RBPs or proteins function cooperatively to bind to target RNAs. To fill this gap, we have previously developed RNA Immunoprecipitation in Tandem

Methods in Enzymology, Volume 655
ISSN 0076-6879
https://doi.org/10.1016/bs.mie.2021.03.019

followed by high-throughput sequencing (RIPiT-seq) to characterize RNA targets of compositionally distinct RNP complexes by sequentially immunoprecipitating two proteins from the same RNP and sequencing the co-purifying RNA footprints. Here, we provide an updated and improved protocol for RIPiT-seq. In this protocol, we have used CRISPR-Cas9 to introduce affinity tag to endogenous protein of interest to capture a more representative state of an RNP complex. We present a modified protocol for library preparation for high-throughput sequencing so that it exclusively uses equipment and reagents available in a standard molecular biology lab. This updated custom library preparation protocol is compatible with commercial PCR multiplexing systems for Illumina sequencing platform for simultaneous and cost-effective analysis of large number of samples.

1. Introduction

All steps in RNA metabolism—from RNA synthesis to RNA degradation—are mediated by proteins that bind RNA to control its biogenesis and function (reviewed in Hentze, Castello, Schwarzl, & Preiss, 2018; Müller-McNicoll & Neugebauer, 2013; Singh, Pratt, Yeo, & Moore, 2015). Such proteins include RNA-binding proteins (RBPs) and their interacting proteins, which often act in concert within multisubunit ribonucleoprotein (RNP) complexes. Identification of RNAs bound to an individual RBP or within a multisubunit RNP provides insight into functions of RBPs, RNPs, and RNAs themselves. Thus, methods for identifying RBP/RNP cargo RNAs serve as an important tool to understand gene regulation.

The past decade has seen tremendous progress in identifying RNA targets of individual RBPs via coupling of ultra-violet (UV) light Cross-Linking and Immunoprecipitation (CLIP) with high-throughput sequencing (CLIP-seq) (reviewed in Lee & Ule, 2018). In CLIP-seq, RNAs are UV-crosslinked to their bound RBPs and RNA segments bound by a specific RBP are recovered after stringent immunoprecipitation (IP) ofthe crosslinked RBP-RNA complex. These RNA segments are then converted into DNA libraries for identification via high-throughput sequencing. Depending on the library preparation strategy employed, some CLIP-seq variants can identify the RBP crosslinking site on RNA, and hence reveal RBP binding site at nucleotide resolution. However, CLIP-seq is disadvantageous when used for certain RBPs or RBP-associated factors which poorly crosslink to RNA with UV-light (Patton et al., 2020; reviewed in Wheeler, Nostrand, & Yeo, 2018). In addition, when studying multisubunit RNP complexes (e.g., cleavage and polyadenylation specificity factor (CPSF) complex involved in mRNA 3′ processing, multifactor mRNA decapping complex),

CLIP-seq can only identify targets of individual RBP but not the RNP complex, thus might lose the valuable information about the sites where two or more proteins are potentially acting in a synergistic manner (Patton et al., 2020).

Our lab studies exon junction complex (EJC), an RNA-binding protein complex that is deposited on mRNA exon-exon junctions during pre-mRNA splicing by the spliceosome (reviewed in Boehm & Gehring, 2016; Hir, Saulière, & Wang, 2016; Woodward, Mabin, Gangras, & Singh, 2017). EJC is composed of a trimeric core and more than a dozen peripheral factors. The EJC composition is heterogeneous due to incorporation of distinct peripheral factors that carry out different functions (Mabin et al., 2018; Wang, Ballut, Barbosa, & Le Hir, 2018). To study the RNA substrates of these individual EJC compositions, we have previously developed and employed an approach termed RNA immunoprecipitation in tandem (RIPiT) followed by high-throughput sequencing (RIPiT-seq) to identify the RNA targets of such multisubunit RNP complexes (Mabin et al., 2018; Singh et al., 2012). In the RIPiT procedure, one subunit of an RNP complex is first immunoprecipitated via an affinity tag, which enables non-denaturing elution of the RNPs containing this protein by preserving protein-protein associations. During the first IP, the RNPs are also treated with RNases to digest away RNAs that are not directly bound within RNPs. The RNase-digested RNPs thus obtained are then subjected to second IP of another subunit of the RNP complex. Following the second IP, the RNA footprints are isolated and converted into cDNA libraries for high-throughput sequencing. RIPiT-seq enables identification of RNA targets of compositionally distinct EJCs and it should be applicable for identification of RNA targets of other multisubunit RNP complexes in general (Fig. 1). While RIPiT-seq is less technically challenging as compared to CLIP-seq and can identify RNA targets of an RNP complex, it is important to note that it likely has reduced specificity as compared to CLIP-seq due to substantial carryover of contaminating RNAs (e.g., ribosomal RNA) during the nondenaturing immunoprecipitations.

Here, we describe an updated RIPiT-seq method, where we use CRISPR-Cas9 to introduce an affinity tag at the endogenous gene locus thereby reflecting the RNA cargoes of an RNP complex more accurately than via overexpression of a tagged protein. We also present an updated kit-free RNA library preparation method that, compared to its previous versions (Gangras, Dayeh, Mabin, Nakanishi, & Singh, 2018; Heyer, Ozadam, Ricci, Cenik, & Moore, 2015; Sterling, Veksler-Lublinsky, & Ambros, 2015; Woodward, Gangras, & Singh, 2019), is more generally

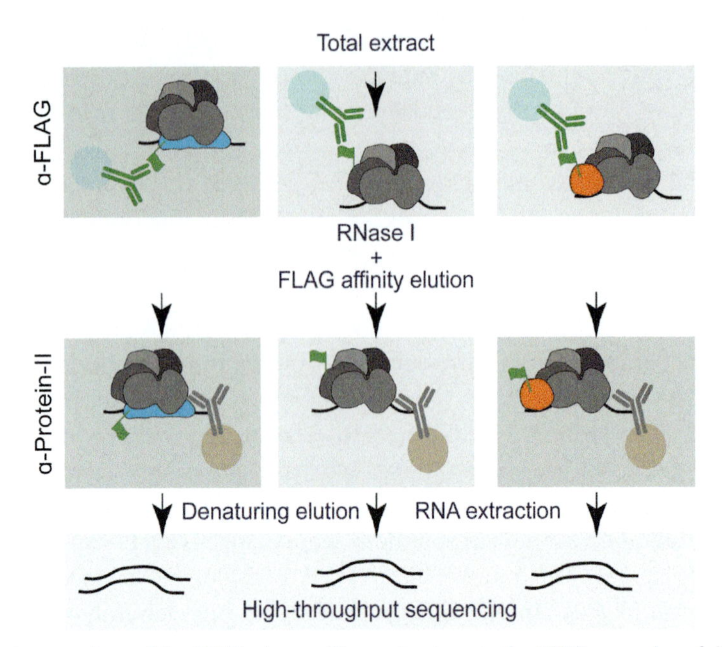

Fig. 1 An overview of the RIPiT scheme. The main steps in the RIPiT procedure following cell lysis are shown. From total extract, RNP complex is first subjected to immunoprecipitation of the tagged protein (shown here to be FLAG-tagged). The immunoprecipitated complexes are then subject to RNase I digestion while still bound to FLAG-affinity resin. The RNP complexes are then eluted under nondenaturing condition with the 3 × FLAG peptide and subject to a second immunoprecipitation using primary antibody targeting second protein of the RNP complex. RNAs are then eluted from the protein complex and converted into cDNA libraries for high-throughput sequencing. RNA bound by a specific composition in an otherwise heterogeneous RNP complex can be determined by choosing different pairs of proteins for the two immunoprecipitations.

accessible as it uses common lab equipment and reagents, and is compatible with commercial Illumina indexing primers, which enables high-level multiplexing at lower cost.

2. Before you begin

The RIPiT-seq procedure described here uses HCT 116 cells, a near-diploid colorectal cell line that is amenable to CRISPR-Cas9-mediated engineering of an affinity tag at endogenous gene locus and can be grown in medium-scale ($1–3 \times 10^9$ cells) cultures in standard tissue culture dishes. Other cell lines satisfying both these criteria can also be used for RIPiT-seq. Alternatively, affinity-tagged proteins can be expressed at near-endogenous levels from exogenously introduced DNA copies (e.g., we have successfully used TRex 293 cells and the Flp-In system for Tetracycline-inducible expression of affinity-tagged proteins for RIPiT-seq (Mabin et al., 2018)).

For the first IP in RIPiT, we have used FLAG tag, a short eight amino-acid tag for which a high-affinity antibody resin and competitor peptide for gentle elution are commercially available for scalable affinity purifications. The short nature of this tag allows it to be easily introduced into an endogenous gene locus via genome engineering to tag a polypeptide at either terminus and minimizes any chance of interference of protein function by the tag. Numerous other small affinity tags (e.g., HA tag, 3 × FLAG, Strep tag II) can also be similarly used in the procedure with minor modifications.

All home-made buffers can be prepared ahead of time. Components that should be added fresh just before use are noted. All buffers should be prepared with stocks that have been sterilized by autoclaving unless otherwise noted. One exception is Milli-Q water used to make large volume of isotonic wash buffer (IsoWB) used during RIPiT. While we have successfully carried out RIPiT using unautoclaved Milli-Q water to make IsoWB; to minimize concerns of RNase contamination autoclaved (and DEPC-treated, if necessary) water can be used.

For all library preparation steps, we recommend using pre-sterilized filter tips.

3. Materials and equipment

3.1 Equipment

- Gene Pulser Xcell Electroporation Systems (or another electroporation system for mammalian cells) (Bio-Rad #1652660)
- Sonifier Cell Disruptor (e.g., Branson benchtop sonifier SFX250)
- Blue Light Transilluminator
- Automated Cell Counter (e.g., Countess II Automated Cell Counters) or a Hemocytometer
- Magnetic Rack Separator (Sergi Lab Supplies)
- Gene Pulser Electroporation Cuvettes, 0.2 cm gap (Bio-Rad #1652082) or similar products

3.1.1 Cell line

- HCT 116, Human Colorectal Carcinoma (ATCC CCL-247)

3.1.2 Reagents

- 5′ DNA Adenylation Kit (NEB #E2610S)
- T7 Endonuclease I (NEB #M0302S)
- 3 × FLAG peptide (APExBIO #A6001)
- Protease Inhibitor Cocktail (APExBIO #K1007)
- Phosphatase Inhibitor Cocktail (APExBIO #K1015)
- Recombinant Cas9 (QB3 MacroLab, UC Berkeley)
- Aphidicolin (Fisher Scientific #AC611970010)
- ANTI-FLAG M2 Affinity Gel (Sigma #A2220)
- RNase I (Lucigen #N6901K)
- EIF4A3 Antibody (Bethyl Laboratories #A302-980A)
- SYBR Gold Nucleic Acid Gel Stain (Thermo Fisher Scientific #S11494)
- 10% TBE-Urea gel (Bio-Rad #4566036) (or equivalent product from other vendors)
- Maxima H Minus Reverse Transcriptase (Thermo Fisher Scientific #EP0753)
- McCoy's 5A (Modified) Medium (Fisher Scientific #16-600-108)
- Fetal Bovine Serum (Sigma-Aldrich #F2442)
- Penicillin-Streptomycin (10,000 U/mL) (Fisher Scientific #15-140-122)
- Trypsin-EDTA (0.05%), phenol red (Fisher #25300-062)
- Ingenio Electroporation Solution, or another mammalian electroporation solution (Mirus Bio #MIR50114)
- Dynabeads Protein A (Thermo Fisher Scientific #10002D)
- Dynabeads Protein G (Thermo Fisher Scientific #10004D)
- 5PRIME Phase Lock Gel (QuantaBio #2302820) or similar product such as Corning high-vacuum silicone grease (Sigma-Aldrich #Z273554)
- Quick CIP (NEB #M0525)
- T4 RNA Ligase 2, truncated K227Q (NEB #M0351L)
- T4 RNA Ligase Buffer (NEB #B0216L)
- Biotin-11-dATP (AAT Bioquest #17014 or PerkinElmer #NEL540001EA)
- Biotin-16-Aminoallyl-2′-dCTP (TriLink BioTechnologies #N-5002-1)
- Low Molecular Weight DNA Ladder (NEB #N3233S) or similar DNA ladder
- RNasin Plus Ribonuclease Inhibitor (Promega #N2615)

- Hydrophilic Streptavidin Magnetic Beads (NEB #S1421S)
- CircLigase ssDNA Ligase and CircLigase 10 × Reaction Buffer (Lucigen #CL4115K)
- HighPrep PCR beads (MagBio #AC-60005) or similar product such as NucleoMag (Macherey-Nagel #744970.5) and AMPure XP (Beckman Coulter #A63880)
- NEBNext Q5 Master Mix (NEB #M0544S)
- 5 M Betaine solution (Sigma-Aldrich #B0300)
- SYBR Green I Nucleic Acid Gel Stain, 10,000 × concentrate (Thermo Fisher Scientific #S7563)

3.1.3 Chemicals

- Phenol/Chloroform/Isoamyl Alcohol 125:24:1, pH 4.3 (PCIA pH 4.3; Fisher Scientific #BP1754I-400)
- Phenol/Chloroform/Isoamyl Alcohol 25:24:1, pH 8.0 (PCIA pH 8.0; Fisher Scientific #BP1752I-100)
- 50% PEG8000
- Deionized Formamide (Sigma-Aldrich #4650-500ML)
- Cycloheximide (Sigma-Aldrich #C7698-1G)

3.1.4 Home-made buffers

- 1 × Phosphate Buffered Saline (PBS)
  137 mM NaCl; 2.7 mM KCl; 10 mM Na_2HPO_4; 1.8 mM KH_2PO_4
- Hypotonic Lysis buffer
  20 mM Tris-HCl pH 7.5; 15 mM NaCl; 10 mM EDTA (Substitute with 1 mM $MgCl_2$ if interactions are Mg dependent); 0.1% Triton X-100; 1 × Protease Inhibitor Cocktails*; (optional) 1 × Phosphatase Inhibitor Cocktail* (* add fresh every time)
- Isotonic Wash Buffer (IsoWB)
  20 mM Tris-HCl pH 7.5; 150 mM NaCl; 0.1% IGEPAL CA-630
- 2 × Dilution Buffer
  20 mM Tris-HCl pH 7.5; 150 mM NaCl; 0.2% Triton X-100; 10 mM EDTA; 0.2 mg/ml BSA*
  2 × Protease Inhibitor Cocktail*; (optional) 2 × Phosphatase Inhibitor Cocktail* (* add fresh every time)
- Conjugation Buffer
  1 × PBS; 0.01% Tween-20
- Clear Sample Buffer
  100 mM Tris-HCl pH 6.8; 4% SDS; 10 mM EDTA

- 5 × First strand (FS) w/o $MgCl_2$ Buffer
  250 mM Tris-HCl pH 8.3; 375 mM KCl
- 1 × Urea Load Buffer
  1 × TBE; 12% ficoll; 7 M Urea; 0.01% bromophenol blue; 0.02% xylene cyanole FF
- 100 mg/mL cycloheximide stock
- DNA Elution Buffer:
  300 mM NaCl; 1 mM EDTA
- 50 × Denhardt's solution:
  1% Ficoll 400; 1% Polyvinylpyrrolidone (PVP); 1% Bovine serum albumin (Fraction V)
- Strep Bead Wash Buffer:
  10 mM Tris-HCl pH 7.5; 1 mM EDTA; 0.3 M NaCl; 0.1% Tween-20
- Strep Bead Resuspension Buffer:
  10 mM Tris-HCl pH 7.5; 0.1 mM EDTA; 0.3 M NaCl
- 1 mM biotin-dNTPs
  0.25 mM dGTP; 0.25 mM dTTP; 0.175 mM dATP; 0.075 mM biotin-dATP; 0.1625 mM dCTP; 0.0875 mM biotin-dCTP

3.1.5 Oligos

- sgRNA or cr:tracrRNA: Guide RNA can be designed using various platforms (e.g., https://benchling.com/). Guide RNA cut site should be as close to the tag insertion site as possible. Synthetic guides can be ordered from multiple vendors such as Synthego and IDT. If necessary, editing efficiency of individual guide can be evaluated from T7 Endonuclease I assay according to the manufacturer protocol.
- Single-stranded oligo donor (ssODN): ssODN contains an affinity tag sequence flanked by 35–50 nt homology arm (Fig. 2A). We include a short GGGS linker following the tag sequence that also contains BamHI (encodes GS) site for the screening purpose.
- mirCat33 adapter: /rApp/TGGAATTCTCGGGTGCCAAGG/ddC/. If adenylated adapter is not commercially available, adenylation can be performed using 5′ DNA Adenylation Kit following the manufacturer protocol. We have reduced RNA:enzyme ratio down to 4:1 without any obvious decrease in adenylation efficiency. The adenylated adapter can be purified with silica column or with Urea-PAGE gel.
- RT Primer: /5Phos/GGNNNNNNNNAGATCGGAAGAGCGT CGTGTAGGGAAAGAGTGT/iSp18/GTGACTGGAGTTCAGAC GTGTGCTCTTCCGATCTCCTTGGCACCCGAGAATTCCA:

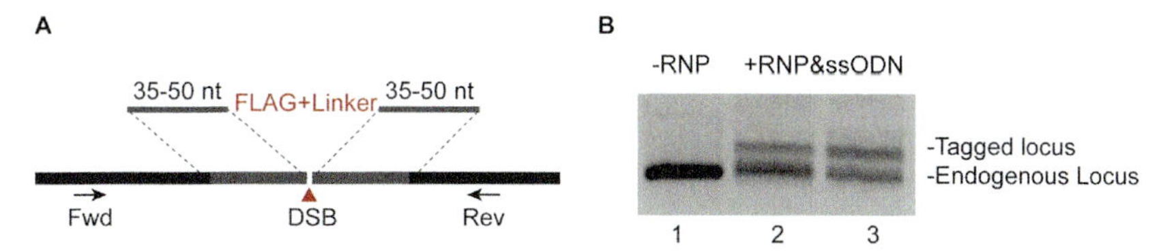

Fig. 2 CRISPR-Cas9-mediated insertion of affinity tag and bulk assessment of tag insertion. (A) Scheme of CRISPR-Cas9-mediated insertion of affinity tag (FLAG) into endogenous locus of a gene. Single-stranded oligo donor (ssODN) for homology-directed repair is on the top where FLAG tag sequence and a flexible linker (amino acid sequence GGGS) is flanked by 35–50nt DNA sequence with perfect homology with the DNA sequence surrounding the cut site (triangle) induced by guide RNA-Cas9 RNP complex. Primers (labeled Fwd and Rev) flanking the insertion site are used in the PCR reaction with genomic DNA as a template to identify the insertion event. (B) A bulk-level (i.e., in a pool of cells) analysis of CRISPR-mediated insertion efficiency by amplifying the insertion site with PCR 48–72h after guide RNA-Cas9 RNP complex electroporation. With a 20–30% insertion efficiency and relatively small amplicon, there should be a clear shift of amplicon containing inserted tag (two replicates shown in lanes 2 and 3, labeled tagged locus) as compared to wild-type endogenous locus (single band in lane 1).

Underlined is an 8-mer unique molecular identifier (UMI) used for removing PCR duplicates. RT primer is ordered via IDT as DNA oligo with HPLC purification.

- I5_uni: AATGATACGGCGACCACCGAGATCTACACACACTC TTTCCCTACACGACGCTCTTCCGAT
  C*T (* Phosphorothioate Bond). Order from IDT as desalted Ultramer.
- I7_uni: CAAGCAGAAGACGGCATACGAGATGTGACTGGAGT TCAGACGTGTGCTCTTCCGATC*T (* Phosphorothioate Bond). Order from IDT as desalted Ultramer.
- Commercial Illumina I5/I7 Indexing Primers: I5/I7_uni primers do not contain barcodes and are used in qPCR for the cycle number determination. For final PCR to prepare library for high-throughput sequencing, commercial Illumina indexing primer set can be ordered from vendors such as Lexogen or IDT. Alternatively, individual indexing primers can be ordered as desalted Ultramers from IDT.

4. Step-by-step method details

4.1 Affinity tag knock-in using CRISPR-Cas9

1. 1 day before CRISPR-Cas9 ribonucleoprotein (RNP) electroporation, seed approximately 2.5×10^6 HCT 116 cells in a 10-cm plate in growth

media (McCoy's 5A (Modified) Medium, 10% FBS, 1% Penicillin-Streptomycin) supplemented with 2 μg/mL aphidicolin for cell cycle synchronization (Lin, Staahl, Alla, & Doudna, 2014; Rivera-Torres & Kmiec, 2017; Rivera-Torres, Strouse, Bialk, Niamat, & Kmiec, 2014). Grow cells overnight under standard growth conditions (5% CO_2 and 37 °C). *Note:* Cell synchronization, as show in the references indicated earlier as well as in our experience, strongly enhances the HDR efficiency.

2. 4 h before electroporation, replace media with fresh growth media without aphidicolin to release the cells from aphidicolin-block and enter S phase (Rivera-Torres & Kmiec, 2017).
3. Prepare RNP complex as described below. RNP complex should be at least 1 μM in the 50 μL final electroporation reaction.

Cas9	50 pmol
sgRNA	50–100 pmol (Cas9:guide RNA ratio can be between 1:1 and 1:2)

Add Ingenio electroporation solution to a total volume of 20 μL. Mix well and incubate at room temperature (RT) for ~20 min.

4. To harvest cells for electroporation, wash cells with 10 mL of 1 × PBS. Add 1 mL of Trypsin-EDTA and incubate at 37 °C for 5 min to dissociate cells. Resuspend cells in 9 mL of growth media. Use a small aliquot to count cells using a cell counter to estimate cell concentration. Aliquot into 1.5 mL centrifuge tubes appropriate volume of cell suspension to get 2.5×10^5 cells per reaction. Harvest cells at 500 × *g* for 3 min.
5. To wash cells with 1 × PBS, resuspend cell pellet in 1 mL 1 × PBS and centrifuge again at 500 × *g* for 3 min.
6. Resuspend cells in 28.5 μL Ingenio electroporation solution. Add 1.5 μL (150 pmol) of ssODN. Mix resuspended cells with RNP complex from step 3 to a final volume of 50 μL.
7. Transfer the 50 μL cell-RNP mix to a 0.2 cm electroporation cuvette. For HCT 116 cells, use the following setting for electroporation: 120 V, 13 ms/per pulse; 2 pulses with 1 s interval.
8. Transfer cells to a 6-well plate containing 2 mL growth media and incubate under standard growth conditions.
9. 48–72 h later, dissociate cells by trypsinization as in step 4 above, and resuspend in 2 mL growth media. Count cells, dilute cells to either achieve 5–10 cell/mL to seed in 96-well plates or seed diluted cells (5–15 cells/mL) in 10-cm plates to isolate single colonies.

Pellet remaining cells to isolate genomic DNA and analyze homology-directed repair (HDR) efficiency via PCR (Fig. 2). *Note*: To confirm efficient insertion of affinity tag into an endogenous gene locus, a bulk-level screening of CRISPR-Cas9 knock-in efficiency can be performed on the RNP-transfected cells. This can be achieved using a pair of primers flanking the insertion site, as shown in Fig. 2. Template DNA for this genotyping PCR can be obtained via a direct lysis protocol described here (Ramlee, Yan, Cheung, Chuah, & Li, 2015), skipping the need to isolate genomic DNA. After 2–3 weeks, pick single colonies (number of colonies to pick will depend on the HDR efficiency) and screen cells for correct insertion. Individual colonies are screened using the same PCR primers in the previous step to identify the cell clones that harbors the tag insertion. Sanger sequencing is required to validate the correct insertion of the tags.

4.2 RNA immunoprecipitation in tandem

1. Grow HCT 116 cells expressing affinity-tagged protein of interest in 15-cm plates under standard growth conditions. (To ensure enough RNA is recovered, at least four 15-cm plates with around 70% cell confluency are needed per RIPiT for an abundant RNP like EJC.)
2. 1 h prior to harvesting cells, add cycloheximide (CHX) to 100 μg/mL final concentration in the media (prepare fresh 100 mg/mL CHX stock each time). *Note*: This step allows accumulation of untranslated mRNAs and hence the EJC-bound pool of mRNAs over the 1-h period. It can be omitted if RNP of interest is not impacted by the translating ribosome.
3. Rinse cells once with 25 mL of 1 × PBS (with 100 μg/mL CHX) and harvest cells in 10 mL 1 × PBS (with 100 μg/mL CHX) using a cell scrapper. Pellet cells at 1000 × *g* for 5 min at 4 °C.
4. Resuspend cells in 4 mL ice-cold gentle hypotonic lysis buffer. To solubilize chromatin associated RNPs, sonicate lysate using Branson tabletop sonifier at 15% amplitude for 30 s in burst of 2 s with 3-s intervals. *Note*: This step is optional and can be omitted if RNPs of interest are primarily in soluble nucleoplasm or cytoplasm.
5. Add NaCl to 150 mM. Keep on ice for 5 min. Spin at 15000 × *g*, 4 °C, 10 min. *Note*: To increase stringency, higher salt concentrations can be used as long as the RNP of interest remains intact at these elevated ionic conditions.

6. Save 40 μL lysate supernatant as input. Store at −20 °C. Rest of the lysate will be used for FLAG immunoprecipitation.
7. Wash anti-FLAG Affinity Gel three times with 1 mL IsoWB. *Note*: 40 μL of the affinity gel (20 μL beads volume) is usually enough to deplete endogenous FLAG-tagged protein from total extract prepared from one 15-cm plate. 200 μL affinity gel is usually sufficient for 4–6 15-cm plates.
8. Add cleared lysate from step 6 to the washed FLAG affinity gel. Nutate at 4 °C for 1 h.
9. Pellet FLAG affinity beads at 1000 × *g*, 1 min at 4 °C. Save 40 μL of supernatant to later check efficiency of depletion of the target protein. Store at −20 °C.
10. Wash beads four times with 1 mL IsoWB (ice-cold).
11. After fourth wash, add 200–300 μL of IsoWB supplemented with RNase I at concentration of 0.01–0.1 unit/μL. Gently vortex at 4 °C for 15 min.
12. Wash beads four times with 1 mL ice-cold IsoWB.
13. Elute FLAG-tagged protein containing RNPs with at least one bed volume (100 μL) of elution buffer containing 3 × FLAG peptide at 500 μg/mL in IsoWB. Gently shake tube contents at 4 °C for 30 min. Repeat the elution twice and combine both the elutions. Save 20 μL of elution to test for efficiency of FLAG IP. *Note*: 4 °C elution is important to preserve protein-protein association.
14. During incubation period for the FLAG elution, prepare antibody-protein A/G DynaBeads conjugates for the second IP. Pre-wash 25 μL protein A or protein G DynaBeads three times with conjugation buffer and resuspend the beads in 1 mL conjugation buffer. Add appropriate amount of the primary antibody against the second protein of the RNP. For EIF4A3 (EJC protein), 5 μg of antibody is enough for one RIPiT from 4 to 6 15 cm plates. Gently vortex for 30 min at RT. Wash antibody conjugated beads three times with 1 mL IsoWB.
15. Add 1 volume of 2 × Dilution Buffer to the FLAG elution and mix the elution with protein A/G conjugated primary antibody for the second protein target.
16. Nutate at 4 °C for 1 h.
17. Wash beads six times with 1 mL ice-cold IsoWB.
18. To elute RNPs, add 20 μL Clear Sample Buffer and incubate at 50 °C for 5 min. Save 2 μL and store at −20 °C for western blot (Fig. 3).
19. To extract RNA from the rest of the elution, add 80 μL water and 100 μL of Phenol/Chloroform/Isoamyl Alcohol (PCIA) (125:24:1; pH 4.3), mix vigorously. Centrifuge at 15000 × *g* for 5 min at RT.

20. Transfer 100 μL aqueous phase to a new tube, add 1/10 volume of 3 M sodium acetate pH 5.2, 1/100 volume of 1 M $MgCl_2$, 2 μL of 5 mg/mL glycogen, and 3 volumes of cold 100% ethanol. Mix well and incubate at −80 °C for 1 h or −20 °C overnight. *Note*: Using vacuum grease or commercial phase locking tubes is helpful in physically separating the organic phase to facilitate the transfer of aqueous phase for every phenol chloroform cleanup step hereafter.

21. Centrifuge for 30 min at 15000 × *g* at 4 °C. Discard supernatant.

22. Wash pellet with ice old 70% ethanol and centrifuge for 5 min at 15000 × *g* at 4 °C. Discard supernatant.

23. Air dry RNA pellet and resuspend RNA in 17 μL RNase-free water.

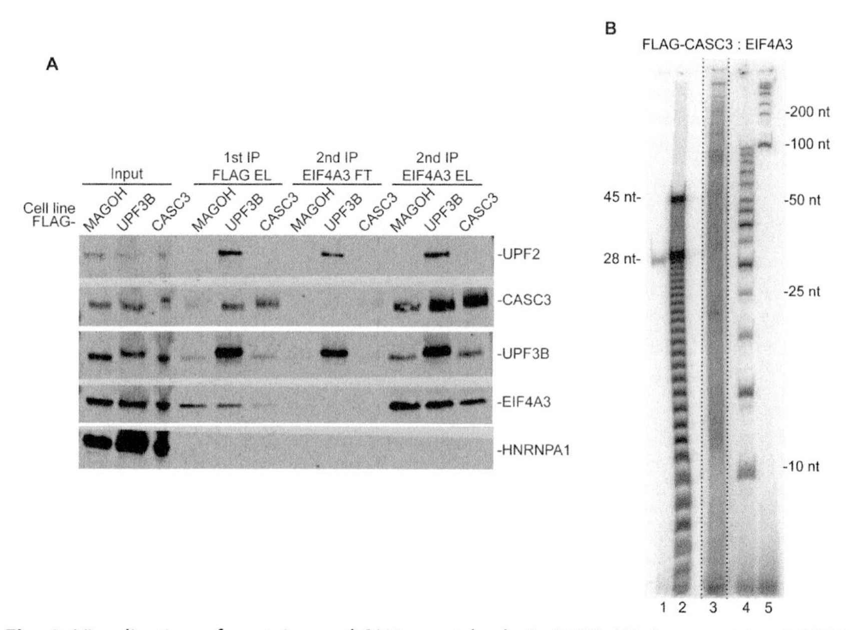

Fig. 3 Visualization of proteins and RNAs enriched via RIPiT. (A) An example of RIPiT validation via western blots. Western blots showing levels of the proteins (on the right) in various fractions labeled on the top (EL, elution; FT, flowthrough). Cell lines expressing the endogenously tagged protein used in RIPiT are indicated above each lane. As seen in the figure, different complexes have different composition, as reflected by different levels of UPF2 protein in the three complexes. (B) An example of RNA footprints from FLAG-CASC3:EIF4A3 containing complexes following 5′-end-labeling with [γ-32P] ATP and resolution on a 26% Urea-PAGE gel. Shown image was spliced at the vertical dotted lines. Size distribution of the RNA footprints can be estimated via comparison with radio-labeled base-hydrolyzed poly-U_{45} RNA ladder (lane 2) and DNA ladders (lane 4 and 5). A known amount (0.1 pmole) of single-stranded RNA oligo (lane 1) is radio-labeled and run on the gel to estimate RIPiT RNA concentration.

24. Dephosphorylate resuspended RNA by adding 2 μL NEB CutSmart Buffer (10 ×) and 1 μL Quick CIP enzyme. Mix well and incubate at 37 °C for 30 min.
25. To the dephosphorylated RNA, add 80 μL water and 100 μL of Phenol/Chloroform/Isoamyl Alcohol (PCIA) (125:24:1; pH 4.3), mix vigorously. Centrifuge at 15000 × *g* for 5 min.
26. Transfer 100 μL aqueous phase to a new tube, add 1/10 volume of 3 M sodium acetate pH 5.2, 1/100 volume of 1 M $MgCl_2$, 2 μL of 5 mg/mL glycogen, and 3 volumes of cold 100% ethanol. Mix well and incubate at −80 °C for 1 h or −20 °C overnight.
27. Centrifuge for 30 min at 15000 × *g* at 4 °C. Discard supernatant.
28. Wash pellet with ice old 70% ethanol and centrifuge for 5 min at 15000 × *g* at 4 °C. Discard supernatant.
29. Air dry RNA pellet and resuspend RNA in 5 μL water.
30. Optional: To visualize length distribution of immunoprecipitated RNA fragments, and to quantify their amount, 1 μL of RNA can be end labeled with [γ-32P] ATP using T4 polynucleotide kinase and visualized on high resolution (20–26% polyacrylamide) gel (Fig. 3).

4.3 RNA footprints library preparation

1. Set up following reaction in a PCR tube:

mirCat33 adapter	7 pmol
RNA (from RIPiT)	4 μL
Water	to 4.8 μL

2. Incubate in a PCR machine at 65 °C 10 min; 16 °C 5 min; 4 °C hold.
3. Transfer tubes to ice and add:

T4 RNA Ligase Buffer	1.5 μL
50% PEG8000	7.5 μL
20 mM DTT	0.75 μL
T4 RNA Ligase 2 (K227Q)	0.45 μL
Total Volume	15 μL

Incubate at 30 °C for 6 h, 65 °C for 20 min and 4 °C hold.

4. Add to the same tube containing 15 μL ligation (on ice):

10 μM RT Primer	1 μL
1 mM biotin-dNTPs	10 μL

5. Incubate at 65 °C 5 min followed by 4 °C for at least a minute.
6. Add the following reagents to the same tube:

5 × FS w/o $MgCl_2$ Buffer	9 μL
100 mM DTT	2.25 μL
Maxima H-RTase	1.2 μL
RNasin Plus	1.25 μL
Water	5.3 μL
Total volume	45 μL

7. Incubate at 55 °C for 45 min; inactivate RTase at 70 °C 15 min; hold reaction at 4 °C if needed.
8. Transfer the reaction to 1.5 mL Eppendorf tube. Add 155 μL water and 200 μL PCIA (25:24:1; pH 8.0) mix vigorously. Centrifuge at 15000 × *g* for 5 min.
9. Transfer 200 μL aqueous phase to a new tube, add 1/10 volume of 3 M sodium acetate pH 5.2, 1/100 volume of 1 M $MgCl_2$, 2 μL of 5 mg/mL glycogen, and 3 volumes of cold 100% ethanol. Mix well and incubate at −20 °C for 15 min.
10. Centrifuge for 15 min at 15000 × *g* at 4 °C. Discard supernatant.
11. Wash pellet with ice old 70% ethanol and centrifuge for 5 min at 15000 × *g* at 4 °C. Discard supernatant.
12. Air dry and resuspend cDNA pellet in 10 μL 1 × Urea Loading Dye. Denature at 65 °C for 5 min before loading on a 10% TBE-Urea gel. Run at 200 V for 40 min. For size markers, load Low Molecular Weight DNA Ladder and 10 pmol of RT primer along with the cDNA samples.

13. Stain gel with SYBR gold following the manufacturer guidelines.
14. Visualize gel on a blue light illuminator. Excise gel segments corresponding to 130–200 nt (or application-dependent) long reverse transcribed products (Fig. 4). Elute with the PAGE elution buffer at 30 °C overnight (300–600 μL). To maximize recovery, elution from gel pieces can be done twice and the two elutions can be combined.
15. Aliquot 10 μL of streptavidin beads for each sample. Wash three times with 0.1 M NaOH.
16. Wash twice with Strep Bead Wash Buffer.
17. Wash once with Strep Bead Resuspension Buffer.
18. Block beads with 30 μL of Strep Bead Resuspension Buffer + 3 μL 50 × Denhardt's solution. Incubate at 30 °C for 20 min with gentle shaking.
19. Wash beads twice with Strep Bead Wash Buffer.
20. Add cDNA elution to the beads, incubate at 30 °C for 30 min with gentle shaking.
21. Wash beads twice with Strep Bead Wash Buffer and once with Strep Bead Resuspension Buffer.
22. Add 5 μL water to resuspend the beads and proceed with cDNA circularization.
23. To resuspended streptavidin beads (volume will be ~5.5 μL), add the following:

CircLigase 10 × Reaction Buffer	1 μL
1 mM ATP	0.5 μL
50 mM MnCl2	0.5 μL
5M betaine	2 μL
CircLigase	0.5 μL
Total Volume	10 μL

24. Incubate 60 °C for 3 h followed by head inactivation at 80 °C for 10 min.
25. Wash beads with twice with Strep Bead Wash Buffer and twice with Strep Bead Resuspension Buffer.
26. Elute circularized cDNA in 15 μL 95% deionized formamide + 10 mM EDTA at 65 °C for 5 min and 90 °C for 2 min.

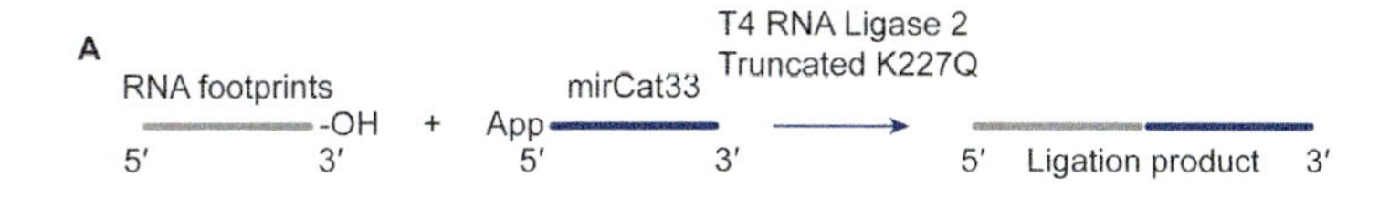

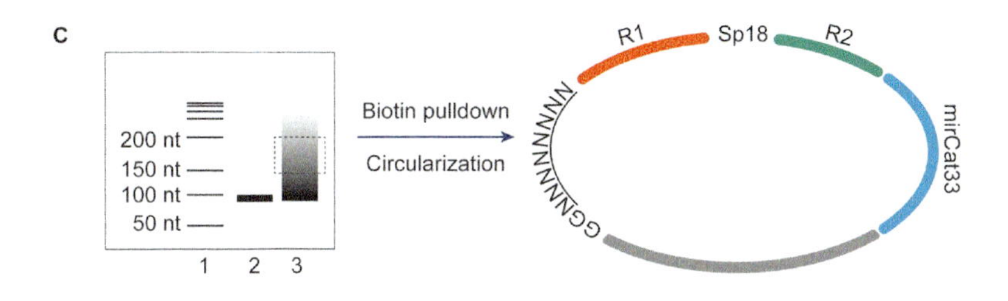

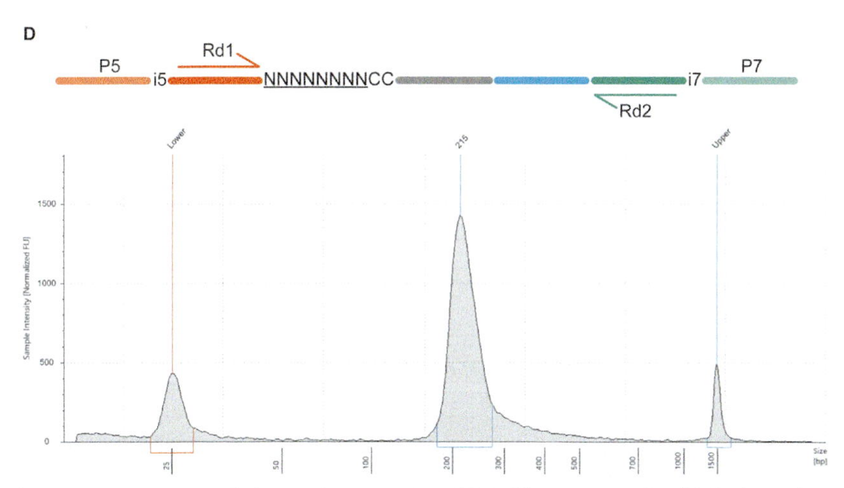

Fig. 4 An overview of the main steps in cDNA library prep for high-throughput sequencing. (A) The first step consists of ligation of a pre-adenylated mirCat33 adapter to the dephosphorylated 3′ end of RNA footprints with T4 RNA Ligase 2, truncated K227Q. (B) Ligation product from A is then used for reverse transcription reaction (in the same tube as ligation) using the special RT primer that includes a region complementary to mirCat33 adapter, binding sites for read two primer binding sequence linked via polyethylene glycol linker (Sp18) to read one primer binding sequence, the 8-nucleotide random sequence that serves as unique molecular identifier (UMI) and two Gs at the 5′-end that conform to the preferred base for the CircLigase used in a subsequent step. The RT reaction includes biotinylated dATP and dCTP to specifically

(Continued)

27. Add 85 μL water and 100 μL PCIA (25:24:1; pH 8.0), mix vigorously. Centrifuge at 15000 × *g* for 5 min.

28. Transfer 100 μL aqueous phase to a new tube, add 1/10 volume of 3 M sodium acetate pH 5.2, 1/100 volume of 1 M $MgCl_2$, 2 μL of 5 mg/mL glycogen, and 3 volumes of cold 100% ethanol. Mix well and incubate at −20 °C for 15 min.

29. Centrifuge for 15 min at 15000 × *g* at 4 °C. Discard supernatant.

30. Wash pellet with ice old 70% ethanol and centrifuge for 5 min at 15000 × *g* at 4 °C. Discard supernatant.

31. Air dry and resuspend cDNA pellet in 11 μL water.

32. Use 1 μL of cDNA for 30 μL qPCR. Set up a parallel reaction using 1 μL water as control:

NEBNext Q5 Master Mix	15 μL
I5_uni (20 μM)	1.5 μL
I7_uni (20 μM)	1.5 μL
2.5 × SYBR Green I	1.2 μL
cDNA	1 μL
Water	9.8 μL

33. Run the following qPCR program:

Initial Denaturation	98 °C	30 s
Denaturation	98 °C	10 s
Annealing/Extension	65 °C	75 s × 35 cycles
Final Extension	65 °C	5 min

Fig. 4—Cont'd label the extended RT products for enrichment in the next step. (C) Left: A schematic of a gel image where RT products (lane 3) are resolved on UREA-PAGE and visualized via SYBR-gold staining along with a low range DNA ladder (lane 1) and unextended RT primer (lane 2). Size-selected cDNAs (marked by dotted box) are eluted from gel, pulled down using streptavidin beads and eluted from beads. Such cDNAs are next subject to circularization. Circularized cDNAs are then ready for PCR amplification with indexing primers. (D) Schematic of the final library product with Illumina I5/I7 dual indexes and a bioanalyzer trace example. Rd1 and Rd2 represent read1 and read2 sequencing primers.

34. To determine cycle number for final PCR, identify cycle number required for qPCR to reach 50% of the maximum fluorescence value, then subtract 3 cycles and this will be used for final PCR using the rest 10 μL cDNA.

35. Set up 30 μL final PCR reaction:

NEBNext Q5 Master Mix	15 μL
I5 Indexing Primers (20 μM)	1.5 μL
I7 Indexing Primers (20 μM)	1.5 μL
cDNA	10 μL
Water	2 μL

36. Run the following qPCR program:

Initial denaturation	98 °C	30 s
Denaturation	98 °C	10 s
Annealing/extension	65 °C	75 s × Cycle number in Step 34
Final extension	65 °C	5 min
Hold	4 °C	∞

37. Add 36 μL (1.2 ×) HighPrep PCR beads or similar product. Mix thoroughly and incubate at RT for 5 min.

38. Place the tube on magnetic rack for ~2 min or until beads are separated. Carefully discard supernatant.

39. While still on magnetic rack, wash beads twice with freshly prepared 80% ethanol for 30 s each.

40. Discard ethanol and air dry for 2–5 min.

41. Add 20 μL water to the dried beads. Mix and incubate for 2 min.

42. Place the tube on magnetic rack until beads are separated, transfer elution to a new tube.

43. 2 μL of the library DNA can be diluted with 2 μL water for High Sensitivity-DNA TapeStation Bioanalyzer assay. As PCR products resulting from unextended circularized RT primer is expected to be

167 bp (with I5/I7 dual 8 bp indices), insert size can be deduced from the Bioanalyzer results. Library molarity of the region containing desired size PCR products as reported by the 2200 TapeStation Software (Agilent) will be used for library pooling.

5. Expected outcomes

5.1 Affinity tag knock-in using CRISPR-Cas9

RNP electroporation and ssODN-mediated HDR regularly yields 20–30% knock-in efficiency in our hands. Certain gene loci can present exceptions and show lower knock-in efficiency. In such cases, a selectable marker-based knock-in strategy can be adopted (Park, Won, Pentecost, Bartkowski, & Lee, 2014). Small scale FLAG IP should be performed to validate that the knock-in protein can be successfully immunoprecipitated and that it can co-IP its associated proteins.

5.2 RNA immunoprecipitation in tandem

After RIPiT of the RNP complex, small fractions saved during the procedure (total extract, depleted extract after first IP, immunoprecipitates from first IP, unbound fraction from the second IP and immunoprecipitates from the second IP) should be analyzed by western blotting to validate the immunoprecipitation of the complex (Fig. 3A). Proteins immunoprecipitated during first and second IP should be of highest abundance on western blots (Fig. 3A). Other factors of the same RNP complex should also be readily observed on the western blots along with the immunoprecipitated proteins. A comparison of total extract with depleted extract after the first IP will indicate the efficiency of the first IP step. Similarly, the unbound fraction after the second IP will contain proteins that associate with the target of the first IP but not with the protein immunoprecipitated second.

RNA footprints isolated from RIPiT should contain a complex mixture of RNAs that should appear as a smear on the RNA gel (Fig. 3B) or BioAnalyzer profile. The presence of prominent distinct-sized bands in the RNA profiles will be indicative of high levels of RNAs of distinct size and sequences, often originating from abundant RNAs such as transfer RNAs or ribosomal RNAs. In such cases, specificity of IPs can be increased by more stringent washing (e.g., higher salt in IsoWB) or by formaldehyde crosslinking RNPs prior to cell lysis followed by denaturing FLAG IP (see below). EJC RIPiTs from typical input material noted in the procedure above regularly yield sufficient amount of RNA for subsequent library preparation. Recommend amount of RNA for input into the downstream library preparation is 1 pmol, although the library preparation method described above works for inputs as low as 0.1 pmol (Gangras et al., 2018).

5.3 RNA footprint library preparation

Following reverse transcription (RT), the extended RT product will appear as a smear above the prominent band at 98 nt, which is the unextended RT primer (Fig. 4C). To ensure that final cDNA library has inserts >30 nt, which is the lower limit for robust mapping of fragment sequences to the human genome, RT products of lengths >130 nt should be size-selected. When input RNA is >1 pmol, the final PCR reaction should require 7–12 cycles. However, if RNA input is higher or lower, PCR cycles outside this range should also be fine for the downstream high-throughput sequencing. It is highly recommended to carefully estimate the number of PCR cycles for the final PCR to be in the linear amplification range. However, if over-amplification occurs, the unique molecular identifiers (UMIs) in the RT primer (Fig. 4C) enable removal of PCR duplicates during downstream processing of sequencing reads. Finally, when the above in-house RNA library preparation procedure is used for the first time, we recommend cloning a small amount of the final PCR product into a plasmid vector (e.g., TA cloning vector provided PCR products carry A overhangs) and identifying insert sequences of 6–12 clones via traditional Sanger sequencing to make sure the library products contain expected sequences.

6. Quantification and statistical analysis

Following Illumina high-throughput sequencing, the first step is to extract the UMI sequences. The UMI-tools package (Smith, Heger, & Sudbery, 2017) offers such an extraction command. Alternatively, first eight nucleotides of sequencing reads can be manually extracted using program such as Awk. UMI sequences should be moved to the end of identifier line in the fastQ file for downstream deduplication purpose as indicated in the manual (Smith et al., 2017). Next, fastQ files are subject to adapter trimming. We select only those reads that start with a CC and end with the mirCat33 adapter sequence. Inclusion of reads missing the mirCat33 adapter might mislead the binding site information of RBP as the 3′ end of the RNA fragment will be unknown.

Clean reads thus obtained after adapter trimming can then be aligned to the reference genome. We highly recommend to first map reads to rRNA or other highly abundant non-coding RNAs that are not of interest, and then align the remaining reads to the genome. Following genome-mapping, PCR duplicates can be removed using UMI-tools dedup function (Smith et al., 2017). The genome-mapped reads should reflect binding site of the RNP complex and it will vary based on the complex being pulled down. For the EJC, we observe strong enrichment on the exon-exon junctions when aligned reads are visualized using tools such as Integrated Genome Viewer or UCSC Genome Browser. Similar quality control should be performed based on the RNP of interest that has been immunoprecipitated. As a first step, read densities can be quantified in the expected binding regions of the targeted RNP complex and compared to average read densities across all regions. For further identification and quantification of specific RNP binding sites, peak calling algorithms can be used (e.g., Kucukural, Özadam, Singh, Moore, & Cenik, 2013; Uren et al., 2012).

7. Advantages

RIPiT can be used to identify footprints of RNP complex as well as RBPs that are inefficiently UV-crosslinked to RNA (Patton et al., 2020). The procedure is less technically demanding as compared to CLIP, which also requires the antibody-target protein interaction to withstand high salt and denaturing detergents. The newly updated RNA footprint library preparation method described here allows a much more streamlined, accessible and cost-effective method to prepare cDNA libraries from small RNA fragments. First of all, most of the reagents and equipment needed for the procedures are readily available in a standard molecular biology lab. Only one step in the procedure calls for size selection of DNA, and this can be accomplished using commercially available precast Urea gels that can be run in regular SDS-PAGE gel running tanks. Estimation of PCR cycle numbers via qPCR offers a rapid and precise method to determine appropriate PCR amplification. Finally, the compatibility of the single RT primer used in the procedure with the commercial Illumina indexing primers allows sample multiplexing for cost-effectiveness.

8. Limitations

One major challenge with RIPiT is that tandem IP requires a relatively large number of input cells. We usually start with at least four 150-mm plates of 70–80% confluent cells to ensure enough RNA is recovered at the end. This is especially true if targeted RBP is substoichiometric to the RNP complex. Thus, RIPiT can be applied when biological starting material is not limiting. As compared to CLIP-seq where crosslinked RNPs are under stringent wash and gel purification, RIPiT-seq only uses the gentle wash conditions for preserving protein-protein and protein-RNA interactions. Due to this reason, RIPiT-seq reads contain higher level of rRNA carryover (and snRNA in the case for EJC due to its spliceosome association). Also, as RIPiT-seq will enrich RNAs bound directly to a protein as well as those bound via other proteins, it cannot differentiate between direct versus indirect RNA binding. Due to lack of any crosslinks between the interacting RNAs and proteins, RIPiT-seq also cannot reveal the binding site information whereas some variants of CLIP-seq can reveal RBP crosslinking sites at nucleotide resolution.

9. Alternative methods/procedures

In cases where RNA-protein interactions within RNPs are labile and/or short-lived in cell extracts, chemical crosslinking of cells prior to cell lysis with formaldehyde (Mabin et al., 2018; Patton et al., 2020) or other crosslinking agents (Obrdlik, Lin, Haberman, Ule, & Ephrussi, 2019) can stabilize such RNPs. Once RNPs are crosslinked, the first IP can be carried out under more stringent conditions in the presence of high salt and/or stronger detergents. Such alternative strategies can also alleviate non-specific enrichment of abundant RNAs such as rRNAs. As the underlying principle of RIPiT is to perform double purification of a compositionally defined RNP, other methods that can achieve two step purifications can also be applied to an RNP of interest, if suitable. Such examples include tandem affinity purification (e.g., Chen et al., 2018) or density gradient fractionation combined with IP (Bohlen, Fenzl, Kramer, Bukau, & Teleman, 2020; Wagner et al., 2020). In case of certain stable RNPs such as spliceosomal complexes, single IP may be sufficient to enrich an RNP and its bound RNAs (e.g., Burke et al., 2018). Nonetheless, the RIPiT-seq procedure described here is an accessible, scalable and facile approach to interrogate genome-wide binding landscapes of multisubunit RNPs, which operate at almost all steps in regulation of gene expression.

References

Boehm, V., & Gehring, N. H. (2016). Exon junction complexes: Supervising the gene expression assembly line. *Trends in Genetics*, *32*(11), 724–735. https://doi.org/10.1016/j.tig.2016.09.003.

Bohlen, J., Fenzl, K., Kramer, G., Bukau, B., & Teleman, A. A. (2020). Selective 40S footprinting reveals cap-tethered ribosome scanning in human cells. *Molecular Cell*, *79*(4), 561–574.e5. https://doi.org/10.1016/j.molcel.2020.06.005.

Burke, J. E., Longhurst, A. D., Merkurjev, D., Sales-Lee, J., Rao, B., Moresco, J. J., et al. (2018). Spliceosome profiling visualizes operations of a dynamic RNP at nucleotide resolution. *Cell, 173*(4), 1014–1030.e17. https://doi.org/10.1016/j.cell.2018.03.020.

Chen, W., Moore, J., Ozadam, H., Shulha, H. P., Rhind, N., Weng, Z., et al. (2018). Transcriptome-wide interrogation of the functional intronome by spliceosome profiling. *Cell, 173*(4), 1031–1044.e13. https://doi.org/10.1016/j.cell.2018.03.062.

Gangras, P., Dayeh, D. M., Mabin, J. W., Nakanishi, K., & Singh, G. (2018). Cloning and identification of recombinant argonaute-bound small RNAs using next-generation sequencing. In K. Okamura, & K. Nakanishi (Eds.), *Argonaute proteins: Methods and protocols* (pp. 1–28). Springer. https://doi.org/10.1007/978-1-4939-7339-2_1.

Hentze, M. W., Castello, A., Schwarzl, T., & Preiss, T. (2018). A brave new world of RNA-binding proteins. *Nature Reviews Molecular Cell Biology, 19*(5), 327–341. https://doi.org/10.1038/nrm.2017.130.

Heyer, E. E., Ozadam, H., Ricci, E. P., Cenik, C., & Moore, M. J. (2015). An optimized kit-free method for making strand-specific deep sequencing libraries from RNA fragments. *Nucleic Acids Research, 43*(1), e2. https://doi.org/10.1093/nar/gku1235.

Hir, H. L., Saulière, J., & Wang, Z. (2016). The exon junction complex as a node of post-transcriptional networks. *Nature Reviews Molecular Cell Biology, 17*(1), 41–54. https://doi.org/10.1038/nrm.2015.7.

Kucukural, A., Özadam, H., Singh, G., Moore, M. J., & Cenik, C. (2013). ASPeak: An abundance sensitive peak detection algorithm for RIP-Seq. *Bioinformatics, 29*(19), 2485–2486. https://doi.org/10.1093/bioinformatics/btt428.

Lee, F. C. Y., & Ule, J. (2018). Advances in CLIP technologies for studies of protein-RNA interactions. *Molecular Cell, 69*(3), 354–369. https://doi.org/10.1016/j.molcel.2018.01.005.

Lin, S., Staahl, B. T., Alla, R. K., & Doudna, J. A. (2014). Enhanced homology-directed human genome engineering by controlled timing of CRISPR/Cas9 delivery. *eLife, 3*, e04766. https://doi.org/10.7554/eLife.04766.

Mabin, J. W., Woodward, L. A., Patton, R. D., Yi, Z., Jia, M., Wysocki, V. H., et al. (2018). The exon junction complex undergoes a compositional switch that alters mRNP structure and nonsense-mediated mRNA decay activity. *Cell Reports, 25*(9), 2431–2446.e7. https://doi.org/10.1016/j.celrep.2018.11.046.

Müller-McNicoll, M., & Neugebauer, K. M. (2013). How cells get the message: Dynamic assembly and function of mRNA–protein complexes. *Nature Reviews Genetics, 14*(4), 275–287. https://doi.org/10.1038/nrg3434.

Obrdlik, A., Lin, G., Haberman, N., Ule, J., & Ephrussi, A. (2019). The transcriptome-wide landscape and modalities of EJC binding in adult drosophila. *Cell Reports, 28*(5), 1219–1236.e11. https://doi.org/10.1016/j.celrep.2019.06.088.

Park, A., Won, S. T., Pentecost, M., Bartkowski, W., & Lee, B. (2014). CRISPR/Cas9 allows efficient and complete knock-in of a destabilization domain-tagged essential protein in a human cell line, allowing rapid knockdown of protein function. *PLoS One, 9*(4), e95101. https://doi.org/10.1371/journal.pone.0095101.

Patton, R. D., Sanjeev, M., Woodward, L. A., Mabin, J. W., Bundschuh, R., & Singh, G. (2020). Chemical crosslinking enhances RNA immunoprecipitation for efficient identification of binding sites of proteins that photo-crosslink poorly with RNA. *RNA, 26*, 1216–1233. rna.074856.120 https://doi.org/10.1261/rna.074856.120.

Ramlee, M. K., Yan, T., Cheung, A. M. S., Chuah, C. T. H., & Li, S. (2015). High-throughput genotyping of CRISPR/Cas9-mediated mutants using fluorescent PCR-capillary gel electrophoresis. *Scientific Reports, 5*(1), 15587. https://doi.org/10.1038/srep15587.

Rivera-Torres, N., & Kmiec, E. B. (2017). A standard methodology to examine on-site mutagenicity as a function of point mutation repair catalyzed by CRISPR/Cas9 and SsODN in human cells. *Journal of Visualized Experiments*, *126*, e56195. https://doi.org/10.3791/56195.

Rivera-Torres, N., Strouse, B., Bialk, P., Niamat, R. A., & Kmiec, E. B. (2014). The position of DNA cleavage by TALENs and cell synchronization influences the frequency of gene editing directed by single-stranded oligonucleotides. *PLoS One*, *9*(5), e96483. https://doi.org/10.1371/journal.pone.0096483.

Singh, G., Kucukural, A., Cenik, C., Leszyk, J. D., Shaffer, S. A., Weng, Z., et al. (2012). The cellular EJC interactome reveals higher-order mRNP structure and an EJC-SR protein nexus. *Cell*, *151*(4), 750–764. https://doi.org/10.1016/j.cell.2012.10.007.

Singh, G., Pratt, G., Yeo, G. W., & Moore, M. J. (2015). The clothes make the mRNA: Past and present trends in mRNP fashion. *Annual Review of Biochemistry*, *84*(1), 325–354. https://doi.org/10.1146/annurev-biochem-080111-092106.

Smith, T., Heger, A., & Sudbery, I. (2017). UMI-tools: Modeling sequencing errors in Unique Molecular Identifiers to improve quantification accuracy. *Genome Research*, *27*(3), 491–499. https://doi.org/10.1101/gr.209601.116.

Sterling, C. H., Veksler-Lublinsky, I., & Ambros, V. (2015). An efficient and sensitive method for preparing cDNA libraries from scarce biological samples. *Nucleic Acids Research*, *43*(1), e1. https://doi.org/10.1093/nar/gku637.

Uren, P. J., Bahrami-Samani, E., Burns, S. C., Qiao, M., Karginov, F. V., Hodges, E., et al. (2012). Site identification in high-throughput RNA–protein interaction data. *Bioinformatics*, *28*(23), 3013–3020. https://doi.org/10.1093/bioinformatics/bts569.

Wagner, S., Herrmannová, A., Hronová, V., Gunišová, S., Sen, N. D., Hannan, R. D., et al. (2020). Selective translation complex profiling reveals staged initiation and co-translational assembly of initiation factor complexes. *Molecular Cell*, *79*(4), 546–560.e7. https://doi.org/10.1016/j.molcel.2020.06.004.

Wang, Z., Ballut, L., Barbosa, I., & Le Hir, H. (2018). Exon junction complexes can have distinct functional flavours to regulate specific splicing events. *Scientific Reports*, *8*(1), 9509. https://doi.org/10.1038/s41598-018-27826-y.

Wheeler, E. C., Nostrand, E. L. V., & Yeo, G. W. (2018). Advances and challenges in the detection of transcriptome-wide protein–RNA interactions. *WIREs RNA*, *9*(1), e1436. https://doi.org/10.1002/wrna.1436.

Woodward, L., Gangras, P., & Singh, G. (2019). Identification of footprints of RNA: Protein complexes via RNA immunoprecipitation in tandem followed by sequencing (RIPiT-Seq). *Journal of Visualized Experiments*, *149*, e59913. https://doi.org/10.3791/59913.

Woodward, L. A., Mabin, J. W., Gangras, P., & Singh, G. (2017). The exon junction complex: A lifelong guardian of mRNA fate. *WIREs RNA*, *8*(3), e1411. https://doi.org/10.1002/wrna.1411.

Generation of 3′UTR knockout cell lines by CRISPR/Cas9-mediated genome editing

Sibylle Mitschka[a], **Mervin M. Fansler**[a,b], **and Christine Mayr**[a,b,*]
[a]Cancer Biology and Genetics Program, Memorial Sloan Kettering Cancer Center, New York, NY, United States
[b]Tri-Institutional Training Program in Computational Biology and Medicine, Weill-Cornell Graduate College, New York, NY, United States
*Corresponding author: e-mail address: mayrc@mskcc.org

Contents

Methods in Enzymology, Volume 655
ISSN 0076-6879
https://doi.org/10.1016/bs.mie.2021.03.014

Abstract

In addition to the protein code, messenger RNAs (mRNAs) also contain untranslated regions (UTRs). 3′UTRs span the region between the translational stop codon and the poly(A) tail. Sequence elements located in 3′UTRs are essential for pre-mRNA processing. 3′UTRs also contain elements that can regulate protein abundance, localization, and function. At least half of all human genes use alternative cleavage and polyadenylation (APA) to further diversify the regulatory potential of protein functions. Traditional gene editing approaches are designed to disrupt the production of functional proteins. Here, we describe a method that allows investigators to manipulate 3′UTR sequences of endogenous genes for both single- 3′UTR and multi-3′UTR genes. As 3′UTRs can regulate individual functions of proteins, techniques to manipulate 3′UTRs at endogenous gene loci will help to disentangle multi-functionality of proteins. Furthermore, the ability to directly examine the impact of gene regulatory elements in 3′UTRs will provide further insights into their functional significance.

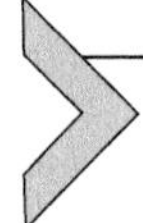

1. Introduction

1.1 The functional significance of 3′UTRs

The 3′ untranslated region (3′UTR) is an integral part of messenger RNAs (mRNAs), encompassing the sequence between the translational stop codon and the poly(A) tail. While the coding region provides cells with the building plan for a particular protein, the 3′UTR can assist in regulating protein abundance, localization, and function (Mayr, 2019). The versatility of 3′UTRs is enabled by a wide array of *trans*-acting factors that interact with the mRNA, including RNA-binding proteins, as well as short and long non-coding RNAs. Together, these interactions support processes that modulate and specify protein function. However, technical limitations have impeded a more detailed understanding of 3′UTR-mediated functions.

Fundamentally, 3′UTRs fulfill two important functions in cells: First, 3′UTRs enable binding of the cleavage and polyadenylation machinery during co-transcriptional mRNA processing. Second, 3′UTRs allow cells to regulate the fate of an mRNA through a variety of post-transcriptional mechanisms.

1.2 Mechanism of mRNA 3′ end processing

Cleavage and polyadenylation are essential steps for the processing of primary transcripts into mature mRNAs, and therefore for protein expression. This generic mechanism of transcript processing occurs co-transcriptionally and is used by all protein-coding genes with the exception of histone genes.

3′UTRs contain the sequences required for the recruitment of the cleavage and polyadenylation machinery. This gene architecture allows the amino acid sequence to be unconstrained by the processing signal. DNA mutations in signals required for 3′ end processing cause a decrease in steady-state mRNA levels with important consequences for human health and fitness (Chang, Yeh, & Yong, 2017; Mariella, Marotta, Grassi, Gilotto, & Provero, 2019). For example, mutations in poly(A) signals in the genes encoding the transcription factor p53 or hemoglobin cause a predisposition to cancer or thalassemia, respectively (Higgs et al., 1983; Orkin, Cheng, Antonarakis, & Kazazian, 1985; Stacey et al., 2011). In addition, more than half of all genes in humans encode more than one functional poly(A) site, thus allowing alternative cleavage and polyadenylation (APA) to occur (Lianoglou, Garg, Yang, Leslie, & Mayr, 2013).

In eukaryotes, 3′ end processing is initiated upon recognition of a poly(A) signal within a favorable sequence context. In vertebrates, the poly(A) signal is a nucleotide hexamer of the sequence A(A/U)UAAA or variants thereof (Gruber et al., 2016; Ulitsky et al., 2012). The canonical hexamers AAUAAA and AUUAAA are usually more efficient at inducing cleavage and polyadenylation than other poly(A) signal variants. Therefore, they can be found at poly(A) sites of single UTR genes as well as distal poly(A) sites of multi-UTR genes. Efficient poly(A) sites also exhibit a higher degree of cross-species conservation (Wang, Nambiar, Zheng, & Tian, 2018; Wang, Zheng, Yehia, & Tian, 2018). In contrast, non-canonical hexamers and weaker sequence contexts are predominantly associated with proximal poly(A) sites of multi-UTR genes.

The polyadenylation machinery is composed of four multiprotein complexes that contact the pre-mRNA through several sequence elements (Tian & Manley, 2017) (Fig. 1A). While the poly(A) signal hexamer is an important element for 3′ end processing in vertebrates, a broader sequence context is required to recruit the cleavage and polyadenylation machinery (Martin, Gruber, Keller, & Zavolan, 2012). U-rich sequences that function as auxiliary elements are found both upstream and downstream of functional cleavage sites (Fig. 1B). The cleavage and polyadenylation specificity factor complex comprised of CPSF-160, CPSF-100, CPSF-73, CPSF-30, FIP1, and WDR33 contacts both the poly(A) signal hexamer as well as U-rich sequences in the vicinity. The auxiliary motif UGUA, which is frequently found upstream of the poly(A) site, is the preferred binding motif for cleavage factor I complex consisting of CFI_m25 plus either CFI_m59 or CFI_m68. The cleavage stimulation factor (CstF77, CstF50, CstF64, and CstF64τ)

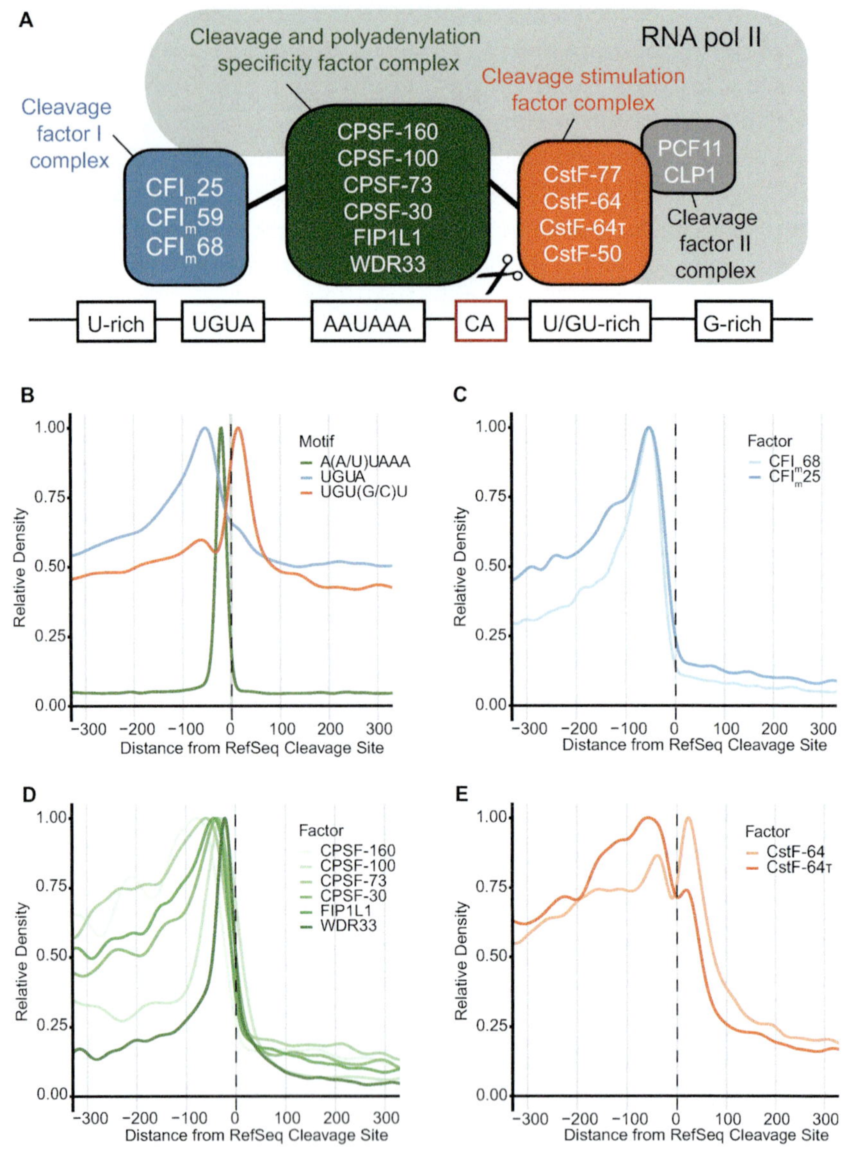

Fig. 1 mRNA cleavage and polyadenylation requires multiple sequence elements surrounding the cleavage site. (A) Schematic of the multiprotein complex responsible for mRNA cleavage and polyadenylation in humans. (B) Sequence context of functional poly(A) sites showing the poly(A) signal hexamer as well as two common auxiliary motifs. The metaplot is aligned to the transcript end in the longest isoform of RefSeq-annotated human genes. (C–E) Metagene analysis shows densities of the binding sites of protein components of the cleavage and polyadenylation machinery determined by CLIP: (C) Cleavage Factor I complex (D) Cleavage and Polyadenylation Specificity Factor complex and (E) Cleavage Stimulation Factor complex. Binding sites were retrieved from the POSTAR2 database and aligned to RefSeq-annotated transcript ends (Zhu et al., 2019).

binds to U- or GU-rich sequences downstream of the cleavage site (Fig. 1C–E). Together, the combination of upstream and downstream sequence elements determines the intrinsic strength of poly(A) sites (Cheng, Miura, & Tian, 2006). In addition, the availability of 3′ end processing factors also impacts usage rates, resulting in diverse 3′UTR isoform expression patterns across cell types and tissues (Lianoglou et al., 2013). APA-mediated differences in 3′UTR isoform expression have also been observed during proliferation, cell stress, immune cell activation, and cancer (Berkovits & Mayr, 2015; Mayr & Bartel, 2009; Sandberg, Neilson, Sarma, Sharp, & Burge, 2008; Zheng et al., 2018).

1.3 3′UTRs as regulators of mRNA stability and protein function

In addition to enabling 3′ end processing, 3′UTRs encode information that impact gene functions through post-transcriptional regulation. Since the sequence elements required for cleavage and polyadenylation are mostly found within the last 100–200 nucleotides upstream of the cleavage site (Fig. 1C–E), additional "non-essential" upstream sequences can evolve to fulfill regulatory functions. Intriguingly, the median 3′UTR length, and thus the available sequence for encoding additional information, positively correlates with morphological complexity in animal evolution (Chen, Chen, Juan, & Huang, 2012; Mayr, 2017).

Traditionally, 3′UTRs have been primarily investigated for their role in regulating protein output. *Cis*-regulatory elements located in 3′UTRs recruit RNA-binding proteins and microRNAs (miRNAs) that can modulate mRNA stability and translation efficiency. The half-life of mammalian mRNAs ranges from a few minutes to several hours. This large range is the result of finely tuned processes that guide post-transcriptional mRNA turnover and translational control. A prime example for these processes is the *Fos* mRNA, whose AU-rich elements located in the 3′UTR cause fast mRNA turnover unless stress-activated cellular pathways antagonize *Fos* mRNA degradation (Otsuka, Fukao, Funakami, Duncan, & Fujiwara, 2019). Similarly, lack of regulation by the AU-rich element in the *Tnf* 3′UTR results in chronic and fatal overproduction of the Tnf-α inflammatory cytokine in mice (Kontoyiannis, Pasparakis, Pizarro, Cominelli, & Kollias, 1999).

Additionally, 3′UTRs are known to control mRNA localization which permits spatially restricted protein synthesis. In yeast, conserved 3′UTR elements have been shown to promote translation of membrane proteins at the endoplasmic reticulum (Chartron, Hunt, & Frydman, 2016; Loya et al., 2008) and mitochondria (Margeot et al., 2002). In eukaryotic cell models, specialized RNA granules that are intertwined with the endoplasmic reticulum enable efficient transport of proteins to the plasma membrane in a 3′UTR-dependent manner (Ma & Mayr, 2018). Similarly, 3′UTR-mediated association of the mRNA with the cytoskeleton has been shown to promote nuclear localization of the encoded protein (Levadoux, Mahon, Beattie, Wallace, & Hesketh, 1999). In neurons, cytoskeletal proteins are transported to the developing growth cone for local translation via a zipcode element located in 3′UTRs (Zhang, Singer, & Bassell, 1999).

Brain cells are known to express the highest proportion of long isoform transcripts among all cell types (Lianoglou et al., 2013). Strikingly, long and short mRNA isoforms of the same gene are often preferentially sorted into different compartments, for example the soma or the neuropil. A recent study in brain tissue found that degradation rates of long and short 3′UTR mRNA transcripts from the same gene are not correlated, suggesting that local mRNA isoform turnover is directed by independent pathways (Tushev et al., 2018). Due to the local environment during and after translation, the protein generated from a particular 3′UTR isoform can be differentially modified, as has been shown for HMGN5 (Moretti et al., 2015).

Specification of protein function without differences in mRNA localization has been demonstrated for the gene encoding the ubiquitin ligase BIRC3 (c-IAP2). The long 3′UTR of *BIRC3* was shown to facilitate cell surface expression of a transmembrane protein involved in chemokine sensing (Lee & Mayr, 2019). In contrast, both long and short 3′UTR isoforms generate BIRC3 protein involved in apoptosis regulation. Due to the unique protein functions conferred by the long *BIRC3* 3′UTR, leukemia cells benefit from upregulating expression of the long 3′UTR isoform at the expense of the short 3′UTR isoform.

Another example of essential information provided by 3′UTRs are selenoproteins. The mRNAs of these proteins rely on a pair of RNA hairpin structures, called SECIS elements, for the incorporation of the rare amino acid selenocysteine into the protein peptide chain (Berry et al., 1991; Kryukov et al., 2003).

Despite these exciting advances, we are just beginning to systematically decipher the genetic information that is encoded in 3′UTRs. The difficulty of this task is rooted in the fact that 3′UTRs rely on fundamentally different principles of encoding information than the universal triplet code found in coding regions. The most basic information layer in untranslated RNAs is the nucleotide sequence. However, since the 4-letter nucleotide alphabet has an inherently low information density, sequence alone is often insufficient to confer specificity. RNA secondary structure incorporates additional information and enables increased specificity for recruitment of *trans*-acting factors (Dominguez et al., 2018). A large number of RNA-binding proteins interact with mRNAs on the basis of structure, or a combination of sequence and structure (Dominguez et al., 2018; Sanchez de Groot et al., 2019). However, RNA structures are believed to be more dynamic and flexible than protein folds and in silico predictions remain challenging (Bevilacqua, Ritchey, Su, & Assmann, 2016; Yu, Lu, Zhang, & Hou, 2020). Experimental data in yeast show that the RNA structure of closely related 3′ end isoforms can be dramatically different; this creates unique constraints for accessibility by *trans*-acting factors (Moqtaderi, Geisberg, & Struhl, 2018). In addition, 3′UTRs can contain a large number of *cis*-regulatory elements creating the potential for modulation and diversification of functions in a combinatorial manner (Iadevaia & Gerber, 2015). Finally, additional mRNA features including poly(A) tail length and a large repertoire of RNA modifications can further impact the fate of an mRNA (Jalkanen, Coleman, & Wilusz, 2014; Roundtree, Evans, Pan, & He, 2017). These challenges highlight the need for a more systematic evaluation of 3′UTR-mediated functions to identify and validate principles that govern mRNA-related processes.

1.4 Existing tools to study regulatory 3′UTR functions have systematic biases

Until now, most functional studies have focused on the role of 3′UTRs as regulators of protein output. Indeed, there are numerous examples showing that RNA-binding proteins and miRNAs that interact with 3′UTRs can modulate mRNA stability and translation rates (Matoulkova, Michalova, Vojtesek, & Hrstka, 2012). However, the notion that abundance regulation is the primary function of 3′UTRs is not so much a reflection of actual biology, but rather the result of the current limitations for studying alternative 3′UTR functions.

In particular, reporter assays have been commonly used as a proxy for endogenous 3′UTR behavior. While easy to perform, 3′UTR reporters are uniquely designed to measure quantitative differences in expression and cannot resolve other 3′UTR-related functions. Moreover, reporter assays analyze the impact of a 3′UTR - or even a part of it - outside of its endogenous sequence context. As part of an mRNA molecule, 3′UTRs co-evolved along with the other mRNA components. By cropping out sequence segments, reporter assays do not consider the potential crosstalk between 3′UTR, coding region and 5′UTR. Folding and accessibility of a 3′UTR sequence can differ in the context of a synthetic reporter gene in comparison to its cognate sequence environment with consequences for post-transcriptional expression regulation (Kristjansdottir, Fogarty, & Grimson, 2015; Lautz, Stahl, & Lang, 2010; Wissink, Fogarty, & Grimson, 2016). The coding region in particular was shown to modulate post-transcriptional expression regulation. Cottrell et al. found that changes in codon optimality impact the degree of repression mediated by miRNAs binding to the 3′UTR (Cottrell, Szczesny, & Djuranovic, 2017). Furthermore, selective testing of the immediate region surrounding the putative target site, while convenient, decreases the sequence length of the 3′UTR. This can introduce an additional bias as previous research has suggested that shorter 3′UTRs exhibit stronger repression in response to overexpression of miRNAs than longer 3′UTRs (Saito & Saetrom, 2012).

Another type of experimental bias is introduced by the overexpression of 3′UTR regulators such as miRNAs and RNA-binding proteins in cells. This approach compares the response of a target gene in the presence or absence of overexpression of a *trans*-acting factor. However, regulators and their target mRNAs exist in a defined concentration equilibrium. As such, the regulatory potential is impacted by the relative concentration of regulator and target, the optimality of the binding site, as well as the abundance of alternative targets (Saito & Saetrom, 2012; Witwer & Halushka, 2016). Overexpression of miRNAs, for example, usually increases their cellular levels to several hundred-fold over their endogenous abundance. As the targeting efficiency is dose-dependent, the pool of mRNA targets may be greatly expanded at concentrations exceeding physiological levels. It is probably for these reasons that many genomic miRNA knockouts have revealed modest or no effects on previously reported target mRNAs (Baek et al., 2008; Miska et al., 2007). This suggests that the number of functionally relevant interactions could be much smaller than previously assumed.

Finally, both miRNAs and RNA-binding proteins target a large number of different mRNAs, which in turn are subject to the regulation by a number of other regulators. Factors implicated in post-transcriptional regulation usually target different mRNAs acting in a common pathway, potentially to amplify a particular biological response (Ben-Hamo & Efroni, 2015; Zanzoni, Spinelli, Ribeiro, Tartaglia, & Brun, 2019). Overexpression of an RNA-binding protein will usually cause expression changes across hundreds of mRNAs, some of them indirectly. Separating these direct from secondary effects and evaluating their individual physiological significance remains challenging.

Given these limitations, it is perhaps not surprising that some genetic models for 3′UTR-mediated expression regulation have presented evidence contradictory to results obtained by non-endogenous studies (Mitschka & Mayr, 2020; Zhao et al., 2017).

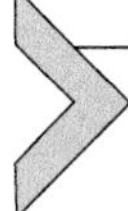

2. Experimental design

2.1 Method overview

In this chapter, we outline strategies that will allow researchers to generate cell models to systematically investigate 3′UTR-dependent functions. Such cell models are created through defined genomic deletions of 3′UTR sequences, while preserving co-transcriptional pre-mRNA processing. We will discuss suitable approaches to target both single- and multi-UTR genes. For multi-UTR genes, we will specifically focus on genes expressing classical tandem 3′UTRs in the last exon that do not alter the amino acid sequence of the encoded protein. Importantly, alternative 3′UTR transcripts can also arise from intronic APA. In contrast to tandem APA, intronic APA events usually affect both the protein coding part as well as the 3′UTR. Usage of intronic APA sites seems to be regulated by a mutual interplay with splicing processes (Lee et al., 2018; Tian, Pan, & Lee, 2007). While genomic deletion strategies can also be applied for deleting these isoforms, we focus here on 3′UTR-dependent functions that do not alter the amino acid sequences of proteins.

CRISPR/Cas9 is now a widely used technique for genome editing. The ability to induce sequence-specific DNA double-strand breaks greatly accelerates the creation of genetically modified cell models and organisms. For traditional protein knockouts, small indel mutations arising at CRISPR/Cas9 cut sites are used to generate frame-shift mutations. However, these regional mutations are not suitable to interrogate functions of non-coding sequences. Alternatively, precision genome editing with designed DNA repair templates remains time-consuming, as the activity of the homology-directed repair (HDR) pathway is generally low in mammalian cells. Instead, our and other previous methods exploit the non-homologous end joining (NHEJ) pathway to generate defined genomic deletions (Bauer, Canver, & Orkin, 2015; Joberty et al., 2020; Mitschka & Mayr, 2020; Thomas et al., 2020; Zhao et al., 2017; Zhu et al., 2016). Specifically, a pair of CRISPR guide

RNAs (gRNA) is used to cut within the 3′UTR, thereby creating a deletion that is flanked by the two cut sites.

2.2 Useful tools for 3′ end annotations

The comprehensive annotation of mRNA 3′ ends is an ongoing project whose progress is most advanced for the human genome. Several useful online tools are available to search for gene-specific 3′ end annotations across different species. Among them are polyASite 2.0 (Herrmann et al., 2020), PolyA_DB3 (Wang, Nambiar, et al., 2018; Wang, Zheng, et al., 2018) and APASdB (You et al., 2015).

Over the last decade, increased sequencing depth has led to an increase in the number of annotated 3′ ends. Of those, very few sites have been validated using orthogonal methods that do not involve sequencing. Artifacts created by internal priming at genomic poly(A)-rich sites remain a challenge for sequencing-based annotation techniques (Gruber et al., 2016). In addition, not all 3′ end sequencing methods have been validated to deliver quantitative results. We therefore advise that any project aiming to delete a 3′UTR or to influence 3′UTR isoform usage should set out to confirm the usage of annotated poly(A) sites. Northern blot analysis can be used to verify and quantify the expression of different 3′UTR isoforms (see also Section 3.6).

2.3 Selection of cell models

While the method described in this chapter can be applied to virtually any cell line, some attention should be paid to the choice of cell model prior to starting the experiment.

2.3.1 Robust expression of the gene of interest

The deletion of the regulatory part of a 3′UTR can result in upregulation, downregulation or no change in expression of the gene of interest compared to the corresponding wild-type cells. In either case, it is the endogenous promoter that drives transcription and the baseline expression level of the gene of interest should be robust enough to enable downstream analysis. For multi-UTR genes, we recommend confirming the expression level of all relevant 3′UTR isoforms in the cell line of choice.

2.3.2 The efficiency of homozygous deletions depends on gene copy numbers

Many commonly used cell lines have aberrant karyotypes with more than two copies per gene. In order to generate a homozygous 3′UTR knockout, all gene copies need to be successfully edited. The presence of additional alleles reduces the chance of obtaining homozygous deletion clones. As is generally true for genome editing, cells with a near-diploid karyotype such as embryonic stem cells, HCT116 (modal chromosome number of 45), U87-MG (44) and WI-38 (46) are usually preferable to cell lines with complex karyotypes including HeLa (82) or MDA-MB-453 (90) cells. Due to local copy number variations, actual gene copy numbers may differ from the overall ploidy state of any given cell line. Therefore, precise karyotype information can help to inform the choice of an appropriate cell model.

In some cases, it might still be desirable to use a cell model with three or more alleles. Screening a larger number of colonies might be necessary to identify homozygous clones. As the deletion efficiency generally decreases with the distance of the two CRISPR/Cas9 cut sites (Bauer et al., 2015), generating larger deletions can further aggravate the problem. Alternatively, homozygous deletion mutants can also be created by performing a second round of transfection and selection after obtaining heterozygous clones.

2.3.3 Requirement for clonal growth

We have noticed that some cell lines are not able to grow as single cells. The addition of cell type-specific conditioned media can help to alleviate this problem. Poor cell survival can also be caused by mechanical stress during the process of cell sorting. As an alternative to growing cells in individual wells, transfected cells can be seeded sparsely in a large culture dish. In this case, individual clones need to be picked under a light microscope for subsequent screening and expansion.

2.4 Developing a gene-specific deletion strategy

CRISPR/Cas9 is now the most widely used tool for genome editing in both cell lines and organisms. Recruitment of the Cas9 nuclease by a programmable guide RNA (gRNA) makes this tool customizable to the specific needs of investigators. Stable binding of the Cas9/gRNA ribonucleoprotein (RNP) complex requires the presence of the cognate protospacer adjacent motif (PAM) downstream of the DNA region complementary to the gRNA. The canonical PAM sequence of the most commonly used Cas9 nuclease

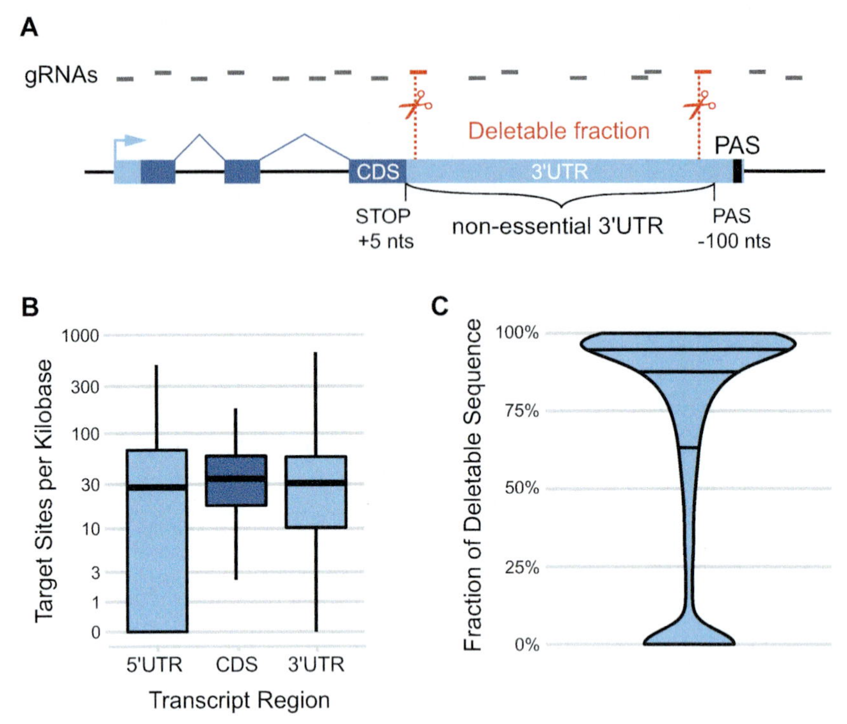

Fig. 2 Density of unique and efficient Cas9 gRNA sequences in human genes. (A) Gene model with unique and efficient gRNAs depicted above. Cleavage sites of a suitable gRNA pair for 3′UTR deletion are highlighted with dotted lines. PAS, poly(A) site; CDS, coding region; nts, nucleotides. (B) Frequency of unique and efficient *Sp*Cas9 CRISPR gRNAs in human RefSeq annotated genes separated by transcript region. Efficient gRNAs were categorized as having a Doench/Fusi score of $\geq$30. (C) Prediction of the deletable, non-essential 3′UTR sequence portion using gRNA criteria used in B. We defined the non-essential 3′UTR as the region between the translational stop codon +5 nucleotides and -100 nucleotides upstream of the annotated transcript end of the longest RefSeq transcript per gene, excluding 3′UTRs with a total length of less than 200 nucleotides.

variant derived from *Streptococcus pyogenes* (*Sp*Cas) is NGG. In addition, the gRNA sequence should be unique in the genome to ensure target specificity. The fact that 3′UTRs are generally more A/T-rich than coding region sequences creates some constraint on PAM site availability (Fig. 2A and B). However, human 3′UTRs still contain a median density of about 31 unique and efficient Cas9 gRNAs per kilobase of sequence (Concordet & Haeussler, 2018; Doench et al., 2016). Given the availability of gRNAs, our analysis of human 3′UTRs using the longest RefSeq-annotated transcript per gene showed that the median percentage of removable non-essential 3′UTR sequence is still close to 90% (Fig. 2C).

2.4.1 Deletion of entire 3′UTRs in single- or multi-UTR genes to generate minimal 3′UTRs

For this type of deletion, a gRNA pair is used to delete the entire 3′UTR except for the region involved in 3′ end processing which results in mRNA transcripts with a minimal 3′UTR (Fig. 3).

Upstream gRNAs should have predicted cleavage sites close to the translational stop codon (TAA/TAG/TGA). The Cas9 cleavage site is expected to be located between nucleotide three and four upstream of the PAM site. While NHEJ can be very precise, there is the possibility that additional nucleotides will be deleted due to microhomology-mediated end joining (MMEJ). In order to ensure that this does not result in mutations of the protein coding region, the theoretical Cas9 cleavage site should not be within five nucleotides of the stop codon (Canver et al., 2014; Owens et al., 2019).

In order to avoid interference with 3′ end processing, cleavage by the downstream gRNAs should not impair cleavage and polyadenylation by deleting auxiliary sequence elements (Fig. 1). Unfortunately, the exact sequence context required for 3′ end processing is not precisely defined

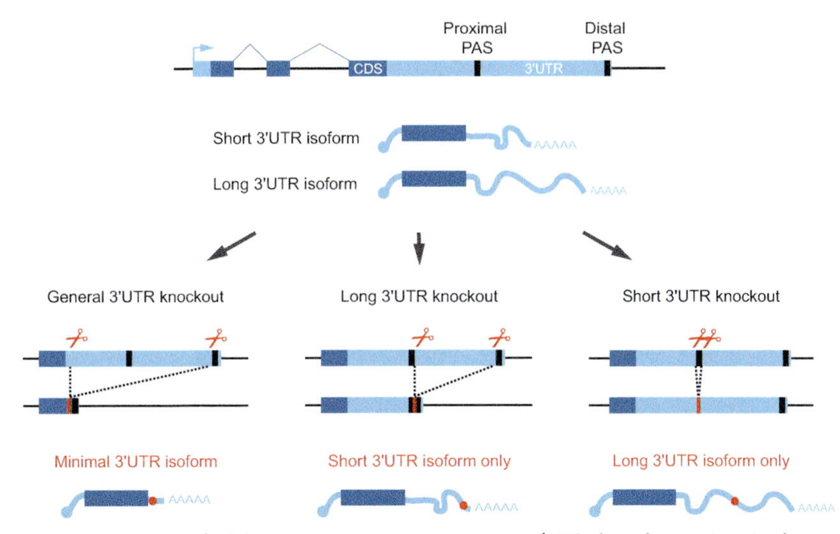

Fig. 3 Overview of deletion strategies to create 3′UTR knockouts in single and multi-UTR genes. Shown are deletion strategies at genomic gene loci as well as the results in the processed mRNAs (indicated by AAAAA, as symbol for the poly(A) tail to denote mRNAs).

for individual genes. As a general rule, we do not recommend positioning the downstream gRNA closer than 100 nucleotides upstream of the cleavage site as auxiliary elements for cleavage factor recruitment are highly enriched in these regions (Fig. 1C–E).

2.4.2 Deletion of the short 3′UTR isoform in multi-UTR genes

This type of mutation is effectively a knockout of the short 3′UTR isoform, while the overall gene-specific transcriptional output is preserved. To this end, two gRNAs with binding sites flanking the poly(A) signal used to produce the short 3′UTR isoform are selected. After the deletion, cleavage and polyadenylation can no longer occur at this site. RNA polymerase II read-through will result in exclusive usage of the downstream poly(A) site(s) and generate only mRNAs with long 3′UTRs. First, the hexamer serving as the putative poly(A) signal needs to be identified which is usually found about 25 nucleotides upstream of the cleavage site (Fig. 1B). Next, non-overlapping pairs of unique gRNAs flanking the poly(A) signal are chosen. Because the poly(A) signal is essential for successful cleavage and polyadenylation, it is not required that the deletion includes the additional sequence context surrounding the poly(A) signal. In fact, designing gRNA pairs with small distances to each other is recommended, as they are more efficient in generating deletions and minimize removal of other regulatory sequence elements.

2.4.3 Deletion of the long 3′UTR isoform in multi-UTR genes

For deleting the long 3′UTR isoform, a gRNA pair is used to delete the genomic region encoding the extended 3′UTR downstream of the proximal poly(A) site (Fig. 3). As a result, the distal, usually stronger poly(A) site will effectively move into close proximity to the proximal site. The combined effect of both poly(A) signals will likely enable highly efficient cleavage and polyadenylation at this location and is expected to preserve overall mRNA expression of the gene. After identifying both poly(A) signals, upstream gRNAs located 3′ of the proximal poly(A) signal and downstream gRNAs located 5′ of the distal poly(A) signal are selected. Because of the substantial length of some 3′UTRs, the desired deletion can be several kilobases in length. Since the deletion efficiency is inversely correlated with the distance of the Cas9 cleavage sites (Bauer et al., 2015), prior testing of individual gRNA efficiencies is particularly important for the success of this strategy (see Section 2.6). In cases with a large distance between the cleavage sites, an alternative approach is to exclusively delete the distal poly(A) site with a pair of gRNAs flanking the site as described above. However, this approach will reduce total mRNA expression by the amount that the long 3′UTR isoform contributed to it.

2.5 Design of CRISPR gRNAs

A number of different online tools can be used to identify unique and efficient gRNAs in the target region (recently reviewed in Hanna and Doench (2020)). In addition, the UCSC genome browser also enables visualization of all possible Cas9 gRNA locations across the human (hg38) genome (Concordet & Haeussler, 2018; Haeussler et al., 2016). Whenever possible, it is recommended to choose gRNAs with two or more mismatches to other putative target sites to minimize the chances of off-target mutations and translocations. Importantly, our protocol employs vector-based expression systems in which poly(T)-stretches act as a termination signal for the endogenous Pol III. Therefore, the gRNA sense oligonucleotide, located on the strand containing the PAM sequence, is not allowed to contain four or more consecutive T's (Gao, Herrera-Carrillo, & Berkhout, 2018). Finally, gRNAs that will be transfected in pairs should not overlap in their target sequence.

Once the gRNA sequences are selected, adapter nucleotides are added to the gRNA sense and antisense sequences:

gRNA sense oligo: 5′-CACCGNNNNNNNNNNNNNNNNNNNN-3′ ($N=20$).

gRNA antisense oligo: 5′-AAACNNNNNNNNNNNNNNNNNNNNC-3′ ($N=20$).

Oligonucleotides for gRNA cloning can be ordered from any commercial vendor.

2.6 Testing of gRNAs

Ideally, two to three gRNAs should be chosen for each side flanking the desired deletion and tested for their respective cleavage efficiency. Suitable methods for determining cleavage efficiencies include the T7 endonuclease E1 mismatch detection assay, Sanger sequencing followed by tracking of indels by decomposition (TIDE) or targeted deep sequencing. A comparison of these methods and detailed protocols can be found elsewhere (Sentmanat, Peters, Florian, Connelly, & Pruett-Miller, 2018). Importantly, the deletion efficiency in cells is probably limited by the least active gRNA.

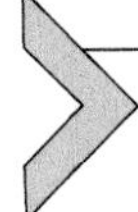

3. Protocol

3.1 Required materials

3.1.1 Reagents

- pX330-U6-Chimeric_BB-CBh-hSpCas9 vector (available from Addgene, #42230) or a suitable alternative vector (see Section 3.3)
- gRNA oligos, designed as described in Section 2.5
- *Bbs*I-HF enzyme (NEB, R3539S)
- T4 DNA Ligase (NEB, M0202S)
- DH5α competent bacteria (homemade or commercial)
- LB liquid medium and LB agar plates
- Ampicillin sodium salt (Fisher Scientific, BP1760–5) and kanamycin sulfate (Fisher Scientific, BP906–5)
- Sequencing primer/CBh_rev: CGTCAATGGAAAGTCCCTATTGGC
- Cell line of interest at an early passage (see Section 2.3)
- Cell line-specific culture medium and additives
- *For adherent cells only:* Trypsin-EDTA (0.05%) with phenol-red (Thermo Fisher/GIBCO, 25300062)
- DAPI (Sigma-Aldrich, D9542)
- PBS
- Lipofectamine 2000 (Thermo Fisher/Invitrogen, 1166830) or reagents required for alternative transfection methods (see Section 3.3)
- Opti-MEM I Reduced Serum Medium (Thermo Fisher, 31985062)
- pmaxGFP (Lonza) or a different fluorescent marker plasmid
- FBS (e.g., Thermo Fisher/GIBCO, 26140079)
- QuickExtract DNA Extraction Solution (Lucigen, QE09050)
- *Taq* DNA polymerase with buffer (NEB, M0273)
- Deoxynucleotide (dNTP) solution mix (NEB, N0447S), diluted in water to 2 mM of each nucleotide
- Gene-specific primers for screening PCR, reconstituted and diluted to 10 μM in water (see Section 3.5)
- Gel Loading Dye, Purple 6 × (NEB, B7024S)
- UltraPure agarose (Invitrogen, 16500500)
- Ethidium bromide solution (Thermo Fisher, 17898)
- Zero Blunt TOPO PCR Cloning Kit (Thermo Fisher, 451245)
- Q5 High-fidelity DNA polymerase (NEB, M0491S)

3.1.2 Equipment

- Tabletop centrifuge
- 37 °C incubator for bacteria, equipped with a shaker

- QIAquick PCR Purification Kit (Qiagen 28104)
- QIAprep Spin Miniprep Kit (Qiagen 27104)
- PCR strips
- PCR thermal cycler
- NanoDrop spectrophotometer
- FACS sorting device capable of sorting single cells into cell culture plates under sterile conditions, e.g., BD FACSAria Cell Sorter
- *Optional:* sterile syringes, with luer-lock (e.g., BD 309653)
- *Optional:* sterile syringe filters, 0.2 μm pores (e.g., Corning, 431219)
- Light microscope
- 96-well cell culture plates (e.g., Costar 3585)
- 5 mL polystyrene round-bottom tubes with cell-strainer cap (Falcon, 352235)
- *Optional:* Multichannel pipettes and disposable reagent reservoirs
- Gel electrophoresis equipment
- GelDoc or similar instrument for gel visualization

3.2 Cloning of gRNAs into the pX330 vector

1. Make 10 × annealing buffer: 100 mM Tris pH 8, 10 mM EDTA, 500 mM NaCl.
2. Reconstitute lyophilized gRNA oligos in water at a concentration of 100 μM.
3. Digest pX330 plasmid DNA:
   - 2 μg pX330 vector DNA
   - 2 μL CutSmart buffer
   - 1 μL *Bbs*I-HF enzyme
   - Ad 20 μL sterile water
4. Incubate reaction for one hour at 37 °C.
5. Purify digested plasmid DNA using the QIAquick PCR Purification Kit according to the manufacturer's instructions. Elute silica-bound DNA in 20 μL of water.
6. Determine DNA concentration with a Nanodrop.
7. For annealing of gRNA DNA duplexes in a thermocycler, combine in a PCR strip:
   - 1 μL of sense DNA oligo
   - 1 μL antisense DNA oligo
   - 2 μL 10 × annealing buffer
   - 16 μL water
8. Incubate at 95 °C for 5 min, then decrease temperature by 0.1 °C/s until room temperature is reached.

9. Set up ligation reaction:
   - 50 μg *Bbs*I-digested pX330 plasmid DNA
   - 1 μL of 1:10 dilution of annealed oligo mix, or water as a control
   - 1 μL 10 × T4 DNA ligase reaction buffer
   - 1 μL T4 DNA ligase
   - Add water to a final volume of 10 μL
10. Incubate ligation reaction for one hour at room temperature.
11. Transform competent bacteria with 3 μL of ligation reaction using standard procedures. Plate the bacteria on ampicillin-containing LB agar plates (100 μg/mL final concentration) and grow overnight at 37 °C.
12. The next day pick 3–5 bacterial colonies per plate and inoculate 4 mL ampicillin-containing LB liquid media with bacteria. Grow overnight at 37 °C in a shaking incubator.
13. Perform plasmid DNA extraction from bacteria cultures using the QIAprep Spin Miniprep Kit according to the manufacturer's instructions. Elute plasmid DNA from spin columns in 50 μL water.
14. Measure DNA concentration using a NanoDrop. Validate correct insertion of gRNA sequences by Sanger sequencing using the CBh_rev as sequencing primer.

3.3 Transfection of target cells with gRNAs

The cell line of choice should be in an early passage and maintained in a state of exponential growth through regular splitting. For best results, cell type-specific transfection protocols should be established prior to starting the experiment. Different methods, including lipofection, calcium phosphate transfection or electroporation are suitable to deliver the CRISPR plasmids into target cells.

The relative ratios of the two gRNA plasmids to fluorescent marker plasmid are 10:10:1 (Fig. 4).

This is an example protocol for the transfection of HEK293 cells using Lipofectamine 2000:

1. The day before the transfection split 6×10^5 cells in each well of a 6-well plate containing 2.5 mL growth media per well.
2. The next day, HEK293 cells should have reached 60–90% confluency. Combine 1.2 μg of each of the two pX330-gRNA plasmids with 120 ng of pmaxGFP plasmid and add 140 μL Opti-MEM I Reduced Serum Medium.

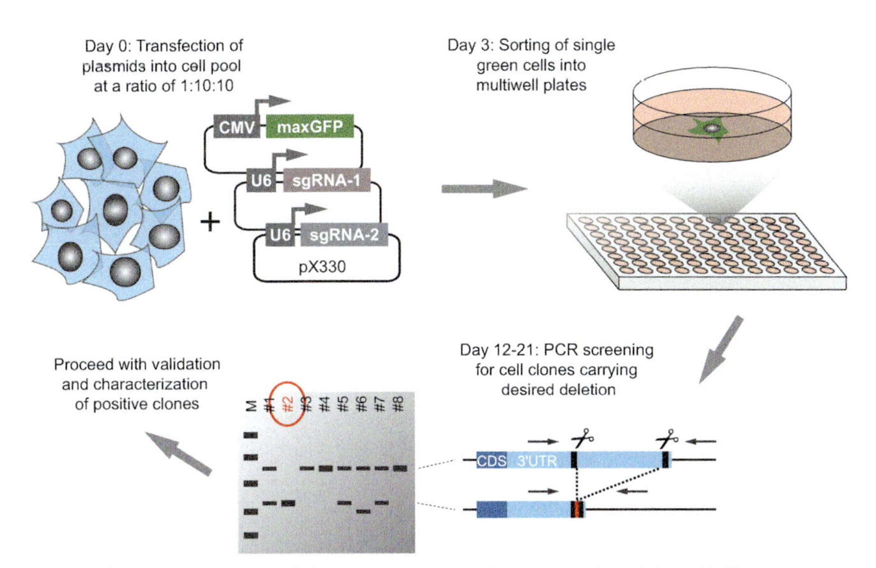

Fig. 4 Schematic overview of the major steps in the protocol to delete 3′UTRs at endogenous gene loci.

3. Add 10 μL of Lipofectamine 2000 to 140 μL Opti-MEM I Reduced Serum Medium, mix briefly, and combine with plasmid-containing solution from previous step. Mix well and incubate for 10 min at room temperature.
4. Add 250 μL of the transfection solution to a 6 well and mix gently by rocking the plate back and forth.
5. The following day change media or split cells if necessary. Continue to grow and split the cells until sorting for GFP-positive cells at day three to five after transfection.

3.4 Seeding of single cells with FACS

Three to five days after transfection, GFP-positive cells are sorted into single wells using FACS (Fig. 4). It is important not to sort the cells too early after transfection, because continued CRISPR/Cas9 activity can lead to further genome editing.

In this protocol, we recommend the use of a FACS sorter to seed transfected single cells into a 96-well culture plate. The ability to select transfected, i.e. fluorescent, cells is particularly useful for cell lines that have low transfection efficiencies. In addition, cell sorting usually yields more accurate seeding of single cells than manual seeding. However, if a cell sorter is unavailable or the cell line of choice is very sensitive to mechanical stress, cells can instead be counted and seeded at an average concentration of 0.33 cells/well. When performing manual seeding, it is particularly important to ensure that cells are properly resuspended in medium to avoid cell clumps. For manual seeding, the fluorescent marker plasmid can be omitted in the transfection mix (Section 3.3).

1. Prepare the required number of 96-well plates by adding 200 μL of culture medium per well. Prewarm and equilibrate the plates in a cell culture incubator with 5% CO_2 (see tip 1).
2. Three days after transfection, detach transfected cells using Trypsin-EDTA (for adherent cells only).
3. Add excess volume of culture medium with 10% FBS to quench Trypsin and collect cells in a 15 mL conical tube. Centrifuge the cells for 5 min at 250 *g* and thoroughly resuspend the cell pellet in 10% FBS in PBS at 1×10^6/mL. To avoid clogging of the FACS sorter, pass cell suspension through the filter top of a 5 mL round bottom FACS tube. Prepare a sample of untransfected cells in parallel to enable accurate gating during cell sorting. Add DAPI at a final concentration of 0.1 μg/mL to the cell suspension (see tip 2). Keep cell samples on ice until cell sorting.
4. Using a FACS sorter, GFP-positive, DAPI-negative single cells are directly sorted into 96-well culture plates containing 200 μL prewarmed and CO_2-equilibrated medium. In short, intact cells are first selected in a forward versus side scatter (FSC/SSC) plot to exclude small debris and particles with high granularity. This population is further sub-gated for doublet exclusion by plotting area versus height or width (e.g., FSC-A versus FSC-H). Finally, cells with low DAPI signal and high GFP signal are selected for sorting into individual 96 wells (see tip 3).
5. After sorting, continue to culture cells at 37 °C/5% CO_2, with partial media changes every 3–5 days. To change medium, remove 100 μL of medium per well and replace with 100 μL of fresh medium.
6. Depending on the cell type, sufficiently large colonies will form within 12 to 21 days after initial cell seeding. Starting from day 10, daily inspections of the culture plates with a light microscope will help to determine the best time to harvest the cell clones. At that time, wells containing colonies will change media color and individual colonies should have reached a size of 2–5 mm in diameter. Importantly, very few wells should contain more than one independent colony. Whenever possible, wells containing multiple colonies should be excluded from the analysis.

Tips and Troubleshooting

1. To reduce cell death upon culture of single cells, the culture media can be supplemented with up to 50% of conditioned media from the same cell line. To this end, condition the medium for 24–48 h with freshly split cells from a culture of exponentially growing cells. At the time of collection, the conditioned medium should not be exhausted, i.e. the phenol-red pH indicator should not have turned orange/yellow. Clear the medium from floating cells and cell debris by centrifuging at 3000 *g* for 20 min. Afterwards, filter the supernatant through a 0.2 μm syringe filter. Store conditioned media for up to seven days at 4 °C or freeze at −20 °C for longer storage.
2. DAPI is a fluorescent DNA stain that is mostly impermeant to intact living cells. It is used to discriminate between dead (DAPI-positive) and live cells (DAPI-negative) by FACS. As not all FACS sorters are equipped with the appropriate short wavelength lasers needed to excite DAPI, propidium iodide (PI) can be used as an alternative to DAPI. However, PI and GFP have strongly overlapping emission spectra, thus requiring signal compensation during setup.
3. Cells with high GFP protein expression levels (e.g., top 25% of GFP-positive population) are more likely to also express high levels of CRISPR/Cas9 plasmids. It was shown that higher Cas9/gRNA expression levels correlate with higher cleavage efficiency (Hsu et al., 2013).

3.5 PCR screening

PCR is a fast and efficient method to screen large numbers of clones for the desired deletion. For small deletions (<500 nucleotides) a single primer pair flanking the deletion site can be used to amplify both the wild-type and edited alleles (Fig. 4). For larger deletions >1000 nucleotides, amplification of the wild-type allele is often not feasible using crude genomic DNA extracts. Therefore, an additional primer pair spanning one of the two cut sites should be used to identify wild-type alleles. For convenience, a three-primer PCR strategy can be employed to simultaneously detect both allele variants. Optimal PCR conditions should be established in advance (see also tips and troubleshooting below).

When working with suspension cells, start with protocol step 4.

1. Once the cell clones have reached a sufficiently large size, aspirate medium and wash cells with 100 μL PBS per well.

2. Add 30 μL of Trypsin-EDTA solution per well and incubate at 37 °C for 5 min or until cells start to detach.
3. Add 200 μL of culture medium to each well to quench the trypsin digest. Carefully resuspend the cells in medium by repeated pipetting.
4. Transfer 50 μL of resuspended cells into a new 96-well plate containing fresh medium to cultivate as a backup.
5. Transfer 100 μL of the cell solution into a new tube (1.5 mL tube or PCR strip/plate) and spin cells down at 1000 *g* for 3 min.
6. Remove supernatant and resuspend cell pellet in 50 μL QuickExtract DNA extraction solution. Adding more QuickExtract solution is advised if the solution appears visibly cloudy (see also tips and troubleshooting).
7. Place tubes in heat block or thermocycler at 65 °C for 6 min, followed by 98 °C for 2 min. Afterwards, place samples on ice to proceed with setup of the PCR reactions or freeze at −20 °C. For best results, use stored genomic DNA within 3 days and avoid freeze/thaw cycles.
8. Add 2 μL of the crude genomic DNA solution to 18 μL of PCR master mix containing gene-specific primers:
   - 2 μL 10 × Standard *Taq* reaction buffer
   - 1 μL screening primer forward (10 μM)
   - 1 μL screening primer reverse (10 μM)
   - 2 μL dNTP mix (10 × stock with 2 mM of each nucleotide)
   - 0.3 μL *Taq* Polymerase
   - 11.7 μL water
9. Run PCR in a thermocycler with template-specific annealing temperatures and extension times (see troubleshooting).
10. Prepare a 1–1.5% agarose gel (depending on the expected product sizes) in Tris-acetate-EDTA (TAE) buffer containing 0.5 μg/mL ethidium bromide. Add 4 μL gel loading dye to each PCR reaction. Load 12 μL on the agarose gel and run at 10 V/cm.
11. Visualize DNA bands under UV light using a GelDoc or equivalent device.

Tips and troubleshooting:

The convenience of this one-step method for DNA extraction makes it ideal for medium and high throughput screening applications. However, this procedure does not purify genomic DNA like classical genomic DNA extraction methods. Therefore, the PCR reaction needs to be optimized to work well for crude genomic DNA extracts. In general, the amplification of smaller PCR products is more efficient, while amplification of larger DNA regions can be challenging. If the screening PCR does not work reliably the following steps can be optimized:

1. Choosing alternative primer pairs can often help to improve PCR efficiency.
2. High concentrations of denatured protein in the extraction solution can inhibit the PCR reaction. Diluting the crude genomic DNA mixture with water in ratios from 1:1 to 1:10 can solve this problem. Of note, due to residual RNA and protein contamination, quantification of genomic DNA in extracts using spectroscopy (e.g., Nanodrop) is highly inaccurate.
3. Optimization of PCR reaction conditions should be optimized if the PCR reaction continues to fail. An adjustment of PCR annealing temperatures and testing of PCR reaction additives (e.g., $MgCl_2$, DMSO) can improve yields with structured DNA templates.
4. While a standard *Taq* polymerase is usually sufficient for screening applications, the use of polymerases with higher processivity such as *Pfu* and KOD polymerase can lead to better results with difficult DNA templates.

In some clones, the interpretation of screening PCR results can be complicated when more than the expected number of allele variants are detected. These implausible genotyping results can occur due to a number of different reasons:

1. The assumed gene copy number is incorrect. Check cell line information for local allele copy numbers.
2. Multiple cell clones grew in the same well. When using FACS sorting, this should only happen in a small proportion of wells, but it is more likely to occur when manual seeding was performed. Before seeding into wells, it is essential that cell clumps are thoroughly dissociated. In addition, conditioned medium that has not been filtered can contaminate wells with cells.
3. The cell clones underwent further editing after seeding, leading to a mixed chimeric genotype. Genetic chimerism appears occasionally due to continued CRISPR/Cas9 activity. Make sure to wait at least 3 days (up to 5) after transfection before sorting single cells into individual wells.
4. DNA contamination of PCR reagents can cause additional bands to appear. If the band corresponds to DNA from the wild-type allele, the problem can be difficult to identify. Always run a PCR control without genomic DNA along with genotyping samples.

3.6 Expansion of cell clones and sequence validation

On the basis of the screening PCR, cell clones harboring the desired deletion are selected for expansion. Aliquots of edited cell clones should be frozen as backups at an early passage. In order to minimize the risk of clonal artifacts, phenotypical characterization experiments should always be performed using several cell clones. We also strongly recommend that all alleles of edited clones used for experiments are validated by Sanger sequencing. For this purpose, a target site-specific primer pair flanking the deleted region should be used to amplify the edited region by PCR using a high-fidelity polymerase (e.g., Q5 polymerase from NEB). The screening primers from Section 3.5 can be used at this step. After confirming that the PCR was successful by agarose gel electrophoresis, a small amount of the PCR product is directly sub-cloned using the Zero Blunt TOPO PCR cloning kit according to the manufacturer's instructions. Pick up to ten bacteria transformants to expand in liquid LB culture containing kanamycin. Perform a plasmid miniprep and sequence the PCR insert using one of the available primer sequences in the vector (e.g., T7, M13_for, M13_rev).

3.7 Additional validation by northern blot (recommended)

In order to confirm a change in 3′UTR isoform usage as a result of the genomic deletion, we highly recommend a validation experiment using northern blot analysis. Northern blotting is a reliable method to detect and quantify mRNAs isoform expression from a gene of interest. This is particularly important when multi-UTR genes are analyzed. Northern blot analysis can also help to detect unexpected outcomes of the 3′UTR deletion, including sequence inversions, allele heterogeneity, deficiency in poly(A) site usage due to the deletion, and activation of cryptic downstream poly(A) sites.

We have created a detailed protocol for the detection of 3′UTR isoforms using an optimized northern blot protocol (https://dx.doi.org/10.17504/protocols.io.bqqymvxw). The DNA probe should be complementary to the part of the mRNA common to both wild type and deletion cell lines, such as the coding region. For larger deletions, a second probe targeting the deleted part of the 3′UTR can be used in addition.

4. Related techniques

4.1 Alternative delivery systems for CRISPR/Cas9 gRNA pairs

The CRISPR/Cas9 toolbox continues to evolve rapidly and we anticipate further improvements in the near future. Already today, a number of different vector designs can be used to create genomic deletions similar to the approach outlined here. For example, a single vector for simultaneous expression of a tandem pair of gRNAs can be used instead of the two-vector system described here. These systems are especially suitable for genome-wide screening approaches, as they allow expression of two gRNAs from a single lentiviral vector (Gasperini et al., 2017; Thomas et al., 2020; Vidigal & Ventura, 2015). CRISPR/Cas9 vector systems that already incorporate a fluorescent marker or an antibiotic resistance cassette can be used as well.

Notably, many labs now transfect cells with in vitro assembled Cas9-gRNA ribonucleoprotein (RNP) complexes. These RNPs can be introduced by electroporation or cationic lipid carriers and can help to circumvent the problem of low transfection efficiencies in some cell systems. In addition, RNPs have been found to be especially effective for DNA editing in a number of model organism through direct delivery by microinjection or electroporation (Chen, Lee, Lee, Modzelewski, & He, 2016; Farboud et al., 2018). We believe that these methods will provide useful variations to our protocol.

4.2 Alternative CRISPR nucleases for 3′UTR editing

A new generation of engineered CRISPR nucleases might soon be able to substitute for *Sp*Cas9 and expand the pool of targetable unique genomic sequences through different PAM-sequence requirements (Chatterjee et al., 2020; Chatterjee, Jakimo, & Jacobson, 2018; Kleinstiver et al., 2015; Legut et al., 2020). The most studied alternative to Cas9 today is called Cpf1. Unfortunately, there is only limited published information regarding the efficiency of Cpf1-mediated genomic deletions (Dumeau et al., 2019). However, we envision potential advantages of the Cpf1 nuclease over Cas9. First, the PAM sequence for Cpf1 is defined as 5′-NTTT-3′, which could be beneficial for targeting A/T-rich regions including 3′UTRs. Second, in contrast to Cas9, Cpf1 cleaves distal to its PAM site (Zetsche et al., 2015). Accordingly, small indel mutations do not automatically preclude further cleavage by Cpf1, which could potentially enhance deletion efficiencies.

4.3 Related genome editing approaches for the analysis of 3′UTR functions

Investigation of 3′UTR-dependent functions through genetic models is still rare and common standards have not been developed. Nevertheless, different approaches that are related to our procedure have been established. For example, gene "knock up" has recently been presented as a gene editing approach for bypassing 3′UTR-dependent *Gdnf* expression regulation

(Mätlik et al., 2019). Here, a CRISPR/Cas9-mediated knock-in of a strong exogenous poly(A) site is used to prematurely induce cleavage and polyadenylation. By preventing incorporation of the original 3′UTR containing repressive *cis*-regulatory elements, this method has been shown to elevate GDNF protein expression *in vivo*. Notably, the addition of Cre-inducible loxP sites creates the potential for a conditional "knock up" allele. However, the use of a strong unrelated poly(A) site can by itself increase mRNA and protein expression by making pre-mRNA cleavage and polyadenylation more efficient than in the endogenous gene independently of additional *cis*-regulatory 3′UTR elements. Nevertheless, gene "knock up" could be a useful tool to investigate the impact of elevated mRNA expression in a cell type-specific manner.

Recently, an elegant study by Bae et al. (2020) provided genetic evidence for the role of the long *Calm1* 3′UTR isoform in the development of mouse dorsal root ganglia and hippocampus. This study investigated a mouse model carrying a concise deletion of the distal *Calm1* poly(A) signal (similar to our strategy for proximal poly(A) sites). Due to the small size of the required deletion, such a strategy is expected to be very efficient in producing homozygous cell clones or embryos. However, while this approach succeeds in eliminating production of the long UTR isoform, it can reduce total mRNA and protein expression to the extent that the long 3′UTR isoform previously contributed to it.

Finally, instead of deleting the entire 3′UTR, a library containing all possible unique gRNA sequences targeting the region of interest can be introduced into target cells. The resulting pool of cells harboring small indel deletions can then be screened for a particular phenotype. Wu et al. (2017) used this type of gRNA "tiling screen" to functionally dissect 3′UTR sequences of several *Drosophila* genes. The main advantage of this method is that it can deliver precise information regarding the spatial distribution of *cis*-regulatory elements. However, in contrast to *Drosophila* genes, most human genes do not provide the necessary gRNA density to sufficiently interrogate most 3′UTR sequences (Pulido-Quetglas et al., 2017). We therefore imagine that this approach could be expanded by using tandem gRNAs that produce an array of small deletions within the 3′UTR sequence.

Acknowledgments

This work was funded by a postdoctoral fellowship from the DFG to S.M., an NIH training Grant (T32GM083937) to M.M.F. and by the NIH Director's Pioneer Award (DP1-GM123454) and the Pershing Square Sohn Cancer Research Alliance to C.M. as well as by the NCI Cancer Center Support Grant (P30 CA008748).

References

Bae, B., Gruner, H. N., Lynch, M., Feng, T., So, K., Oliver, D., et al. (2020). Elimination of Calm1 long 3'-UTR mRNA isoform by CRISPR-Cas9 gene editing impairs dorsal root ganglion development and hippocampal neuron activation in mice. *RNA*, *26*(10), 1414–1430.

Baek, D., Villén, J., Shin, C., Camargo, F. D., Gygi, S. P., & Bartel, D. P. (2008). The impact of microRNAs on protein output. *Nature*, *455*(7209), 64–71.

Bauer, D. E., Canver, M. C., & Orkin, S. H. (2015). Generation of genomic deletions in mammalian cell lines via CRISPR/Cas9. *Journal of Visualized Experiments*, *95*, e52118.

Ben-Hamo, R., & Efroni, S. (2015). MicroRNA regulation of molecular pathways as a generic mechanism and as a core disease phenotype. *Oncotarget*, *6*(3), 1594–1604.

Berkovits, B. D., & Mayr, C. (2015). Alternative 3' UTRs act as scaffolds to regulate membrane protein localization. *Nature*, *522*(7556), 363–367.

Berry, M. J., Banu, L., Chen, Y., Mandel, S. J., Kieffer, J. D., Harney, J. W., et al. (1991). Recognition of UGA as a selenocysteine codon in type I deiodinase requires sequences in the 3′ untranslated region. *Nature*, *353*(6341), 273–276.

Bevilacqua, P. C., Ritchey, L. E., Su, Z., & Assmann, S. M. (2016). Genome-wide analysis of RNA secondary structure. *Annual Review of Genetics*, *50*(1), 235–266.

Canver, M. C., Bauer, D. E., Dass, A., Yien, Y. Y., Chung, J., Masuda, T., et al. (2014). Characterization of genomic deletion efficiency mediated by clustered regularly interspaced short palindromic repeats (CRISPR)/Cas9 nuclease system in mammalian cells. *The Journal of Biological Chemistry*, *289*(31), 21312–21324.

Chang, J. W., Yeh, H. S., & Yong, J. (2017). Alternative polyadenylation in human diseases. *Endocrinology and Metabolism (Seoul)*, *32*(4), 413–421.

Chartron, J. W., Hunt, K. C. L., & Frydman, J. (2016). Cotranslational signal-independent SRP preloading during membrane targeting. *Nature*, *536*(7615), 224–228.

Chatterjee, P., Jakimo, N., & Jacobson, J. M. (2018). Minimal PAM specificity of a highly similar SpCas9 ortholog. *Science Advances*, *4*(10), eaau0766.

Chatterjee, P., Lee, J., Nip, L., Koseki, S. R. T., Tysinger, E., Sontheimer, E. J., et al. (2020). A Cas9 with PAM recognition for adenine dinucleotides. *Nature Communications*, *11*(1), 2474.

Chen, C. Y., Chen, S. T., Juan, H. F., & Huang, H. C. (2012). Lengthening of 3'UTR increases with morphological complexity in animal evolution. *Bioinformatics*, *28*(24), 3178–3181.

Chen, S., Lee, B., Lee, A. Y., Modzelewski, A. J., & He, L. (2016). Highly efficient mouse genome editing by CRISPR ribonucleoprotein electroporation of zygotes. *The Journal of Biological Chemistry*, *291*(28), 14457–14467.

Cheng, Y., Miura, R. M., & Tian, B. (2006). Prediction of mRNA polyadenylation sites by support vector machine. *Bioinformatics*, *22*(19), 2320–2325.

Concordet, J.-P., & Haeussler, M. (2018). CRISPOR: Intuitive guide selection for CRISPR/Cas9 genome editing experiments and screens. *Nucleic Acids Research*, *46*(W1), W242–W245.

Cottrell, K. A., Szczesny, P., & Djuranovic, S. (2017). Translation efficiency is a determinant of the magnitude of miRNA-mediated repression. *Scientific Reports*, 7(1), 14884.

Doench, J. G., Fusi, N., Sullender, M., Hegde, M., Vaimberg, E. W., Donovan, K. F., et al. (2016). Optimized sgRNA design to maximize activity and minimize off-target effects of CRISPR-Cas9. *Nature Biotechnology*, *34*(2), 184–191.

Dominguez, D., Freese, P., Alexis, M. S., Su, A., Hochman, M., Palden, T., et al. (2018). Sequence, structure, and context preferences of human RNA binding proteins. *Molecular Cell*, *70*(5), 854–867 (e859).

Dumeau, C. E., Monfort, A., Kissling, L., Swarts, D. C., Jinek, M., & Wutz, A. (2019). Introducing gene deletions by mouse zygote electroporation of Cas12a/Cpf1. *Transgenic Research*, *28*(5–6), 525–535.

Farboud, B., Jarvis, E., Roth, T. L., Shin, J., Corn, J. E., Marson, A., et al. (2018). Enhanced genome editing with Cas9 ribonucleoprotein in diverse cells and organisms. *Journal of Visualized Experiments*, *135*, 57350.

Gao, Z., Herrera-Carrillo, E., & Berkhout, B. (2018). Delineation of the exact transcription termination signal for type 3 polymerase III. *Molecular Therapy. Nucleic Acids*, *10*, 36–44.

Gasperini, M., Findlay, G. M., McKenna, A., Milbank, J. H., Lee, C., Zhang, M. D., et al. (2017). CRISPR/Cas9-mediated scanning for regulatory elements required for HPRT1 expression via thousands of large, programmed genomic deletions. *American Journal of Human Genetics*, *101*(2), 192–205.

Gruber, A. J., Schmidt, R., Gruber, A. R., Martin, G., Ghosh, S., Belmadani, M., et al. (2016). A comprehensive analysis of 3′ end sequencing data sets reveals novel polyadenylation signals and the repressive role of heterogeneous ribonucleoprotein C on cleavage and polyadenylation. *Genome Research*, *26*(8), 1145–1159.

Haeussler, M., Schonig, K., Eckert, H., Eschstruth, A., Mianne, J., Renaud, J. B., et al. (2016). Evaluation of off-target and on-target scoring algorithms and integration into the guide RNA selection tool CRISPOR. *Genome Biology*, *17*(1), 148.

Hanna, R. E., & Doench, J. G. (2020). Design and analysis of CRISPR-Cas experiments. *Nature Biotechnology*, *38*(7), 813–823.

Herrmann, C. J., Schmidt, R., Kanitz, A., Artimo, P., Gruber, A. J., & Zavolan, M. (2020). PolyASite 2.0: a consolidated atlas of polyadenylation sites from 3′ end sequencing. *Nucleic Acids Research*, *48*(D1), D174–D179.

Higgs, D. R., Goodbourn, S. E., Lamb, J., Clegg, J. B., Weatherall, D. J., & Proudfoot, N. J. (1983). Alpha-thalassaemia caused by a polyadenylation signal mutation. *Nature*, *306*(5941), 398–400.

Hsu, P. D., Scott, D. A., Weinstein, J. A., Ran, F. A., Konermann, S., Agarwala, V., et al. (2013). DNA targeting specificity of RNA-guided Cas9 nucleases. *Nature Biotechnology*, *31*(9), 827–832.

Iadevaia, V., & Gerber, A. P. (2015). Combinatorial control of mRNA fates by RNA-binding proteins and non-coding RNAs. *Biomolecules*, *5*(4), 2207–2222.

Jalkanen, A. L., Coleman, S. J., & Wilusz, J. (2014). Determinants and implications of mRNA poly(A) tail size- -does this protein make my tail look big? *Seminars in Cell & Developmental Biology*, *34*, 24–32.

Joberty, G., Falth-Savitski, M., Paulmann, M., Bosche, M., Doce, C., Cheng, A. T., et al. (2020). A tandem guide RNA-based strategy for efficient CRISPR gene editing of cell populations with low heterogeneity of edited alleles. *The CRISPR Journal*, *3*(2), 123–134.

Kleinstiver, B. P., Prew, M. S., Tsai, S. Q., Topkar, V. V., Nguyen, N. T., Zheng, Z., et al. (2015). Engineered CRISPR-Cas9 nucleases with altered PAM specificities. *Nature*, *523*(7561), 481–485.

Kontoyiannis, D., Pasparakis, M., Pizarro, T. T., Cominelli, F., & Kollias, G. (1999). Impaired on/off regulation of TNF biosynthesis in mice lacking TNF AU-rich elements: Implications for joint and gut-associated Immunopathologies. *Immunity*, *10*(3), 387–398.

Kristjansdottir, K., Fogarty, E. A., & Grimson, A. (2015). Systematic analysis of the Hmga2 3' UTR identifies many independent regulatory sequences and a novel interaction between distal sites. *RNA*, *21*(7), 1346–1360.

Kryukov, G. V., Castellano, S., Novoselov, S. V., Lobanov, A. V., Zehtab, O., Guigó, R., et al. (2003). Characterization of mammalian Selenoproteomes. *Science*, *300*(5624), 1439.

Lautz, T., Stahl, U., & Lang, C. (2010). The human c-fos and TNFalpha AU-rich elements show different effects on mRNA abundance and protein expression depending on the reporter in the yeast Pichia pastoris. *Yeast*, *27*(1), 1–9.

Lee, S. H., & Mayr, C. (2019). Gain of additional BIRC3 protein functions through 3'-UTR-mediated protein complex formation. *Molecular Cell*, *74*(4), 701–712 (e709).

Lee, S. H., Singh, I., Tisdale, S., Abdel-Wahab, O., Leslie, C. S., & Mayr, C. (2018). Widespread intronic polyadenylation inactivates tumour suppressor genes in leukaemia. *Nature*, *561*(7721), 127–131.

Legut, M., Daniloski, Z., Xue, X., McKenzie, D., Guo, X., Wessels, H. H., et al. (2020). High-throughput screens of PAM-flexible Cas9 variants for gene knockout and transcriptional modulation. *Cell Reports*, *30*(9), 2859–2868 (e2855).

Levadoux, M., Mahon, C., Beattie, J. H., Wallace, H. M., & Hesketh, J. E. (1999). Nuclear import of Metallothionein requires its mRNA to be associated with the perinuclear cytoskeleton. *Journal of Biological Chemistry*, *274*(49), 34961–34966.

Lianoglou, S., Garg, V., Yang, J. L., Leslie, C. S., & Mayr, C. (2013). Ubiquitously transcribed genes use alternative polyadenylation to achieve tissue-specific expression. *Genes & Development*, *27*(21), 2380–2396.

Loya, A., Pnueli, L., Yosefzon, Y., Wexler, Y., Ziv-Ukelson, M., & Arava, Y. (2008). The 3'-UTR mediates the cellular localization of an mRNA encoding a short plasma membrane protein. *RNA (New York, N.Y.)*, *14*(7), 1352–1365.

Ma, W., & Mayr, C. (2018). A Membraneless organelle associated with the endoplasmic reticulum enables 3'UTR-mediated protein-protein interactions. *Cell*, *175*(6), 1492–1506 (e1419).

Margeot, A., Blugeon, C., Sylvestre, J., Vialette, S., Jacq, C., & Corral-Debrinski, M. (2002). In Saccharomyces cerevisiae, ATP2 mRNA sorting to the vicinity of mitochondria is essential for respiratory function. *The EMBO Journal*, *21*(24), 6893–6904.

Mariella, E., Marotta, F., Grassi, E., Gilotto, S., & Provero, P. (2019). The length of the expressed 3' UTR is an intermediate molecular phenotype linking genetic variants to complex diseases. *Frontiers in Genetics*, *10*, 714.

Martin, G., Gruber, A. R., Keller, W., & Zavolan, M. (2012). Genome-wide analysis of pre-mRNA 3′ end processing reveals a decisive role of human cleavage factor I in the regulation of 3′ UTR length. *Cell Reports*, *1*(6), 753–763.

Mätlik, K., Olfat, S., Garton, D. R., Montaño-Rodriguez, A., Turconi, G., Porokuokka, L. L., et al. (2019). Gene Knock Up via 3'UTR editing to study gene function in vivo. *bioRxiv*, 775031.

Matoulkova, E., Michalova, E., Vojtesek, B., & Hrstka, R. (2012). The role of the 3′ untranslated region in post-transcriptional regulation of protein expression in mammalian cells. *RNA Biology*, *9*(5), 563–576.

Mayr, C. (2017). Regulation by 3′-untranslated regions. *Annual Review of Genetics*, *51*(1), 171–194.

Mayr, C. (2019). What are 3' UTRs doing? *Cold Spring Harbor Perspectives in Biology*, *11*(10).

Mayr, C., & Bartel, D. P. (2009). Widespread shortening of 3'UTRs by alternative cleavage and polyadenylation activates oncogenes in cancer cells. *Cell*, *138*(4), 673–684.

Miska, E. A., Alvarez-Saavedra, E., Abbott, A. L., Lau, N. C., Hellman, A. B., McGonagle, S. M., et al. (2007). Most Caenorhabditis elegans microRNAs are individually not essential for development or viability. *PLoS Genetics*, *3*(12), e215.

Mitschka, S., & Mayr, C. (2020). Endogenous p53 expression in human and mouse is not regulated by its 3'UTR. *bioRxiv*, 2020.2011.2023.394197.

Moqtaderi, Z., Geisberg, J. V., & Struhl, K. (2018). Extensive structural differences of closely related 3′ mRNA isoforms: links to Pab1 binding and mRNA stability. *Molecular Cell*, *72*(5), 849–861.e846.

Moretti, F., Rolando, C., Winker, M., Ivanek, R., Rodriguez, J., Von Kriegsheim, A., et al. (2015). Growth cone localization of the mRNA encoding the chromatin regulator HMGN5 modulates neurite outgrowth. *Molecular and Cellular Biology*, *35*(11), 2035–2050.

Orkin, S. H., Cheng, T. C., Antonarakis, S. E., & Kazazian, H. H., Jr. (1985). Thalassemia due to a mutation in the cleavage-polyadenylation signal of the human beta-globin gene. *The EMBO Journal*, *4*(2), 453–456.

Otsuka, H., Fukao, A., Funakami, Y., Duncan, K. E., & Fujiwara, T. (2019). Emerging evidence of translational control by AU-rich element-binding proteins. *Frontiers in Genetics*, *10*, 332.

Owens, D. D. G., Caulder, A., Frontera, V., Harman, J. R., Allan, A. J., Bucakci, A., et al. (2019). Microhomologies are prevalent at Cas9-induced larger deletions. *Nucleic Acids Research*, *47*(14), 7402–7417.

Pulido-Quetglas, C., Aparicio-Prat, E., Arnan, C., Polidori, T., Hermoso, T., Palumbo, E., et al. (2017). Scalable design of paired CRISPR guide RNAs for genomic deletion. *PLoS Computational Biology*, *13*(3), e1005341.

Roundtree, I. A., Evans, M. E., Pan, T., & He, C. (2017). Dynamic RNA modifications in gene expression regulation. *Cell*, *169*(7), 1187–1200.

Saito, T., & Saetrom, P. (2012). Target gene expression levels and competition between transfected and endogenous microRNAs are strong confounding factors in microRNA high-throughput experiments. *Silence*, *3*, 3.

Sanchez de Groot, N., Armaos, A., Grana-Montes, R., Alriquet, M., Calloni, G., Vabulas, R. M., et al. (2019). RNA structure drives interaction with proteins. *Nature Communications*, *10*(1), 3246.

Sandberg, R., Neilson, J. R., Sarma, A., Sharp, P. A., & Burge, C. B. (2008). Proliferating cells express mRNAs with shortened 3' untranslated regions and fewer MicroRNA target sites. *Science*, *320*, 1643–1647.

Sentmanat, M. F., Peters, S. T., Florian, C. P., Connelly, J. P., & Pruett-Miller, S. M. (2018). A survey of validation strategies for CRISPR-Cas9 editing. *Scientific Reports*, *8*(1), 888.

Stacey, S. N., Sulem, P., Jonasdottir, A., Masson, G., Gudmundsson, J., Gudbjartsson, D. F., et al. (2011). A germline variant in the TP53 polyadenylation signal confers cancer susceptibility. *Nature Genetics*, *43*(11), 1098–1103.

Thomas, J. D., Polaski, J. T., Feng, Q., De Neef, E. J., Hoppe, E. R., McSharry, M. V., et al. (2020). RNA isoform screens uncover the essentiality and tumor-suppressor activity of ultraconserved poison exons. *Nature Genetics*, *52*(1), 84–94.

Tian, B., & Manley, J. L. (2017). Alternative polyadenylation of mRNA precursors. *Nature Reviews. Molecular Cell Biology*, *18*(1), 18–30.

Tian, B., Pan, Z., & Lee, J. Y. (2007). Widespread mRNA polyadenylation events in introns indicate dynamic interplay between polyadenylation and splicing. *Genome Research*, *17*(2), 156–165.

Tushev, G., Glock, C., Heumuller, M., Biever, A., Jovanovic, M., & Schuman, E. M. (2018). Alternative 3' UTRs modify the localization, regulatory potential, stability, and plasticity of mRNAs in neuronal compartments. *Neuron*, *98*(3), 495–511 (e496).

Ulitsky, I., Shkumatava, A., Jan, C. H., Subtelny, A. O., Koppstein, D., Bell, G. W., et al. (2012). Extensive alternative polyadenylation during zebrafish development. *Genome Research*, *22*(10), 2054–2066.

Vidigal, J. A., & Ventura, A. (2015). Rapid and efficient one-step generation of paired gRNA CRISPR-Cas9 libraries. *Nature Communications*, *6*, 8083.

Wang, R., Nambiar, R., Zheng, D., & Tian, B. (2018). PolyA_DB 3 catalogs cleavage and polyadenylation sites identified by deep sequencing in multiple genomes. *Nucleic Acids Research*, *46*(D1), D315–D319.

Wang, R., Zheng, D., Yehia, G., & Tian, B. (2018). A compendium of conserved cleavage and polyadenylation events in mammalian genes. *Genome Research*, *28*(10), 1427–1441.

Wissink, E. M., Fogarty, E. A., & Grimson, A. (2016). High-throughput discovery of post-transcriptional cis-regulatory elements. *BMC Genomics*, *17*, 177.

Witwer, K. W., & Halushka, M. K. (2016). Toward the promise of microRNAs - enhancing reproducibility and rigor in microRNA research. *RNA Biology*, *13*(11), 1103–1116.

Wu, Q., Ferry, Q. R. V., Baeumler, T. A., Michaels, Y. S., Vitsios, D. M., Habib, O., et al. (2017). In situ functional disSection of RNA cis-regulatory elements by multiplex CRISPR-Cas9 genome engineering. *Nature Communications*, *8*(1), 2109.
You, L., Wu, J., Feng, Y., Fu, Y., Guo, Y., Long, L., et al. (2015). APASdb: a database describing alternative poly(A) sites and selection of heterogeneous cleavage sites downstream of poly(A) signals. *Nucleic Acids Research*, *43*(Database issue), D59–D67.
Yu, B., Lu, Y., Zhang, Q. C., & Hou, L. (2020). Prediction and differential analysis of RNA secondary structure. *Quantitative Biology*, *8*(2), 109–118.
Zanzoni, A., Spinelli, L., Ribeiro, D. M., Tartaglia, G. G., & Brun, C. (2019). Post-transcriptional regulatory patterns revealed by protein-RNA interactions. *Scientific Reports*, *9*(1), 4302.
Zetsche, B., Gootenberg, J. S., Abudayyeh, O. O., Slaymaker, I. M., Makarova, K. S., Essletzbichler, P., et al. (2015). Cpf1 is a single RNA-guided endonuclease of a class 2 CRISPR-Cas system. *Cell*, *163*(3), 759–771.
Zhang, H. L., Singer, R. H., & Bassell, G. J. (1999). Neurotrophin regulation of β-actin mRNA and protein localization within growth cones. *Journal of Cell Biology*, *147*(1), 59–70.
Zhao, W., Siegel, D., Biton, A., Tonqueze, O. L., Zaitlen, N., Ahituv, N., et al. (2017). CRISPR-Cas9-mediated functional disSection of 3’-UTRs. *Nucleic Acids Research*, *45*(18), 10800–10810.
Zheng, D., Wang, R., Ding, Q., Wang, T., Xie, B., Wei, L., et al. (2018). Cellular stress alters 3′UTR landscape through alternative polyadenylation and isoform-specific degradation. *Nature Communications*, *9*(1), 2268.
Zhu, S., Li, W., Liu, J., Chen, C.-H., Liao, Q., Xu, P., et al. (2016). Genome-scale deletion screening of human long non-coding RNAs using a paired-guide RNA CRISPR–Cas9 library. *Nature Biotechnology*, *34*, 1279–1286.
Zhu, Y., Xu, G., Yang, Y. T., Xu, Z., Chen, X., Shi, B., et al. (2019). POSTAR2: Deciphering the post-transcriptional regulatory logics. Nucleic Acids Research, 47(D1), D203-D211.

Modulation of alternative cleavage and polyadenylation events by dCas9-mediated CRISPRpas

Jihae Shin[a], Ruijia Wang[a], and Bin Tian[a,b,*]

[a]Department of Microbiology, Biochemistry, and Molecular Genetics, Center for Cell Signaling, Rutgers New Jersey Medical School, Newark, NJ, United States

[b]Program in Gene Expression and Regulation, Center for Systems and Computational Biology, The Wistar Institute, Philadelphia, PA, United States

*Corresponding author: e-mail address: btian@wistar.org

Contents

Abstract

The CRISPR/Cas9 technology is revolutionizing genomic engineering. The high efficiency and selectivity of the system have inspired the development of various derived tools for gene regulation at different levels, such as transcriptional activation or inhibition,

Methods in Enzymology, Volume 655
ISSN 0076-6879
https://doi.org/10.1016/bs.mie.2021.04.016

epigenetic modification, splicing, and base editing. Cleavage and polyadenylation (CPA) is an essential 3′ end maturation step for almost all eukaryotic mRNAs. CPA is tightly coupled with transcriptional termination, and its activity impacts gene expression. Over half of all human genes display alternative polyadenylation (APA), where multiple cleavage and polyadenylation sites (PASs) lead to mRNA isoforms with variable termini. APA isoforms often have distinct metabolisms, and their relative abundance can change drastically in different cells. Here, we describe a method based on delivering a catalytically dead Cas9 (dCas9) to genomic regions nears the PAS, which alters APA site usage in 3′UTRs or introns. This method, named CRISPRpas, allows investigators to examine functional significance of APA isoforms of individual genes. We also describe using the bioinformatics program APAlyzer to examine APA events of interest with RNA-seq data.

1. Introduction

Cleavage and polyadenylation (CPA) is essential for 3′ end maturation of almost all eukaryotic mRNAs (Shi & Manley, 2015). CPA is carried out by the 3′ end processing machinery at the CPA site, commonly referred to as poly(A) site (PAS). The strength or processing efficiency of a PAS is governed by its surrounding sequence motifs such as the AAUAAA hexamer or its variants, upstream UGUA elements, upstream U-rich elements, and downstream U- or GU-rich elements (Tian & Graber, 2012; Tian & Manley, 2017). Mutations changing the PAS sequence have been reported in a growing number of human diseases, such as thalassemia and systemic lupus erythematosus (Danckwardt, Hentze, & Kulozik, 2008; Hollerer, Grund, Hentze, & Kulozik, 2014). Recent studies have also identified human single nucleotide polymorphisms (SNPs) near the PAS that can alter PAS usage and impact gene expression (Shulman & Elkon, 2020; Yang et al., 2020).

Most mammalian genes have multiple PASs, resulting in expression of the alternative polyadenylation (APA) isoforms (Hoque et al., 2013; Shepard et al., 2011). APA sites in 3′ untranslated regions (3′UTRs) shorten or lengthen the 3′UTR sequence. Because 3′UTRs are often enriched with binding motifs for *trans*-activating regulators, such as RNA-binding proteins, miRNAs, and long non-coding (lnc) RNAs, 3′UTR APA alters aspects of mRNA metabolism, including mRNA stability, translation, and subcellular localization (Mayr, 2017). In addition, APA sites in introns lead to expression of transcripts containing distinct open reading frames, including non-functional truncated ones (Dubbury, Boutz, & Sharp, 2018; Lee et al., 2018). While intronic polyadenylation (IPA) sites are generally considered "cryptic" and are suppressed in normal conditions, emerging studies suggest that their usage is dynamically regulated in many developmental and pathological conditions (Dubbury et al., 2018; Kamieniarz-Gdula & Proudfoot, 2019; Lee et al., 2018; Li et al., 2015; Williamson et al., 2017). While the biological importance of APA is increasingly appreciated, experimental strategies to modulate PAS usage are still limited.

The CRISPR/Cas9 system has emerged as a powerful tool for genome editing (Doudna & Charpentier, 2014). The Cas9-mediated genomic editing of PAS was employed in several recent studies to examine the importance of specific APA isoforms (Kamieniarz-Gdula & Proudfoot, 2019; Liu et al., 2017; Wang et al., 2018; Wang, Zheng, Wei, Ding, & Tian, 2019). In most cases, the approach involves deletion of a specific PAS, leading to isoform expression changes. Also notable is that sub-sequence deletion of 3′UTR by CRISPR/Cas9 has been used to investigate post-transcriptional regulation (Zhao et al., 2017) when precise 3′UTR motif editing is difficult due to lack of suitable guide RNAs (gRNAs). One drawback of Cas9-mediated genome editing is that it permanently changes DNA sequence and requires extensive cell manipulation and isolation of single clones for each targeting site. A rapid and efficient programmable tool that impact APA without genomic sequence changes would therefore be desirable.

Owing to its high efficiency, the CRISPR/Cas9 system has also been harnessed for gene regulation beyond genome sequence alterations. For example, endonuclease deficient Cas9 (catalytically dead Cas9, dCas9)-based CRISPR systems have been used for transcriptional inhibition (CRISPRi) and activation (CRISPRa) (Gilbert et al., 2013; Konermann et al., 2015; Qi et al., 2013). We recently developed a non-genomic editing method, named CRISPRpas, to alter APA in 3′UTRs or introns by delivering dCas9 to the downstream region of a proximal PAS (pPAS) (Shin, Ding, Baljinnyam, Wang, & Tian, 2021). By blocking the progression of RNA polymerase II (Pol II), the dCas9 promotes usage of the pPAS and blocks usage of distal PAS (dPAS). In addition to gRNA targeting efficiency, important parameters for CRISPRpas efficiency also include the DNA strand for targeting, distance from the target site to the PAS, and the PAS strength. Here, we describe experimental design and methods in detail of our approach of CRISPRpas in a mammalian cell line. We also discuss the advantage, limitation, and potential applications for this method. Users can employ CRISPRpas to modulate APA of their gene of interest in order to examine functional consequences of APA for the gene.

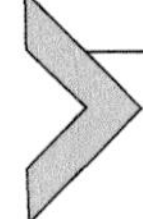

2. Experimental design

2.1 Bioinformatic analysis of APA

Before the experimental setup, it is important to gather enough information about the mRNA APA isoforms expressed from the target gene. Several 3′ end annotation databases are available including PolyA_DB3 (Wang, Nambiar, Zheng, & Tian, 2018), polyASite 2.0 (Herrmann et al., 2020), and APASdB (You et al., 2015). The data from these databases can be visualized on the UCSC genome browser as custom tracks. The UCSC genome browser also has PolyA-Seq tracks (hg19 assembly) from Merck Research Laboratories for multiple human tissues (Derti et al., 2012), which users can toggle for visualization. Other individual experimental data sets from 3′ end sequencing experiments such as 3′READS (Hoque et al., 2013), 2P-Seq (Spies, Burge, & Bartel, 2013), 3P-Seq (Jan, Friedman, Ruby, & Bartel, 2011; Nam et al., 2014), PAS-Seq (Shepard et al., 2011), A-seq (Gruber et al., 2014), and QuantSeq (Moll, Ante, Seitz, et al., 2014) are also rich sources of information for PAS usage in specific cell types. In addition, computational tools, such as DaPars (Xia et al., 2014), APAtrap (Ye, Long, Ji, Li, & Wu, 2018), APAlyzer (Wang & Tian, 2020), and QAPA (Ha, Blencowe, & Morris, 2018) can be used to infer changes of APA isoform expression using regular RNA-Seq data. Accordingly, the Genotype-Tissue Expression (GTEx) project provides a comprehensive resource to study tissue-specific APA in non-diseased human populations (GTEx Consortium, 2013) and data from the Encyclopedia of DNA Elements (ENCODE) project are informative for APA in various cell types (Davis et al., 2018; The ENCODE Project Consortium, 2012).

2.1.1 PolyA_DB3

PolyA_DB3 (https://exon.apps.wistar.org/polya_db/v3/index.php) catalogs PASs in several genomes using data from the 3′ end region extraction and sequencing (3′READS) method (Wang, Zheng, Yehia, & Tian, 2018). The 3′READS method utilizes a direct ligation of an adapter to the 3′ end of poly(A) + RNAs, which removes mispriming events at internal A-rich regions and enables accurate interrogation of the PAS (Hoque et al., 2013). The current version of the PolyA_DB is based on ~1.2 billion 3′READS reads from over 360 human, mouse, and rat samples. Users can visit the PolyA_DB website and query PAS usage of individual genes in these genomes (hg19, mm9, and rn5). In addition, data for chicken genes (galGal4) are available for conservation analysis. The UCSC hub files consisting of PAS positions (bigBed format) and expression levels, based on mean RPM (reads per million mapped reads) of PAS-supporting reads (bigWig format) are also available through the website.

2.1.2 PolyASite 2.0

PolyASite 2.0 (https://polyasite.unibas.ch) is based on publicly available human, mouse, and worm 3′ end sequencing datasets generated by several methods including 3'READS (Hoque et al., 2013), 2P-Seq (Spies et al., 2013), 3P-Seq (Jan et al., 2011; Nam et al., 2014), PAS-Seq (Shepard et al., 2011), A-seq (Gruber et al., 2014), and PolyA-Seq (Derti et al., 2012). The current version encompasses ~2.3 billion reads from over 400 samples (Herrmann et al., 2020). Users can search for PASs in the gene of interest or add custom tracks for human (hg38), mouse (mm10), and *C. elegans* (ce11) using aggregated data. Importantly, PolyASite 2.0 provides readily accessible UCSC genome browser tracks for individual RNA-Seq data that are included in the database. For example, the 'samples' tab has a description of each sample (with its NCBI GEO number) to check experiment and sample details. Custom tracks can be added to the UCSC genome browser through simple clicks. Sample sources are also well documented on the website, allowing users to obtain information about isoform expression in specific cell or tissue types.

2.1.3 Publicly available, large RNA-seq datasets

The GTEx project provides a comprehensive public resource to study tissue-specific gene expression and regulation, with samples collected from nearly 1000 non-diseased individuals (GTEx Consortium, 2013). GTEx has a Transcript Browser under the Expression tab, enabling visualization of transcript expression and isoform structures in over 50 tissues. The isoform structure information is particularly helpful for designing targeting gRNAs and validation primer sets. The ENCODE project provides extensive transcriptomic data from various cell types (Davis et al., 2018; The ENCODE Project Consortium, 2012). Under the Functional genomics data tab, users can filter data based on assays, organisms, and cell types, etc. After selection of interested RNA-Seq experiments, transcript expression can be directly visualized in the Files section, or through UCSC genome browser tracks in File details tab. These tissue- and cell type-specific RNA-Seq datasets are of great value to assessing isoform expression levels, mitigating biases in expression analysis that are often associated with 3′ end sequencing.

Note: 3′ end sequencing methods often have biases in expression analysis. This is because reads from 3′ end sequencing are confined to a small sequence region. Biases can come from sample preparation or read mapping. By contrast, reads from regular RNA-Seq methods cover wider regions, and thus contain less biases. Cross-examination of expression data between RNA-Seq and 3′ end sequencing datasets could help the users infer isoform expression patterns more accurately.

2.1.4 APA isoform stability datasets

Because APA isoforms can have different half-lives, consideration of isoform stability differences in experimental design is advisable. For this purpose, users can resort to RNA isoform stability datasets for guidance. There are several publicly available datasets measuring RNA stability with 3′ end sequencing methods, such as half-life calculation based on transcriptional inhibition in mouse fibroblast NIH3T3 cells (Spies et al., 2013) and rodent hippocampal neurons (Tushev et al., 2018), and comparison of metabolic labeled (with 4-thiouridine, 4sU), newly made RNAs with pre-existing RNAs to derive stability of RNA isoforms in mouse fibroblast NIH3T3 cells (Zheng et al., 2018). We have also deposited a similar dataset for human HEK293T cells into the GEO database under the accession number GSE161727.

2.2 CRISPRpas

CRISPRpas takes advantage of dCas9 in blocking the progression of Pol II, thereby promoting the usage of the pPAS and inhibiting the usage of the distal PAS (dPAS). A schematic of the system is shown in Fig. 1. When a PAS is present on the pre-mRNA, CRISPRpas helps CPA at the PAS. However, because the pre-mRNA is also subject to degradation after Pol II is stalled or falling off from the template (Fig. 1), some considerations of gRNA design are needed to ensure CPA of pre-mRNA is the major pathway in CRISPRpas.

2.2.1 gRNA design for CRISPRpas

Our previous study showed that CRISPRpas efficiency is sensitive to gRNA location as well as PAS strength (Shin et al., 2021). The gRNAs for CRISRPpas should be designed to target non-template strand of the DNA sequences (Fig. 1). A distance of $\geq$500 nucleotides downstream of the target PAS often appears necessary in our studies (Shin et al., 2021). In general, multiple gRNAs should be tested for their efficiencies and

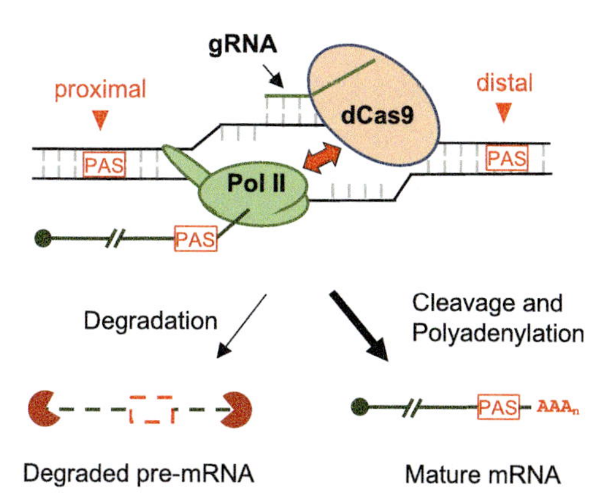

Fig. 1 Schematic of CRISPRpas. A model for dCas9-mediated APA regulation for a hypothetical gene with two PASs, namely, proximal and distal PASs. When a dCas9/gRNA targets a region between the proximal PAS and the distal PAS, RNA Pol II is stalled near the target site. The nascent transcript could be degraded after falling off from RNA Pol II (a minor pathway), or be subject to cleavage and polyadenylation (a major pathway), resulting in expression of an APA isoform using the proximal PAS.

sensitivities for a given gene. Sometimes it might not be possible to find high-specificity gRNAs due to design restrictions, in which multiple suboptimal gRNAs might be necessary. We recommend testing at least 3–5 gRNAs for any given APA event. To calculate gRNA specificity score, one can use CRISPOR (Haeussler et al., 2016) or similar programs to design gRNA oligos (Hanna & Doench, 2020). In the case of CRISPOR, a web-based interface allows users to paste genomic region of interest and outputs possible gRNA target sequences ranked by the specificity score. Selection of a gRNA with a high specificity score can minimize off-target effects. Moreover, because poly(T)-stretches act as a termination signal for Pol III, gRNA candidates with four or more consecutive T's should be avoided (most gRNA designing tools already consider it in their algorithms). The annealed oligos can be cloned into U6 promoter plasmid with *Bbs*I sites followed by Cas9 scaffold sequence such as pGR9 (Shin et al., 2021), phU6-gRNA (addgene # 53188), and pDonor_hU6 (addgene # 69312). Alternatively, 5′ and 3′ end 2′-O-Methyl and phosphorothioate modified synthesized single gRNAs can also be ordered from a commerical vendor. Testing gene-specific gRNAs should be compared to the non-specific control gRNA, which can be selected from validated sequences. For example, 5′-TTCTCTTGCTGAAGCTCGA-3′ is a validated non-specific sequence from a previous study (Tanenbaum, Gilbert, Qi, Weissman, &

Vale, 2014). The addgene.org has a list of validated negative control gRNA sequences (https://www.addgene.org/crispr/reference/grna-sequence).

2.2.2 dCas9 expression system

Robust dCas9 protein expression is critical for efficient CRISPRpas activity. The transient dCas9 expression vector can be co-transfected with a gRNA-encoding plasmid. Alternatively, stable cell lines expressing dCas9 can be established using various methods such as lentivirus transduction, especially when the type of cell being used is refractory to transfection. In this chapter, we describe a protocol using a stable cell line that has PiggyBac transposase-mediated genomic integration of dCas9 sequence to ensure its consistent expression at high levels. The PiggyBac transposon system facilitates preferential integration of transposons at "TTAA" sites that are randomly dispersed in the genome (Woodard & Wilson, 2015). Also, the PiggyBac system offers a large cargo-carrying capacity, which is amenable for large proteins like dCas9 (MW = 163 kDa). As mentioned above, when the cell type is known to have a low efficiency for traditional transfection methods (such as liposome-based transfection of plasmid DNA), a lentivirus-mediated transduction is recommended.

2.3 Experimental validation of APA changes

Successful APA changes can be confirmed by a variety of molecular biology techniques including northern blot analysis and RT-qPCR. For the validation of APA by RT-qPCR, users should carefully design specific primer sets amplifying common UTR (cUTR) and alternative UTR (aUTR) sequences, respectively. The standard $\Delta\Delta$ Ct method for comparative expression analysis can be used to calculate the relative expression of each isoform vs the expression of house-keeping genes, such as *GAPDH* or *ACTIN*. For IPA modulation, primer sets amplifying IPA isoforms and full-length isoforms (using PASs in the last exon) should be designed. In both cases, it is important to ensure RNA quality, especially free of contamination of genomic DNA, which may come from the RNA purification step. It is recommended to treat RNA samples with DNase prior to the reverse transcription reaction. Also, a negative control in which reverse transcriptase is not added can serve to indicate RNA quality, as any PCR product in this reaction should be the result of contaminant genomic DNA. When designing and validating primers, primers should follow the guidelines for standardized primer design, e.g., 15–20 bp long, 50–55% G/C, and amplicon size between 70 and 200 bp. Many user-friendly web-based primer design tools are available, including Primer3Plus (Untergasser et al., 2012). Importantly, the designed primer sequences should also be checked for specificity in the transcriptome using alignment tools such as BLAST. It is advisable that primer sets are to be tested with reference cDNA samples to confirm specificity prior to experiments. Samples generated by reverse transcription without RT enzyme should be used as a negative control. A single PCR product is expected for high specificity primer sets.

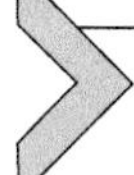

3. Materials

3.1 Reagents

- Plasmids
  - HyPB7 transposase plasmid (PB210PA-1 or equivalent, System Biosciences)
  - PiggyBac dCas9 expressing plasmid (e.g., PB-CAG-dCas9-10x GCN4-P2A-BFP)
  - U6 promoter gRNA expressing plasmid (e.g., pGR9 (Shin et al., 2021), phU6-gRNA (Addgene # 53188), and pDonor_hU6 (Addgene # 69312))
- Oligonucleotides (non-specific control oligos and gene-specific oligos ordered from commercial vendors)
  - Sense oligo for gRNA (5'-CACCGNNNNNNNNNNNNNNNNNN NNN-3′)
  - Antisense oligo for gRNA (5'-AAACNNNNNNNNNNNNNNNN NNNNC-3′)
  - Sequencing primer for gRNA plasmid (e.g., U6 forward 5'-GAGG GCCTATTTCCCATGATT-3′)
  - Gene specific PCR primers targeting different isoforms
- Cell line of interest
- Cell line-specific culture media and additives
- Fetal Bovine Serum (FBS)
- Antibiotic-Antimycotic (100X) (Gibco, 15240096)
- 0.05% Trypsin-EDTA (Gibco, 25-300-054)
- Phosphate-buffered saline (PBS)
- RNase-free water
- OPTI-MEM (Thermo Fisher, 31985062)
- Lipofectamine 3000 or similar transfection reagents (Invitrogen, L3000001)
- Hygromycin B (Millipore Sigma, 10843555001)

- Round bottom FACS tube (Falcon, 352235)
- 96-well cell culture plate
- 24-well cell culture plate
- 12-well cell culture plate
- 6-well cell culture plate
- FastDigest BpiI (FD1014, Thermo Scientific)
- TE buffer: 10 mM Tris-Cl, 0.1 mM EDTA, pH 8.5
- QIAquick Gel Extraction Kit (Qiagen, 28706)
- T4 Polynucleotide kinase (PNK, NEB, M0201)
- T4 DNA ligase (NEB, M0202)
- LB liquid medium
- Ampicillin (or other antibiotics resistance gene expressing in the gRNA plasmid)
- LB agar plates
- DH5α competent bacterial cells (homemade or commercial)
- Plasmid DNA miniprep kit (NEB, T1010)
- TRIzol (Invitrogen, 15596026)
- 3 M sodium acetate (pH 5.2)
- TURBO DNA-free Kit (Invitrogen, AM1907)
- Chloroform (Sigma-Aldrich, 366927)
- Isopropanol (Sigma-Aldrich, I9516)
- 100% Ethanol
- UltraPure agarose (Invitrogen, 16500500)
- Ethidium bromide solution (Thermo Fisher, 17898)
- 10 mM dNTP solution mix (NEB, N0447)
- M-MLV reverse transcriptase (Promega, M170)
- Oligo (dT)15 primer
- Recombinant RNasin Ribonuclease inhibitor (Promega, N2511)
- Luna universal qPCR master mix (NEB, M3003)

3.2 Equipment

- 37 °C water bath
- Cell culture incubator
- Cell culture hood
- Shaking incubator for bacterial culture
- Tabletop centrifuge for conical tubes
- Refrigerated microcentrifuge
- Fluorescence/light microscope (e.g., Invitrogen EVOS system)
- FACS sorting device (e.g., BD FACSAria Cell Sorter)

- Thermocycler
- Heat blocks
- Agarose gel electrophoresis system
- Nanodrop spectrophotometer
- Real-time PCR systems

4. Methods

4.1 APA event analysis using APAlyzer

Here we use the bioinformatic tool APAlyzer to examine RNA-seq data from GTEx to identify APA events altered by Single Nucleotide Polymorphism (SNP). We show one example analyzing intronic APA of the gene *ANKMY1* (Fig. 2).

1. The raw RNA-seq data for GTEx data was downloaded from dbGaP (phs000424.v7) (GTEx Consortium, 2017) including 5032 RNA-seq samples.
2. The SNP calling genotype data and phenotype data of GTEx were also downloaded from the same resource, only individuals covered by the RNA-seq were kept for the analysis.
3. Gene expression and intronic APA analysis were done using Bioconductor package APAlyzer (v1.4.0) (Wang & Tian, 2020).

Code:

```
BiocManager::install("APAlyzer")
library(APAlyzer)
URL= "https://github.com/RJWANGbioinfo/PAS_reference_RData/blob/
master/"
file= "hg19_REF.RData"
source_data(paste0(URL,file,"?raw= True"))
PASREF=REF4PAS(refUTRraw_hg19,dfIPA_hg19,dfLE_hg19)
UTRdbraw=PASREF$UTRdbraw
DFUTRraw=PASEXP_3UTR(UTRdbraw, bamfiles)
```

4. For gene expression analysis, reads mapped to CDS were used. For multi-exon genes, CDS region in the last exon was excluded. The read count was then normalized by CDS length and sequencing depth as transcript per kilobase million (TPM).

Code:

```
library(APAlyzer)
library("GenomicFeatures")
library("org.hs.eg.db")
```

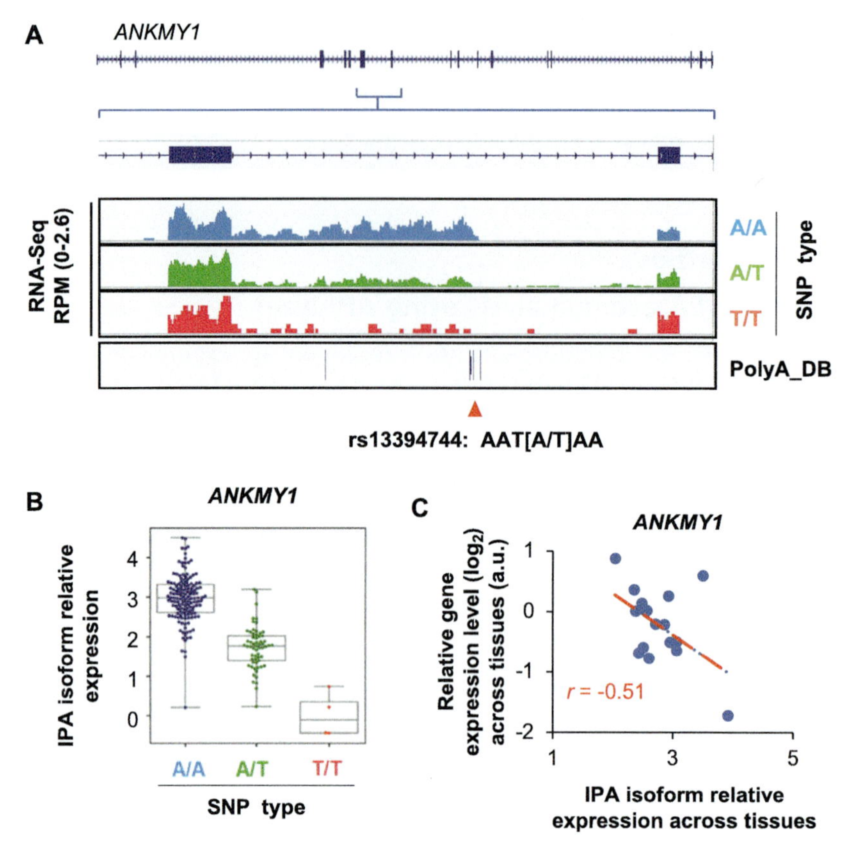

Fig. 2 Analysis of an APA event of *ANKMY1* using GTEx data. (A) An SNP in human *ANKMY1* gene affects cleavage and polyadenylation (CPA) in one of its introns. The *ANKMY1* gene is shown in a UCSC genome browser track and the region of interest is shown in a blowup image. GTEx RNA-seq data are segregated by three SNP types, as indicated. PASs from polyA_DB3 are also shown. The SNP variants are A or T in the AAT[T/A]AA motif for CPA. Because AATAAA is a stronger motif for CPA than AATTAA, there is more intronic CPA in cells with the A allele than those with the T allele. As such, RNA-seq reads covering the intronic region before the PAS are more abundant in A/A samples than A/T or T/T samples. (B) Relative expression (RE) of IPA isoform in individuals with different alleles. RE value is based on comparison of intronic read density with read density in other regions of the gene, and is calculated by APAlyzer. (C) Scatter plot for *ANKMY1* IPA-RE vs normalized gene expression in 17 different tissues. This result indicates that IPA of *ANKMY1* negatively impacts its expression. a.u., arbitrary unit.

```
extpath = system.file("extdata", "hg19.refGene.R.DB", package = "APAlyzer")
txdb = loadDb(extpath, packageName = 'GenomicFeatures')
IDDB = org.hs.eg.db
CDSdb = REFCDS(txdb,IDDB)
GENE_TPM = GENEXP_CDS(CDSdb, bamfiles)
```

5. For IPA analysis, IPA sites with percent of samples with expression (PSE) > 5% were first extracted from PolyA_DB version 3.2 (Wang, Nambiar, et al., 2018). The upstream and downstream regions of each IPA were then defined as the region between the site to its closest upstream 5′ or 3′ splicing sites (SS), and the region between the site to its closest downstream 3′SS. 5′SS and 3′SS information was obtained from RefSeq and Ensembl. The relative expression of IPA isoform (IPA-RE) = $\log_2(a-b)/c$, where a and b are read densities in IPA upstream and IPA downstream regions, respectively, and c is read density of the 3′ most exons. Only IPAs with at least five reads in each of the three regions were used for further analysis.

Code:

```
library(APAlyzer)
dfIPA=PASREF$dfIPA
dfLE=PASREF$dfLE
IPA_OUTraw=PASEXP_IPA(dfIPA, dfLE, bamfiles)IDDB=org.hs.eg.db
CDSdb=REFCDS(txdb,IDDB)
GENE_TPM=GENEXP_CDS(CDSdb, bamfiles)
```

6. For global APA analysis, IPA-RE of each gene were first standardized across samples, and then the median value of IPA-RE were used to represent the global IPA level for each tissue.

4.2 Establish a dCas9 expressing stable cell line

4.2.1 Determine hygromycin B kill curve

1. Place Hygromycin B powder at RT for 30 min before use.
2. Dissolve 100 mg of Hygromycin B in 1 mL of distilled water. Solution may be stored at 4 °C. Alternatively, aliquot and store at −20 °C. Do not re-freeze the aliquots after thawing.
3. Seed cells of the parental cell line in a 24-well plate at different densities (0.5–2×10^5 cells per well).
4. On the next day, remove medium and add fresh medium with varying concentrations of Hygromycin B (0, 100, 200, 400, 600, 800 μg/mL final concentration).
5. Examine viability every 2 days. Culture for 14 days. Replace the media containing antibiotics every 3 days.
6. Determine the lowest concentration of antibiotics that begins to show massive cell death in 3 days and kills all cells within 2 weeks. For HEK293T cells, we used 400 μg/mL Hygromycin B for 6 days.

4.2.2 *Transfection of PiggyBac plasmids*

1. (Day 1) Before transfection, split 3×10^5 cells in each well of a 12-well plate with 1 mL of growth media.
2. (Day 2) On the day of transfection, HEK293T cells should be ~70% confluent.
3. For one well of a 12-well plate combine:
   - 166 ng HyPB7 transposase expressing vector (~7.6 kb)
   - 833 ng PiggyBac dCas9 expressing vector (e.g., PB-CAG-dCas9-10xGCN4-P2A-BFP, 14.6 kb)

   A ratio of 1:2.5–1:5 for transposase to transposon vector ratio is recommended.
4. Combine DNA with P3000 in OPTI-MEM. Include several control wells such as no transposase control (transposon vector only) and no DNA control for Hygromycin B resistance control.
   - 1 μg plasmids
   - 2 μL P3000
   - 50 μL OPTI-MEM
5. Prepare another tube for 2 μL Lipofectamine 3000 in 50 μL of OPTI-MEM.
6. Mix tubes from steps 4 and 5. Incubate 10 min at RT.
7. Add 100 μL of the transfection mix drop-wise to one well in a 12-well plate and mix by gentle rocking.
8. (Day 3) On the next day, change media or split cells if necessary.
9. (Day 5) Start hygromycin selection for 6 days with 400 μg/mL Hygromycin B. No transposase or no DNA control cells should die off within 3 days, as determined in Section 4.2.1.
10. Check for positive integrations after 6 days. In case of PB-CAG-dCas9-10xGCN4-P2A-BFP, nuclear BFP signals for positive cells should be clearly visible (the BFP has a nuclear localization signal). Confirm dCas9 expression with western blot analysis of protein lysates.

4.3 FACS sorting and monoclonal selection of positive cells (optional)

While Hygromycin-resistant bulk cells are good enough for initial analysis, FACS sorting followed by monoclonal selection for cells expressing BFP at high levels would be desirable for more robust experimental analysis.

1. Prepare 96-well plates with 100 μL of pre-warmed media.
2. Wash dCas9 expressing stable cells with PBS.
3. Detach cells with trypsin-EDTA.
4. Add an excess volume of growth media to stop trypsin activity and collect cells in a 15 mL conical tube. Spin down the cells for 5 min at 500 RCF and resuspend the cell pellet with 0.5 mL media.

5. Pass cell suspension through a 50 μm filter top into a 5 mL round bottom FACS tube.
6. Using a FACS sorter, individually sort properly gated (FSC/SSC) BFP-positive cells in a 96-well culture plate containing 100 μL of pre-warmed media in each well.
7. For the remaining sorted cells (bulk), plate them into a 6 well plate and recover after 2–7 days. Freeze half of the cells for future plating and manually dilute remaining cells for seeding of single cells into a 96 well plate. Plate cells at two different densities, for example, 0.5 cells/well and 1.5 cells/well.
8. For single clones, continue to culture cells at 37 °C and do not disturb cells for 7–10 days.
9. From day 7, inspect the cells daily with a light microscope for single colony growth and BFP signals. Wells containing multiple colonies should be discarded. Change media for wells containing a single colony if necessary.
10. Check BFP signal and dCas9 expression level in each established cell lines.

Note: Depending on the cell type, a direct single-cell sorting into a plate can sometimes be too harsh and cells may not recover well. In this case, recover sorted cells after 2–7 days and manually perform single-cell expansion by serial dilution as described in step 7.

4.4 Cloning of gRNAs

Users need to clone a non-specific control gRNA in the same empty vector, and use it as a negative control for CRISPRpas experiments. Previous studies using the Cas9/dCas9 system have listed many non-specific control sequences, which can be handy for negative control gRNA design. The addgene.org website also has a list of negative controls with validated sequences. We show one example using CRISPRpas to perturb APA of the gene *ANKMY1* (Fig. 3A).

1. Reconstitute dry gRNA oligos in water to 100 μM.
2. Digest 1 μg of U6 promoter gRNA plasmid with FastDigest BpiI (*Bbs*I) for 30 min at 37 °C.
   - 1 μg of gRNA plasmid DNA
   - 1 μL BpiI (BbsI) (10 U/μL)
   - 2 μL 10 × FastDigest buffer
   - DW up to 20 μL
3. Run digested linear plasmid on an agarose gel and excise the correct size band.
4. Extract linear plasmid DNA using a gel extraction kit.
5. Elute digested plasmid and resuspend in TE buffer.

6. Determine DNA concentration using Nanodrop spectrophotometer.
7. Phosphorylate gRNA oligos for 30 min at 37 °C.
   - 1 μL sense oligo (100 μM)
   - 1 μL antisense oligo (100 μM)
   - 1 μL 10 × T4 PNK reaction buffer
   - 0.5 μL T4 PNK
   - DW up to 10 μL
8. Anneal oligos in a thermocycler. Incubate for 5 min at 95 °C and ramp down to 25 °C at 5 °C/min.
9. Dilute annealed oligos from step 8 by 1:100.
10. Ligate BbsI digested gRNA plasmid DNA and annealed oligos for 1 h at RT.
    - 50 ng BbsI digested plasmid from step 2
    - 1 μL annealed and diluted oligo duplex from step 9
    - 1 μL 10 × T4 ligase reaction buffer
    - 1 μL T4 DNA ligase
    - DW up to 10 μL
11. Add ligation mix to competent bacterial cells. Incubate on ice for 20 min.
12. Heat shock at 42 °C for 40 s and return to on ice. Add 1 mL of LB media.
13. Recover 1 h in the 37 °C incubator.
14. Plate transformants on ampicillin + LB agar plates (100 μg/mL).
15. Pick 3–5 colonies.
16. Inoculate 5 mL of LB + Amp bacterial culture and grow overnight at 37 °C in a shaking incubator.
17. Prepare plasmid DNA miniprep followed by measurement of DNA concentration by using a Nanodrop.
18. Validate correct insertion of gRNAs by Sanger sequencing using an upstream sequencing primer (e.g., U6 forward).

4.5 Transfection of gRNA plasmid

In this section we describe the usage of a stable cell line expressing dCas9 ($HEK293T^{dCas9}$) from step 4.1. Transient transfection of plasmid expressing dCas9 can also be used to drive the expression of dCas9, using methods such as lipofection, calcium phosphate, electroporation, etc. In these cases, optimization of transfection efficiency is recommended prior to experiments.

1. (Day 1) The day before the transfection, split 3×10^5 cells in each well of a 12-well plate with 1 mL of growth media.

2. (Day 2) On the day of transfection, HEK293T^{dCas9} cells should be ~70% confluent. Dilute 1 μg of pGR9 gRNA expressing plasmid and 2 μL of P3000 in 50 μL of OPTI-MEM.
3. Prepare another tube for 2 μL Lipofectamine 3000 in 50 μL of OPTI-MEM.
4. Mix tubes from steps 2 and 3. Incubate 10 min at RT.
5. Add 100 μL of the transfection mix to a 12 well and mix by gentle rocking.
6. (Day 3) On the next day, change media or split cells if necessary.
7. (Day 4–5) Proceed to validation and functional assays.

4.6 Validation of CRISPRpas by RT-qPCR

4.6.1 DNase-free RNA purification

1. Remove growth media and wash cells with PBS.
2. Add 0.5 mL of TRIzol reagent directly to the culture dish to lyse the cells.
3. Homogenize samples by pipetting up and down several times, and transfer to a microcentrifuge tube.
4. Incubate cells for 5 min to permit complete dissociation of the ribonucleoprotein complex.
5. Add 0.1 mL of chloroform and incubate at RT for 5 min.
6. Centrifuge the sample for 15 min at 12,000 × g at 4 °C.
7. Transfer the aqueous phase containing the RNA to a new tube. Avoid transferring any of the interphase.
8. Add 0.25 mL of isopropanol to the aqueous phase and incubate for 10 min.
9. Centrifuge for 10 min at 12,000 × g at 4 °C.
10. Discard the supernatant with pipetting.
11. Wash the RNA pellet in 1 mL of cold 75% ethanol and centrifuge for 5 min at 12,000 × g at 4 °C.
12. Discard the supernatant.
13. Centrifuge for 1 min at 12,000 × g at 4 °C and discard liquid.
14. Resuspend the pellet in 50 μL of RNase-free water.
15. Determine RNA concentration.
16. Add 5 μL of 10 × TURBO DNase buffer and 1 μL of TURBO DNase (2 U) to the RNA and mix gently.
    - 10 μg RNA
    - 5 μL 10 × TURBO DNase buffer
    - 1 μL of TURBO DNase (2 U)
    - DW up to 50 μL

17. Incubate 30 min at 37 °C.
18. Resuspend DNase inactivation reagent.
19. Add 5 μL DNase inactivation reagent. Incubate 5 min at RT, mixing occasionally. Flick the tube 2–3 times during the incubation period.
20. Centrifuge at 10,000 × *g* for 1.5 min and transfer 30 μL of supernatant to a new tube.

4.6.2 Reverse transcription

1. Thaw 5 × RT buffer, 10 mM dNTP mix, 1 μg/μL Oligo (dT)15 primer.
2. Prepare RT mix.
   - 5 μL 5 × reaction buffer
   - 1.25 μL dNTP mix solution
   - 25 U RNasin ribonuclease inhibitor
   - 0.75 μL (200 U) M-MLV RT
   - DW up to 19.5 μL
3. Take 5 μL of RNA and add 0.5 μL of 1 μg/μL Oligo (dT)15 primer. Mix well and heat the tube at 70 °C for 5 min. Cool the tube immediately on ice for 5 min. Briefly spin the tube to collect the solution at the bottom of the tube. Transfer to a PCR tube.
4. Add 19.5 μL of RT mix. Also prepare a negative control without RT to be used to check genomic DNA contaminants.
5. Mix gently and incubate in the thermocycler at 42 °C for 1 h.
6. Briefly spin the tube, collect the supernatant, and dilute RT product up to 200 μL.

4.6.3 Quantitative PCR reaction

Typically, qPCR can be set up with technical replicates (3 ×) and biological replicates (e.g., three or more samples). Primers should be tested for specificity before the experiment. The standard ΔΔ Ct method for comparative expression analysis can be used to calculate the relative expression of each isoform compared to the expression of a house-keeping gene, such as *GAPDH* or *ACTIN*.

1. Thaw qPCR master mix at RT, then place it on ice. Briefly mix by inversion.
2. Determine the total volume for the appropriate number of reactions, plus 10% overage. Use 9 μL of diluted RT reaction for 20 μL PCR reaction.
   - 9 μL cDNA
   - 0.5 μL forward primer (10 μM)
   - 0.5 μL reverse primer (10 μM)
   - 10 μL qPCR master mix

3. Aliquot the assay mix into a qPCR plate. Seal the plate with optically transparent film.
4. Spin the plate at 2500 rpm for 2 min to remove bubbles and collect supernatant.
5. Proceed to a real-time instrument. Initial denaturation for 60 s at 95 °C followed by 40 cycles of 15 s at 95 °C and 30 s at 60 °C. Add a melt curve step at the end for quality check.
6. Proceed to qPCR data analysis using a standard method.

We show one example analyzing APA changes of the gene *ANKMY1* by CRISPRpas (Fig. 3B).

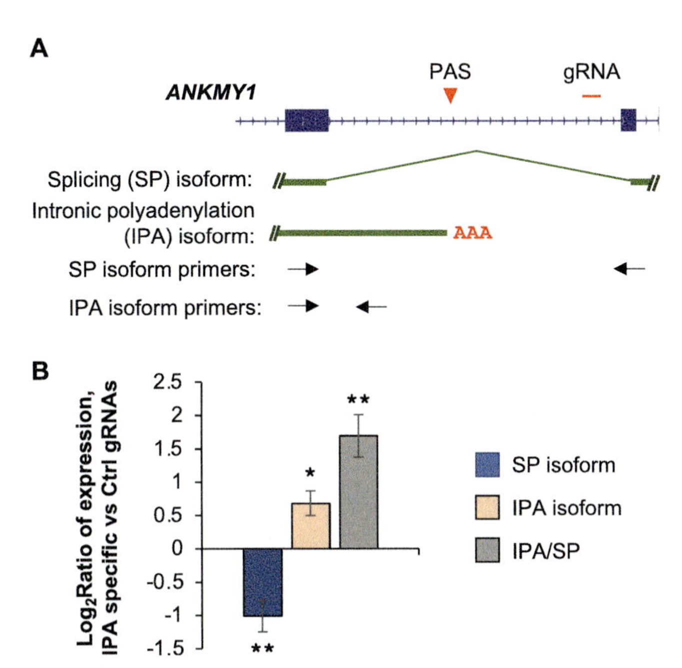

Fig. 3 Alternation of APA of *ANKMY1* by CRISPRpas. (A) Design of gRNA. The human *ANKMY1* gene structure is shown for the region encompassing the intronic PAS afftected by an SNP (Fig. 2). Blue rectangle is coding exon and blue solid line with arrows is intron. The intronic PAS is denoted with a red arrowhead. The gRNA for CRISRPpas is designed to target the non-template strand of DNA. The primers for RT-qPCR are also indicated, which interrogate the expression of splicing (SP) and IPA isoforms. (B) RT-qPCR analysis of relative amounts of SP (blue) and IPA (orange) isoforms. *GAPDH* was used as an internal control. The ratio of IPA/SP isoforms (gray) is also shown. Control (Ctrl) or specific gRNAs were transfected in HEK293T^{dCas9} cells, and RNA was extracted for RT-qPCR. Log_2 expression ratio was normalized to Ctrl. Error bars are standard error of mean based on biological replicates ($n = 4$). *P* value was calculated with the Student's *t*-test. ($*P < 0.05$; $**P < 0.01$).

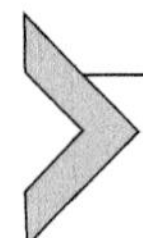

5. Discussion

5.1 Advantages

A major advantage of CRISPRpas is that 3′UTR shortening or IPA promotion (by the suppression of distal PAS isoform) can be easily achieved via a simple transfection without extensive genome engineering. CRISPRpas elicits immediate changes in APA and the functional significance of the change can be tested within a few days. In our hands, at least one out of three designed gRNAs can successfully modulate APA for over half of the genes we have tested. It is also reasonable to speculate that using multiple gRNAs might further increase the success rate of the CRISPRpas experiments.

Functional consequences of APA events can be studied by using traditional CRISPR/Cas9-mediated gene editing. For example, either the 3′UTR or the PAS can be edited to allow APA isoform changes. However, these approaches prove challenging in primary cells because CRISPR/Cas9 manipulation requires cell proliferation. Hence, non-genomic approaches that can alter APA is desirable in primary cells. CRISPRpas can be adapted to lentiviral or other viral vector system to efficiently express dCas9 and gRNAs in primary cells. For example, lentiCRISPR v2-dCas9 vector (Addgene plasmid # 12233) can be used to clone specific gRNAs into the lentiviral vector. Lentiviral transduction of such vector would lead to expression of dCas9 and gRNA in cell types that are refractory to regular transfection reagents (e.g. lipofectamine). In addition, with the currently available dCas9 transgenic or knock-in mice (Duan et al., 2018; Fujita et al., 2018), primary cells that are already expressing dCas9 proteins can be isolated, cultured and transduced with lentivirus expressing gRNAs for CRISPRpas.

5.2 Limitations

A limitation of the current CRISPRpas version is that it can only be used to inhibit distal PAS usage. Because transcription elongation is coupled with PAS selection, further experiments to target dPASs can be tested. For example, delivering dCas9 downstream of dPAS may promote the dPAS usage and expression of long isoform.

Another limitation at the present is a possible side effect of pre-mRNA degradation (shown in Fig. 1). We found that the distance between PAS and target site is critical for gene expression level changes, suggesting that CPA requires some time to complete. Whether this step can be optimized is a work for the future.

6. Summary

Here we describe the CRISPRpas method to change APA in 3′UTRs or introns. We show its efficacy in a human cell line engineered to stably express the dCas9 proteins. We have found the method simple and robust for modulation of APA.

Acknowledgments

We thank Qingbao Ding and Erdene Baljinnyam for technical support. This work was supported by the National Institutes of Health grant number GM084089 to B.T.

References

Danckwardt, S., Hentze, M. W., & Kulozik, A. E. (2008). 3′ End mRNA processing: Molecular mechanisms and implications for health and disease. *The EMBO Journal*, *27*(3), 482–498. https://doi.org/10.1038/sj.emboj.7601932.

Davis, C. A., Hitz, B. C., Sloan, C. A., Chan, E. T., Davidson, J. M., Gabdank, I., et al. (2018). The encyclopedia of DNA elements (ENCODE): Data portal update. *Nucleic Acids Research*, *46*(D1), D794–D801. https://doi.org/10.1093/nar/gkx1081.

Derti, A., Garrett-Engele, P., Macisaac, K. D., Stevens, R. C., Sriram, S., Chen, R., et al. (2012). A quantitative atlas of polyadenylation in five mammals. *Genome Research*, *22*(6), 1173–1183. https://doi.org/10.1101/gr.132563.111.

Doudna, J. A., & Charpentier, E. (2014). Genome editing. The new frontier of genome engineering with CRISPR-Cas9. *Science*, *346*(6213), 1258096. https://doi.org/10.1126/science.1258096.

Duan, J., Lu, G., Hong, Y., Hu, Q., Mai, X., Guo, J., et al. (2018). Live imaging and tracking of genome regions in CRISPR/dCas9 knock-in mice. *Genome Biology*, *19*(1), 192. https://doi.org/10.1186/s13059-018-1530-1.

Dubbury, S. J., Boutz, P. L., & Sharp, P. A. (2018). CDK12 regulates DNA repair genes by suppressing intronic polyadenylation. *Nature*, *564*(7734), 141–145. https://doi.org/10.1038/s41586-018-0758-y.

Fujita, T., Kitaura, F., Oji, A., Tanigawa, N., Yuno, M., Ikawa, M., et al. (2018). Transgenic mouse lines expressing the 3xFLAG-dCas9 protein for enChIP analysis. *Genes to Cells*, *23*(4), 318–325. https://doi.org/10.1111/gtc.12573.

Gilbert, L. A., Larson, M. H., Morsut, L., Liu, Z., Brar, G. A., Torres, S. E., et al. (2013). CRISPR-mediated modular RNA-guided regulation of transcription in eukaryotes. *Cell*, *154*(2), 442–451. https://doi.org/10.1016/j.cell.2013.06.044.

Gruber, A. R., Martin, G., Muller, P., Schmidt, A., Gruber, A. J., Gumienny, R., et al. (2014). Global 3' UTR shortening has a limited effect on protein abundance in proliferating T cells. *Nature Communications*, *5*, 5465. https://doi.org/10.1038/ncomms6465.

GTEx Consortium. (2017). Genetic effects on gene expression across human tissues. *Nature*, *550*(7675), 204.

GTEx Consortium. (2013). The genotype-tissue expression (GTEx) project. *Nature Genetics*, *45*(6), 580–585. https://doi.org/10.1038/ng.2653.

Ha, K. C. H., Blencowe, B. J., & Morris, Q. (2018). QAPA: A new method for the systematic analysis of alternative polyadenylation from RNA-seq data. *Genome Biology*, *19*(1), 45. https://doi.org/10.1186/s13059-018-1414-4.

Haeussler, M., Schonig, K., Eckert, H., Eschstruth, A., Mianne, J., Renaud, J. B., et al. (2016). Evaluation of off-target and on-target scoring algorithms and integration into the guide RNA selection tool CRISPOR. *Genome Biology*, *17*(1), 148. https://doi.org/10.1186/s13059-016-1012-2.

Hanna, R. E., & Doench, J. G. (2020). Design and analysis of CRISPR-Cas experiments. *Nature Biotechnology*, *38*(7), 813–823. https://doi.org/10.1038/s41587-020-0490-7.

Herrmann, C. J., Schmidt, R., Kanitz, A., Artimo, P., Gruber, A. J., & Zavolan, M. (2020). PolyASite 2.0: A consolidated atlas of polyadenylation sites from 3′ end sequencing. *Nucleic Acids Research*, *48*(D1), D174–D179. https://doi.org/10.1093/nar/gkz918.

Hollerer, I., Grund, K., Hentze, M. W., & Kulozik, A. E. (2014). mRNA 3'end processing: A tale of the tail reaches the clinic. *EMBO Molecular Medicine*, *6*(1), 16–26. https://doi.org/10.1002/emmm.201303300.

Hoque, M., Ji, Z., Zheng, D., Luo, W., Li, W., You, B., et al. (2013). Analysis of alternative cleavage and polyadenylation by 3′ region extraction and deep sequencing. *Nature Methods*, *10*(2), 133–139. https://doi.org/10.1038/nmeth.2288.

Jan, C. H., Friedman, R. C., Ruby, J. G., & Bartel, D. P. (2011). Formation, regulation and evolution of *Caenorhabditis elegans* 3'UTRs. *Nature*, *469*(7328), 97–101. https://doi.org/10.1038/nature09616.

Kamieniarz-Gdula, K., & Proudfoot, N. J. (2019, Aug). Transcriptional control by premature termination: A forgotten mechanism. *Trends in Genetics*, *35*(8), 553–564. https://doi.org/10.1016/j.tig.2019.05.005.

Konermann, S., Brigham, M. D., Trevino, A. E., Joung, J., Abudayyeh, O. O., Barcena, C., et al. (2015). Genome-scale transcriptional activation by an engineered CRISPR-Cas9 complex. *Nature*, *517*(7536), 583–588. https://doi.org/10.1038/nature14136.

Lee, S. H., Singh, I., Tisdale, S., Abdel-Wahab, O., Leslie, C. S., & Mayr, C. (2018). Widespread intronic polyadenylation inactivates tumour suppressor genes in leukaemia. *Nature*, *561*(7721), 127–131. https://doi.org/10.1038/s41586-018-0465-8.

Li, W., You, B., Hoque, M., Zheng, D., Luo, W., Ji, Z., et al. (2015). Systematic profiling of poly(A)+ transcripts modulated by core 3′ end processing and splicing factors reveals regulatory rules of alternative cleavage and polyadenylation. *PLoS Genetics*, *11*(4), e1005166. https://doi.org/10.1371/journal.pgen.1005166.

Liu, Y., Han, X., Yuan, J., Geng, T., Chen, S., Hu, X., et al. (2017). Biallelic insertion of a transcriptional terminator via the CRISPR/Cas9 system efficiently silences expression of protein-coding and non-coding RNA genes. *The Journal of Biological Chemistry*, *292*(14), 5624–5633. https://doi.org/10.1074/jbc.M116.769034.

Mayr, C. (2017). Regulation by 3'-untranslated regions. *Annual Review of Genetics*, *51*, 171–194. https://doi.org/10.1146/annurev-genet-120116-024704.

Moll, P., Ante, M., Seitz, A., et al. (2014). QuantSeq 3′ mRNA sequencing for RNA quantification. *Nature Methods*, *11*. https://doi.org/10.1038/nmeth.f.376.

Nam, J. W., Rissland, O. S., Koppstein, D., Abreu-Goodger, C., Jan, C. H., Agarwal, V., et al. (2014). Global analyses of the effect of different cellular contexts on microRNA targeting. *Molecular Cell*, *53*(6), 1031–1043. https://doi.org/10.1016/j.molcel.2014.02.013.

Qi, L. S., Larson, M. H., Gilbert, L. A., Doudna, J. A., Weissman, J. S., Arkin, A. P., et al. (2013). Repurposing CRISPR as an RNA-guided platform for sequence-specific control of gene expression. *Cell*, *152*(5), 1173–1183. https://doi.org/10.1016/j.cell.2013.02.022.

Shepard, P. J., Choi, E. A., Lu, J., Flanagan, L. A., Hertel, K. J., & Shi, Y. (2011). Complex and dynamic landscape of RNA polyadenylation revealed by PAS-Seq. *RNA*, *17*(4), 761–772. https://doi.org/10.1261/rna.2581711.

Shi, Y., & Manley, J. L. (2015). The end of the message: Multiple protein-RNA interactions define the mRNA polyadenylation site. *Genes & Development*, *29*(9), 889–897. https://doi.org/10.1101/gad.261974.115.

Shin, J., Ding, Q., Baljinnyam, E., Wang, R., & Tian, B. (2021). (2021-01-01 00:00:00). CRISPRpas: Programmable regulation of alternative polyadenylation by dCas9. *bioRxiv*.
Shulman, E. D., & Elkon, R. (2020). Systematic identification of functional SNPs interrupting 3'UTR polyadenylation signals. *PLoS Genetics*, *16*(8), e1008977. https://doi.org/10.1371/journal.pgen.1008977.
Spies, N., Burge, C. B., & Bartel, D. P. (2013). 3' UTR-isoform choice has limited influence on the stability and translational efficiency of most mRNAs in mouse fibroblasts. *Genome Research*, *23*(12), 2078–2090. https://doi.org/10.1101/gr.156919.113.
Tanenbaum, M. E., Gilbert, L. A., Qi, L. S., Weissman, J. S., & Vale, R. D. (2014). A protein-tagging system for signal amplification in gene expression and fluorescence imaging. *Cell*, *159*(3), 635–646. https://doi.org/10.1016/j.cell.2014.09.039.
The ENCODE Project Consortium. (2012). An integrated encyclopedia of DNA elements in the human genome. *Nature*, *489*(7414), 57–74. https://doi.org/10.1038/nature11247.
Tian, B., & Graber, J. H. (2012). Signals for pre-mRNA cleavage and polyadenylation. *Wiley Interdisciplinary Reviews: RNA*, *3*(3), 385–396. https://doi.org/10.1002/wrna.116.
Tian, B., & Manley, J. L. (2017). Alternative polyadenylation of mRNA precursors. *Nature Reviews. Molecular Cell Biology*, *18*(1), 18–30. https://doi.org/10.1038/nrm.2016.116.
Tushev, G., Glock, C., Heumuller, M., Biever, A., Jovanovic, M., & Schuman, E. M. (2018). Alternative 3' UTRs modify the localization, regulatory potential, stability, and plasticity of mRNAs in neuronal compartments. *Neuron*, *98*(3), 495–511. e496. https://doi.org/10.1016/j.neuron.2018.03.030.
Untergasser, A., Cutcutache, I., Koressaar, T., Ye, J., Faircloth, B. C., Remm, M., et al. (2012). Primer3--new capabilities and interfaces. *Nucleic Acids Research*, *40*(15), e115. https://doi.org/10.1093/nar/gks596.
Wang, Q., He, G., Hou, M., Chen, L., Chen, S., Xu, A., et al. (2018). Cell cycle regulation by alternative polyadenylation of CCND1. *Scientific Reports*, *8*(1), 6824. https://doi.org/10.1038/s41598-018-25141-0.
Wang, R., Nambiar, R., Zheng, D., & Tian, B. (2018). PolyA_DB 3 catalogs cleavage and polyadenylation sites identified by deep sequencing in multiple genomes. *Nucleic Acids Research*, *46*(D1), D315–D319. https://doi.org/10.1093/nar/gkx1000.
Wang, R., & Tian, B. (2020). APAlyzer: A bioinformatics package for analysis of alternative polyadenylation isoforms. *Bioinformatics*, *36*(12), 3907–3909. https://doi.org/10.1093/bioinformatics/btaa266.
Wang, R., Zheng, D., Wei, L., Ding, Q., & Tian, B. (2019). Regulation of intronic polyadenylation by PCF11 impacts mRNA expression of long genes. *Cell Reports*, *26*(10), 2766–2778. e2766 https://doi.org/10.1016/j.celrep.2019.02.049.
Wang, R., Zheng, D., Yehia, G., & Tian, B. (2018). A compendium of conserved cleavage and polyadenylation events in mammalian genes. *Genome Research*, *28*(10), 1427–1441.
Williamson, L., Saponaro, M., Boeing, S., East, P., Mitter, R., Kantidakis, T., et al. (2017). UV irradiation induces a non-coding RNA that functionally opposes the protein encoded by the same gene. *Cell*, *168*(5), 843–855. e813. https://doi.org/10.1016/j.cell.2017.01.019.
Woodard, L. E., & Wilson, M. H. (2015). piggyBac-ing models and new therapeutic strategies. *Trends in Biotechnology*, *33*(9), 525–533. https://doi.org/10.1016/j.tibtech.2015.06.009.
Xia, Z., Donehower, L. A., Cooper, T. A., Neilson, J. R., Wheeler, D. A., Wagner, E. J., et al. (2014). Dynamic analyses of alternative polyadenylation from RNA-seq reveal a 3'-UTR landscape across seven tumour types. *Nature Communications*, *5*, 5274. https://doi.org/10.1038/ncomms6274.
Yang, Y., Zhang, Q., Miao, Y. R., Yang, J., Yang, W., Yu, F., et al. (2020). SNP2APA: A database for evaluating effects of genetic variants on alternative polyadenylation in human cancers. *Nucleic Acids Research*, *48*(D1), D226–D232. https://doi.org/10.1093/nar/gkz793.

Ye, C., Long, Y., Ji, G., Li, Q. Q., & Wu, X. (2018). APAtrap: Identification and quantification of alternative polyadenylation sites from RNA-seq data. *Bioinformatics*, *34*(11), 1841–1849. https://doi.org/10.1093/bioinformatics/bty029.

You, L., Wu, J., Feng, Y., Fu, Y., Guo, Y., Long, L., et al. (2015). APASdb: A database describing alternative poly(A) sites and selection of heterogeneous cleavage sites downstream of poly(A) signals. *Nucleic Acids Research*, *43*(Database issue), D59–D67. https://doi.org/10.1093/nar/gku1076.

Zhao, W., Siegel, D., Biton, A., Tonqueze, O. L., Zaitlen, N., Ahituv, N., et al. (2017). CRISPR-Cas9-mediated functional dissection of 3'-UTRs. *Nucleic Acids Research*, *45*(18), 10800–10810. https://doi.org/10.1093/nar/gkx675.

Zheng, D., Wang, R., Ding, Q., Wang, T., Xie, B., Wei, L., et al. (2018). Cellular stress alters 3'UTR landscape through alternative polyadenylation and isoform-specific degradation. *Nature Communications*, *9*(1), 2268. https://doi.org/10.1038/s41467-018-04730-7.

Printed in the United States
by Baker & Taylor Publisher Services